MULTICHIP MODULE TECHNOLOGIES AND ALTERNATIVES:

THE BASICS

Cover Photo
Courtesy of Unisys photographer, Paul Robinson. The top MCM is a hermetic processor module utilizing the *latest* MCM-D technology with fine pitch, high lead count, flip TAB connections. The bottom three packages are MCM-C modules with conventional wire bond connections. These modules are described in Chapter 14.

MULTICHIP MODULE
TECHNOLOGIES AND ALTERNATIVES:

THE BASICS

Edited by:
Daryl Ann Doane
Paul D. Franzon

VNR VAN NOSTRAND REINHOLD
New York

Library of Congress Catalog Card Number 92-2779
ISBN 0-442-01236-5

Printed in the United States of America.

Van Nostrand Reinhold
115 Fifth Avenue
New York, New York 10003

Chapman and Bell
2-6 Boundary Row
London, SE1 8HN, England

Thomas Nelson Australia
102 Dodds Street
South Melbourne 3205
Victoria, Australia

Nelson Canada
1120 Birchmount Road
Scarborough, Ontario M1K 5G4, Canada

16 15 14 13 12 11 10 9 8 7 6 5 4 3 2 1

Library of Congress Cataloging-in-Publication Data

Multichip module technologies and alternatives : the basics / [edited] by Daryl Ann
    Doane and Paul D. Franzon.
        p. cm.
    Includes Index.
    ISBN 0-442-01236-5
    1. Microelectronic packaging. 2. Integrated circuits — Very large scale
integration — Design and construction. I. Doane, Daryl Ann. II. Franzon, Paul D.
TK7874.M864 1993
621.39'5 — dc20   92-2779
                                                                      CIP

DEDICATION

This book is dedicated with sincere gratitude
to the 42 authors who contributed their knowledge,
suggestions and many hours of writing
to form the chapters we requested.
Without their diligence, timeliness of response and
patience with our requests
for additional or alternative information,
there would be no book.

ABOUT THE EDITORS

Daryl Ann Doane, President, DAD Technologies, Inc., is a specialist in the manufacturing of IC devices and packages, and in the marketing of materials, equipment and processes for such manufacturing. She has published more than 40 reports, papers and reviews documenting new process development and yield optimization. DAD Technologies, Inc. is a technology-based consulting organization serving both producers and users. Clients have included both government and commercial organizations connected with IC fabrication and packaging.

Dr. Doane previously served as Director, Microelectronics R&D for Hunt Chemical (now Olin Hunt Specialty Products). Prior to that she was a Member of the Technical Staffs of AT&T Bell Laboratories, RCA and Hewlett-Packard Laboratories.

Dr. Doane is a member of the American Physical Society (APS), Materials Research Society (MRS), International Society for Hybrid Microelectronics (ISHM), International Electronics Packaging Society (IEPS), and is a Senior Member of the IEEE. She received the 1980 Distinguished New Engineer Award from the Society of Women Engineers. She also served as President of the New Jersey Section of that Society.

Dr. Doane presently serves as Director of Technical Marketing for the IEEE-CHMT (Components, Hybrids & Manufacturing Technology) Society. She also is an elected member of the Board of Governors of the National CHMT Society. She served previously as Chairman of the San Diego Chapter of the CHMT Society.

Dr. Doane received the BA with Distinction in Chemistry and Physics, an MA in Chemistry, and an MS and PhD in Metallurgy and Materials Science from M.I.T. and the University of Pennsylvania, respectively.

Paul D. Franzon is currently an Assistant Professor in the Department of Electrical and Computer Engineering at North Carolina State University. He has over eight years experience in electronic systems design and design methodology research and development. During that time, in addition to his current position, he has worked at AT&T Bell Laboratories in Holmdel NJ, at the Australian Defense Science and Technology Organization, as a founding member of a successful Australian technology start-up company, and as a consultant to industry, including positions on Technical Advisory Boards.

Dr. Franzon's current research interests include design sciences/methodology for high speed packaging and interconnections systems and also for high speed and low power chip design. In the past, he has worked on problems and projects in wafer scale integration, IC yield modeling, VLSI chip design and communications systems design. He has published over 45 articles and reports.

His teaching interests focus on microelectronic systems building including package and interconnection design, circuit design, processor design and the gaining of hands-on systems experience for students.

Dr. Franzon is a member of the IEEE, ACM and ISHM. He serves as the Chairman of the Education Committee for the National IEEE-CHMT Society.

Dr. Franzon received a BS in Physics and Mathematics, a BE with First Class Honors in Electrical Engineering and a PhD in Electrical Engineering all from the University of Adelaide, Adelaide, Australia.

TABLE OF CONTENTS

FOREWORD

Far from being the passive containers for semiconductor devices of the past, the packages in today's high performance computers pose numerous challenges in interconnecting, powering, cooling and protecting devices. While semiconductor circuit performance measured in picoseconds continues to improve, computer performance is expected to be in nanoseconds for the rest of this century - a factor of 1000 difference between on-chip and off-chip performance which is attributable to losses associated with the package. Thus the package, which interconnects all the chips to form a particular function such as a central processor, is likely to set the limits on how far computers can evolve.

Multichip packaging, which can relax these limits and also improve the reliability and cost at the systems level, is expected to be the basis of all advanced computers in the future. In addition, since this technology allows chips to be spaced more closely, in less space and with less weight, it has the added advantage of being useful in portable consumer electronics as well as in medical, aerospace, automotive and telecommunications products. The multichip technologies with which these applications can be addressed are many. They range from ceramics to polymer-metal thin films to printed wiring boards for interconnections; flip chip, TAB or wire bond for chip-to-substrate connections; and air or water cooling for the removal of heat.

While there are several books now on packaging, these books deal with the subject of multichip modules as part of packaging in general, or they treat a particular multichip module technology or they are at an advanced level. What is needed, therefore, is a *comprehensive book at the basic level*, structured so that anyone entering the field can quickly learn about the technologies, understand the tradeoffs, review the product examples, an make systems level decisions.

Such a book has been provided by Daryl Ann Doane and Paul D. Franzon. They have worked with an outstanding team of packaging experts from industry and universities. Together they have produced **Multichip Module Technologies and Alternatives: The Basics**, an outstanding book for both industry and university use. It is equally appropriate as an introduction to the multichip module technologies for those just entering the field, and as an up-to-date basic technical book for those currently practicing in it.

The books deals with the subject of multichip modules along three parts: systems level perspectives including packaging technology options and costs, the basics of ceramic, thin film and printed wiring board technologies as well as chip and module level connections; thermal and electrical design considerations including electrical testing; and finally product examples illustrating how multichip modules have been useful.

The basic and integrated nature of the book clearly reflects the dedication and the hard work of the editors and the authors.

Rao R. Tummala
IBM Fellow

PREFACE

Welcome!

Welcome to our book. We feel it is a unique book in the field of packaging and we hope you find it both useful and enjoyable. We (editors and authors) have certainly enjoyed bringing it to you!

Uniqueness of the Book

This is a very unique book! Its uniqueness comes about in two ways: first in the approach to the subject, and second, in the approach to the writing of the book.

First, we feel that this book helps define a turning point in the discipline of packaging. The "bottleneck" to increased systems performance is now more often the package than the chip. One effect of this is that suddenly a whole "breed" of engineers need to gain an understanding of how package design affects their systems performance and cost goals. Another effect is the widespread recognition that multichip module packaging technologies are possible solutions to this performance limitation. The attendant explosive growth in the MCM technology alternatives available is a testimony to this recognition.

Until recently, packaging was mainly the domain of mechanical and materials specialists, and furthermore was rarely taught at universities. *This has changed.* Almost overnight, the pressure of high on-chip speeds and high transistor counts meant that packaging became a subject that must be understood by just about any engineer involved in designing a system. When these engineers tried to establish their understanding they found themselves in a virtual "Tower of Babel." First, often the same terms were used for different things or, just as bad, different terms were used for the same thing. Second, as this was the domain of specialists, significant background was required in each discipline in order to understand it. This book turns the discipline of packaging into a subject accessible to the generalist rather than just one accessible to the specialist. Common terms are defined and crosslinked to the terms in current usage. Each discipline is presented in such a way that is both understandable to all and is useful. In providing this book, we help turn packaging into a discipline in which anyone can participate. We also present a book highly suited for teaching within a university.

The book also is unique in how it was created! Packaging is a multidisciplinary subject. We realized that no two people can claim mastery of all the disciplines needed and involved in the subjects we wished to cover. We also realized that rapid writing was important. Thus an edited text was called for. But edited texts are often poorly lacking in terms of understandability and flow. Usually an editor relies on selecting appropriate experts and then accepting what each expert writes with minor alterations.

This book is very different. It could be called a "closely edited" text. Each chapter has had the heavy hand of development of both editors in it. This happened in several forms. Often we spent many hours with individual authors defining what we thought was appropriate and guiding them in the actual writing. In all cases, as well as seeking outside reviewers, we wrote "anonymous" reviews ourselves. In some cases, we directly adjusted the text after it was submitted. Thus we are responsible for the final product as much as each chapter author is.

The result, we feel, is a book that has the authority that comes with a book written by a team of experts, but has the understandability, completeness and flow of a book written by a single author, At least we hope that the book comes as close to that ideal as possible!

Audience

For whom is the book intended? **This book is for everyone!** The emphasis in the book is on understanding the fundamentals and the reporting of real experiences rather than including a huge amount of data. It is intended for those who need a broad exposure to the concepts underlying the design, fabrication, packaging, assembly, manufacturing of multichip modules and the costs associated with alternative packaging technologies. The book is intended for applications, manufacturing and design engineers as well as for technical decision makers and managers who are confused about MCM issues but who wish to understand the fundamentals and basics. Specifically, it will be useful to

- **engineers** in design, processing, fabrication, manufacturing, assembly and test who need to choose a packaging technology for specific product and application goals;

- **managers** determining technology alternatives for new systems needs;

- **marketing; sales technologists** who need a working knowledge of the alternative MCM technologies.

The book also is a suitable text for advanced undergraduate and graduate students in design, electrical, mechanical and systems engineering as well as for students in the applied sciences as discussed further on in this Preface.

Philosophy of the Book

What are the key decisions needed when considering using MCM technologies? There are four perspectives: materials, manufacturing, systems performance and cost requirements. The book presents the basics of MCM technologies from these four perspectives with an emphasis on decision-making. How do you choose a packaging technology from a systems performance, cost perspective? What are the fundamental materials and manufacturing issues that need to be considered? Is the packaging strategy appropriate to the design goal?

This book is applications-oriented in the sense that it discusses

- Examples of MCMs and the products and systems in which they have been used,
- Examples of equipment and processes used to design, build and test MCMs,
- Actual case studies and insights provided by leading companies themselves rather than a summary of reported results.

Organization of the Book

The book is divided into four parts which we call

- Part A - The Framework
  (Making Decisions: The Big Picture)
- Part B - The Basics
- Part C - Case Studies
- Part D - Closing the Loop

Each Part begins with what we call a "frame" in which we summarize the purpose or goals of the Part. The frames alert the reader to what theme dominates in the Part, and what can be learned from it. The frames are set apart from the rest of the book by the border around their pages.

For example, in Part A, the goal is to present a basis for understanding multichip module technology possibilities for packaging. We call this a *framework for understanding*. The basic definitions needed for understanding are presented. Some alternative packaging approaches are discussed, and the themes of decision-making and the design process that run through the book are also introduced. The reader is alerted to look for the global picture in Part A, an overview that a breadth of knowledge is required for understanding, and that an awareness of the basic concepts introduced is Part A will enable him or her to follow the detailed technical information in Part B.

Part B provides an understanding of the technical fundamentals in the design, fabrication and testing of multichip modules. The tutorials in Part B are unique in the published literature. For example, there is detailed coverage of MCM-L, the fabrication technology based on laminate structures. Another special feature of Part B is the discussion of multichip module-to-printed wiring board (second level) connections in Chapter 10. The test chapter (Chapter 13) provides specific guidelines on how to design MCM products for testing and how to reduce test cost and effort. In each Chapter, the reader is alerted to the options associated with the technologies presented so that these options can be related to product applications.

Part C consists of "reports" from some companies that have created and are selling successful MCM products. The development of their MCM technologies, internally from existing expertise, is described. Here the reader will benefit from the insights shared by these companies.

Finally, Part D "closes the loop." It focuses on what aspects of all the technologies presented are likely to have the greatest impact on meeting future needs from a systems perspective. It also presents a *forward view* by describing some open, unsolved problems, open challenges that must be met if future systems are not to be constrained in performance by limitations in packaging technologies.

Use as a Textbook

We feel that this book is highly usable as a packaging textbook. It covers this interdisciplinary topic at a level that is accessible to all while providing concrete learning material. This book emphasizes decision-making and design, and thus, relates directly to what practicing engineers do for a living. It would be useful in both graduate courses and advanced undergraduate courses. It has the tightness and continuity of a textbook in contrast to a typical edited book. In our opinion, this book is superior to any current packaging "textbook" for use as a text mainly because of the style of presentation.

Despite its title, it could be used for many courses that do not feature only MCMs. Throughout this book, single chip packaging is presented as an alternative to be considered. The design chapters apply equally well to single chip and multichip package design. The only single chip package elements that are missing are a discussion of how plastic packages are made and a specific discussion of how boards are assembled. Chapter 9 covers some of these related points. Board manufacturing is covered in Chapter 5. Single chip ceramic package manufacturing uses similar processes as those for making ceramic MCMs which are discussed in Chapter 6.

There are several types of courses that can use this book. One of us (PDF) uses the book in a semester long graduate course on packaging design for electrical engineers. That course starts with an overview of packaging alternatives (Chapters 1 and 2, with 5 to 10 as references), discusses systems level decision-making (Chapter 3 and parts of the Case Studies), and briefly discusses cost (parts of Chapter 4). Almost half of the course is concerned with electrical design of packaging (Chapter 11). A small amount of time is spent on thermal design (Chapter 12), to the level of detail needed by an electrical engineer, and on the appropriate impact of testing on design (Chapter 13).

The book also would be useful for a more technology and manufacturing oriented course. There Chapter 1 and 2 would be used to introduce the course. Establishing the reasons for the multiple packaging alternatives available and how to choose among them would be covered with the aid of Chapter 3. Chapter 4 provides a detailed look into the impact of manufacturing costs. Chapters 5 through 10 would be used to teach the bulk of the course. Brief references to Chapters 11 through 13 would provide technologists with an understanding of how their decisions affect design. Chapter 17 would offer a detailed manufacturing viewpoint.

Finally, this book also could be useful to mechanical and chemical engineers and materials science-based courses on packaging.

Daryl Ann Doane
Paul D. Franzon

ACKNOWLEDGMENTS

With the publication of this book, it is a special pleasure for one of us (DAD) to express gratitude and appreciation to Paul B. Wesling, Tandem Computers. His longstanding and patient role as mentor and resource person has provided an immeasurable educational opportunity and insight regarding the value of technical publications. He currently serves as Vice President of Publications for the IEEE Components, Hybrids and Manufacturing Technology (CHMT) Society. He has been a *pioneer* in the area of technical publications related to peer-reviewed journals, reprint as well as conventional books and Conference Proceedings. His innovative ideas in publications have been adopted by Societies outside of the IEEE, and also are reflected in the value and vision for this book. Thank you, Paul!

Several people advised us on technical and organizational topics. They generously contributed their time and valuable expert knowledge in extended discussions. In particular, Dr. Robert C. Sundahl, Intel, Dr. Walter J. Bertram, Alcoa Electronic Packaging and Professor R. Wayne Johnson, Auburn University, shared their ideas on what was needed in the book, as well as providing detailed reviews for several of the chapters. We wish to thank them.

John H. Lau, Hewlett-Packard was helpful in sharing his experience as an editor of three books for Van Nostrand Reinhold. He never tired of answering questions; we appreciate his patience and helpfulness very much.

John Nelson, Unisys, is very special! A faithful colleague and friend, he graciously responded positively to so many requests, providing the value of his longstanding experience and insight in the packaging industry in answering questions pertaining to the book. It has been a wonderful opportunity to discuss the technical aspects of the book, as well as its human side, with him. Many of the unique aspects of the book are due to his suggestions. *We thank him for everything!*

Each chapter in the book was reviewed by at least three individuals whose areas of expertise covered the chapter topic. We also deliberately asked individuals who worked in areas peripheral to a chapter topic to review the chapters for comprehension and clarity. Although we cannot list these more than 54 individuals, we are very appreciative of their efforts, thoroughness and helpful comments. We took them all very seriously.

Still, with all the technical expertise brought to form the contents of this book, there would have been no way to complete a book of this scope, size and complexity had it not been for the unique combinations of dedication and perseverance of the staff and management at Van Nostrand Reinhold (VNR), our publisher. We are very grateful for their efforts. Especially noteworthy was the standard of inspiration set by Mr. Stephen Chapman, Electrical Engineering Editor. He shared our "vision" for the book and persevered on our behalf for the conventional as well as the nonconventional requests we made. He arranged for an accelerated publication schedule so the book could be available when requested. Without his dedication to the book it would not have appeared in such a timely manner.

Sue Alexander of Seaborn Enterprises, Chandler AZ, was responsible for producing the galleys with artwork for the entire book. She demonstrated a high level of skill and innovation in adapting busy tables and complex equations to a text already filled with an unusual amount of photos and artwork. All the while she was faced with unprecedented time constraints. We are very much in awe of the final excellence of her creation in the layout and appearance of the book.

Most of all, we would like to thank those who are close to us. One of us (DAD) wishes to express special thanks to Dr. John L. Doane, Timothy A. ("TAD") Doane and Stevie Doane for their love, support and patience. Their faithful encouragement and sustaining, as well as helpful, assistance contributed significantly to the existence of the book. The other of us (PDF) gives loving thanks to Debra L. Ray in acknowledgment of her forebearance during the many long evenings and weekends when he was not there.

Part A—The Framework
Making Decisions: The Big Picture

"Cheshire Puss," she began, rather timidly,
"Would you tell me, please,
which way I ought to walk from here?"
"That depends a good deal
on where you want to get to,"
said the Cat.

Alice in Wonderland
by Lewis Carroll

In the first part of this book, we provide a *framework (or basis) for understanding* multichip module packaging possibilities. Construction of our framework begins by defining the alternative approaches available with respect to multichip module structures and the packaging strategies involving such structures. It also includes such issues as manufacturability, performance, cost and time-to-market common to any packaging strategy - single chip or multichip. Finally, in the construction of the framework, we suggest a design methodology or decision-making process by which a packaging strategy can be coupled to a particular product goal. There is no single "right" or "wrong" packaging technology, but only relative to a specific application goal.

By constructing our framework of understanding, it is possible to cover a broad base of information in a short time. Such an overview, or global picture, also emphasizes the multidisciplinary nature of the multichip module technologies involved. Because of the breadth of understanding required in the packaging "sciences," there is some confusion about multichip module packaging as a viable, practical technology alternative. Suddenly people who did not have to think at all about packaging must consider complex packaging alternatives. Managers and engineers are being forced to make decisions without a complete understanding of packaging alternatives. In Part A, we bring together the multiple disciplines involved in such decision making considerations in a way understandable to the nonspecialist and to the student, as well as informative to a worker in the field.

In Chapter 1 of Part A, for example, the multichip module structure is defined and the alternative multichip module technologies are described. The themes of decision-making and the design process that run through the book also are introduced. The challenges presented to growth in the use of MCM systems, such as their performance, cost and associated infrastructure, as well as single chip alternatives, are discussed.

Chapter 2 provides an overview of MCM alternatives concentrating on materials and manufacturing issues. Specific examples of how some multichip modules have been manufactured are discussed.

We also consider (in Chapter 3) the relationships between the "needs" of the system being packaged and the choices of packaging technologies in terms of performance and cost, needs being defined in terms of application or product goals.

Packaging decisions are very sensitive to both direct and indirect costs which are discussed in all the chapters. Specifically, Chapter 4 of Part A focuses on cost modeling, at the component level, and from a manufacturing and design perspective.

In the Framework, the factors that influence high level packaging decisions are explored. Thus, Part A is particularly useful to the technical manager who needs to make decisions on the best applications of this new technology to company products, but does not need to understand the detail required to design these products. The practicing engineer will gain the breadth of knowledge required to understand the full impact of the new technology. The basic concepts introduced will enable an understanding of the detailed technical information provided in Part B. Marketing and sales people also will find this part useful in communicating to customers how the company packaging decisions help meet customer needs.

1

INTRODUCTION

Daryl Ann Doane

1.1 BACKGROUND AND DEFINITIONS

1.1.1 Purpose and Perspective of the Book

This is a book about multichip modules (MCMs). And the strategies involved in implementing them *successfully*! The book discusses the technology alternatives appropriate for MCMs, and the decision-making processes required for choosing the "best" technology for a particular application. The book helps to answer the question: *Is the technology applicable (or appropriate) to the design goal?*

Multichip modules are not new! The basic science and technologies involved in the design, materials, fabrication and manufacturing of MCMs have been known generally for over 30 years. These packaging technologies span a wide range of technical disciplines, such as chemistry, physics, materials science, electrical and mechanical engineering. The broad range of technologies which now must be understood by an MCM designer, for example, has expanded beyond a conventional experience base. One goal of this book is to provide information so that such an individual can learn about the various available technologies, and how and which one to choose for his or her particular packaging needs.

In the past 30 years, there has been an exponential growth in transistor speed and in the number of transistors per chip. For example, both have doubled over the last three years. Chip interconnection speeds and chip I/O count have not increased at the same rate. As a result, chip interconnections play a more dominant and limiting role in determining overall system speed and performance. Packaging of the chips has become a more significant factor in performance. Systems level performance improvements are now being limited more by the packaging and interconnection technologies, and less by the chip technology itself.

Advanced packaging and interconnection technologies have been mainly the domain of high performance computers whose performance otherwise would have been limited by the package. MCMs are viewed by some as the *only suitable technology* for packaging such systems. Soon the performance of many more systems will be limited by the package, so it is anticipated that MCMs will appear in a wider range of systems in the future.

Some fundamental advantages of MCMs are:

- Increased system speed
- Reduced overall size
- Ability to handle chips with large numbers of input/outputs (I/Os)
- Increased number of interconnections in a given area
- Reduced number of external connections, for a given system function

Every external package connection degrades the system performance somewhat by its parasitics (undesired circuit elements). Each package lead has some parasitic capacitance and inductance that distort the shape of the signals passing through it. Reducing the number of external connections not only reduces these parasitics, but also generally contributes to system reliability. Thus, one way MCMs would reduce the parasitic effects is through the reduction in the number of second level connections they afford.

This book discusses selected technologies and alternatives for MCM design and fabrication. A multidisciplinary approach is used to evaluate the interactions between materials, processes, designs and assembly techniques. Contributions by different authors provide perspectives from many different viewpoints. The emphasis throughout the book is on making decisions among the technology alternatives for specific applications and products.

This chapter has two main purposes. First it provides the definitions needed to understand the book and relates these definitions to terms used in other chapters. These terms are common "jargon" in various disciplines, technologies or corporate cultures, and thus, may vary from chapter to chapter depending on the perspective. By relating these terms to each other and to the basic concepts, we hope to remove some stumbling blocks in the way of the reader's understanding.

The second purpose of this chapter is to introduce some of the themes and perspectives presented in the book. Major themes include decision-making as it relates to designing MCM-based products and as it relates to implementing MCMs in the marketplace, MCM structures and technologies, and MCM manufacturing. These themes are introduced so that the reader can start relating to the multiple disciplines and approaches that are concerned with packaging right from the start of the book.

1.1.2 ARCHITECTURE - Building a Multichip Module Structure

In this section we define what is meant by a multichip module, and introduce the alternative technologies used to build a MCM structure.

What is a Multichip Module?
In this book we define a multichip module (MCM) as a structure consisting of two or more integrated circuit chips electrically connected to a *common circuit base* and interconnected by conductors in that base. The conductors in the base are usually patterned in multiple layers separated by a suitable insulating dielectric material, with vias for interconnections between layers. The base also provides the required mechanical support for the chips. Figure 1-1 shows schematically an MCM structure.

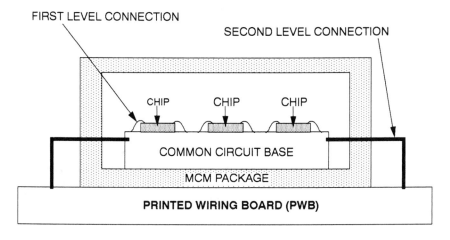

Figure 1-1 MCM architecture (schematic).

Based on this definition of an MCM structure, *multichip modules are not new!* In the very early days, MCMs were called hybrid circuits. Hybrid circuit assemblies are still in use, and usually contain lower lead count analog and digital integrated circuits (ICs), as well as active or passive discrete components. They may contain many layers of metallic conductors and ceramic dielectrics, but they classically have been formed one layer at a time. Chip-on-board (COB) assemblies, where bare chips are wire bonded to conventional printed wiring boards, also have been in use for some time.

Another definition for MCMs is based on the interconnection technologies that can be used in their structures. In particular, the IPC (Institute for Interconnecting and Packaging Electronic Circuits), according to its standard IPC-MC-790 [1], defines three types of technologies that can be used to make MCM structures:

- **MCM-L:** Modules using advanced forms of printed wiring board (PWB) technologies to form the copper conductors on plastic laminate-based dielectrics.

- **MCM-C:** Modules constructed on cofired ceramic substrates using thick film (screen printing) technologies to form the conductor patterns. ("Cofired" means that the conductors and the ceramic are all heated in an oven at one time.)

- **MCM-D:** Modules whose interconnections are formed by the thin film deposition of metals on deposited dielectrics, which may be polymers or inorganic dielectrics.

Some of the authors in this book refer to these structures as laminate MCMs, cofired ceramic or thick film MCMs, and thin film MCMs, respectively. All of these structures satisfy the definition of containing multiple chips electrically connected to, and interconnected on, a single base structure that is eventually connected to a printed wiring board.

A third approach to defining an MCM is based on the packaging efficiency achieved by the technology. Packaging efficiency (or "silicon density") is defined as the percentage of area occupied by silicon ICs [2]. Using this approach, an MCM can be defined as follows: An MCM is a structure in which a packaging efficiency of greater than 30% is achieved. This definition implies a particular degree of technology which allows the chips to be packed closely together. It excludes many hybrid and COB designs. For the purposes of this book, we seek to understand the broader construction of a multichip module structure without the 30% packaging efficiency restriction.

What is the MCM Architecture?

The basic idea of MCMs is to decrease the average spacing between ICs in an electronic system. The fundamental aspect of MCM technology is therefore the chip interconnection technology. This technology includes the method of connecting the I/O conductors on the chips to the common circuit base, and also the method of sending signals through that base between the chips. The main goals are higher performance resulting from reduced signal delays between chips, reduced overall size and reduced numbers of external connections.

The elements that make up an MCM structure, or architecture, are shown in Figure 1-1. These elements are the chips, the first level connections, the common circuit base, the second level connections and the package. The remainder of this section functionally defines these elements and introduces related terminology. Succeeding sections briefly present some of the available technology alternatives.

We use the term *common circuit base*, because it is common to all the chips, because it contains the signal interconnection circuits and the power and ground distribution circuits, and because it provides a mechanical support base for all the chips. Sometimes the top of the common circuit base is called the "substrate" or the "MCM substrate". See, for example, Figure 2-1. In that figure, the lower "substrate base" is that part of the common circuit base that provides mechanical stiffness to the MCM. In Figure 1-2, the "substrate/package circuitry" and the "substrate" correspond, respectively, to the "MCM substrate" and the "substrate base" of Figure 2-1.

Sometimes the substrate and the substrate base are the same. For example, in an MCM-C, the interconnections are formed in a solid ceramic. On the other hand, in an MCM-D, the interconnections are formed in a plastic-like polyimide dielectric. For mechanical support this substrate then needs to be attached to a substrate base, typically ceramic or silicon. (In other places, a silicon substrate base may be called "the substrate," based on its similar function in an IC.)

The first level connections consist of the signal and power/ground wires provided between the chips and the common circuit base. Not shown in Figure

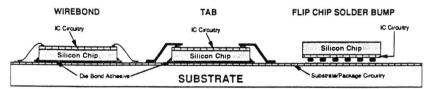

Figure 1-2 Common types of first level connections. Chip to common circuit base. (Courtesy E. Larson).

1-1 is the die attach. The terms "die attach," "chip attach" or "chip connect" are used to describe the physical anchoring of the die to the common circuit base. The wires in the first level connection usually form only part of the die attach. Often the chips are bonded to the common circuit base as well.

Note the distinction between the use of the words "connection" and "interconnection." In this book, the term "connection" is used to refer to the electrical connection between two levels of the packaging architecture in an electronic system. The term "interconnection" (signal interconnection, interconnection structure, or interconnection circuit) refers to the conductors provided within the common circuit base.

The second level connections are the connections between the MCM and the PWB on which it is mounted. Again, these paths refer to the current carrying conductors. Separate structures are sometimes used to fix the MCM physically to the PWB. Second level connection alternatives are explored in Chapter 10.

The package is the final part of an MCM, which can be seen, touched and felt. It provides the following functions:

- Physical protection from environmental (corrosion, humidity) and mechanical (vibrations, shock) stresses and from handling by automatic machinery used during manufacturing processes
- Part of the second level connection elements for the distribution of signals, power, and ground
- A means for removing the heat dissipated in the chips
- A space transformation (fanout) from the fine conductor spacings on chips to the coarser spacings on PWBs

Thus, the electronic package is an electromechanical structure that supports, protects and connects (electrically and thermally) the devices contained within it, independent of the actual number of chips involved - single chip or multichip. The package serves as the link between the component or circuit level functions (such as ICs, discrete devices, resistors, capacitors) and system level functions (such as circuit board assembly and board to board connections). In addition, it often allows testing and burn-in of its contained parts prior to further assembly.

In Table 1-1, we summarize the basic MCM architecture outlined above, and provide some examples of functions and technologies for each level in the architecture. (Sometimes this architecture is called the "hierarchy.") Notice that the order of the levels from top to bottom in Table 1-1 is the same as the physical order, shown schematically in Figure 1-1. In the next sections we define in more detail some of the technologies shown in that Table.

Table 1-1 Basic MCM Architecture.

LEVEL	FUNCTIONS	TECHNOLOGIES
Chips	Digital Analog	Si: CMOS, bipolar GaAs
1st level connection	Conductor connection from chips to common circuit base	Peripheral: wire bond, TAB, flip TAB Area: flip chip solder bump
Common circuit base	Signal interconnection Power and ground conductors	Hybrid circuits MCM-L, MCM-C, MCM-D MCM-Si, MCM-D/C
MCM package	Hermeticity Heat removal Physical protection Conductors	Peripheral conductors: DIP, QFP Area array conductors: PGA, PAC
2nd level connection	Conductor connection to PWB	Plated through-hole Surface mount

1.1.3 First Level Connection and
Common Circuit Base Alternatives

Most of the methods for connecting the I/O conductors on the chips to the common circuit base also are used within single chip packages. See Figure 1-2. Wire bonding and tape automated bonding (TAB), for example, are used to make the connections between peripheral conductors on single chips and their packages. With "flip chip" arrangements, such as flip TAB, the chip is flipped over so that its conductors are on the bottom, facing the common circuit base. Flip chip solder bump (FCSB) connections are analogous to the connections beneath single chip surface mount area array packages to conductors on PWBs.

Physical attachment ("chip attach" or "die attach") of the chips to the common circuit base is a distinct operation from wire bonding or TAB conductor connections, but occurs simultaneously with the solder connections in FCSB. With flip TAB, there is no metallurgical bond between the chip and the common circuit base. Sometimes epoxy is used, and sometimes the chip is held in place by the connected TAB leads.

The common circuit base is that part of the MCM structure that defines the MCM technology used: MCM-L, MCM-C or MCM-D. However, MCM signal interconnection technologies themselves have developed from other technologies, such as patterning ICs on silicon, packaging ICs and mounting ICs on PWBs. The names for these technologies are relatively new [1], but the basic technologies are fairly mature.

For example, decreasing the sizes of conductor lines and spaces on laminate PWBs leads to the MCM-L technology ("L" stands for laminate). MCM-L technology also is an outgrowth of the COB technology, since the packages around individual chips are omitted. The laminate normally includes several layers of patterned metal conductors pressed together (laminated) with interspersed polymer dielectric insulating layers. The adhesive joining of all these layers is done at a relatively low temperature.

MCM-C technology ("C" stands for cofired ceramic, but also may include thick film) is a development of traditional hybrid circuit technology, where printed thick film conductors are used to provide the signal interconnections. To increase the circuit density, multiple thick film layers of conductor and dielectric are used, just as in PWBs. The layers originally had to be built up with sequential ceramic firing operations, since the dielectric was applied as a thick film of ink. Later developments allowed conductors and ceramic layers of alumina to be fired all at once, provided this was done at a higher temperature ("high temperature cofired ceramic" technology or HTCC). One early application of this cofired MCM-C structure, with over 30 ceramic layers, was for mainframe IBM computers [3].

The glass-ceramic dielectrics used originally in ink form were later made into tapes. (The layers of unfired ceramic generically are called "green tape." "Green" is a ceramicist's term for the unfired material, even though the sheets of tape actually may be other colors.) Simultaneous firing of all the layers was then possible at the lower temperatures used in original hybrid circuit technology. This is called "low temperature cofired ceramic" technology or LTCC.

To obtain the highest density MCM signal interconnection circuits, other technologies use techniques such as photolithography developed for patterning circuits on silicon chips. With MCM-D ("D" stands for deposited), both the conductors and the insulating dielectric layers that separate them are deposited. The dielectric is usually an advanced (organic) polymer such as polyimide, although (inorganic) silicon dioxide (SiO_2) is sometimes used. Multilayer conductors are deposited as thin films, such as are used on single layer hybrid circuits for resistors and bonding pads. Vacuum metal deposition processes such as sputtering, polymer deposition processes such as spin and spray coating, chemical vapor deposition (CVD) of SiO_2, and patterning processes such as wet chemical, or plasma (dry) etching and laser etching are familiar chip processing techniques.

The dielectrics used in MCM-D usually are not strong enough to stand alone. Thus, in this technology a separate, different material is used for mechanical stiffness and support. It may be silicon, metal, ceramic containing no conductors, or a PWB or multilayer cofired ceramic that contains conductors. The identification of this material is sometimes included in the technology label. For example, when MCM-D technology is used with a multilayer cofired ceramic substrate base containing the MCM power and ground circuits and signal I/O vias, the technology is called "MCM-D/C." When silicon is used, the MCM technology is called "MCM-Si." MCM-Si technologies may use either thermally grown SiO_2 or CVD SiO_2.

1.1.4 MCM Packaging Alternatives

Packaging technologies involve a consideration of the materials to be used, the geometry of the package, as well as the thermal and electrical design parameters. This section begins by considering the basic functions that a package must perform. Since most MCM packages have been, until now, similar or identical to single chip packages, the classifications of these packages also are summarized. Then, some features of packages specific to MCMs are examined.

Entire books [G1] have been written on microelectronics packaging. For details, the reader is referred to these general references at the end of the chapter. The intent in this section is to provide a working knowledge of the language of packaging and some aspects of packaging that will be useful and relevant to MCM design and technology selection.

Classifications of Packages
Packages can be classified by their main material, the second level connection technique, the means used to remove heat, the method and degree of protection (encapsulation) of the chips, first level connections, and common circuit base. In this part, we explore certain aspects of these classifications. Second level connection classifications are explained more fully in Chapter 10. Available options for heat removal are discussed in Chapter 12.

The most common materials used for the package body are ceramic and plastic. If ceramic is used, it is made using the same approaches used to make ceramic substrates, and then a metal lid is brazed to the package. Plastic bodies usually are made through a process called injection molding, where the chip is placed in a mold into which plastic is injected.

Packages may contain internal cavities, or recesses, for mounting chips or MCM substrates. These cavities provide room for many connections to the package leads.

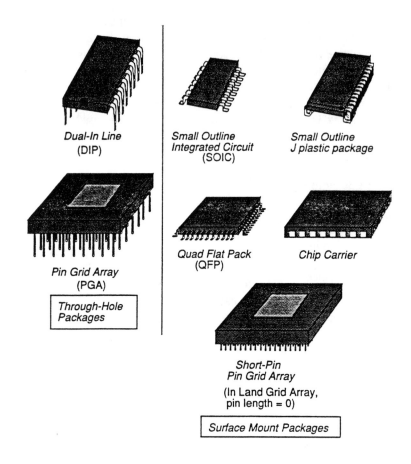

Figure 1-3 Common through-hole and surface mount packages. (Courtesy P. Franzon.)

Some of the single chip types of package also used for MCMs are sketched in Figure 1-3. One way to classify these packages is in terms of the method of mounting to the next level in system assembly: through-hole mounting or surface mounting. The conductors are called leads or (straight) pins; in the case of surface mounted packages they may be pads.

Plated through-holes (PTH) in a printed wiring board (PWB) refer to precision holes drilled through the board and plated with copper. These copper-plated holes form electrical interconnections between bond pads on the top of the board and the conductors within the board. They also provide interconnections between the different conductor planes within the board.

Current PWB design rules and manufacturing capability limit the use of through-hole packages with high pin counts. For example, current PWB designs can accommodate holes on a 0.100" (2.5 mm) grid. Each package requires one hole for every package pin; each pin is typically 0.035" (0.9 mm) in diameter. Advanced PWBs can accommodate holes with 0.050" spacing. Surface mounted packages, on the other hand, can be connected to pads of the surface of PWBs with 0.025" spacing.

Packages can be grouped further according to physical arrangement of the leads (peripheral or area array), and package lead geometries, such as J-shaped and gull wing. The SOIC (small outline IC) and QFP (quad flat pack) shown in Figure 1-3, for example, have gull wing leads.

In Table 1-2, we classify the common packages, provide their abbreviations, and indicate typical lead (conductor) count and spacing.

The progression in Table 1-2 is from the simple to the advanced. The upper package types in this table all have peripheral leads. We start with dual in-line packages (DIPs), which have leads on only two sides and which are mounted in PTHs. Of the surface mounted packages with peripheral leads, SOJ stands for a small outline package with J- shaped leads, TSOP stands for thin small outline package, and PLCC stands for plastic leaded chip carrier. Next come the area array packages, with PTH mounted pin grid arrays (PGAs) first. If the PGA pins are shortened as shown in Figure 1-3, then PGAs can be surface mounted. Since surface mount pads on a PWB can be spaced more closely than PTHs, the possible lead pitch is reduced.

Area array pad array carriers (PACs) with solder bumps on the underside may be soldered directly to the PWB in a process analogous to connection of the solder bump for flipped chips (FCSB) to their MCM substrate. The process sometimes is called C5 (Controlled Collapse Chip Carrier Connection) for the PAC, and C4 (Controlled Collapse Chip Connection) for the FCSB. PACs are presently in the prototype stage; future lead spacings may drop to 0.02". A PAC is similar to a PGA with zero-length leads. Such a PGA is called a land grid array (LGA).

Properties of MCM Packages

Table 1-2 is only a starting point for MCM packaging: there are many alternatives.

MCM packages often look like regular-sized versions or large versions of well known single chip packages with pins or leads emerging from them. Sometimes the MCM has a highly customized package that does not resemble an existing single chip package. Often the substrate base (mechanical support

Table 1-2 Common Package Classifications.

PACKAGE TYPE	BODY		LEAD		TYPICAL LEAD	
	Plastic	Ceramic	Arrange	Connect	Pitch (in.)	Count
Dual In-Line	PDIP	Cer/DIP	Peripheral	PTH	0.05 - 0.1	64
Small Outline	SOIC SOJ TSOP	Flat Pack	Peripheral	Surface	0.025 - 0.05	28
Chip Carrier	PLCC		Peripheral	Surface	0.05	84
Quad Flat Pack	PQFP		Peripheral	Surface	0.01 - 0.05	256
Pin Grid Array	PPGA	CPGA	Area	PTH	0.05 - 0.1	144; 299
Short Lead PGA	yes	yes	Area	Surface	0.05	to 500
Pad Array Carrier = Land Grid Array	yes	yes	Area	Surface	0.02 - 0.05	to 1000

structure of the common circuit base) is used to form part of the package. In other cases, a package housing is not used. Instead, leads or pins are attached to the common circuit base for second level connection, and the exposed chips and wires are sealed against moisture by using epoxy or gel.

MCM packages must be mated carefully to the total MCM architecture. For example, the MCM common circuit base technology determines the possible lead geometry of the package. If there are electrical vias through the substrate bases for MCM-D or MCM-Si, the package leads then must be peripheral (as shown schematically in Figure 1-1). To allow area array conductors under the package with MCM-D, a multilayer ceramic package can be added (as in MCM-D/C).

The first level connection determines the possible thermal paths. In the case of flip chip mounting, much of the heat can be taken through the top of the package. Then rather poor thermal conductivity in the MCM substrate can be tolerated. Since there are no electrically sensitive conductors on the backsides of the chips, the chips may be connected through thermal gel or elastic spacers directly to the package top, to which a heatsink or cold plate can be attached.

Hermeticity describes the relative effectiveness, or extent, of sealing of a package. Hermetic packages have very low leak rates in helium testing ($< 10^{-7}$ atm-cc/sec typically). Hermeticity or near-hermeticity is important in packaging, since moisture can cause electrical circuit failures. Water absorption in organic dielectrics results in swelling that can cause separation of conductors, and other electrochemical activity such as copper plating, for example. Also, water may induce migration of silver from conductors, leading to short circuits. HTCC MCM-Cs may be used without a separate package, since the fired ceramics are hermetic. As discussed in Chapter 6, most LTCC ceramics probably are hermetic also. The lowest cost MCM-Ls have simple epoxy encapsulation for protection, without a separate package [4], and thus are not hermetic. Recently, it has been realized that many nonhermetic sealing approaches, including plastic packaging, epoxy and gel encapsulation, are sufficiently resistant against contaminants and moisture to provide good reliability in many, even military, applications. This subject is currently under investigation. See Figure 1-4 for some possible encapsulation options.

The performance of a package is discussed further in Chapter 3.

1.2 FINDING YOUR WAY

The other chapters in **Part A - The Framework** provide further perspective on MCMs. They also provide bases for choices between MCM technologies from the point of view of manufacturability, performance and cost, in Chapters 2, 3 and 4, respectively. Chapter 2 provides an overview related to the materials and the manufacturability of MCMs. Specific examples of how some MCMs have been manufactured are presented. Chapter 3 covers the relationships between the needs of the system being packaged and the alternative technologies that exist ("need" being defined in terms of application or product goals, and being coupled with performance and cost as tradeoffs).

Chapter 4 focuses on MCM cost at the component level in keeping with the book's emphasis on MCM components. Two types of cost models are described. One is manufacturing activity-based (called technical cost modeling), useful when in-house manufacturing details are known. The other is a design activity-based model useful when the only manufacturing-related information available to the designer are vendor quoted prices, for example. Although many companies

Hermetic:

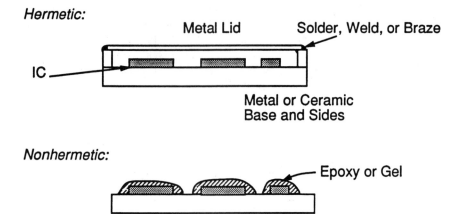

Nonhermetic:

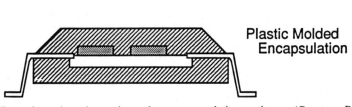

Figure 1-4 Hermetic and nonhermetic package encapsulation options. (Courtesy P. Franzon).

pursue a technology based primarily on previous experience, these chapters may provide the incentive to examine less familiar technologies.

In **Part B - The Basics**, MCM-L, MCM-C, and MCM-D signal interconnection technologies are discussed in some depth in Chapters 5, 6 and 7, respectively. Chapter 8 examines the dielectrics used with MCM-D. The several parts of Chapter 9 treat the physical die attach methods as well as the conductor connection technologies such as wire bonding, TAB, and flip chip methods similar to FCSB for the interior of the MCM. External connection of MCMs to PWBs is considered in Chapter 10.

The maximization of electrical (signal propagation) performance and thermal performance is considered in Chapters 11 and 12, respectively. MCM electrical testing issues are a big factor in overall MCM costs and are considered in Chapter 13.

The Case Studies in Part C consider various developments of actual products based on particular MCM technologies. The Unisys experience with MCM-C and MCM-D is examined in Chapter 14. Hughes MCM-C and MCM-D products are covered in Chapter 15. Experience with MCM-Si is considered in Chapter 16, and the DEC MCM-D experience is considered in Chapter 17. For examples of MCM-L products, the end of Chapter 5 may be consulted.

1.3 THE IMPORTANCE OF MATERIALS

Even a cursory reading of the various chapters in Parts A and B reveals the central importance of materials properties for MCM technologies. Many of the most important properties are highlighted in Table 1-3, which also shows some of the symbols and units. We mention some examples briefly in this section.

Numerical values for many of these properties are given in Chapters 5 through 8. For example, Table 5-1 lists the properties of dielectrics used in MCM-L, Tables 6-5, 6-7 and 6-8 show the properties of MCM-C materials, including HTCC and LTCC materials, and Table 7-3 gives the properties of dielectrics and conductors used in MCM-D. Tables 8-2, 8-4, 8-5 and 8-6 list important properties for various MCM-D thin film dielectrics, and Table 8-13 summarizes these properties for polyimides.

The electrical properties are most important for high frequency signal interconnections, where they affect the signal delay between chips and the spacing between conductors. These properties are introduced in Chapters 2 and 3, and are discussed in considerable detail in Chapter 11. For example, the speed of signal propagation in a dielectric is inversely proportional to $\sqrt{\varepsilon_r}$, so higher dielectric constants, ε_r, cause longer delays.

Performance is degraded not only by slower propagation between chips, but also by distortion of square-shaped signal pulses. Major sources of this degradation are unintentional resistances, inductances and capacitances in the circuits between chips. These unintentional elements are called parasitic circuit elements. Some of these parasitics are caused by the first level connections; others are related to the materials in the common circuit base; still others are caused by second level connections.

Parasitic resistances, for example, are associated with losses in the metal conductors and in the dielectric of the common circuit base. These resistances increase the rise time of the pulse, which effectively increases the signal delay. Dielectric losses are proportional to the dielectric loss tangent, $\tan \delta$. The loss tangent (dimensionless) is sometimes called the dissipation factor, in which case it is normally given in percent.

Parasitic capacitance between conductors is due to the dielectric, and can cause crosstalk coupling of one signal into another. Parasitic mutual inductance is associated with the magnetic fields around current-carrying conductors and can

Table 1-3 Principal MCM Materials Properties.

ELECTRICAL PROPERTIES	THERMAL PROPERTIES	MECHANICAL PROPERTIES	PHYSICAL PROPERTIES	CHEMICAL PROPERTIES
Dielectric constant, ε_r	Coefficient of thermal expansion CTE (ppm/°C)	Young's modulus E (GPa or kpsi)	Microstructure (grain size)	Metal oxidation
				Metal migration
			Flatness and Planarization	
Loss tangent tan δ	Thermal conductivity (W/m-K)	Poisson's ratio ν	Hermeticity	Reactivity
				Adhesion
Resistivity (Ω-cm)	Shrinkage	Elongation	Melting point	Toxicity
	Thermal stability	Flexural strength (GPa)	Glass transition temperature, T_g	Environmental

also cause crosstalk. To keep the crosstalk acceptably low in high frequency interconnections, the conductor spacing must increase with increased ε_r. These factors are explained further in Chapters 3 and 11.

Thermal properties are important during the processing that forms the MCM and/or during the thermal cycling that occurs in MCM operation. Ceramic shrinkage after processing ("firing") and the thermal stability of organic dielectrics at processing temperatures are examples of dielectric properties important mostly for process design. A high thermal stability means that there is no significant outgassing or change in the mechanical dimensions that could lead to delamination.

Thermal conductivity is most important with high power single and multichip module operation. The thermal conductivity of each path from the chip to the outside world must be considered to maintain the junction temperatures on the chip at acceptable levels during operation. Poor heat transfer through materials with low thermal conductivity can be overcome by placing special conductors ("thermal vias") through the material, or by sending most of the heat through another path, such as a conductor attached to the "backside" (side away from the I/Os) of a flip chip connected die. Optimal design of these paths and also of heat transfer to cooling media such as air and water is examined in Chapter 12.

Differences in the coefficient of thermal expansion (CTE) between chips, common circuit base or package can lead to breakage either during cool down from a processing step or after extended thermal cycling. Whenever the temperature changes, connected parts with different CTEs expand or contract by different amounts, resulting in strain on the interface between parts. With large temperature changes, such as occur during processing, this might result in immediate failure, such as conductor breaks, or conductor peeling from a dielectric. Repeated smaller temperature changes, such as on/off operation in the field, eventually might result in thermal fatigue failures.

The units of CTE are parts per million per °C (ppm/°C). For example, with a CTE of 100 ppm/°C, the length of an unconstrained dielectric originally 10 mm long increases by 0.1 mm in a temperature rise of 100°C. Note particularly that the CTE itself can be a strong function of temperature, and can also be anisotropic (direction dependent). The CTE (also called the thermal coefficient of expansion, or TCE) for some ceramics and organic dielectrics are tabulated in Chapters 6 and 8.

Some mechanical properties are important for their role in determining the response of materials to temperature cycling. Consider, for example, the bowing or camber of a silicon substrate beneath a thin dielectric. Excessive bowing following cool down from processing can make photolithographic conductor patterning difficult in MCM-Si. This bowing results from the different CTEs of the substrate and dielectric and the substrate, and is proportional to the ratio of the Young's modulus of the dielectric to that of the substrate. A large Poisson's ratio corresponding to an almost incompressible dielectric can also cause large bowing. These issues are discussed Chapter 7.

Conductors with high elongation and ceramics with high flexural strength can make a design more robust to thermal and mechanical cycling. In copper, the substantial plastic flow or elongation before breakage is a definite advantage. Similarly, solders that can plastically deform to accommodate CTE mismatch between chips and the MCM substrate can make flip chip solder bump (FCSB) technology more reliable.

Several units of stress for quantities such as Young's (elastic) modulus and flexural strength are in common use. Some equivalences are: 1 GPa = 1000 N/mm^2 = 145 kpsi, where GPa is gigapascal, N is newtons, and kpsi is 1000 pounds per square inch.

The physical and chemical properties listed in Table 1-3 are important mainly in processing. For example, a smooth and flat surface is important for high resolution conductor patterning. A ceramic with a large grain size may have unacceptably large surface roughness. In a multilayer structure, a polymer with good planarization can smooth out the features, or height variations, of layers that it covers.

Adhesion between metal conductors and polymer dielectrics is important in MCM-L and MCM-D processing. The glass transition temperature, T_g, of a glassy polymer is its softening temperature. Generally a lower T_g is beneficial for adhesion of the polymer to other layers and for stress relaxation, but is detrimental to dimensional stability. See the discussion in Chapter 8.

Copper reacts with polyimides and oxidizes at high temperatures. Therefore, copper in multilayer structures often is protected with another metal, such as nickel or chromium, which in turn should not react with the neighboring dielectric. Good conductors, such as copper, silver and gold, melt at temperatures required in HTCC processing, so lower conductivity metals must be used in HTCC structures. These and many other processing issues are covered in detail for each of the MCM signal interconnection technologies in Chapters 5, 6 and 7.

Hermeticity or near-hermeticity is important for long term reliability. This is particularly important for MCM-C devices that may have no separate package. Verification of LTCC ceramic hermeticity allows the use of relatively inexpensive silver conductors, which are susceptible to metal migration when exposed to moisture. Absorption of moisture in polymers may cause swelling and consequent breakage of attached conductors.

Toxicity of compounds of beryllium and alloys of lead under certain conditions, and environmental concerns over chlorinated fluorocarbons and some solvents, has led to searches for alternatives to these materials.

New materials also can make new high performance and/or cheaper technologies practical. For example, porous cordierite ceramic was developed for the interior of MCM-L laminates because it has a relatively high thermal conductivity and Young's modulus, and a relatively low dielectric constant and CTE. Some other new materials possibilities have been touched upon in this short review. Hopefully, readers of this book will be inspired to find new materials solutions to problems that impede the introduction of present and future MCM technologies.

1.4 THE IMPORTANCE OF MANUFACTURING PROCESSES

New MCM manufacturing processes to reduce costs are under active investigation, as can be seen from a reading of the chapters in Part B. These include blurring of the various MCM technology definitions described above.

For example, MCM-L may approach the thin film capability of MCM-D to pattern fine lines by using additive conductor processing instead of the subtractive etch processing traditionally used to pattern conductors on PWBs. The common circuit base is still a low cost laminate. Similarly, new developments of thick film screen printing emulsions and wires may allow fine conductor lines to be patterned on LTCC dielectrics, while at the same time new LTCC dielectrics have been developed with dielectric constants as low as those of the organic dielectrics used in MCM-D. In these ways, performance may approach that of MCM-D but at lower cost.

On the other hand, MCM-D costs can be reduced by increasing yields and using larger substrates. For example, patterning of somewhat wider lines and spaces on MCM-D substrates can lead to higher yields, since there are fewer large defects that can cause shorts and opens than there are small defects. Larger wafers (for MCM-Si) can become useable also. This is a simple example of improved manufacturability, where the yield is less sensitive to process imperfections. This type of manufacturability improvement is important particularly because it reduces costs even without the reduction in process imperfections that normally occur in proceeding up the learning curve to high volume production.

Just as with IC patterning, transfer of a new technology from a prototype lab to the production floor is critical. The production environment offers new challenges to high yields. Superior R&D efforts by U.S. companies have not been matched with equal efforts in transferring technology to production. Fast turnaround and accurate monitoring of factory production flow are essential.

High performance with high yields and accompanying low costs then requires creative and competent people from process design engineers to technology transfer experts to production technicians. An adequate supply of such people presupposes that manufacturing is viewed as an attractive alternative for students making career choices. Manufacturing competence is critical to a healthy economy [5], and we hope that readers of this book will even find that manufacturing can be glamorous and rewarding!

1.5 THE IMPORTANCE OF INDUSTRY INFRASTRUCTURE

In order for a product (the "final" product) to be designed and produced successfully, a number of intermediate products and services must be available. For MCMs, for example, these include the substrates, the chips, CAD tools and training in how to use them, test machines, etc. In principle, a company could provide such intermediate products and services from internal resources. Such companies are referred to as "vertically integrated companies." However, for most companies this would be inefficient. Thus, a portion of those intermediate products and services must exist outside the company as common, shared resources. These intermediate products and services provided to the industry as a whole are referred to as the "industry infrastructure."

Naturally, the infrastructure needs of any one company depend on what it does not wish to provide from its internal resources. The most common case is the company whose only asset is design expertise. Such a company relies on others to manufacture the ICs, MCMs, PWBs and other assemblies that make up their designs, and to provide the information and tools necessary to carry out their designs. In order to implement high performance MCM-based products successfully, such a company needs the following infrastructure elements:

- One or, preferably, two or more suppliers who can manufacture the same "unpopulated" (without chips) MCM in quantity with high yields and with sufficient manufacturing capability to be cost competitive. Two suppliers are preferred to reduce risk of product starvation if one supplier has a temporary problem, for example. This is referred to as "multi-sourcing." True multi-sourcing is rare now, but most MCM designers are coping adequately with single suppliers. MCMs need to be delivered fully tested and guaranteed.

- One, preferably two or more suppliers of qualified bare die suitable for MCM mounting. By "qualified," it is meant tested, burnt in, and guaranteed to the degree that packaged die are. Generally, it is more efficient for the chip manufacturer to perform these tests than for the chip user. By "suitable," it is meant that die are provided in the following forms:

- Bare, suitable for wire bonding
- With a pad layout suitable for attachment to a standard TAB frame, or with a TAB frame
- With solder bumps, suitable for flip connection. Since die are easiest to solder bump wafer form, this should be done before dicing and shipping.

One side effect of this requirement is that multiple suppliers provide chips with the same size, pad layout and electrical specifications. One supplier might suffice as long as the extra risk is acceptable. This is rare for chips, however. Chip production is very susceptible to production problems.

- At least two suppliers of the other required package components, such as TAB frames (if not provided by the chip supplier) and heatsinks.

- Manufacturers who can assemble all of these components, determine if they work correctly, and repair them, if needed. Providers of unpopulated MCMs also could offer this service. Standards are required so that automated assembly equipment can be developed.

- Computer-aided design (CAD) tools. These tools produce the MCM physical layout (locations of the chips and the interconnecting circuits, for example) and turn this layout into a suitable format for feeding to production machines. Ideally these tools automatically produce a design that functions correctly.

- Computer-aided engineering (CAE) tools to help in early design decisions, such as which MCM technology to use and to verify that the design will work. The verification step requires simulators to verify the logic design, the electrical properties of the interconnection and package structures, the thermal design, and the test plan. Other verification tools check that the manufacturing design rules have been followed and that the design will have no yield or other manufacturing problems.
    All of these tools should be seamless in that the designer should not have to re-enter information that is described already in another tool. For example, the designer should be able to simulate an interconnection structure by "clicking" on it in the layout; he or she should not have to type in a simulation file.

- Computer-based libraries. These libraries should provide physical information, including die size and pad location, as well as information needed to simulate the design, such as logic models of the circuits and package structures, thermal models of heat dissipation and package properties, test models of the chip, and the MCM design rules. Again, these libraries should be seamless. The designer should not need to type in or move data between programs in order to do the design. For example, clicking of the interconnection to be simulated should allow appropriate models to be pulled from the libraries automatically.

- Information and training. *This book is an element of the infrastructure!*

As of today (mid-1992), not all of this infrastructure is in place, particularly in the United States and Europe. (Japanese companies tend to be large and vertically integrated, or to have such close relationships with supplier companies that they are effectively vertically integrated.) This picture is changing very rapidly. Next we summarize the current infrastructure and how it is changing.

Though many MCM foundries exist, only a few can manufacture MCMs in high volume. Those that can often are using lines that were previously internal to vertically integrated companies. In the future, more of the smaller companies need to be able to provide volume manufacturing of MCMs once the need for such high volume exists. The work of standards committees to develop standard MCM sizes and second level packages hopefully will enable a designer to obtain the same MCM package from multiple sources. This activity also should enable equipment manufacturers to produce equipment that can handle and assemble these MCMs automatically.

Until very recently, guaranteed MCM-suitable die were available only at a high price premium over the equivalent packaged part. Several service companies are providing these parts, buying wafers from the chip manufacturers, and testing them. One chip manufacturer recently announced the availability of tested, guaranteed bare die, at the same prices as the equivalent packaged part [6]. Hopefully, as demand increases and the cost of at-speed bare die testing and burn-in decreases, more manufacturers will offer this service at a price *less than* the equivalent packaged part.

Currently, obtaining multiple sources of bare die with the same pad locations and in the same size is difficult. Manufacturers may need to collaborate. For TABed (or TABable) die to become available, for example, standard pad sizes and locations are needed, as well as standard TAB frame sizes (see Chapter 9). For solder bumped die to become pervasive, more chip companies will need to license or develop a solder bump technology. One helpful development would be for a foundry to add solder bumps to die in wafer form obtained from chip suppliers.

CAD and CAE vendors provide many good computer tools. However, seamless operation with each other and with libraries is not quite there. More agreement between the tool vendors is needed. Part of this agreement could be in the form of common frameworks for information exchange between tools. A small number of frameworks are emerging that provide this commonality. Seamless libraries require that the CAE vendors and the library providers (the chip and MCM manufacturers) agree to common formats for this information.

Achievement of much of the above infrastructure requires agreement in a competitive environment. Most companies recognize that this agreement is necessary for their own commercial success. A number of forums exist for the required consensus building effort. These forums have been provided through professional and industry service organizations, such as the IEEE, ISHM, IEPS, IPC, EIA, JEDEC, SRC; by the government (particularly DARPA, DOD-Air Force, Army); and by Consortia (particularly MCC and MCNC).

MCM vendors may copy profitably the approach of vendors of ASICs (Application Specific Integrated Circuits). Such vendors provide standard packages, high level CAD descriptions of their gate arrays, and quick turnaround. MCM equivalents would include: industry standards for new MCM packages; high level CAD descriptions of microprocessors, logic and memory chips; CAD aids for analyzing propagation delay, reflections, crosstalk, and final noise margin; and CAD thermal design tools. Much of the engineering can be done in advance of specific customer requirements, not only to produce faster turnaround, but also to improve reliability. Increased volume relative to total custom products also will provide a cost advantage.

Last but not least, training and information is needed. This book fulfills part of that mission. Conferences and seminars also provide part of this information. Universities have been slow in providing interdisciplinary courses in packaging technologies. This book will help overcome this by serving as a possible course text. Training courses for faculty, with government and industry-provided support to attend such courses, have been successful in the past for other important emerging technologies. The chip design industry, in particular, has benefited from this activity in the past. Now it is the turn of the electronics packaging industry!

1.6 DECISION-MAKING AS A PROCESS

With the introduction of MCMs, the number of packaging alternatives available to the designer, and the number of decisions that must be made, have increased substantially. The case studies presented in Part C of this book provide unique examples of the decision-making process. The descriptions there of the decision-making process emphasize the decision to introduce an MCM technology and then designing into it rather than just describing the reported results. In this section, we describe some general features of the decision-making process.

Engineering, managerial and marketing personnel together have to decide on the packaging alternative that best meets the need of their customer. An MCM technology provides a significant performance premium for the customer, but possibly at additional cost and schedule risk. Together with engineering staff, the manager needs to determine the impact of the different packaging technologies on performance and cost. Marketing and sales staff need to be provided with a clear understanding of what these new technologies are providing for their customers.

Knowing what packaging alternatives are available, we first investigate possible applications for MCM technologies by looking at the market.

1.6.1 Determining the Application: Possible MCM Markets

Though any such classification is somewhat arbitrary, it is possible to identify six categories of electronic systems:

1. **Consumer Products:** Included are consumer entertainment products, home appliances, and personal communications products, such as wireless (cellular) telephones
2. **Aerospace and Military Products:** Included are avionics, satellites, and military communications equipment
3. **Computers:** Classified broadly into the following subcategories according to applications and relative importance of performance and cost factors:
   - *Low-end computers*, such as PCs used for general office applications for example, word processing. Low-end computers can be defined as computers designed for users who wish to minimize cost.
   - *Mid-range computers*, such as workstations and servers used mainly for technical applications, for example, circuit simulation. Mid-range computers can be defined as computers intended for users who wish to maximize the ratio of performance to cost.
   - *High-end computers*, including supercomputers and mainframes used respectively for specialized scientific applications such as climate modeling and for applications requiring the rapid processing of large amounts of data such as airline reservation systems. High-end computers can be defined as computers intended for users who wish to maximize performance.

- *Portable computing products*, including notebooks
- *Peripherals* including input/output devices (scanners, printers, disks, screens)
- *Embedded computers*, such as machinery controllers and automobile computers
4. **Biomedical Systems:** Included are items such as office ultrasound machines, large CAT scanners and human-embedded devices
5. **Telecommunications:** Included are centrally provided equipment, such as switches, PBXs and their line cards, for example
6. **Instrumentation:** Included are oscilloscopes and test equipment

Each of these categories has somewhat different packaging needs.

The electronics parts usually found in consumer electronics do not have high pin counts (64 pins is a maximum typical count) and do not require high speed interconnection delays. Thus these systems usually use plastic single chip packages or, if small size creates a selling advantage, chip-on-board packages (the cheapest form of a bare chip mounted package). For example, most hand held calculators use a chip-on-board technology.

The majority of aerospace and military electronics systems are either signal processing or communications systems. Several examples are given in Chapter 15. In a signal processing system such as a radar processor, high circuit speeds are not usually required. Instead, additional system performance is achieved by having many chips working on the same signal at the same time. (However, there has been a recent trend to greater use of high speed chips in military electronics.) Though these chips may have only moderate I/O pin counts, the system size and weight can benefit enormously from MCM technology, and thus their use in this domain is common.

To obtain peak performance out of a computer, high I/O count parts must be connected with the shortest possible interconnection delay. (This is discussed in Chapter 3.) For example, the DEC 21064 RISC microprocessor, found in workstation products, is packaged currently in a 431-pin ceramic PGA and runs at 150 MHz. Even small computers tend to have high pin counts. For example, the Intel 386SL microprocessor, found in the current generation of notebook computers, is packaged currently as a 132-pin (LGA) and runs at 25 MHz. To pack these high pin count chips in a small space and to achieve small interconnection delays, most mid-range computer manufacturers are considering the use of MCM technology. Notebook computer manufacturers are seriously considering laminate (MCM-L) technology, as it provides considerable performance improvement, mainly through reduced size. Workstation manufacturers are considering both laminate and thin film (MCM-D) technologies, and several have constructed prototypes. Very low-end computers, such as PCs costing under $1000 (1992 dollars), seek performance advantages but cost dominates and thus they use plastic single chip packaging.

High-end computer manufacturers have been using MCM technology, mainly ceramic (MCM-C) technology, for many years. As it is not currently possible to manufacture a high speed mainframe processor as a single chip, MCM technology enables high-end computer manufacturers to interconnect many chips and make a "virtual" large chip as an MCM. They require the highest performance out of MCM technology. Examples are given in Chapters 14 and 17.

Embedded computer applications, such as a washing machine controller, a printer controller, a fuel injection controller, or even an automobile navigation computer, tend to be low performance applications. As pin counts and clock speeds are not high, and space is not usually at a premium, advanced MCM technology has little to offer. However, ceramic hybrid circuits have been used for a long time in automobiles as a means to cope with the harsh environmental conditions. Plastic packages provide insufficient protection in this hot, hydrocarbon-filled environment, and hermetic hybrid packages provide a reasonable cost alternative.

Typical biomedical systems would be a computer aided tomography (CAT) scanner, or an implanted defibrillator. Much of the electronics in a CAT scanner is used for signal processing, similar to applications in military systems. However, since size is relatively unimportant, these systems do not need advanced MCM technology. Size is important in implanted systems, but the electronics parts in such systems have been neither high pin count nor high speed. Nevertheless, high pin count electronics parts are appearing in new implanted systems, and hybrid circuit packaging may no longer suffice.

Telecommunications equipment, such as large switches, tend to require both high speeds and large amounts of wiring. The telephone industry is looking very seriously at MCM technology for the provision of these functions.

Instrumentation such as oscilloscopes and test systems places unique demands on its electronics in that it must operate faster than the electronics in the system it is used to observe or test. Hybrid circuits often have been used in such systems and the use of newer MCM technologies is anticipated.The performance and cost considerations associated with these different systems are discussed also in Section 3.3.

1.6.2 Determining the MCM Technology: Business Decisions

Once a particular market application seems attractive for further consideration, many decisions must be made. Fundamental business issues include whether to develop an internal capability or to use the contract services of another company, whether to try to change the technology with each new product, or to try to identify and develop a technology that will apply to several products.

Customer requirements for electrical performance, package size, unit cost and time-to-market must always be met. Such considerations dictate many of the choices in choosing an MCM technology. At the same time, there are always alternative choices for some aspects of every technology. It is important to recognize the cost and time-to-market advantages gained by exploiting any design or manufacturing infrastructure (and its accompanying expertise) that exists already, and that may be available within a company. For example, in a company with ready access to silicon IC design and fabrication resources, it probably will be cost effective to utilize an MCM-Si architecture. Similarly, assuming that the same desired performance specifications can be met, a company with skills in ceramics and thick film chemistry might initially pursue development of an MCM-C architecture.

On the other hand, if it is desired to do more than demonstrate the MCM concept, or to provide a limited number of application specific MCM prototypes, it is necessary to look at the longer range capabilities of a given MCM design to avoid being locked into an inflexible technology. It is important also to consider outside vertically integrated companies (IBM, DEC, Hughes, HP to name only some) that are willing and able to sell designs, components and functional subsystems. A readily accessible infrastructure greatly eases entry into the MCM business and provides a technology capable of meeting all the initial performance requirements. Such an infrastructure may be ill-suited for scaling up to large volume production or may be incapable of meeting expected future performance requirements. For example, a wire bond MCM-L architecture may represent a good route into the business, but an investment in flip chip MCM-D technology will have to be made eventually if exceptional high frequency performance is a long-range objective.

The importance of automated production using standard parts needs to be emphasized. As long as MCM technology development is focused on highly customized forms for small production runs, MCM costs will be high. Automation and standardization are necessary for cost-effective medium and large MCM product runs. This means the selection and refinement of MCM processes and designs appropriate for high speed automated batch processing, assembly and testing. As far as possible, these MCM technologies also must exploit the generic assembly and testing tools and standards that can be shared across a broad (industry wide) design and manufacturing base. In this case, even MCM-D may become a high volume cost effective alternative to VLSI integration and conventional packaging, and quickly claim its rightful share of the market!

1.6.3 Designing the Product: Multidisciplinary Engineering

With the growing need for aggressive packaging technologies, an interdisciplinary approach is needed for the engineering decision making process. Until recently, electronic package engineering was a discipline concerned mainly with manufacturing and reliability issues. Other engineers, separately, could take the package types offered by the package engineers and design products that met their required specifications. Today unfortunately, this strategy is no longer possible.

Key decisions in packaging involve detailed material, manufacturing and reliability issues as well as electrical and thermal knowledge. Now design and systems engineers need to understand packaging technologies because package performance has become critical to their design function. Until recently, the choice of packaging technology was obvious once the chips were defined. Now this is not the case.

Even if a strategic direction in packaging technology has been selected, there are other subchoices and decisions still to make. Thermal design has become more intensive in nature. A different perspective is needed in electrical design. Managing the test process, for example, must be given a higher priority than previously. More effort must be spent on understanding materials and manufacturing issues as now their effect on cost, and their interaction with system function, is greater.

"Concurrent engineering" refers to cases where design and manufacturing decisions have to be made early and with the interaction of many different kinds of engineers. MCM product design clearly is one of these cases.

1.7 OVERALL PROSPECTS FOR MCMs

In this book, we consider not only various MCM alternatives, but also the other alternative: *"None of the Above."* Why MCMs at all? Two major perceived impediments to the widespread production of MCMs are:

- **Design Time**: Designing an MCM generally takes longer than designing the equivalent PWB. Then why bother with the new complications of MCMs if single chips will have the same performance and density by the time an MCM system is finally brought to market?
- **Cost**: Even if the design time can be reduced, will sufficient investment ever be made in U.S. production of MCMs to make their costs competitive?

With regard to design time considerations, a number of points should be made. First, as MCM technology becomes more pervasive, the additional design time overhead will decrease to the point where it is not a serious factor, particularly with proper application of concurrent engineering techniques. Second, many MCMs contain a mix of semiconductor technologies (such as digital, analog, memory, silicon and GaAs) and cannot be replaced by a single chip. Third, large MCMs contain so many transistors that a single chip equivalent will take a long time to develop. Fourth, it does not make sense to design a custom chip for a low volume application. An equivalent MCM will often make sense, however. Finally, design time is not an important factor in many applications where product lifetimes are long, for example, in military applications.

Even in today's commercial markets, however, relatively long life cycles are possible. Systems based on popular microprocessors sometimes fall into this category, and it is encouraging to see that Intel, for example, is beginning to offer bare die of its popular chips [6]. Secondly, even large computer systems are gravitating towards parallel processing based on microprocessors. The size advantage of MCMs is attractive here, and a single modular design might be used over and over to add on processing capability. Another possibility for MCMs lies in designing systems such that the MCMs can be replaced later with improved chips having equivalent performance and I/Os without affecting the rest of the system [8].

With regard to costs, there is serious concern that MCMs will never be commercially viable in the U.S. without the commitment of long term investment. The case of Group 3 Fax technology is cited [9], where investment in the U.S. was withdrawn prior to the development of a large volume market that ultimately drove the costs down. Suggested solutions to this problem include direct government support and management enthusiasm for an initial high volume capability [9], an improved government climate for long term investment and better marketing forecasts [10]. These solutions, however, are not under the control of the individual small or medium size manufacturer.

Not only are initial MCM costs high because the technology is new, but also because there are new design and testing interfaces between chip makers, MCM producers, and systems houses [11]. Long term working relationships need to be initiated within the industry to reduce costs associated with these interfaces [10]-[11]. Costs may be lower in the long run if systems people pick qualified suppliers and give them some confidence in a stable market while helping them develop the needed standards and interfaces. Trying to maintain the usual lowest bidder approach can be counterproductive, because then no supplier may wish to invest in the volume production capability that will drive the cost down.

Several chip makers (Intel, Motorola to name only a few) have taken the initiative, without waiting for support from systems houses. They have not only offered bare die of their popular chips, but they have started offering fully tested units ("known good die") [6]. Bare die testing eliminates an uncertainty leading to possible low yield of more expensive MCM units with several chips.

Ramp-up of MCM production may also occur fairly gradually. Since there are different possible systems and different MCM technologies, there is not the all-or-nothing quantum barrier of the Group 3 Fax technology cited above. MCM production may start out meeting market needs for "few chips packages" [12].

MCMs also offer the unique capability for optimized chip technology. No longer must different types of circuits be built on the same chip. Instead a single IC fabrication process can be used for an entire chip. For example, RAM processes can be used for RAMs, logic processes for CPUs, analog processes for CODECs and GaAs, if needed.

A promising niche for commercial MCMs in the short term is in systems where small size is absolutely essential, but the ultimate in performance is not [6]. One possible example is portable communication products [13]. The reduced interconnection complexity associated with MCMs also is attractive in these applications.

High performance systems where the speed is limited by the number of I/Os available on a single chip package are a longer term candidate for MCMs [14]. Reductions in the number of second level connections also should lead to higher reliability in these applications. MCMs have another unique advantage in their huge array contacting capability (see Chapter 18). Finally, MCMs can provide optimal functional performance, if signals only have to travel at high speed between chips within the same MCM package.

High performance alone, however, is not now a sufficient driving force for volume MCMs. Intel recently dropped its production of an advanced MCM for the reason cited at the beginning of this section: the next generation chip device came out with performance comparable to the MCM [6].

Many of these issues are discussed in more detail in Chapter 18. We hope that this cursory view sparks your interest!

1.8 SUMMARY

MCMs offer the potential for increased chip density leading to reduced size of electronic systems. Together with reduced size, MCMs offer a number of advantages. The speed performance is improved due to smaller chip spacings and reduced parasitics. Reliability is improved due to the reduction of the number of second level connections.

The successful use of MCM technology requires careful application. The alternatives and the economic issues that affect its use must be considered. Currently, MCM components are more expensive than the equivalent collection of single chip components and PWBs. Sometimes, an advanced technology custom chip will make sense over a small MCM, particularly a high volume application. If that solution does not make sense, then MCM technology must be considered seriously in any performance-driven application. In any case, a large system built out of MCMs might be less expensive actually than the equivalent single chip package system due to substantial savings in total size.

If MCMs are to be used in an application, full consideration should be given also to the effects of infrastructure, standardization and automated assembly on the part to be sued. Careful, concurrently engineered design is required.

The main challenge to keep in mind while reading this book is how to realize these potential advantages of MCMs for applications you may have. Technologies must be developed and chosen so that the MCMs are not hindered from reaching their potential. In particular, an MCM should maintain the performance of its component chips with minimal degradation. The material in Part B of this book will help in developing the required technology. The examples in Part C will help in making the right technology choices. Part D will help to verify whether you have made the right choices. *We wish you success in an exciting adventure!*

Acknowledgments

The author expresses sincere gratitude to her co-editor, Dr. Paul D. Franzon for his helpful technical discussions and encouragement pertaining to this chapter and to the overall scope of the book. His technical collaboration in the goals of this book has been a most rewarding opportunity! She also acknowledges with sincere appreciation the informative, technical discussions and helpful suggestions of John Nelson, Unisys, John Segelken and T. Dixon Dudderar, AT&T Bell Laboratories and Walter Bertram, Alcoa Electronic Packaging. Their encouragement and support in completing this chapter have been of particular value also.

General References

These following books provide excellent reference material and more detailed information on topics related to packaging and covered in the book.

G1   R. R. Tummala, E. J. Rymaszewski, eds., *Microelectronics Packaging Handbook*, New York: Van Nostrand Reinhold, 1989.

G2   D. P. Seraphim, R. Lasky, C. Y. Li, *Principles of Electronic Packaging*, New York: McGraw-Hill, 1989.

G3   R. W. Johnson, R. K. Teng, J. W. Balde, eds., *Multichip Modules: System Advantages, Major Construction, and Materials Technologies*, New York: IEEE Press Reprint Book, 1991.

G4   G. Messner, I. Turlik, J. Balde, P. Garrou, eds., *Thin Film Multichip Modules*, Reston VA: ISHM, 1992.

G5   J. J. Licari, L. R. Enlow, *Hybrid Microcircuit Technology Handbook*, Park Ridge NJ: Noyes Publications, 1988.

G6   C. A. Harper, ed., *Electronic Packaging and Interconnection Handbook*, New York: McGraw-Hill, 1991.

G7   C. A. Harper, ed., *Handbook of Thick Film Microelectronics*, New York: McGraw-Hill, 1982.

G8   M. L. Dertouzos, R. K. Lester, R. M. Solow, *Made In America: Regaining The Productive Edge*, New York: Harper Perennial, 1989.

G9   K. B. Clark, T. Fujimoto, *Product Development Performance-Strategy, Organization, and Management in the World Auto Industry*, Boston MA: Harvard Business School Press, 1991.

G10   J. W. Dally, *Packaging of Electronic Systems (A Mechanical Engineering Approach)*, New York: McGraw-Hill, 1990.

G11   H. B. Bakoglu, *Circuits, Interconnections, and Packaging for VLSI*, Reading MA: Addison-Wesley, 1990.

G12   J. S. Hwang, *Solder Pastes in Electronics Packaging*, New York: Van Nostrand Reinhold, 1989.

G13   J. H. Lau, ed., *Solder Joint Reliability: Theory and Applications*, New York: Van Nostrand Reinhold, 1991.

G14   C. F. Coombs, Jr., *Printed Circuits Handbook-3rd Ed.*, New York: McGraw-Hill, 1988.

The Materials Research Society holds week-long symposia on Electronic Packaging Materials Science at some of their meetings (April and November of each year). The papers from each Symposium are published in a hard cover volume. Some recent volumes, for example, are:

G15   E. D. Littie, P. S. Ho, R. Jaccodine, K. Jackson, eds., *Electronic Packaging Materials Science V*, MRS Symposium Proceedings Nov. 1990, vol. 203, Pittsburgh PA: Materials Research Society, 1991.

G16   E. D. Littie, R. C. Sundahl, R. Jaccodine, K. A. Jackson, eds., *Electronic Packaging Materials Science IV*, MRS Symposium Proceedings April 1989, vol. 154, Pittsburgh PA: Materials Research Society, 1989.

References

1   Publication IPC-MC-790, "Guideline for Multichip Technology Utilization", IPC (Institute for Interconnecting and Packaging Electronic Circuits), Lincolnwood, IL, 1990.

2   J. J. Reche, "High Density Interconnect for Advanced Packaging," *Proc. NEPCON West*, (Anaheim CA), pp. 1308-1318, Feb. 1989.
See also:
W. H. Knausenberger and L.W. Schaper, "Interconnection Costs of Various Substrates - The Myth of Cheap Wire," *IEEE Trans. CHMT*, vol. CHMT-3, no. 4, pp. 634-637, Dec. 1980.

3   A. J. Blodgett, Jr., "Microelectronic Packaging," *Scientific American*, vol. 249, pp. 86-96, July 1983.
A. J. Blodgett, Jr., "A Multilayer Ceramic Multichip Module," *IEEE Trans. CHMT*, vol. CHMT-3, no. 4, pp. 634-637, Dec. 1980.
For a more recent article giving the background as well as the latest information on the glass ceramic/copper multilayer substrates for the IBM 390/9000 high performance computers see:
A. H. Kumar, R. R. Tummala, "State-of-the-Art, Glass-Ceramic/Copper Multilayer Substrate for High Performance Computers," *Int. J. Hybrid Microelectronics*, vol. 14, no. 4, pp. 137-150, Dec. 1991.

4   L. M. Higgins III, *et al.*, "Glob-Top Encapsulant Suitability for Large Wire Bonded Die on MCM-L," *Proc. Int. Conf. Multichip Modules*, (Denver CO), pp. 482-484, April 1992.

5   M. L. Dertouzos, R. K. Lester, R. M. Solow and The MIT Commission on Industrial Productivity, *Made in America: Regaining the Productive Edge*, New York: Harper Perennial (A Division of Harper Collins Publishers, 1989.

6   *Electronic Engineering Times*, May 4, 1992

7   W. Blood, "ASIC Design Methodology for Multichip Modules," *Hybrid Circuit Techn.*, vol. 8, pp. 21-27, Dec. 1991.

8   J. L. Hennessy, "Trends in Processor and System Design and the Interaction with Advanced Packaging," *Proc. IEEE Multichip Module Conference MCMC-92*, (Santa Cruz CA), pp. 1-3, March 1992.

9   J. W. Balde, "Crisis in Technology: The Questionable Ability to Make Thin Film Multichip Modules," *Proc. of the IEEE*, Dec. 1992.

10 E. Jan Vardiman, "Economic and Political Implications of Disinvestment in the MCM Industry," *Proc. Int. Conf. Multichip Modules*, (Denver CO), pp. 71-73, April 1992.

11 F. Bachner, *et al.*, "Defining the Interface between Suppliers and Users in the Multichip Module Marketplace," *Proc. Int. Conf. Multichip Modules*, (Denver CO), pp. 52-55, April 1992.

12 L. M. Higgins III, "Perspectives on Multi-Chip Modules: Substrate Alternatives," *Proc. IEEE Multichip Module Conference MCMC-92*, (Santa

Cruz CA), pp. 12-15, March 1992.

13 B. Freyman, B. Miles, "A DSP-based Multichip Module employing Advanced Multilayer Printed Circuit Board Technology," *Proc. Int. Conf. Multichip Modules*, (Denver CO), p. 490, April 1992.

14 W. M. Siu, "MCM and Monolithic VLSI: Perspectives on Dependencies, Integration, Performance and Economics," *Proc. IEEE Multichip Module Conference MCMC-92*, (Santa Cruz CA), pp. 4-7, March 1992.

2

MCM PACKAGE SELECTION: A MATERIALS AND MANUFACTURING PERSPECTIVE

Allison Casey Dixon and Edward G. Myszka

2.1 INTRODUCTION

Multichip packaging is receiving increased attention as electronic equipment manufacturers drive toward smaller, faster and less expensive products. By connecting several chips together in a single package:

- Board size can be reduced by up to a factor of 10 or more
- Signal propagation between chips can be up to three times faster
- The number of solder connections in a system can be reduced

Even so, multichip modules (MCMs) will be utilized only where they are the least expensive method of meeting system requirements. The choice of MCM materials and manufacturing processes greatly influences the cost of a multichip module technology in terms of piece part cost, manufacturing yield, manufacturing cycle time and repairability. Materials choices are also dominant factors in the electrical and thermal performance of a module. There is no single "right" choice; rather, different choices are appropriate for different applications.

This chapter presents the technology choices available today for the various parts of an MCM described in Table 2-1 - signal interconnect, substrate base, MCM substrate, package body, chip mounting and module level connection.

Different combinations of these parts are used for different MCM technologies. Some of the typical combinations are sketched in Figure 2-1. Comparisons of manufacturing flows, materials properties, performance and cost are emphasized in this chapter, along with the concept that there are no "right" choices except in light of a specific application.

After the technology choices are described, global material and manufacturing related issues are discussed - cost, performance, thermal path, rework and manufacturability. The chapter concludes with some examples of modules designed for given applications.

Table 2-1 Multichip Module Parts.

PART	DESCRIPTION	EXAMPLES
Chip Mounting	Electrical and mechanical connection of chip to substrate.	Die attach/wire bond, TAB, flip chip
Package Body	Additional structural support, environmental protection and signal connection to outside world.	Ceramic packages
Substrate Base	Structural support for the signal interconnect. Signal interconnect may be deposited in the substrate base or the substrate base may be an integral part of the signal interconnect.	Silicon, ceramic, organic laminate, metal
Signal Interconnect	Metal and dielectric patterns forming the circuitry between chips.	Copper/polyimide, tungsten/alumina, aluminum/silicon dioxide.
MCM Substrate	Signal interconnect plus the substrate base, may require an additional package.	MCM-C, MCM-L, MCM-D, MCM-D/C, MCM-Si
Module Level Connection	Electrical and mechanical connection of module to motherboard. Integral part of either the MCM substrate or the package body.	PGA, PAC, gull wing lead

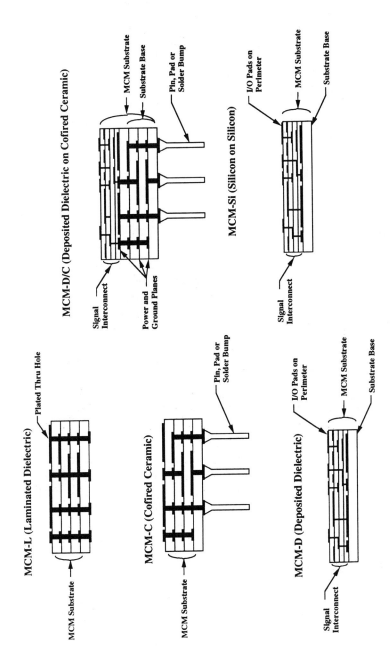

Figure 2-1 Typical MCM design configurations.

2.2 PACKAGE BODY AND SUBSTRATE BASE CHOICES

The package body and substrate base are the structurally robust piece parts that form mechanical support for the MCM. The package body is simply a housing for the module. Some module technologies require use of a package body, and others do not. In Figure 2-1, the MCM-D and MCM-Si modules will need a package body to house the fragile module and to provide electrical connection to the outside world. Typically, an MCM-D or MCM-Si module is wire bonded to a ceramic package body. The package body provides mechanical and environmental protection and a means to attach pins or leads to transfer electrical signals to the outside world. In some cases, a package body is not necessary because the substrate base, the structural support for the metal and dielectric patterns forming the circuitry between chips (signal interconnect), provides sufficient mechanical support and a means of electrically connecting the module to the outside world.

The various types of MCMs are sketched in Figure 2-1. Figures 2-2 shows photos of an assembled module and the MCMs packaged parts used to fabricate it. The MCM-L, MCM-C, and MCM-D/C modules make use of integrated MCM substrates that do not require use of a separate substrate base or package body. MCM-D and MCM-Si, on the other hand, utilize all three parts: substrate base, MCM substrate, and package body.

Three general classes of materials may be used for package bodies or substrate bases: ceramics, organic laminates and metals. Important characteristics of these materials are summarized in Table 2-2 and are discussed below [1].

2.2.1 Ceramics

Ceramics used in MCMs have several advantages. Foremost of these is that a properly designed ceramic piece part can serve as the module package body, the MCM substrate, and the module level connection — all in a single integrated part. Ceramics are electrically non-conductive, simplifying module design. They can form part of a hermetic package easily.

Alumina (also known as aluminum oxide or Al_2O_3) is the most common, inexpensive and widely used ceramic substrate material. Alumina may be used in an MCM as either an MCM substrate or as a stand alone package body for MCM-D or MCM-Si type modules. There are two general forms of alumina used – multilayer cofired alumina, which can contain conductive metal traces or planes, and single-layer pressed alumina. Multilayer cofired alumina is made from individual layers of unfired material (called "green tape") which can be patterned with conductive metal traces. Many layers (up to 50 or more) can be stacked together and fired at high temperatures to form a single unit. Further

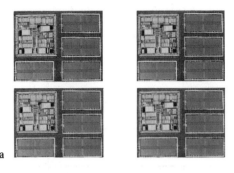

a

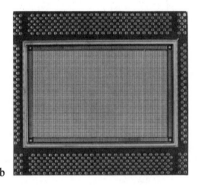

b

c

Figure 2-2   MCM-Si package showing (a) chips on silicon substrate forming module, (b) ceramic package base and (c) assembled MCM package.

Table 2-2 Properties of Package Bodies/Substrate Bases.

	Formula	Cost	Availability	CTE $(10^{-6}/°C)$	Thermal Conductivity $(W/m\text{-}K)$
Pressed Alumina	Al_2O_3	X	Excellent	8.1	30
Aluminum Nitride	AlN	4X	Poor	4.3	260
Silicon Carbide	SiC	2X	Poor	3.7	270
Mullite	$3\ Al_2O_3/$ $2\ SiO_2$	X	Poor	4.5	2-6
Glass Ceramic	Various	X	Fair	3.0-4.2	5
Organic Laminate	Various	0.5X	Excellent	12.0-17.0	0.2-0.3
Metals	Various	0.3X	Good	6.0-20.0	200-400

Note: Silicon, CTE = $3.5 \times 10^{-6}/°C$.

processing details are discussed in Section 2.3.1 of this chapter and in Chapter 6. Cofired alumina is a mature technology with highly automated factories in place. Its major disadvantages are shrinkage and warpage. These factors are controlled tightly in the manufacturing process and compensated for in the design process. Shrinkage and warpage limit the density of conductive metal traces on a cofired ceramic MCM due to the additive tolerance of each layer. Shrinkage and warpage also cause problems with mounting high lead count, tight pitch chips in cases where the planarity of connection points is critical. Figure 2-14b (discussed in Section 2.7.2) is an example of a prototype workstation module utilizing cofired ceramic as a substrate base.

Pressed alumina is a single sheet of ceramic material that is independently fired. Pressed alumina does not contain conductive traces or planes and serves strictly as a package body for MCM-D or MCM-Si modules or as a substrate base for MCM-D/C modules.

As shown in Table 2-2, alumina is a poor conductor of heat relative to other ceramics. It is, however, orders of magnitude better than organic laminates. Alumina has a coefficient of thermal expansion (CTE) more than twice that of silicon. When silicon chips are mounted on alumina, thermal expansion mismatch creates stress that can cause material fatigue, leading to premature failure.

New ceramic materials aimed at improving CTE and thermal conductivity are currently in prototype production. Some of the most promising materials such as aluminum nitride, silicon carbide and mullite, are listed in Table 2-2 for comparison with standard alumina.

2.2.2 Organic Laminates

Organic laminates, commonly referred to as printed circuit board (PCB) or printed wiring board (PWB), are used in many different forms for electronics packaging. In MCMs, organic laminates are used most often as an integrated package body and MCM substrate. In general, sheets of polymer materials (polyimide, FR-4, BT resin plus glass reinforcement) are sandwiched between layers of metal (usually copper) traces. The stack up is then bonded by lamination in a hydraulic press or autoclave under heat and pressure (350°F and 325 psi for standard FR-4). Further processing details for laminate-based MCM technologies are described in Chapter 5.

Like ceramics, organic laminates have the advantage of combining package base and MCM substrate functions into a single piece part. Another advantage of organic laminates is the match of their CTE to the mother board. Any MCM will be mounted eventually on a PWB. The organic laminate-based technology typically has the lowest manufacturing cost of all module technologies because it is mature and employs both batch and parallel processes (see Section 2.3.2). The major disadvantages are very poor thermal conductivity, a CTE four to five times that of silicon, and warpage. Figure 2-3 is a photograph of an organic laminate MCM substrate, consisting of seven copper layers, BT resin dielectric and an embedded leadframe.

2.2.3 Metals

Metals also are used as package bodies for MCMs [2]. Often metals are chosen for high power MCMs because of their excellent thermal conductivity. Metals also provide a very rigid and flat surface for signal interconnect layers within the module. The major disadvantage of using metals as package bodies is that signal connections cannot be made directly through the metal to the outside world.

2.3 SIGNAL INTERCONNECT AND MCM SUBSTRATE CHOICES

Module signal interconnect refers to the pattern of metal and dielectric that forms the circuitry between chips and from the chips to the outside world. As

Figure 2-3 Organic laminate MCM.

discussed in Chapter 1, there are many MCM substrate types available. MCM-C, MCM-L, MCM-D, MCM-D/C and MCM-Si, as well as conventional thick film hybrids, are all available today. The MCM designer is faced with many choices for the fundamental technology, as well as in the selection of materials used within each of these technologies. The selection of the dielectric material, as well as the conductor material, also strongly influences the overall performance of the module. This is true especially for modules containing high speed digital and mixed analog/digital components with switching frequencies in excess of 50 MHz or switching rise times less than 2 ns. This section focuses on the description of the core MCM technologies and processes and the materials which are available to manufacture such MCMs.

There are many materials used as conductors for MCM applications. Copper (Cu) and aluminum (Al) are utilized as conductors for MCM-D, MCM-D/C and MCM-Si applications. Copper is preferred for its lower resistivity. Copper is being used also in experimental MCM-C modules fabricated at low cofiring processing temperatures. Gold (Au) is used occasionally for MCM-D and MCM-D/C, but its relatively high cost prohibits its use in most commercial applications. Refractory metals such as tungsten (W), molybdenum (Mo) and manganese (Mn) are used in high temperature cofired ceramic processing. These metals are selected for their high melting points rather than for their electrical properties. Metals such as W, Mo and Mn are electrical conductors able to withstand the high (1500°C) firing temperatures required to sinter the ceramic layers together to form a monolithic multilayer MCM-C substrate.

2.3.1 MCM-C (Cofired Ceramics)

Multilayer ceramic structures have been designated as MCM-C. Further details on ceramic-based technologies are given in Chapter 6. This type of MCM can be categorized in two groups: those processed at high temperature (HTCC) and those at low temperature (LTCC). HTCC has been available for decades and is used most commonly [3]. Refractory metals such as W, Mo, and Mn are used as electrical conductors in HTCC processing. These metals are selected because of their inherent high melting point. These metals remain stable and do not decompose during the sintering process where temperature extremes can reach (1500°C). Low temperature cofiring, although fairly new to the industry, has generated a great deal of interest. The ability to use highly conductive noble metals such Ag, Au and, most recently, Cu as the conductor offers many advantages over HTCC modules [4]. Unfortunately, since this technology is quite young and its penetration into the market relatively small, the substrate cost can be as much as 70 – 100% higher than HTCC substrates. As the market penetration increases, this disparity is anticipated to decrease. In the cofire processing for both the high-and low-temperature technologies, a liquid slurry is formed from ceramic particles and organic binders and then cast into a solid sheet. This sheet is often referred to as the "green tape" because of its unfired state. Typical costs for moderate production volumes associated with green tape are $0.06 and $0.11 per square inch for HTCC and LTCC, respectively [5].

Next, holes for vias are generated in the green tape. The most commonly used via diameter sizes range from 0.015 – 0.008" with more aggressive designs reaching 0.004". Vias can be drilled, but are formed more commonly by a punching operation. Punching can be accomplished by a single punch head positioned sequentially at each site by computer numerical control (CNC). This often is referred to as "soft tooling." Alternatively, a custom die head that simultaneously punches all the vias on a single layer is referred to as "hard tooling."

Soft Tooling versus Hard Tooling

The nonrecurring engineering cost (NRE) for hard tooling is higher than soft tooling, but the unit price is typically lower. The decision on which type of tooling to be purchased is a function of the anticipated volume. For low volume runs, soft tooling NRE, which *can average $30,000 for an eight layer module, can be cost effective. Conversely, for high volume applications the cost of hard tooling, which can exceed $60,000, can easily be amortized over the life of the product to produce a cost effective solution. One disadvantage of soft tooling is that there typically is a maximum run rate associated with the tool. That is, there exists a maximum number of modules that are built and shipped each month. This is typically associated with the existing fabrication throughput capabilities.*

After the green tape sheets are formed and via holes drilled or punched, a conductive ink (a refractory metal for HTCC MCM substrates or a noble metal for LTCC MCM substrates) is applied to each layer through a screen printing process (see Chapter 6). 10 mil lines are standard and 4 mil lines are readily fabricated using this process. Vacuum can be applied to the underside of the punched sheet to pull the ink into the hole to coat the side walls or, in many cases, to fill the via. Solid vias are advantageous for high density applications because they can be stacked directly on top of each other, otherwise the vias will be staggered and hence require additional area.

After patterns are screened on each layer, the layers are stacked and the assembly is fed into a furnace and fired at temperatures above 1400°C for the high temperature process and approximately 800°C for the low temperature process. The result is a monolithic structure containing all the interconnects.

During the firing process the organic binder decomposes, the ceramic densifies and, unfortunately, the structure reacts by shrinking. This shrinkage becomes a problem during the subsequent assembly of chips where precise positioning of the chip connections is required. This is true especially of high density tape automated bonding (TAB) and flip chip interconnects where the I/Os are fixed. If wire bonding is used as the chip connection method, an automatic wire bonder with look-ahead vision (an optical pattern recognition system that determines the precise location of bonding pads used to make small adjustments to accommodate substrate shrinkage) can be used. Another solution is to deposit a thin film layer of metal or copper/polyimide after firing the ceramic to facilitate chip connection of high I/O devices. This postfire processing can be provided only for the outermost surface of the module.

Thermal solutions in an MCM are accommodated typically by placing vias strategically under the chips (thermal vias) and/or by brazing a Cu/W heatsink to the ceramic structure. Figure 2-4 shows a single chip multitiered cofired ceramic package with an integral Cu/W heat spreader designed to dissipate up to 30 watts. These features can be implemented readily for multichip packaging, but they can come with both cost and electrical performance penalties.

Figure 2-4 Single chip multitiered cofired ceramic PAC package.

One of the major disadvantages of cofired alumina is its relatively high dielectric constant (ε_r) which can limit the performance capability of MCMs (Section 2.6.2). Since the performance of a module is influenced heavily by design layout as well as by material selection, no rigid cut off exists for using cofired ceramic technology. Modules have been built and tested at clock frequencies in excess of 50 - 75 MHz. New ceramic materials with lower dielectric constants, such as glassy ceramics (Table 2-2), have been developed for modules operating in excess of this. The next section presents a brief description of several of these materials.

Low Dielectric and Glassy Ceramics
The most well known glassy ceramic was developed by IBM. These ceramics have low cofiring temperatures which make them co-sinterable with either Cu or Au conductors. Additionally, the materials have a low dielectric constant and mechanical properties which may be tailored for a particular application. Others are attempting to modify the material properties by introducing particles suspended in the ceramic matrix. Upon sintering, or through a post annealing

treatment, porosity is introduced to the matrix [6]. Since the dielectric constant of the gaseous material produced by the porosity is very low, the effective dielectric constant of the assembly is also low. Experimental data on research grade materials suggest that the effective dielectric constant can be reduced to approximately 2.5. These materials hold great promise but first must prove compatibility with metal films and the ability to withstand the environmental testing required to become a reliable substrate material.

High Thermal Conductivity Ceramics

Attention is being focused also on the development of ceramic materials with higher thermal conductivities than alumina. Besides alumina, beryllia (BeO) is the most widely used ceramic in the microelectronics industry, primarily because of its high thermal conductivity. Although cofired structures have not been developed for this material, it has been used extensively as a heatsink for semiconductor lasers and as a substrate base for high power RF applications. The greatest disadvantage of BeO is in its processing. BeO dust and particulates have been designated as carcinogens and, thus, extreme care must be taken when cutting or grinding this material. As a result, aluminum nitride (AlN) has been introduced as an alternative to BeO for MCMs (Table 2-2). AlN has 80% of the thermal conductivity of BeO and can be cofired in a multilayer structure [7]. With these beneficial properties, it is anticipated that AlN will become a cost effective alternative to BeO in the near future.

2.3.2 MCM-L (Organic Laminates)

The MCM-L structures can be regarded as a laminated PWB scaled to meet the requirements and dimensions of a MCM as discussed in Chapter 1. Fabrication methods vary among manufacturers, but, in practice, most techniques closely follow the infrastructure previously developed by the PWB industry. The following section is intended to provide the reader with a brief description of the most widely used fabrication methods, materials and their associated advantages and disadvantages.

Many organic laminate types are used to manufacture modules for various applications. Epoxy glass (such as FR-4) is most common. For applications where material stability is required at higher temperatures, materials such as Bismaleimide triazine (BT Resin) and polyimides with higher glass transition temperatures (T_g) can be used. Laminates can be double clad, single clad or not clad at all. Cladding is thin copper foil applied to the sheet of dielectric material. It is this foil that is patterned to make the conductors. Double clad and single clad refer to the number of sides of the dielectric material which are coated with copper foil. The clad sheets are called "laminates." Several clad sheets stacked and bonded together are also called "laminates." Copper foil thickness is measured in ounces. For example, a one ounce copper clad laminate will measure 0.0014" thick. One ounce and one-half ounce copper are typical for the PWB industry and one half, one quarter and even less than one eighth ounce can be found in laminates for the MCM-L industry.

Laminate Sheet Types

Copper clad laminates are available in widths in excess of 24" and are fabricated most commonly by either of two methods. Adhesive laminates rely on the application of a copper foil to a sheet dielectric (bare laminate with no metal) with an adhesive layer (typically acrylic in nature). These materials are clad in a rolling press at very high volumes. Conversely, adhesiveless laminates are fabricated by applying the copper directly to the dielectric with no bonding agent. Adhesiveless laminates can be formed in one of two ways: liquid dielectric may be deposited to the foil and cured, or the metal may be electroplated or vacuum deposited to the fully cured laminate material.

Additive versus Subtractive Processing

Copper conductors are patterned using either a subtractive or an additive process. The subtractive process is used typically with thicker copper films. In the subtractive process the pattern is etched directly into the existing metal. Subtractive processing is used predominantly in the PWB industry. Minimum feature sizes typically are limited to 0.003" - 0.004" due to the isotropic nature of the conductor etching process and the dielectric surface roughness.

Alternatively, in the additive process, a photoresist coating is applied and patterned onto the existing thin metal foil. The actual conductor pattern is electroplated onto this foil, the photoresist is stripped and the thin foil is etched away revealing the electroplated conductor. Refer to Figures 7-12 and 7-13. Typical conductor patterns can range in size from 0.001" - 0.003". It is predicted that additive processing in conjunction with smoother laminate surfaces, will enable conductor features less than 0.001".

After the clad laminates are patterned, vias are drilled on a CNC machine. To increase process efficiency, many laminates are often stacked during the via drilling process on machines with multiple spindles. Currently, minimum via size is approximately 0.004" - 0.006" in diameter and is limited by the availability of reliable drill bits. The copper layers can be patterned sequentially on a single clad laminate or two adjacent layers may be patterned in parallel on a free standing double clad film. The ability to pattern both sides of the laminate simultaneously is a major advantage of MCM-L because it effectively reduces the manufacturing cycle time.

After vias are formed and the Cu layers patterned, each layer can be inspected, an advantage of parallel processing. The inspection of layers prior to final lamination minimizes the possibility of the transfer of defects into the completed module. This early inspection for defects will increase the likelihood of receiving a functionally working substrate. The yielded copper layers are stacked with prepreg (partially cured laminate material) inserted between each layer. The stack is laminated in a press or autoclave with heat and pressure as mentioned previously.

The major disadvantages with organic laminate substrates used today are low routing density, poor thermal conductivity through the substrate and a high CTE, compared to silicon. New materials are being developed to alleviate many of these problems. These materials include advanced polyimides, aramids and fluoropolymers with homogeneous matrices and composite laminates. The electrical (dielectric constant) and mechanical (CTE) properties of many of these materials can be tailored for specific applications [8]-[9]. As the industry progresses to larger ICs with more demanding chip connection methods, such as flip chip, these materials will draw more attention.

For MCM-L applications one noticeable change in the fabrication of the laminate structure is the use of thinner dielectric films. These thin (25 - 50 μm) films allow controlled impedance, high density modules to be fabricated with transmission line design criteria. These fine line capabilities contribute also to the increase in routing density. However, large metallized pads on each layer are required to aid in the via drilling operation. To truly achieve high density modules novel approaches to via generation also will be required. Processes under consideration include wet chemistry etching, reactive plasma etching and laser ablation techniques. Each of these approaches to the fabrication of small vias is anticipated to break the current 0.004" barrier set by the mechanical drilling process and will increase significantly the routing density of MCM-L substrates. Unfortunately, many of the conventional laminates used today are not compatible with these advanced via processing techniques. For example, when using an excimer laser, since the ablation threshold of the glass weave used in conventional laminates is more than an order of magnitude greater than that for a polyimide matrix, processing of vias is complicated. The glass weave thermally decomposes at the high energy densities required to remove it. Similarly, the acrylic adhesives used to bond copper foils to dielectric sheets also have high ablation thresholds. Laminate suppliers have recognized the limitations of conventional laminate materials and have begun to offer new, more advanced laminates specifically targeted for such applications. Many of these new laminates are adhesiveless and do not contain a glass fiber weave. Instead, these new materials are made up of the pure matrix material or contain small particulates dispersed within the matrix to control its properties and processability. As the industry progresses to more advanced via fabrication techniques, we predict increased use of new laminate materials.

In most cases, laminates are fabricated in a large panel. At the end of production, the layers are cut out of the panel. The ability to fabricate multiple modules in a large panel format is cost effective. However, as the panel size increases, the layer-to-layer registration becomes more difficult. Parallel processing offers the potential to integrate two distinctly different laminate patterning processes: conventional PWB for low density layers (power and ground planes), and advanced fabrication processing, such as laser ablation for high density (signal) layers. Since layer-to-layer registration is less of an issue with low density layers, large panel sizes can be processed. These large panels can be cut into smaller, more manageable panels and mated to panels containing the higher density layers during the lamination process. Additionally, recent advances in the lamination of the multilayer organic structures to leadframes has made it possible to produce leaded MCM-L substrates configured as quad flat packs (QFPs).

2.3.3 MCM-D (Deposited Dielectric)

MCM-Ds are fabricated by a sequential deposition of conductor, typically Cu or Al, and dielectric layers (typically polyimide) on a substrate base made of ceramic, silicon or metal. As the fabrication of MCM-L parallels the PWB industry, MCM-D most closely parallels the processing techniques of the semiconductor industry. Semiconductor type processing provides for very fine lines with high routing densities. However, the semiconductor processing equipment used to manufacture MCM-D substrates is expensive and low volume runs make it difficult to depreciate the high capital costs. MCM-D substrates are fabricated in 4", 5" or 6" round or square "wafers." Fabrication costs are associated primarily with the wafer and not with finished substrate size; that is, a single substrate on a wafer will cost about twice as much as a design where two substrates can be fit onto a single wafer. The cost driver is the number of substrates that can fit onto a given wafer size (number "up").

Another disadvantage of MCM-D technology is that it often requires the use of a separate package body to house the MCM substrate and to make electrical and mechanical connection to the mother board. This results in additional

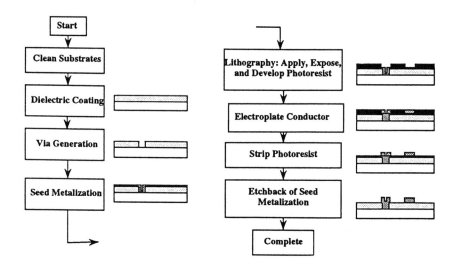

Figure 2-5 Process flow: MCM-D.

assembly operations and higher manufacturing costs. A more cost effective approach to MCM-D is its integration with cofired ceramic. These structures are detailed in Section 2.3.4 on MCM-D/C.

Although MCM-D fabrication processes differ from one manufacturer to another, a common method is presented to give an idea of the processing required to build such a module. Figure 2-5 is a flow chart of a typical process.

A liquid polymer (typically polyimide, but also benzocyclobutene (BCB) and other fluoropolymers) dielectric layer is deposited on the substrate base by a conventional spin coating process. The liquid dielectric material is dispensed onto the center of the substrate and spun until the material spreads uniformly to cover the substrate. Although this is an effective process in terms of uniformity, it is far from efficient. As much as 50 - 80% of the material is spun off and wasted. As typical costs for the polymers can range from $1.00 to $2.00 per gram, and up to 4 grams are required to form a 12.5 µm thick film layer on a 4" × 4" substrate, more efficient, alternative dispensing techniques are needed. Alternative deposition methods under investigation include spraying and extrusion. Dielectric thicknesses of 25 µm or less are typical, and multiple coats usually are required to obtain this thickness with good planarization.

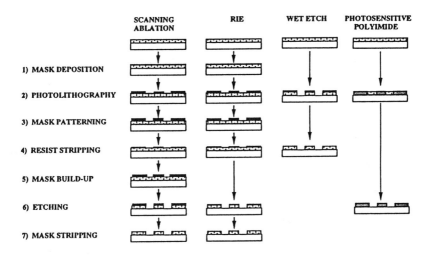

Figure 2-6 MCM-D via formation techniques.

Vias can be formed by scanning laser ablation, reactive ion etching or wet etching of the dielectric, as well as through the exposure of photosensitive polymers. Figure 2-6 shows the comparison of process steps required to form vias. No one method has been accepted as a standard in the industry [10]. Reactive ion etching of vias is commonly used in the U.S. and wet etching in Japan. These techniques are discussed further in Chapter 7.

After the dielectric layer is deposited and vias formed, metal conductors are formed using either additive or subtractive processing. In additive processing, a bus layer metallization for electroplating is first formed either by sputtering or by electroless copper plating. Photoresist is then spun on, the pattern imaged and the bulk pattern is electroplated. The photoresist then is stripped and the bus layer metallization etched away, leaving the electroplated conductor pattern behind. In subtractive processing, the full surface is coated with metal and the conductor pattern etched. The entire process is repeated for each layer. Both subtractive and additive processing are practiced today.

2.3.4 MCM-D/C (Deposited Dielectric on Cofired Ceramic)

Deposited dielectric on cofired ceramic, MCM-D/C, can cost effectively combine the advantages of MCM-C and MCM-D. This technology utilizes deposited dielectric layers on top of a multilayer cofired ceramic substrate base to produce MCM substrates with all of the high density and high frequency attributes of MCM-D coupled with the flexible physical attributes of MCM-C. This combination provides a single substrate that functions as the signal interconnect, substrate base, package body and module level I/O connection. Vias may be brought out the bottom of the substrate, the package I/Os may be terminated in a pad array carrier (PAC) format or mated with pins to simulate a standard, single chip pin grid array (PGA) package. The cofired substrate is part of the package body, eliminating the cost associated with repackaging of the MCM substrate. The photo in Figure 2-14b is an example of a module that uses a MCM-D/C substrate. In this module there are four layers of polyimide dielectric and thin film copper deposited on a standard 299 pin cofired ceramic substrate.

Another advantage of MCM-D/C technology is that signal lines can be embedded in the low dielectric constant deposited layers to minimize crosstalk, while power and ground planes can be contained in the higher dielectric constant ceramic layers. With the development of high dielectric constant ceramic materials ($\varepsilon_r > 200$), it may be possible to incorporate much (if not all) of the decoupling capacitance within the substrate and eliminate the secondary assembly operation required to mount external bypass capacitors.

2.3.5 MCM-Si (Inorganic Thin Film)

Yet another solution to multichip packaging utilizes semiconductor fabrication techniques directly. This technology, often referred to as silicon-on-silicon (MCM-Si), is processed in a similar fashion to conventional silicon ICs. As discussed in Chapter 16, a silicon wafer is used as the substrate base, Al or Cu as the conductor, and silicon dioxide as the inorganic dielectric media. Several microns of the metal conductor and the subsequent silicon dioxide dielectric layers are deposited by a vacuum deposition technique (such as evaporation or sputtering). Layer thicknesses are controlled quite uniformly over areas as large as 6" - 8" diameters. Photoresist is applied and patterned with the support of a wafer spinner and mask aligner, respectively. The metal is etched away by either wet etching or, more typically, by dry etching techniques. Because the silicon wafer substrates are extremely smooth and flat, very fine feature sizes are possible. Although submicron interconnects are possible, typical features of 15 μm are adequate for MCM interconnects.

MCM-Si offers the advantages of the highest signal interconnect density, excellent CTE match to the silicon die and the utilization of existing semiconductor fabrication infrastructure. The equipment used is readily available and the processes have been established and well characterized by the semiconductor manufacturing community. The utilization of this infrastructure has contributed to a reduction in fabrication cycle time (and learning curve) for many entering the MCM field along with the security of the predictable reliability of conventional ICs. Additionally, the inherent low density of silicon makes it ideal for applications in which light weight is critical, such as in the aerospace and military industries.

However, MCM-Si technology also has some significant disadvantages. Foremost among these are that silicon substrates are not a suitable MCM package base. In some cases, the high resistivity of the aluminum conductor precludes its use in high frequency applications. In addition, as with MCM-D, the equipment costs are high. Batch sputtering systems can range from $150,000 to $350,000 and in-line systems from $1,500,000. In combination with low volume manufacturing, typical module costs will be high. Silicon-based technologies are discussed further in Chapter 16.

Other Inorganic MCM Materials
The majority of the industry has focused on the development of organic dielectric materials (MCM-D and MCM-D/C). For the short term these will be the materials of choice. However, in the long term, some inorganics may offer performance benefits that surpass that of today's organics. Research workers have focused attention on synthetic diamond and diamond-like coatings [11]. Although the development of cost effective deposition methods requires additional work, the superior properties of the diamond-like films warrant investigation as dielectric layers for high density interconnect structures. These materials typically combine desirable properties such as a low dielectric constant, extremely high thermal conductivity (more than four times that of copper) and a low CTE. These properties are desired for high frequency, high power and low residual stress applications, respectively.

2.3.6 Thick Film Hybrid MCM

Ceramics have been used in the microelectronics field for many decades. Various ceramic materials exist, but alumina (aluminum oxide) is most widely used as a substrate for the hybrid microelectronics industry. Some argue that the development of thick film hybrids was the origin of the first MCM-type interconnect. In this type of interconnect system, a flat sheet of alumina is typically used as the substrate material. Each layer is applied sequentially by screen printing, as described in Chapter 6. Although this is not a complicated process, one disadvantage with any sequentially fabricated module is layer-to-layer inspection. Since these layers can only be inspected once they have been applied to the module, there exists a potential for a defect in the final layer causing the entire module to be defective. This is unlike parallel processing (for

MCM-L, MCM-C) where all the layers can be fabricated independently and inspected. Many new processes under development either use parallel processing guidelines or minimize the total number of layers.

The substrate is then placed into a furnace where the ink is dried and fired. Conductor geometries are typically limited by the fabrication of the screens used for the printing process to 0.005" lines and spaces.

The dielectric layer is applied in a similar fashion. No dielectric material is applied in regions where vias lead to the above layers. When the subsequent conductor layer is printed, the ink flows into the openings of the dielectric and make contact to the conductor below. The balance of the module is completed by repeating this process.

Thick film hybrid is a mature technology with good infrastructure. A major advantage of this technology is that a large supply of resistive inks are available that can be screened onto the module to integrate resistors. Additionally, there exists an abundant supply of surface termination inks with metallurgies that facilitate wire bonding, soldering or component attachment through conductive epoxies.

Thick film hybrid technology used for MCMs has some serious disadvantages in the areas of performance and routing density. The dielectric constant of most dielectric inks is as high as 8. While high dielectric constant is not in itself a disadvantage, it can make a desired substrate characteristic impedance difficult or impossible to attain. Thick film hybrids are very limited in routing density - typically 0.005" - 0.010" lines and spaces and via sizes of approximately 0.008". From a module yield point of view, it is advantageous to minimize the number of layers. One way would be to increase the routing density so that all signal interconnects could be made on a minimum number of layers. Recent research and development efforts in transfer printing have claimed results of 0.001" feature sizes. Although this process has yet to be implemented in a production environment, it holds promise for the future.

2.4 CHIP MOUNTING CHOICES

The purpose of any chip mounting technique is to provide a suitable path for electrical signals from chip to MCM substrate and to provide a means of mechanically attaching the chips to the substrate. All chip mounting techniques also provide a path for heat generated in the chip to dissipate. The common of chip mounting choices are wire bond/die attach, TAB, flip TAB, and flip chip. Further details of these chip mounting or chip connecting techniques are presented in Chapter 9.

The method of connecting chips to a substrate is probably the most critical choice in module physical design. This choice largely defines the technical

Table 2-3 Chip Mounting Choices.

	Die Attach Wire Bond	TAB	Flip TAB	Flip Chip
Cost	X	> 2X	> 2X	0.8X
Chip Availability	Excellent	Fair	Fair	Poor
Reworkability	Poor	Fair/Poor	Fair	Good
Probe Test	DC	AC	AC	AC
Lead Inductance (nH)	2.0 - 3.5	4.0 - 5.0	4.0 - 5.0	< 1.0
Footprint (Chip +)	20 - 100 mil	80 - 600 mil	80 - 600 mil	Clearance
Peripheral Bond Pitch	4 - 7 mil	3 - 4 mil	3 - 4 mil	10 mil
Area Array Bond Pitch	N/A	N/A	N/A	≥ 10 mil
Max I/O Count	300 - 500	500 - 700	500 - 700	> 1,000

Notes:
- Cost based on high volume production of a nine chip module with over 2000 chip-to-substrate connections
- Wire bond lead inductance for 1.0 mil Al wire with pad-to-pad spacing from 2.0 - 4.0 mm.
- Wire bond footprint depends on cavity or no cavity.
- TAB footprint depends on OLB pitch.
- Chip availability may be poor for high performance devices such as DSP, microprocessor and ASIC.

capabilities and product characteristics of the module. The chip mounting technique dictates what thermal path must be provided for the heat generated in the chip and, in many cases, dictates the type of substrate to be used. The chip mounting or chip connection choice also has heavy implications on module cost, chip availability, chip testability, module reworkability, module size and module performance. These issues, compared in Table 2-3, will be discussed for each of the major chip mounting choices.

2.4.1 Die Attach/Wire Bond

Die attach/wire bond chip mounting of semiconductor chips has been used for over 20 years. The chip is attached mechanically to the substrate by a variety of means including organic adhesives (silver filled epoxy is common) and metal eutectics (gold/silicon and various solders are common). The die attach material can be thermally and/or electrically conductive depending on the material selected. Then the chip is connected electrically, wire bonded, to the substrate by gold or aluminum wires. Figure 2-7a shows chips wire bonded in an MCM. Over years of process development, this technique has been thoroughly characterized and refined to its present production limit of about 4.8 mil pitch (a "mil" being a milli-inch and "pitch" referring to the center-to-center spacing of the wires). Figure 2-7b shows a portion of a chip wire bonded with 464 wires on a 4.8 mil pitch using 1.25 mil aluminum wire - a good example of state of the art production.

The die attach/wire bond technique minimizes MCM risk. High volume precision equipment is readily available in existing semiconductor assembly lines and experienced personnel are available. Yields (number of good wire bonds per number of wire bonds attempted), and thus costs, are predictable. All of this adds up to known variables and low risk.

Another unique advantage of die attach/wire bond is availability of ready to use chips. Bare silicon chips with wire bondable aluminum bond pads have been sold for years by semiconductor manufacturers to hybrid manufacturers. Techniques, specifications and applications information for electrical test, visual inspection, packing, shipping and assembling exist in a somewhat established infrastructure. It is important to note, however, that while many thousands of devices are routinely available in chip form, many state of the art devices such as microprocessors, digital signal processors and application specific integrated circuits (ASICs) are not. These devices are fully tested and sold typically in single chip package form - an important issue for MCM manufacturers.

In addition to pitch limitation, the die attach/wire bond technique has two major disadvantages for MCMs. The first is that die attaching a chip to a substrate complicates thermal management within the module. As shown in Figure 2-7a, heat generated by the chip must be conducted through the die attach material to the MCM substrate, to the package body and, finally, to the outside of the package. But most MCM substrates and package bodies are poor conductors of heat. They also may have CTE (the material property defining the of expansion or contraction at different temperatures, usually expressed in ppm/°C) sufficiently different from that of the silicon chip to create mechanical stress problems.

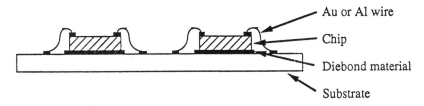

Au or Al wire

Chip

Diebond material

Substrate

Figure 2-7a Die attach/wire bond.

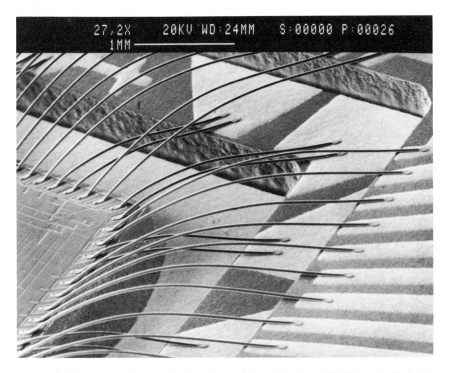

Figure 2-7b State of the art wire bonding (1.25 mil Al wire, 465 I/O, 4.8 mil pitch).

The second major disadvantage of using the die attach/wire bond technique for MCMs is that bare silicon chips cannot today be tested at-speed (normal operating AC frequency) because the large inductance of probe needles masks the test results. At-speed testing is normally done by the semiconductor manufacturer after chips are packaged. From a MCM manufacturing perspective, the scenario is grim: partially tested chips are assembled together in a module and then the entire module is tested at operating frequency. Any one chip that is not "up to speed" causes the entire module to fail test. Module yield is affected dramatically by the number of chips in the module. Even if the module design and test techniques are clever enough to locate a bad chip, the problem is compounded by the fact that die attach/wire bonded chips are difficult to remove and replace. Thus, the MCM manufacturer is faced with poor yields and little chance of rework. In addition, technology is not commonly available today to burn-in bare silicon chips; burn-in must be done at the module level (see Chapter 13).

The die attach/wire bond technique is often cited for its low cost - about $0.25 for die attach and approximately $0.01 per wire bond. But as we have seen, the total module manufacturing costs could be quite high due to low yields. This technique is suited best for modules consisting of a few, low power, inexpensive chips and an inexpensive substrate. Wire bond techniques are discussed in Section 9.3.

2.4.2 TAB

Tape automated bonding (TAB) is a method of electrically and mechanically connecting a chip to the substrate that has many possible advantages over conventional die attach/wire bond techniques for MCMs. The tape is an etched-out piece of metal consisting of tiny beam leads. The end of the leads connecting to the chip are typically on a small pitch and are fanned out to a larger pitch at the end that connects to the substrate. Chips are first bonded to the tape (inner lead bond or ILB) with each chip pad connected up to a beam lead by thermocompression (heat and pressure) bonding. The ILB unit is then excised from the tape and die attached to the substrate using conventional means described in Chapter 9. As a last step, the other end of the beam leads are bonded, using heat and pressure, to the substrate (outer lead bond or OLB). There are many techniques and metallization schemes for accomplishing inner and outer lead bonding. From a manufacturing point of view, the two main types are gang bonding and single point bonding. In gang bonding, all leads are connected at the same time by means of a hot bar thermode. The process is fast and uniform, but lead planarity is a difficult issue. In single point bonding, each lead is connected individually by means of a hot point or laser bonder. Figure 2-8a shows a schematic diagram of chips TAB bonded in a MCM.

TAB has been developed in the United States primarily to handle more chip connections at tighter pitches than wire bonding. TAB works well for high lead count, small pitch chips where wire bond yields become unpredictable and costly. Production TAB ILB bonding is done today at 4 mil pitch and has been demonstrated in the laboratory at 3 mil pitch and less. Figure 2-8b is a photograph of a 385 mil square chip with 360 pads TAB inner lead bonded at a 4 mil pitch. The beam leads are fanned out for outer lead bonding at a pitch of 8 mil. The large fanout is required for two reasons. First, tight pitch bonding requires leads to be very precisely coplanar and this is less likely to be the case by the time the unit reaches the OLB process step. And second, the substrate or board technology limits the pitch capability of pads on the substrate. If an advanced substrate is used, then non-fanout TAB (where the OLB pitch is the same as the ILB pitch) can be used.

TAB offers several advantages when used to bond chips to substrates in MCMs. Foremost is that ILB units can be tested at ac operating speeds. This means that bad chips can be culled and only known good chips need be assembled into the module, thus dramatically increasing module yields. The value in testing chips before module assembly increases with the number of chips in the module. In addition, the type of chip in a module can increase or decrease the need to test them prior to assembly. For example, ASIC chips typically are not speed sorted at final test; they are designed to fall within a range of specified performance. A module containing all ASIC devices can be expected to perform properly even without at-speed testing of the chips prior to assembly. On the other hand, high speed memory components are speed sorted at final test and graded for sale. A module containing these devices could not be expected to yield well at final test unless the chips had been at-speed tested prior to assembly.

A related advantage of TAB is that inner lead bonded chips can actually be burned-in (operated at-speed and sometimes at elevated temperatures for a period of time to cull out bad units) prior to module assembly. Such practice would probably eliminate the need for burn-in at the module level, reducing the cost of scrapping or reworking finished modules.

Although TAB offers the advantages of assembling only known good chips and the capability of bonding large complex chips, the disadvantages of using TAB in MCMs are numerous. Foremost among these is cost. Two factors can make TAB expensive: cost of the TAB tape and the number of difficult manufacturing process steps. Complex double metal layer (one metal layer acts as the ground plane to reduce lead inductance) tape can run as high as $30 in a 35 mm format. Single metal layer tape is significantly cheaper - approximately

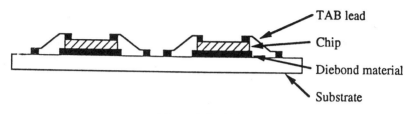

Figure 2-8a TAB.

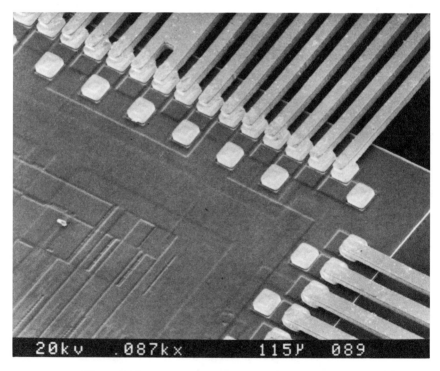

Figure 2-8b State of the art TAB (360 I/O, 4 mil pitch).

$4 in a 35 mm format. Once the chip is inner lead bonded, it cannot be reworked or the tape salvaged. A typical TAB manufacturing flow consists of multiple difficult processes: inner lead bonding, testing, burn-in, testing, die attaching and outer lead bonding. Each of these processes can be low yielding on complex components, and together, they can add up to unacceptably low yields and high cost.

Another disadvantage of TAB is that die attaching a chip to a substrate complicates thermal management within the module. The problem here is identical to that discussed in Section 2.4.1.

Finally, TAB technology often means long leads and large component footprints. Even moderate OLB pitch can easily mean long beam leads, with corresponding high inductance. A very long beam lead may require an integrated ground plane to reduce lead inductance. At this level, TAB tape becomes very difficult and expensive to manufacture. Another disadvantage of TAB is that OLB at reasonable pitches can mean very large device footprints. In one case, a 15 mil OLB drove TAB lead length to over 880 mils. Large footprints mean the electrical signal must travel greater distances, something MCMs are designed to avoid.

In general, TAB technology is best suited for large MCMs containing complex and expensive chips and substrates. For this type of application, the advantages of testability and burn-in at the ILB stage probably outweigh the disadvantages.

2.4.3 Flip TAB

Flip TAB, sketched in Figure 2-9a, is a variation on TAB where ILB bonded units are mounted face down on the substrate for OLB. Flip TAB has all the characteristics, positive and negative, of regular TAB, with two major differences. First, in flip TAB, heat generated by the chips can be removed from the backside of the chip to a module lid, thus simplifying substrate design. And second, flip TAB is easier to repair than regular TAB since no die attaching is done. Instead the inner lead bonded chip simply rests on a spacer material and is held in place by the OLBs as shown in Figure 2-9b. The mainframe computer module shown in Figure 2-13 is another good example of non-fanout flip TAB.

2.4.4 Flip Chip

Flip chip mounting offers the best possible electrical connection between chip and substrate because it eliminates leads altogether. As sketched in Figure 2-10, solder bumps are placed in an array pattern across the chip. The chips are

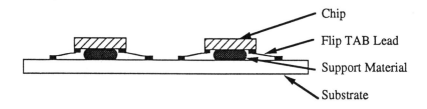

Figure 2-9a Flip TAB chip mounting.

Figure 2-9b State-of-the-art flip TAB.

mounted face down on the substrate and the solder is reflowed (heated to the melting point and then cooled) to form the connection. The chips can be placed as close together as 10 mils with no additional space needed for connections. The technique offers the best performance and smallest footprint at potentially the lowest manufacturing cost.

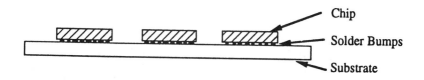

Figure 2-10 Flip chip mounting.

Flip chip electrical performance is superior to the other techniques because the interconnect length is extremely short - usually only about 50 - 100 μm, thus, the inductance is extremely low, as shown in Table 2-3. This can be a significant factor if designing a module for high speed applications.

Small chip footprint is also a performance plus for the flip chip technique. Small footprints mean shorter distances between chips on the module and shorter signal propagation delay. (Propagation delay is the time it takes a signal to leave the output buffer of one chip, travel across the substrate and enter the input buffer of another chip). Product miniaturization is one of the market drivers for MCMs. Obviously, small footprints also mean smaller modules.

From a manufacturing point of view, flip chip has several advantages. There are few process steps and those steps are batch rather than individual. Additionally, because the solder bumps can be placed in an array pattern across the entire surface of the chip, the pitch of these bumps need not be extremely small. As the number of signal connections increases on complex chips, wire bond and TAB techniques must continually strive for smaller and smaller pitches. Once developed, the flip chip technique should be the highest yielding and lowest cost of all.

Finally, flip chip, like TAB, offers the ability to probe test chips at-speed prior to mounting in the module. This means that bad chips can be culled and only known good chips need be assembled into the module, dramatically increasing module yields. The value in testing chips before module assembly increases with the number of chips in the module. In addition, the type of chip in a module can increase or decrease the need to test prior to assembly.

Also, like TAB, a flip chip mounted IC can be removed from the module and replaced should a defect be identified. This repair process is accomplished by locally heating the faulty die to reflow the solder and remove the die. The site is then refreshed by either reflowing or removing the existing solder.

Flip chip has been developed and utilized almost exclusively by IBM for over 10 years. The basic technology is well characterized and production qualified within IBM, and almost universally not characterized or qualified outside of IBM. This fact gives rise to a host of infrastructure issues for the non-IBM module designer selecting flip chip technology. First and foremost is the lack of chips available with solder bump array connections. Basically, they do not exist outside IBM. The outside module designer is faced with procuring standard wire bond compatible chips or wafers and somehow converting them to flip chip compatibility. The problem is compounded by the fact that these chips will have peripheral connection locations, possibly at a very tight pitch. Conversion to a large pitch array pattern of connections, and then bumping them, is a formidable task. Simply asking semiconductor manufacturers to provide flip chip compatible chips may not work. Infrastructure is lacking at this level also. Semiconductor manufacturers do not have computer aided design (CAD) tools or rules for designing flip chips. They lack bumping processes and equipment, probe-test equipment, trained personnel, etc.

The disadvantages of flip chip clearly are related to infrastructure and economics rather than technical. Current efforts at IBM to disseminate the technology to the general industry may eventually alleviate these issues and flip chip could become the preferred interconnect for a wide range of MCMs. For today, the MCM designer must carefully weigh the risks and benefits of using flip chip for a specific application.

2.5 MODULE LEVEL CONNECTION CHOICES

Numerous choices exist for the connection of the MCM to the mother board, but, unlike the vast standards implemented for single chip packages, none presently exist for MCMs. Without industry collaboration, the recent broad activity in MCMs will result in many different package sizes and formats. Fortunately, a cooperative venture has been initiated for the standardization of MCM packaging sizes. A task force operating under the auspices of the IEEE Computer Packaging Committee recently proposed a series of standard MCM package sizes to the EIC/JEDEC Committee [12]. Until this is reconciled, many module sizes will continue to be fabricated. Some of them will adopt the existing single chip packaging formats and benefit from the existing standards implemented by such organizations as JEDEC. The following section provides the basis to better understand the different options that exist and the performance penalties imposed.

2.5.1 Peripheral I/O

Peripheral I/Os are the most commonly used single chip package connection arrangement. The appearance of many MCM designs have been styled to resemble these single chip packages. By doing so the assembly and testing infrastructure can be utilized with little or no tooling modifications. The user does not know whether the package contains multiple chips or a single IC without dissecting the package. These packages can be implemented with virtually all of the MCM substrate types stated in Section 2.3, but are most prevalent in MCM-L where leadframes can be easily laminated to the module or plated through-holes can be used as castillation joints to form leaded and leadless modules, respectively. If leaded, the leads can protrude from one or all four sides. For most applications, the most efficient way to distribute the I/O warrants the use of all four sides. The leads are trimmed and formed to QFP specifications. If no leads are required, metallization is placed on the underside of the module body to facilitate direct soldering. Although this style of package utilizes less space because the solder joints are on the underside of the package perimeter, inspection of the solder joints by conventional means is difficult. Conversely, the lead adds inductance and contributes to the propagation delay of the package. For high frequency applications, leadless modules are preferred.

In both cases, as the number of I/Os terminating from the module increases, either the lead pitch decreases or the module size increases. As the lead pitch decreases, the ability to reliably solder the module to a circuit board becomes progressively more difficult. As this pitch is reduced, soldering defects, such as bridging (solder flows between adjacent I/O and forms an electrical short), also increases. In contrast, if the package body is increased, more space of the mother board will be required for the attachment process. Thus, the use of perimeter I/Os is self limiting. Typical I/O pitches found are 0.050", 0.025" and 0.020", and I/O counts can reach several hundred. For increased interconnection density, one must consider the use of array I/O connections.

2.5.2 Pin Grid Array

Pin grid arrays (PGAs) can be utilized as a solution to overcome the geometrical constraints imposed by the peripherally terminated modules. By distributing the leads on the full area of the underside of the package, the module is more efficiently used. Figure 2-11 demonstrates the module I/O plot versus the area of the package. For the same pitch, significantly more I/Os exist in the PGA format. One hundred mil, 70 mil staggered and 50 mil pitches are commonly used. Package lead counts from several hundred to over 1000 have been fabricated. These pins can be press fit and soldered onto the MCM, as with plastic packages, or brazed in place, as with ceramic. The approximate cost of adding pins ranges from $0.01 to $0.02 per I/O. The PGAs can be through-hole mounted, placed into sockets or the pins may be surface mounted and soldered directly to the board. In many high performance applications, surface mounting

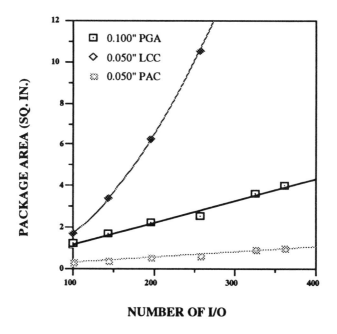

Figure 2-11 Array package efficiency.

will be required. In these cases, as with any leaded module, the inductance of the lead must be taken into consideration. For applications where this inductance is too great, area array formats such as pad array carriers should be considered.

2.5.3 Pad Array Carrier (PAC)

Pad array carriers (PACs) are often referred to as area array packages, land grid arrays (LGAs) or surface mount arrays (SMA). They are similar to PGAs with the exception that the pins are replaced by metallized pads. This package offers all of the efficient I/O termination benefits of the PGA without the negative attributes associated with lead inductance. Additionally, costs are typically less than PGAs since no leads are required. Attachment to the motherboard is through a socket or by a direct soldering process known as C5 (Controlled Collapse Chip Carrier Connection). In this process, solder bumps are deposited onto the metallized pads on the underside of the module and on the mating pads on the motherboard. The PAC then is placed onto the board and reflowed. As with most surface mount applications, small misalignments in component placements are remedied during the reflow cycle by the natural wetting characteristics and surface tension of the solder. This phenomena contributes to a high yielding process.

2.6 GLOBAL MATERIALS AND MANUFACTURING CONSIDERATIONS

2.6.1 Cost

The choice of materials and manufacturing processes greatly influences the final module cost and its suitability for a particular application. Because many MCM technologies are new (such as thin film substrates, flip chip connections and TAB), capital equipment start-up costs can be very high. The immaturity of these technologies also means low initial yields. Although capital costs and initial low yields can easily undermine a start up MCM development effort, this chapter is concerned with cost comparisons based on a full-scale production situation.

The largest contributions to cost for MCM production are module final test yield, substrate cost, and chip cost. Module final test yield is based on the expected yield of the individual components (chips plus substrate) plus an estimate of the damage done by the assembly process. Expected yield of the chip components is a tricky subject, depending heavily on the type of testing done at probe (in wafer form) and at final test for a particular device. For example, a RAM (random access memory) chip is normally tested only for basic functionality at probe. Then the RAM is packaged and sorted at final test for speed and temperature performance. As a last step, the packaged RAM is burned in to cull early life failures. Unless the die is speed and temperature sorted at probe and burned in at wafer level, which is not the normal practice today, the speed, temperature, and burn-in fallout occur at module final test and module burn-in.

If the circuit does not contain redundancy, the module can be considered a "series system" [13]. The expected yield of a series system is the product of the individual expected yields of the components. For example, consider a module containing three different semiconductor chips with expected yields of 95%, 90%, and 97%. The substrate expected yield is 95% and the assembly related yield is estimated at 99%. The expected module final test yield is then:

$$.95 \times .90 \times .97 \times .95 \times .99 = 78\%.$$

It is easy to see that the module yield quickly deteriorates as a function of the number of components and the expected yield of each of those components. Yield can be greatly improved and module costs dramatically reduced by testing module components - chips and substrate - prior to assembly.

For a given chip set, module yield will dominate total costs for small plastic MCM-L type modules where substrate costs are low and module repair may be difficult or impossible. If a repairable technology is used along with an inexpensive substrate, the silicon cost is the major contributor to total cost. At the other end of the spectrum, substrate cost may be the largest contributor to total cost for large, high end, high power ceramic or MCM-D/C modules where module repair is practiced [14].

Substrate cost is driven by three major factors: capital equipment required to fabricate the substrate, the number of module substrates that can be obtained from a single process unit (such as a ceramic wafer) and substrate yield. More mature technologies such as thick film hybrid and wire bond will have the lowest capital equipment costs and the highest substrate yield. Larger process units (such as 12" × 18" PWB panels) will lead to lowest individual substrate costs. Substrate yield is substantially influenced by basic manufacturing approaches, such as parallel versus sequential processing. In parallel processing (most commonly used for PWB and cofired ceramic type substrates), substrate layers are individually fabricated, inspected and yielded. The layers are then stacked to form a single substrate.

2.6.2 Electrical Performance[1]

Electrical performance can be affected significantly by choices made in the areas of signal interconnect and chip, MCM substrate and module level connections. Thermal performance of a package can also affect electrical performance, most likely through degraded performance at excessive junction temperatures. Other than thermal issues, the chief performance factors to be considered during package selection are:

1. Material properties of insulator layers and interconnect conductors
2. Transmission line parameters such as impedance and crosstalk
3. Package parasitics (such as pad capacitance and lead inductance)

Ideally, in any electronic system, the signal leaving the output of some transmitter should arrive, with no changes, at the input of some receiver. However, signals experience attenuation (loss of amplitude) and distortion (noise)

---

[1] Contributed by Eugene Heimbecher.

as they travel through a packaging system. Loss is caused by material properties and noise is caused by numerous factors broadly broken into transmission line concerns and package parasitics. Noise control considerations generally lead to package designs which are large, difficult to manufacture, and expensive. Manufacturing and cost factors, as well as size and weight considerations, usually lead to package designs with poor electrical performance. Balancing these factors against one another is essential for successful MCM system design.

Insulator and Conductor Material Properties

Insulator layers effect electrical performance through two parameters. These are dielectric constant (ε_r or K) and loss tangent (tan δ), also known as dissipation factor (DF). Dielectric constant, ε_r, is a key factor in determining most transmission line parameters; both ε_r and tan δ contribute to what is called dielectric loss in insulator layers. Dielectric loss, like conductor loss (below), results in signal attenuation. Generally, for MCMs with short line length and frequency content below 1 GHz, dielectric loss is not of concern. For these cases, loss due to resistivity of interconnect metal almost always masks dielectric loss. At frequencies much higher than 1 GHz, certain materials must be avoided because of their high dielectric loss. At microwave frequencies, preferred materials are ceramic or teflon; fiberglass epoxy (FR-4) is to be avoided.

The high dielectric constant of a material in and of itself is not necessarily a problem. It is not an automatic limitation to high frequency usage. There are usually undesirable consequences to using high dielectric constant materials for signal interconnect structures. These are primarily: increased signal propagation delay and increased physical size (for identical transmission line parameters and performance).

Interconnect material (such as copper) affects electrical performance primarily through internal resistance resulting in signal losses. (Voltage drop due to these losses is usually negligible in signal interconnect, but may be significant in power and ground lines. This is one reason for use of wide planes to distribute power and ground). Resistance is generally broken down into two categories. These are standard DC resistance and high frequency AC resistance due to skin effect. Skin effect occurs because alternating currents do not flow evenly throughout the cross section of a conductor, but tend to flow only near the outer surfaces. For example, it is a good approximation to say that, for pure copper, a current with a frequency of 300 MHz only flows in the outer 0.15 mil of the conductor cross section [15].

DC resistance is calculated using standard equations; AC resistance is calculated by a ratio of full cross section to skin effect area [16]. Line resistance is not always important. For example, MCM-L conductors are on the order of 3 - 4 mils wide and about 1 mil thick and both DC and AC losses are usually

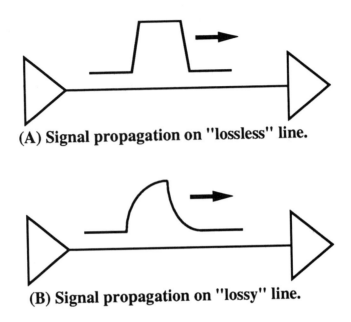

(A) Signal propagation on "lossless" line.

(B) Signal propagation on "lossy" line.

Figure 2-12 Effect of line loss.

negligible. On the other hand, MCM-D conductors may be on the order of 0.4 × 0.15 mil and losses need to be carefully assessed. Measured data (Motorola) of some typical MCM-D substrates show DC resistance on the order of 3 - 5 Ω/cm for copper interconnect and 15 Ω/cm for aluminum.

Large line resistance causes a low pass filter (RC) type roll off on signal edges (Figure 2-12). This roll off is small near the sending end of the line and becomes progressively more pronounced as the signal propagates down the line. The edge degradation is beneficial to noise generation, but can be a significant problem on long lines with critical timing requirements, for example clock distribution lines. Also, the signal rise and fall times may be slowed down so much that CMOS devices become subject to spurious oscillations. The MCM system designer must consider these issues carefully.

Transmission Line Parameters
The understanding of transmission line theory and its application to MCM design is a text in itself [16]. However, it may be sufficient to say that at higher

frequencies and faster edge speeds it always is necessary to maintain the best possible controlled impedance environment for both signal lines and power and ground distribution. For optimum performance in high speed, high frequency systems, signal interconnect always needs to be done in a transmission line manner (such as stripline, with good AC ground and reference planes both above and below the signal). Additionally, voltage distribution should always use full planes well coupled to ground references. A near ideal approach is available in the MCM-D/C. Signal lines can be embedded in low dielectric constant deposited stripline layers with optimized signal performance while power and ground return planes can be contained in a high ε_r, very thin (and hence highly capacitive) layer in the ceramic base. With sufficient power plane capacitance, it would not be necessary to use discrete decoupling capacitors.

Dielectric constant is important in two ways. It affects both transmission line impedance and speed of signal propagation. As an example, for a given stripline geometry, if the impedance is 50 Ω and the delay is 2.03 ns/foot (ε_r = 4.0), doubling the dielectric constant without changing any other factor results in impedance decreasing to 35%) while delay increases to 2.87 ns/foot. (Note the square root of two relationship.) The situation is not as clean for microstrip lines or other multimedia line geometries, but the general trend of decreased impedance and increased delay (as ε_r increases) always holds.

For several reasons it is desirable to have transmission line characteristic impedance (Z_o) fall into the general range of 50 - 150 Ω. Line impedance is generally a complex function of several parameters, but it can be increased (from 35 - 50 Ω) by a combination of making line width smaller and/or making the vertical distance between the signal lines and ground planes larger. When dielectric constants are large, as for ceramic, designs must use either thinner line widths or thicker insulator layers than for designs with smaller dielectric constants, such as polyimide. There are two problems with trying to use thinner lines. First is the fact of running into the manufacturing process limit, as discussed in previous sections. Second is a parameter tolerance issue: a line width of 10 mils ±1 is easier to make consistently than a line width of 2 mils ±1. Thus, the usual strategy is to increase the thickness of the insulator.

The chief physical parameters for controlling impedance are: dielectric constant and line width and vertical distance between signal line and ground plane. It is recommended that a full dimensional analysis of all parameters of all layers in a substrate stack up be performed to identify the worst case maximum and minimum impedances to guarantee acceptable electrical performance. Also, it is often desirable to separate or isolate certain signals from other signals. This can result in the need for additional routing layers as compared with a design with no routing restrictions. These extra layers are undesirable from a manufacturing and cost point of view, but are necessary to obtain required electrical performance.

Crosstalk is the remaining transmission line parameter to be discussed. Crosstalk is the unwanted coupling of energy from an active (signals occurring) section of some circuit or system to some other unconnected and nominally quiet (no signals) section. This can result in a receiver seeing an input signal that is, in fact, only noise. Crosstalk can arise form a number of sources:

- Electro-magnetic interference (EMI)
- Power and ground plane noise
- Capacitive affects when signal lines cross over each other
- Mutual coupling when signal lines run parallel to each other

Generally, this last mechanism is considered the most significant source of crosstalk in well designed systems. (But crossover affects can be significant if several or more occur along the length of the victim line.)

For stripline, maximum crosstalk is not affected by the dielectric constant of the insulator material. If ε_r is doubled, all other factors being unchanged, there will be no change in crosstalk. However, if the dielectric thickness is increased (to increase line impedance from 35 - 50 Ω), then crosstalk will increase due to the reason discussed below. Therefore, with higher ε_r materials, to obtain comparable electrical performance, the line separations must be larger and the overall physical size of the design will probably increase.

There are a number of strategies for reducing crosstalk, but the most useful is simply to keep things far apart. Microstrip transmission lines are more prone to crosstalk problems and should be avoided in noise sensitive designs. When microstrip is used, separating two lines by a distance equal to 8H (where H is the vertical distance between the bottom of the signal lines and the top of the ground plane) reduces crosstalk to about zero. Stripline is a more effective approach for lines in a noise sensitive application.

As an example, consider an MCM-D/C design, as mentioned above. If the distance between ground planes is about 40 μm (1.6 mils), striplines separated by only 80 μm would have no crosstalk. In practice, it is found that even a separation of 40 μm is enough to achieve close to zero crosstalk. On the other hand, for an MCM-L design using microstrip, the situation can be much worse. For example, to guarantee no crosstalk in a structure with a dielectric thickness of 4 mils, signal lines would have to be separated by over 30 mils. Thus, there is a 10 to 1 routing density advantage in MCM-D (stripline) as opposed to current MCM-L (microstrip) for identical levels of crosstalk.

Crosstalk does not occur only in MCM substrates. A potential problem in flip chip MCM design is redistribution of I/O pads. Most ICs are not designed for flip chip. Hence, the typical procedure is to take a standard chip, with I/O pads on the periphery, and use an extra metal layer to redistribute the I/O pads over the surface of the chip. This must be done with some care. It is possible to route a very noisy output over the top of some very sensitive input. This is an issue that is not well understood currently.

Crosstalk is an issue to be considered in package selection as well. Three primary factors to reduce chip connection and packaging crosstalk are: to keep leads as far apart as possible, to keep parallel leads as short as possible and to use as many ground leads as possible (mixed in with signal leads) to reduce coupling between signal leads.

2.6.3 Thermal Path

Heat produced within semiconductor chips will dissipate through the MCM package and out to ambient air and to the board on which the module is mounted. If heat is not dissipated efficiently, the chip junction temperature (T_j) rises beyond the reliable operating range and operating life is decreased significantly.

Like single chip electronic packages, module thermal performance is measured in terms of the difference in temperature between the chip junction and ambient air per watt of power produced by the chips:

$$\theta_{ja} = (T_j - T_a) / W.$$

The module must provide thermal pathways that conducts heat expected from the chips mounted in the module. Heat is conducted through a combination of four general paths: chip to substrate to ambient, chip to lid to ambient, chip to leadframe to ambient or chip to substrate to heatsink to ambient. The chip to substrate to ambient path is most common in electronic packaging, but for MCMs this path can be complicated by a substrate with mediocre thermal conductivity and very dense circuitry. Designing structures to improve substrate conductivity, such as thermal vias and metal heat spreaders, usually reduces the routing channels available in the substrate. The designer is faced with a difficult tradeoff between routing density and thermal conductivity. The chip to lid to ambient path has been demonstrated on several high power MCM designs, but is workable only when the backside of the chip faces the module lid - flip chip or flip TAB mounting. Many low cost, single chip packages use the chip to leadframe to ambient approach. This path has limited thermal capability (35°C/W in still air) and may complicate the routing of signal interconnect on the module.

2.6.4 Rework

The choice of materials and manufacturing processes determines whether a module can or should be reworked; that is, a faulty chip removed from the substrate and replaced with a new chip. Rework is both a technical and business issue. The cost of the rework process must be less than the cost of a finished module, otherwise rework does not make economic sense. When considering the cost of rework, one must consider costs associated with designing fault location circuitry, isolating a fault, performing the repair, retesting the part and, perhaps, cycling through another rework iteration.

Module level rework begins with fault detection by either visual or electrical tests. Next, the faulty chip is accessed. The chip is demounted, normally by applying heat to the chip connection points. The substrate connection points are then prepared for a new chip, the new chip is mounted, the module encasement is restored and the module is retested. Materials and manufacturing choices are obviously critical at every step of rework. For example, visual detection may be difficult or impossible with flip chip or flip TAB type chip mounting. Electrical detection is successful only if the module circuitry is designed for such testing, using JTAG boundary scan, for example, as discussed in Chapter 13. Chip accessibility can be difficult or impossible in modules with plastic mold encasement or with certain lid attach schemes. Chip demounting is difficult or impossible in modules where a die attach is employed. Substrate connection point preparation is difficult in modules that contain tight pitch chip connection pads, or in those that rely on solder pads on the substrate.

2.7 MODULE DESIGN EXAMPLES

MCM materials and manufacturing choices should be optimized for a given application. In this section, we describe some real examples of MCMs employing various combinations of technologies to suit specific product requirements.

2.7.1 Mainframe Computer Module

Figure 2-13 is a photograph of an MCM designed and built by Siemens-Nixdorf Information Systems for a high speed mainframe computer. It is 4" on a side and contains bipolar ASIC and memory components. The MCM substrate is a multilayer organic laminate (MCM-L). Substrate manufacturing is additive and sequential. Each layer is visually inspected and repaired as it is fabricated [17]. Chip mounting is flip TAB using thermode gang bonding. An elastic spacer on

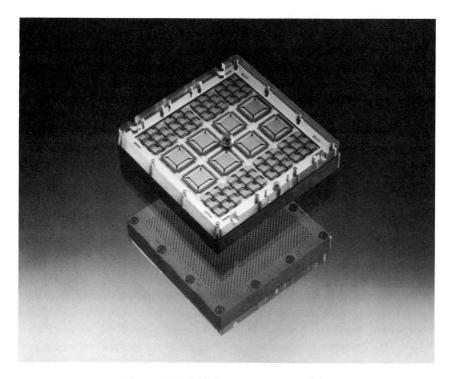

Figure 2-13 Mainframe computer module.

the front or active side of the chip presses the back of the chip directly against an anodized aluminum module lid forming an excellent path for heat transfer. The module can handle more than 500 watts of power. The substrate is mounted in a plastic housing that contains the unique module level I/O, with over 1,000 signals connecting to the mother board. The module I/O is an array of holes in the plastic housing with bifurcated springs designed to accept a matching array of pins mounted on the mother board. The entire module is repairable.

This MCM design is optimal for its intended application of packaging a large number of high speed, high power, expensive bipolar chips. High cost can limit the use of this technology for the general market.

2.7.2 Workstation Module

The MCM sketched in Figure 2-14a is aimed at the performance and density driven CMOS/BiCMOS marketplace, particularly, the departmental computer, communications, peripherals, workstation and PC segments. It is a cost effective, high density MCM developed to accommodate high speed complex ASIC logic devices, microprocessors and memories.

The package body and MCM substrate are combined in a single substrate made of a ceramic base with deposited thin film copper/polyimide interconnect layers (MCM-D/C) [18]. The substrate provides very high density routing with 20 μm lines and 30 μm spaces on two layers. The chip mounting is flip chip, with a thermal gel material providing the heat conduction path between the backside of the chips and the thermally conductive lid. The module I/O is a JEDEC standard 299 pin grid array or a space optimized 50 mil pad array carrier [19] as pictured in Figure 2-14b.

This module is well suited for its intended application. It is capable of connecting 15 to 20 complex chips in a very small area. The power handling capability of 20 - 30 watts is matched to the expected power dissipation of CMOS components. The module offers optimum electrical performance at a cost that serves the market for performance oriented systems.

2.7.3 Low Cost Module

A very inexpensive module commercially available today is pictured in Figure 2-15a. The module, which offers advantages in system miniaturization and system performance, is a direct extension of a single chip QFP package. It utilizes a multilayer PWB substrate (package body and MCM substrate are integrated (MCM-L) with an embedded leadframe that forms the module level peripheral connection (Figure 2-15b). The chips are mounted using standard die attach/wire bond techniques, and the entire assembly is overmolded with plastic.

The major limitations of the module are thermal dissipation and routing density. Heat is primarily conducted from the chip to the copper leadframe The package can dissipate up to 4 watts under forced convection at 500 feet per minute airflow. Routing density is limited primarily by very large vias - 0.8 mm on the internal layers and 0.5 mm on the outer layers.

The technology offers low cost and low risk because it uses a minimum number of inexpensive piece parts and mature processing technologies. It is generally applicable to consumer and communications products where small size and low cost are important.

2.7.4 Data Communications Module

The module shown in Figure 2-16 is a prototype developed for a data communication application. The module consists of two host microprocessors

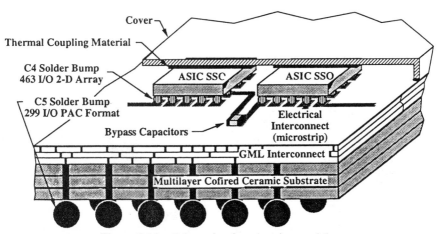

Figure 2-14a Schematic of workstation module.

Figure 2-14b Six chip C4/C5 workstation module: Thin film copper/polyimide on multilayer cofired ceramic, 463 I/O flip chip connections per chip.

Figure 2-15a Low cost module.

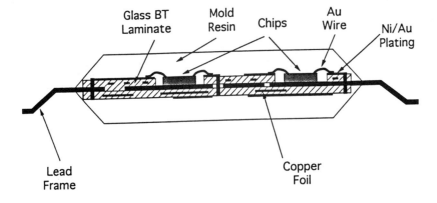

Figure 2-15b Schematic of low cost MCM with integral leadframe.

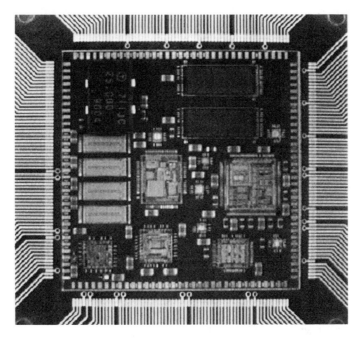

Figure 2-16 Data communications module.

with associated memory, a proprietary chip set and various passive devices. A total of 18 ICs and approximately 45 passive devices are assembled to a 1.8" × 1.8" four layer MCM-D substrate incorporating 75 μm lines and 30 μm vias. Die and wire bond assembly techniques are utilized for chip mounting. In this application MCM processing has made it possible to reduce the size from the original 30.0 square inch PWB to 3.24 square inch.

2.8 FUTURE TRENDS IN MCM MATERIALS MANUFACTURING

MCMs are a solution to the size, parasitic and path length effects of semiconductor packages and their interconnect. The multichip module (MCM) interconnection technology has been maturing as a packaging technology since the first articles were published in the early 80s. The first applications using MCMs used multilayer cofired ceramic substrate technology (HTCC). Low dielectric constant polymers with thin film conductor technology applied to silicon, metal, alumina and cofired multilayer ceramic came later as more package and or chip I/Os were needed.

During this same period plastic quad flat package (PQFP) technology improved significantly with 0.3 mm pitch and up to 500 I/Os being planned for manufacture by 1993. PWB technology and MCM-L are also improving with 4 mil line and space and 6 mil via technology currently available and manufacturers talking about 2 mil line and space PWBs with 4 mil vias by the mid 90s.

With the exception of the very high end computer and military applications and the very low end consumer market, the use of MCM has been relatively modest. Most of the successful commercial users of MCMs have been vertically integrated mainframe computer companies. The successes have come from products derived from a system viewpoint, where the company has control of the subsystem and the components to achieve performance of the product at an acceptable cost.

If MCM technology is to approach the degree of acceptance in the marketplace to that of surface mount technology (SMT) today, issues relating to die availability, low cost substrate technology and the testing of bare chips and modules must be addressed.

An abundant supply and variety of good bare die must become available at an acceptable cost. Off chip performance will determine whether wire bond, TAB or C4 die configurations are suitable for use. As chip clock frequencies rise, the thrust will be to C4 die attach.

2.8.1 Substrates

Today, typical high density MCM-D substrates can cost anywhere from $50 to $100 per square inch, meaning that the substrate cost alone will deter many product designers from considering this packaging approach. A low cost MCM technology must evolve within the next several years if MCMs are to have a major impact across a broad spectrum of products. A high density MCM substrate cost below $5 per square inch must be achieved. A hybrid version of current MCM-L and MCM-D processing is a potential solution. For example, high density signal layers containing small vias fabricated by advanced processing such as laser ablation can be laminated to low density power and ground planes fabricated by conventional PWB processing. Since there is a cost premium for using the high density layers, utilization should be confined to a minimum of signal layers or for the redistribution from high I/O chips to the circuit board. The cost of such a module is expected to approach that of conventional PWB and current MCM-L technologies.

2.8.2 Optical Multichip Modules (MCM-O)

MCM prototypes intended for use at system clock rates in excess of 3 GHz have been fabricated by conventional copper/polyimide processing. Design guidelines for these high frequency modules have been proposed [20]. A potential next step in the evolution of MCMs that could overcome many of the difficulties associated with high frequency, hard wired MCMs is the use of optics for chip-to-chip interconnection. Optical transmission media exhibit terahertz bandwidth, immunity to electromagnetic interference and optical noninteraction. It has been demonstrated that properly designed thin film optical waveguides, physically intersecting with each other and transmitting optical signals, exhibit negligible crosstalk [21]. Therefore, chip-to-chip optical interconnect technology may not require a multilayer system as is required when using conventional, hardwired interconnections. The need for crossovers is eliminated or minimized. However, in practice, it is likely that hardwired and an optical layer will be used, together with the total number of layers significantly reduced from using hardwired alone.

Preliminary results have shown that high speed chip to chip optical interconnects, compatible with conventional MCM substrate processing, are feasible [21]. A semiconductor laser was coupled to a flip chip photodetector through a photolithographically defined polymer optical waveguide (Figure 2-17a). An infrared photograph of the interconnect taken during operation is shown in Figure 2-17b. Optical propagation is from left to right across a 2" ceramic MCM-D substrate. The resulting optoelectronic module is referred to as an optical MCM (MCM-O).

2.8.3 Test

Test methodology and technology to supply functional bare die and assembled MCMs must be developed to a point where testing is not perceived as an impediment to utilizing MCM technology. To assist in this area, standard MCM sizes and footprints (either perimeter or area array) need to be established. In addition, test and burn-in sockets need to be available.

2.8.4 Thermal Control

Thermal solutions for modules which provide for reliable thermal control at reasonable cost need to be developed. For high power applications, C4 die or flip TAB can offer the lowest thermal resistance path (back of die is exposed for heatsinking) without complicating the substrate layout, quite possibly making it the most viable packaging alternative for high power applications [23]. (See Figures 2.9 and 2.14).

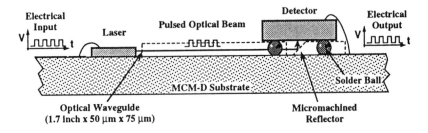

Figure 2-17a Schematic of a chip to chip optical interconnect.

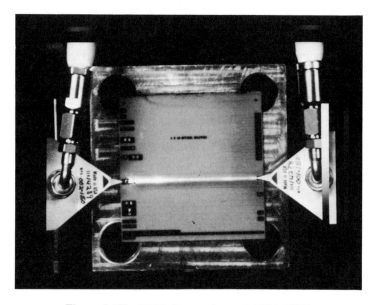

Figure 2-17b MCM-O operating at 1 Gbit/s NRZ.

2.8.5 Environmental Concerns

Finally, more environmentally sound processes need to be developed. The elimination of many toxic solvents and heavy metals from the fabrication of laminate materials and subsequent assembly processes is becoming an issue. The move to eliminate solvents, such as methylene dianaline (MDA) from the manufacture of organic laminates [24] and chlorofluorocarbon (CFC) containing materials, has already begun. In addition, federal legislation is pending that limits or eliminates the use of lead from electronic assemblies [25]. OEMs soon may be responsible for recycling all lead containing products from inception to retirement - cradle to grave. As a result the industry needs to focus on developing non-lead bearing solders and conductive z-axis adhesives and cements.

Acknowledgments

The authors acknowledge and appreciate the ongoing support of J. W. Stafford and S. Bowen. We also wish to thank the numerous staff members of the Motorola Corporate Manufacturing Research Center and ASIC Division for review, discussion and contribution to this manuscript.

A special thanks and acknowledgment to Eugene Heimbecher for his contribution of Section 2.6.2.

References

1   J. L. Sprague, "Multilayer Ceramic Packaging Alternatives," *IEEE Trans. on Components, Hybrids, and Manuf. Tech.*, vol. 13, no. 2, pp. 390-369, June 1990.

2   U. Deshpande, S. Shamouilian, G. Howell, "High Density Interconnect Technology for VAX-9000 System," *Internat. Electr. Packaging Soc.*, IEPS, (Marlborough MA), Sept. 1990.

3   Design Guidelines Multilayer Ceramic, *Kyocera Corporation*.

4   Design Guide: Developing Customized Microelectronic Packaging for High-Performance Applications, *Dupont Electronics*.

5   M. R. Cox, L. Ng, "An Economic Comparison of High and Low Temperature Cofired Ceramic," *Inside ISHM* pp. 31-33, Jan./Feb. 1992.

6   S. J. Stein, R. L. Wahlers, C. Haung, M. Stein, "Interconnection and Packaging of Advanced Electronic Circuitry," *Proc. ISHM*, (Orlando, FL), pp.130-134, Oct. 1991.

7   N. Iwase, K. Anzai, K. Shinozaki, "Aluminum Nitride Substrates Having High Thermal Conductivity," *Solid State Techn.*, pp. 135-138, Oct. 1986.

8   D. J. Arthur, "Advanced Fluoropolymer Dielectrics for MCM Packaging," *First Internat. Conf. on Multichip Modules*, (Denver CO), pp. 270-276, April 1992.

9   P. Fischer, "Materials for a Parallel Approach to High Density Interconnects," *Proc. IEPS*, (Marlborough MA), p. 103, 1990.

10 T. G. Tessier, W. F. Hoffman, J. W. Stafford, "Via Processing Options for MCM-D Fabrication: Excimer Laser Ablation vs. Reactive Ion Etching", *Proc. 41st Electronics Components Techn. Conf.*, (Atlanta GA), pp. 827-834, May 1991.

11 R. C. Eden, "Applicability of Diamond Substrates to Multi-Chip Modules," *Proc. ISHM*, (Orlando FL), pp. 363-367, Oct. 1991.

12 J. W. Balde, "Proposed MCM Standard Sizes - A Report of the IEEE Task Force", *Proc. NEPCON West*, (Anaheim CA), pp. 467-478, Feb. 1992.

13 R. A. Dovich, *Reliability Statistics*, Milwaukee, Wisconsin: ASQC Quality Press, pp. 21-35.

14 A. Krauter, J. Baumann, M. Becker, "Assembly Process for a New Packaging System," *Siemens Research and Development Reports*, Bd. 17, 1988, No. 5, Springer-Verlag.

15 J. Hardy, *High Frequency Circuit Design*, Virginia: Reston Publishing Company, 1979.

16 H. B. Bakoglo, *Circuits, Interconnects and Packaging for VLSI*, New York: Addison-Wesley Publishing Company, 1990.

17 H. Brosamle, B. Brabetz, V. Ehrenstein, F. Bachmann, "Technology for a Microwiring Substrate," *Siemens Research and Development Reports*, Bd. 17, 1988, Nr. 5 Springer-Verlag.

18 E. G. Myszka, A. Casey, J. Trent, "A Multichip Package for High Speed Logic Die," *Proc. 42nd Electronic Components Techn, Conf.*, (San Diego CA), pp. 991-996, May 1992.

19 J. Trent, G. Westbrook, "Fine Pitch Pad Array Carrier Sockets for High Speed Logic Die," *IEEE Multichip Module Conf.*, (Santa Cruz CA), pp. 40-43, March 1992.

20 B. K. Gilbert, W. L. Walters, "Design Options for Digital Multichip Modules Operating at High System Clock Rates," *First Internat. Conf. on Multichip Modules*, (Denver CO), pp. 167-173, April 1992.

21 C. T. Sullivan, "Optical Waveguide Circuits for Printed Wire-Board Interconnections," *Optoelectonic Materials, Devices, Packaging and Interconnects*, SPIE vol. 994, pp. 92-100, 1988.

22 K. W. Jelley, G. T. Valliath, J. W. Stafford, "1 Gbit/s NRZ Chip-To-Chip Optical Interconnect," *Photon. Tech. Lett.*, to be published.

23 R. Darveaux, I. Turlik, "Backside Cooling of Flip Chip Devices in Multichip Modules," *First Internat. Conf. on Multichip Modules*, (Denver CO), pp. 230-241, April 1992.

24 F. Klimpl, "Basics of Multilayer Matter," *Electronic Engineering Times*, Issue 681, Feb. 1992.

25 C. Melton, A. Skipor, J. Thome, "Manufacturing Process Issues of Non Lead Bearing Solder Pastes," *Internat. Conf. on Solder Fluxes and Pastes*, (Atlanta GA), May 1992.

3

MCM PACKAGE SELECTION: A SYSTEMS NEED PERSPECTIVE

Paul D. Franzon

3.1 INTRODUCTION

There are a large number of packaging alternatives available to the design engineer today. This range is not likely to narrow in the near future. The object of this chapter is to provide a framework of understanding for making packaging decisions with the perspective of how best to satisfy the needs of an electronic system.

The design decisions made in any engineering venture are driven by cost and performance. Fundamentally, packaging acts to limit performance and to increase cost. Recent trends in integrated circuit technology suggest that system performance is being limited increasingly by the package. This has resulted in heightened attention to new, more highly customized forms of packaging used to improve system performance. Multichip modules (MCMs) represent a class of packages used to obtain significant improvements in system performance compared with conventional forms of packaging. Highly customized advanced packages are expensive when compared with off the shelf mass produced single chip packages. It is important to realize when advanced packaging is appropriate and when it is not. A primary aim of this chapter is to identify the nature of the tradeoffs involved and to suggest a decision-making process. The aim is not, however, to provide the reader with complete models for that decision-making. Many of the later chapters provide these models.

Fundamentally, there are three types of silicon found in a digital system. First, we find the very large scale integrated (VLSI) logic chips with over 100,000 transistors. Such chips range from general purpose microprocessors to special purpose application specific integrated circuits (ASICs). ASIC styles range from full custom and semi-custom designs to chips whose function is programmed in the manufacturing line. Second, we find the so called "glue chips," the off the shelf chips that provide functions not integrated into the VLSI chips. Today, more and more of the glue logic is being collected into and thus, is being replaced by ASICs. Typically, the only glue chips found are the drivers required to drive large loads and long lines and the receivers and latches that often are at the other end. Finally, we find the memory chips, which are often the most numerous [1].

The need for advanced packaging is driven by the trends in the design and use of these three types of chips. Today's leading CMOS microprocessor chips often contain over 1,000,000 transistors and are clocked at frequencies in excess of 150 MHz. The transistor count and clock speed are expected to continue to grow at a rapid rate, as shown in Figure 3-1. Leading edge VLSI chips create tremendous demands on the package. These chips have high I/O counts with 500+ I/Os being typical for 1992 RISC microprocessors. As the on-chip circuits become faster, the inter-chip package delay, not speeded up by using faster circuit technology, becomes dominant. One of the requirements for reducing this delay is that the chips be placed closer together. The increasingly fast signals produce lots of electrical noise in the package unless the package is carefully designed. The large number of fast circuits also produces lots of heat that must be removed from the system. For example, the DEC Alpha CPU chip dissipates over 30 W when running at a clock speed of 200 MHz.

The purpose of this chapter is to present a framework for making packaging decisions as part of the system design process. The next two sections set the stage for this by describing the system design process and defining the concept of the packaging hierarchy. Following that, the factors through which packaging decisions affect system performance and cost are presented and discussed. A process is then described showing how these performance and cost factors are used to make packaging decisions. An example is given of this decision process.

3.2 SYSTEM DESIGN PROCESS

The phases that make up the design process are summarized in Figure 3-2. In the system specification phase, the system requirements and goals are determined. How this is done is discussed toward the end of the chapter. Most of this chapter is concerned with the high level design phase, in which the

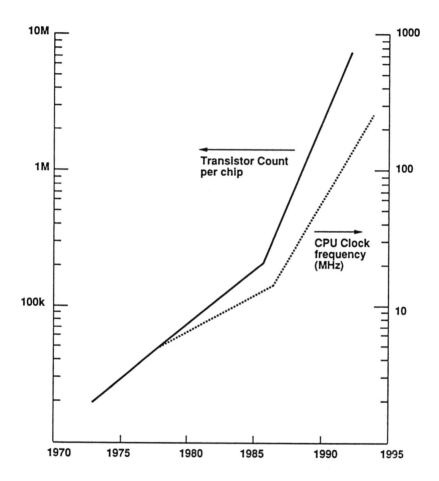

Figure 3-1 Microprocessor clock speed and the number of transistors per die are both growing at an increasing rate. (Adapted from [16].)

organization of the system, and the technologies to be used, are decided. This is described further in Section 3.6. In the low level design phase, the actual circuits are designed schematically. These are turned into layouts of the chips and packages during the prefabrication stage. A layout describes exactly where each transistor and part is placed and where each wire is run. Aspects of these phases are discussed in **Part B - The Basics.**

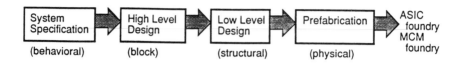

Figure 3-2 The system design process. (Adapted from a chart prepared by Ken Drake of MCC.)

3.3 THE PACKAGING HIERARCHY

Obviously not all systems fit onto one VLSI chip, one MCM or one printed wiring board (PWB). Thus, multiple levels of packaging are needed so that multiple chip packages can be interconnected. This is referred to as a packaging hierarchy, an example of which is given in Figure 3-3. At the lowest level of the hierarchy, the chips are mounted in single chip packages or MCMs. The next level of the hierarchy usually consists of PWBs (sometimes referred to as cards). The PWBs in Figure 3-3 are then connected together via a backplane (also referred to as a back panel, or sometimes board if the PWB was called a card). Backplane to backplane connections are made via a rack, and the racks then might be connected by some means, and so on. This is a very common hierarchy for larger systems though it is by no means the only one available.

Gate to gate interconnections (nets) might not go through this packaging hierarchy (for example, an on-chip interconnection) or might have to go through one or more levels of packaging. As any one net spans more levels of the hierarchy, the length of that net, and also the signal delay, increases substantially. Ideally, the interconnections and connections provided by the packaging hierarchy would be matched to the interconnection needs of the system. For example, Figure 3-4 shows how an electronic system could be considered as a set of interconnected functional blocks. In this case, the interconnection requirements of the system map naturally onto the packaging hierarchy shown previously in Figure 3-3. The nets that need to be short and fast are kept to the low levels of the hierarchy while nets that can be long span all of the levels of the hierarchy.

This ideal situation is rarely met. Most systems require more short nets than usually is provided by the packaging. The physical limitations of the packaging hierarchy (primarily determined by the desire to keep the sizes of individual packages and connectors down to control cost) force a number of these nets to go through more than one level of packaging. Unfortunately, the number of connections that go between package levels is limited by the capacity of the

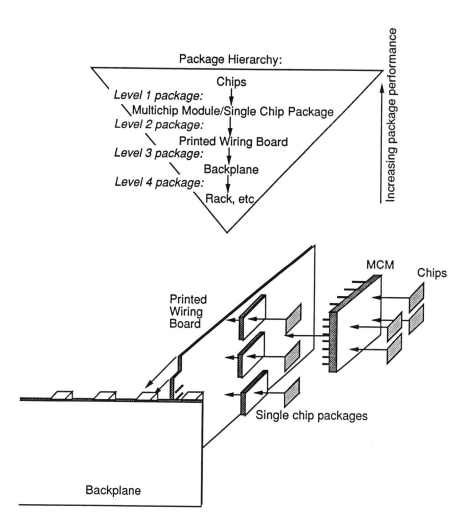

Figure 3-3 The packaging hierarchy.

connectors between levels (and high capacity connectors are expensive). If there are not enough paths available through these connectors then it might be necessary for several nets to share the same path (a bus or multiplexed path).

Interconnect Topology:

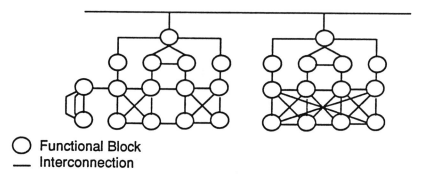

⭕ Functional Block
— Interconnection

Map Interconnect Topology onto Packaging Hierarchy:

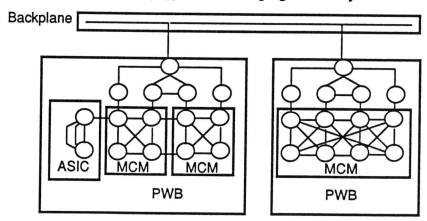

Figure 3-4 The interconnection needs between functional modules in a system often vary across a system. The packaging hierarchy should be structured so as to match these needs as closely as possible.

Thus, in many systems, a large number of nets take a double hit - they must span multiple levels of package and they must share a path with other nets. The job of the designer is to select the package technologies to minimize the number of nets that must run through the high levels of the package, and then to select those nets carefully. A tremendous advantage of MCMs over conventional packaging is that they allow more of the nets to be concentrated in the first two levels of the hierarchy, thus improving the system performance.

There are several alternatives to the hierarchy style illustrated in the bottom of Figure 3-3. For example, sometimes the backplane might be bypassed with dedicated cable connections coming from the other side of the board. This can be extended even further to the situation where most of the interboard connections are dedicated wires. For example, this is done in Cray computers. A slight variation on the hierarchy shown in the bottom of Figure 3-3 is to have a relatively large motherboard with smaller daughterboards mounted on it. This is a common hierarchy in desktop and portable computer systems. It also is used to package the IBM 3081 computer central processing unit. In this case, the main board is 60 cm × 70 cm in size and the daughterboards are MCMs, each containing 100 chip sites [2]. Another hierarchy under active investigation is the use of connections in the third dimension to stack MCMs.

3.4 PACKAGING PERFORMANCE FACTORS

This section describes the factors that relate performance to packaging technology choices. Usually, each performance factor is broken down into the elements that determine it. In some cases, figures of merit used to differentiate package choices are discussed. A figure of merit is a number that attempts to summarize the packaging factor. The usefulness of a figure of merit depends on the accuracy needed to make a decision. Often a full model or evaluation is needed to make a decision.

3.4.1 Size and Weight

Size and weight are often specified as performance goals in many portable and aerospace systems. In the latter this is a primary driving force towards the application of MCMs (see Chapter 15).

A size restriction leading to the use of advanced packaging also might arise artificially. In any system, the size and possibly the weight of each subsystem only is estimated early in the design process. Errors might be made or system goals might change later while detailed design is in progress. In this case it might become necessary to use advanced packaging in a sub-system in order to avoid a complete redesign.

A size limitation might be an area or volume limitation or some mix of the two. For example, in a notebook computer, the height of the computer is limited by the height of the disk drive. Disk drive manufacturers are driven to package the electronics portion of the drive in as small a height as possible, sometimes just a fraction of an inch. On the other hand, a few millimeters of height on the main computer board is not as important as the total area of the board.

If a size limitation is expressed as an area limitation, then a suitable figure of merit for evaluating different packaging approaches is the substrate efficiency. This is defined as the percentage of the substrate covered by silicon [3]. For example, if a 6 cm × 6 cm MCM has 10 0.8 cm^2 die placed on it, the substrate efficiency is given as 10 × 0.8 / (6 × 6) = 22%. Another useful metric is the number of gates per unit of substrate area. These figures sometimes are evaluated on a volumetric, rather than an area, basis.

Particular attention also must be given to the size and weight of the power supply (battery) and the mechanical structures used to remove heat. This might involve a compromise between the desire to use a small system to house the electronics and the need for a large heat removal structure required by a small system with a high heat density.

By replacing multiple single chip packages with an MCM, substantial size and weight reductions are achieved. The closest comparable single chip package solution would be to use ASICs in surface mount packages on a PWB. The magnitude of the reduction possible with an MCM solution depends, in part, on the interconnection capacity needed.

3.4.2 Interconnection Capacity Within Each Level

Interconnection capacity refers to the total amount of wiring provided within a level of the packaging hierarchy. This wiring is provided by layers within the package devoted to signal interconnections. The physical layout of the wires is determined by a routing Computer Aided Design (CAD) tool. Additional layers are usually used for distributing power and ground. Once the details of the parts to be interconnected are known, the package interconnection (or routing) capacity requirement is expressed in the following form [4]:

$$\text{Required Interconnection Capacity} \approx RN_{net} P/E \qquad (3\text{-}1)$$

with the result expressed in units of length. In this equation, E is the efficiency with which the available interconnect capacity can be used (often called routing efficiency and typically takes a value of around 50% [4]), R is the average length of each net (interconnection) in terms of chip pitch, N_{net}, is the number of nets and P is the average chip pitch, the average distance between chip centers.

The available interconnection capacity is given by:

$$\text{Available Interconnection Capacity} \approx \frac{\left(\sqrt{N_{chip}} - 1\right)^2 P^2 \times \text{Number of signal layers}}{\text{Average wire pitch}}$$

$$(3\text{-}2)$$

plus any capacity around the periphery of the board. Here N_{chip} is the number of chips.

A figure of merit often used to compare different packaging technologies is the interconnection density:

$$\text{Average Interconnection Density} = \frac{\text{Number of signal layers}}{\text{Average wire pitch}} \quad (3\text{-}3)$$

Figure 3-5 shows how interconnection density is calculated for a single layer. Figure 3-6 gives a plot of interconnection density and cost per unit area for different interconnection technologies. Wiring capacity is then given by:

$$\text{Available Interconnection Capacity} \approx \left(\sqrt{N_{chip}} - 1\right)^2 P^2$$
$$\times \quad \text{Average Interconnection Density.}$$

$$(3\text{-}4)$$

If the available capacity is less than the required capacity then, short of redoing the high level design, there are six choices:

1. Share (multiplex) signals onto the same interconnections, for examples, reduce N_{net}. This is often done on backplane busses. If it is likely that several functional modules need the same physical line at the same time then performance degrades.

2. Increase the number of signal layers thus increasing weight, size and cost. Typically, the total number of layers increases at a rate almost one and a half times the rate of increase of the number of signal layers. In a high speed system each signal layer has to be next to a power or ground layer (see Section 3.4.4 and Chapter 11). If through-hole components are used on a PWB, the maximum number of layers is limited to 10 or 12 (usually equivalent to a maximum of 8 signal layers).

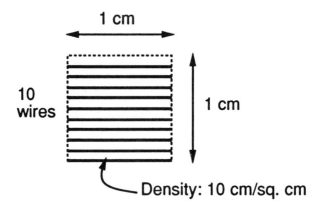

Figure 3-5 How interconnection density is measured.

3. Increase the chip pitch, P, and the size of the board or substrate. This then increases weight, size, cost and delay.

4. Spend additional design time to try to increase the routing efficiency, E, by manually, rather than automatically, routing the nets.

5. Reduce the number of chips by using an ASIC.

6. If the via size is larger than the wire width, thus consuming area that could have been used for wires (see Figure 3-7), then reducing the via size will allow an increase in interconnection density. This is common in PWBs and laminate MCMs.

In general, the finer the average wire pitch and the greater the total interconnection density, the less likely that these undesirable steps are required. In particular, thin film MCMs rarely have insufficient interconnection capacity. The average wire pitch is not always the same as the minimum wire pitch, typical values for which are given in Table 3-1. There are two reasons for this. First, in high speed systems, wires need to be pitched further apart to control crosstalk noise. This is discussed in Chapter 11. Second, via size and style can have a dramatic impact on average interconnect density.

If there are insufficient via sites on the board or MCM, it might be difficult for a router to find a via where it needs one. This is referred to as via

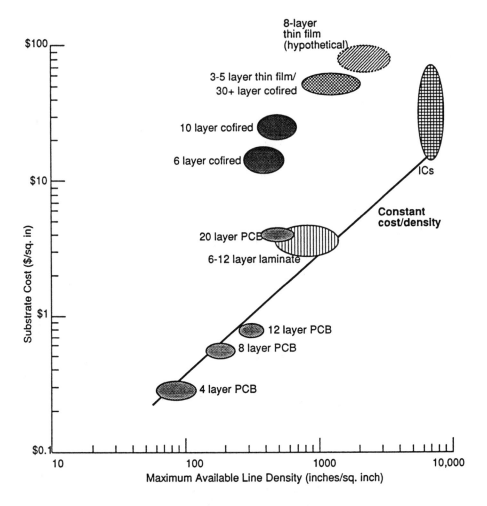

Figure 3-6 Substrate cost (1992) versus interconnection density. (Adapted from [18] with updated prices based on typical 1992 vendor costs.)

starvation. As a rule of thumb, if there are fewer than 1.5 to 2 via sites per signal pin, then the router efficiency, E, decreases [4]. This is particularly a problem with the through hole vias commonly used in PWBs since the number of via sites does not increase as the number of signal layers increases. With high layer count boards and laminate MCMs, buried vias are required to compensate for this. The use of large PWB and laminate MCM vias also can result in via

Table 3-1 Typical Minimum Line Pitches.

TECHNOLOGY	MINIMUM WIRE PITCH	MINIMUM VIA PITCH
Cofired Ceramic MCM	150 μm - 250 μm	250 μm
Thin Film MCM	20 μm - 40 μm	20 μm - 200 μm
Laminate MCM	100 μm - 250 μm	1000 μm - 2500 μm
PWB	250 μm - 300 μm	1000 μm - 2500 μm

starvation. Referring again to Figure 3-7, it can be seen that between every row of vias there are several wires (two or three are typical). This low ratio of wires to vias might also make it difficult for a wire always to find a via where it needs one. In thin film and ceramic technologies the via size can be the same as the conductor size. Buried vias are the norm. Thus, the effects of vias on interconnection properties in these technologies are usually minimal.

Interconnection properties interact with the choice of connection techniques used between levels of the packaging hierarchy. For example, if an edge style of connection is used (such as wire bonding or TAB for the first level connection) then the pads on the chip might have a center to center pitch as small as 75 μm (though 150 μm or more is typical). If the pitch of the wires on the substrate (Table 3-1) is larger, then the chip connect function must include fanout for pitch matching, as shown in Figure 3-8. Making room for this fanout increases the minimum possible chip pitch, P. This is one important reason why conventional packaging tends to have poor substrate efficiencies. Providing this fanout to match a 0.35 - 0.5 mm PWB surface pad pitch makes high pin count surface mount packages large (while a small pin count package, such as a memory, is small). Providing fanout often is necessary even on laminate and cofired ceramic MCMs. It is not necessary on a thin film MCM.

On the other hand, with area connection (such as flip chip solder bump), the pad or pin distance on the chip must be equal to or larger than the minimum via spacing in the package. Providing this fanout to match a 0.1" PWB via pitch makes high pin count pin grid arrays (PGAs) large. The signal layers underneath the chip then are used to bring these signals out from underneath (escape) to be routed to other chips at the required pitch. This consumes some of the interconnection resources underneath the chips. It is important to ensure that sufficient interconnection resources remain underneath the chip for normal interconnections. This is often a problem underneath high pin count PGAs. IBM solves this problem on their flip chip MCMs by assigning special layers (redistribution layers) for the escape function in the Thermal Conduction Module. These layers also perform test and rework functions. Again, the impact is minimized if thin film layers are used.

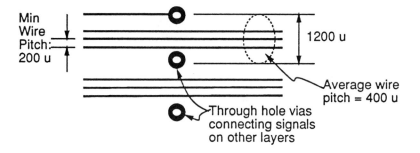

Figure 3-7 One effect of via size on wire pitch.

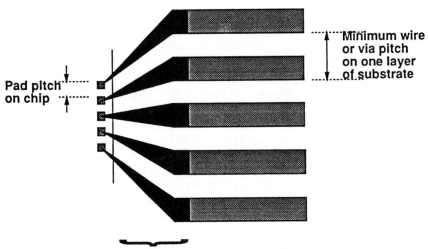

Figure 3-8 Fanout often is required to pitch match the chip pads and the package pads.

3.4.3 Connection Capacity Between Packaging Levels

The connection capacity between packaging levels is much smaller than the interconnection capacity within a level. For example, consider an MCM connected to a PWB. The pin pitch of the connector that goes between the two package levels typically is limited to either the wire pitch of the PWB on one layer only, or to the through-hole via pitch of the PWB, usually the latter. The capacity implied by either of these pitches is much less than the capacity, over a similar cross section, of either the PWB wiring or the MCM wiring. Furthermore, the cost of large high pin count connectors can be very high. (See Chapter 10 for a discussion on MCM to PWB connectors generally and Chapter 18 for a discussion on cost.) This is why richly interconnected systems tend to use large MCMs and large PWBs to minimize the use of connectors and backplanes. By flattening the packaging hierarchy in this way, the bottleneck created by the low capacities of the higher levels of hierarchy and the connectors to them is minimized. For example, the IBM 3081 main CPU board contains nine 9 cm × 9 cm MCMs placed on a 70 cm × 60 cm PWB. However, a large MCM is more expensive to produce than a collection of smaller MCMs. There is a tradeoff between substrate size and connector capacity.

Performance often is compromised to reduce connector capacity requirements between different levels of the hierarchy. For example, the only significant difference between the Intel 386DX microprocessor and the Intel 386SX microprocessor is that the latter provides fewer pins on the single chip package for the memory interface. This reduces the effective performance of the latter but does make it less expensive.

An important factor determining the connection capacity between levels is whether an edge or area connector is used. For example, Figure 3-9 plots the total number of pads versus pad pitch for both area and edge connectors for the IC to MCM interface. With area connection, more connections are made with reduced manufacturing tolerance than with edge connections. This is a major advantage for flip chip with solder bump technology. It is also a major advantage for PGAs and other area array single chip and multichip packages in comparison to peripheral pinned packages such as surface mount. The subject is discussed in more detail in Chapters 10 and 18.

The above discussion assumes the required connection capacity between levels is known. However, often it is not known and must be estimated. A good estimation of the required signal pin capacity, N, for the signals leaving any level of packaging is given empirically by Rent's rule:

$$N = KM^P \qquad (3-5)$$

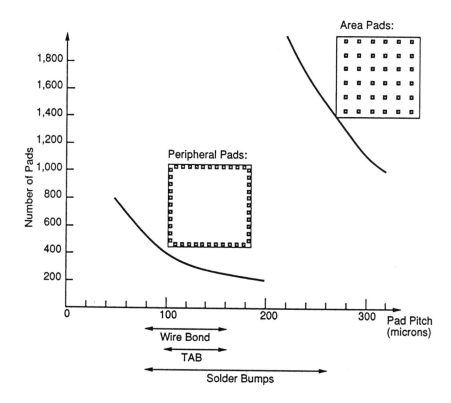

Figure 3-9 Number of I/O pads for a 1 cm^2 chip with peripheral and edge arrangements.

where K is a constant of proportionality, typically about 2.5, M is the number of circuits contained within the lower level of the hierarchy, and p is Rent's constant (typically $0.5 < p < 0.7$). However, it should be noted that K, M and p are determined empirically. Rent's constant tends to decrease with large increases in the number of circuits. (Consider the number of signal pins leaving a personal computer box versus the number leaving a CPU chip; they are roughly comparable indicating a sharp decrease of p with M.) It should be remembered that N does not include the power and ground pins. In high performance systems there can be almost as many power and ground pins as signal pins to control electrical noise.

For modest increases in the number of circuits, M, contained within a package, the number of pins required, N, usually increases. Thus, as further transistor miniaturization occurs, there is an increased requirement for interlevel connectivity. Interlevel connections become an important technological constraint on the package performance, as discussed in Chapter 18.

3.4.4 Delay and Electrical Noise

Delay refers to the time required for a signal to travel between the functional circuit blocks in a system. To a first order approximation, packaging-related delay is broken up into the following contributions (Figure 3-10):

$$t_{delay} = t_{buffer} + t_{flight} + t_{rise-time-degradation} + t_{noise-settle} \qquad (3-6)$$

where t_{buffer} is the delay incurred within the buffer-amplifier, t_{flight} is the time taken for the signal to travel (fly) along the wire at nearly the speed of light, $t_{rise-time-degradation}$ is the extra delay incurred because of an increase in rise time of the signal (the time for the signal to transition between the two different logic voltage levels) on the rise time as compared to the signal at the input of the buffer. $t_{noise-settle}$ is the extra delay that must be incurred while waiting for electrical noise to settle.

If the buffer is small and weak, then t_{buffer} is small but $t_{rise-time-degradation}$ is large. The delay $t_{rise-time-degradation}$, is determined mainly by the ability of the buffer output to charge and discharge the capacitances associated with the interconnection. Part of this capacitance comes from the chip attach leads and $t_{rise-time-degradation}$ improves as the chip attach leads get smaller. Thus smaller buffers are sometimes used in chips intended solely for MCM use. If the line is lossy (resistive), then $t_{rise-time-degradation}$ is increased further. Thus it is desirable to have low loss lines.

The minimum time for the signal to travel down the line, the time of flight, is given by:

$$t_{flight} = \frac{l}{c/\sqrt{\varepsilon_r}} \qquad (3-7)$$

where l is the length of the interconnect, c is the velocity of light in a vacuum and ε_r is the relative dielectric constant of the insulator. Typical values for

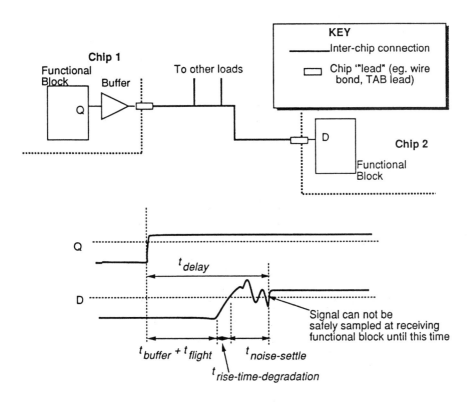

Figure 3-10 Electrical delay.

dielectric constant are given in Chapter 2. In a high speed system, total delay time is limited primarily by the time of flight. This is a driving force behind using packaging types that allow parts to be more closely spaced (reduce l) and for using dielectric materials with lower values for dielectric constant, such as glass-ceramic for cofired ceramic MCMs, polyimide for thin film MCMs and cyanite ester or polyimide for laminate MCMs and PWBs, as discussed in Chapter 1.

Whenever transitions occur on digital signals (0→1 or 1→0), electrical noise is introduced both on the wire (connection or interconnection) carrying the signal and on the wires around it. This noise contributes to the package parasitics defined in Chapter 1. In digital systems, noise considerations indirectly affect system performance. For example, Figure 3-10 shows how increased noise is equivalent to an increase in delay on data lines for digital systems. For clock signals and analog signals, excessive noise will directly compromise correct system function. This is discussed further in Chapter 11.

There are three major sources of noise within a digital system: reflection noise, crosstalk noise and simultaneous switching noise. The system also produces noise that affects the operation of systems around it (and itself is susceptible to such noise produced by other systems). This is referred to as electromagnetic interference (EMI) noise. The magnitude of all of these noise sources depends on the rise time, t_{rise}, of the signal. The faster the rise time, the worse the noise.

Reflection noise is a potential problem if the time of flight becomes comparable with the rise time

$$t_{flight} > t_{rise}/5 \rightarrow t_{rise}/3. \qquad (3\text{-}8)$$

When this is the case, the signal edge needs to see a constant impedance as it travels along the line. Whenever the impedance changes, part of the signal is reflected just as part of a light signal is reflected when it encounters a sheet of glass.

The characteristic impedance, Z_o of a wire is given by:

$$Z_o = \sqrt{L/C} \qquad (3\text{-}9)$$

where L is the inductance per unit length of the line and C is the capacitance per unit length. Maintaining a constant (termed controlled) characteristic impedance requires that L and C remain constant along the length of the line. Doing this requires that the signal line maintain a constant cross sectional geometric relationship with a reference line or plane, either ground or power. Examples of controlled impedance lines are given in Figure 3-11. The most common approach is to use microstrips and striplines, creating the need for reference planes. Typically, offset striplines are used instead of striplines wherein there are two signal layers between each pair of reference planes. Thus in most digital systems there are typically about half as many reference planes as signal planes (reference layers can be shared by signal layers).

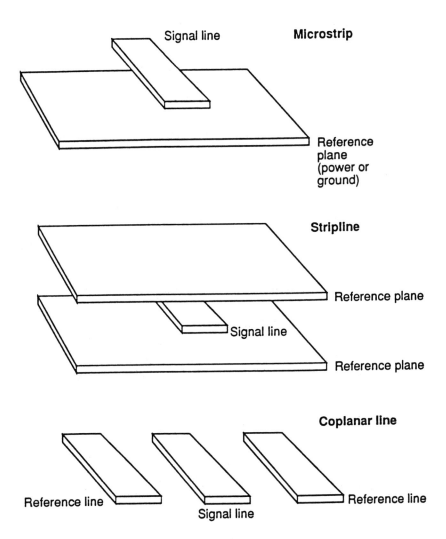

Figure 3-11 Examples of controlled impedance lines.

Also, if a matching terminating resistance, $R = Z_o$ is not placed at one end of the line, then part of the signal is reflected there. This reflection travels up and down the line, interpreted by the receiver as noise. Without a matching termination, reflection noise can be controlled by keeping the line short (see Chapter 11):

$$t_{flight} \; < \; t_{rise}/4. \tag{3-10}$$

Crosstalk noise arises whenever signal lines or chip connect leads run parallel to each other. The signal on the active line couples onto the quiet line as noise. The faster the rise time, the greater the coupled noise. Crosstalk noise is controlled by placing the lines further apart than the minimum line spacing and, sometimes, by limiting the coupled length. Thus, the greater the potential impact of crosstalk noise, the lower the interconnection density is. Chapter 11 discusses how crosstalk noise considerations also lead to the use of reference planes and allow lines to be more closely spaced when the dielectric constant is reduced. One advantage of cofired PGA packages over most plastic packages is the provision for these reference planes within the package.

Simultaneous switching noise occurs whenever a large number of off-chip or on-chip drivers switch at the same time, as shown in Figure 3-12. This switching activity causes a large current spike to flow through the ground and V_{CC} connections. When this current spike flows through the inductance associated with the ground and power circuits a noise voltage will appear on the chip's internal power and ground lines. The magnitude of this noise is given approximately as:

$$V_{SSN-noise} \; = \; L_{eff} N \; \frac{dI}{dt} \tag{3-11}$$

where L_{eff} is the effective inductance of chip ground or V_{CC} connection, N is the number of switching drivers, and dI/dt is the current transient produced by each buffer. The effective inductance, L_{eff}, also can be substantially reduced through the use of shorter chip attach leads and by placing ground planes beneath the leads. As a rule dI/dt increases with decreasing rise time. This noise causes false switching inside the chip, appears as noise on the output leads of any quiet buffers connected to the ground or power rail and increases delay in the switching drivers. One major advantage of MCM technology is that the effective inductance is greatly reduced in comparison to single chip packages. This is particularly true for solder bump and multimetal TAB attachments (TAB with a ground plane).

For the same set of chips, all of these sources of internal noise are usually smaller in an MCM package in comparison to PWB packages. In particular, the shorter connection distances reduce reflection noise as well as the accumulation of crosstalk noise, while the smaller chip connection inductances reduce the amount of simultaneous switching noise. Noise is easier to control in MCMs and its impact on delay is smaller. However, noise usually can be adequately managed in single chip packages. Many packages, particularly ceramic PGAs, provide internal reference planes or ground and power planes for the purposes of controlling noise. Though few plastic packages currently have these planes, some are starting to make limited use of them.

EMI is produced by the package circuits whenever current flows within them. The connections act as antennas producing radiated noise. Meeting required standards for EMI noise can be difficult and often requires the use of metallic screens in the box. One advantage of MCM technology is that it reduces the size of many of these antennas and, potentially, the need for screening.

It is important to note, however, that as transistor speed increases, rise time decreases and noise control becomes more difficult. The subject of electrical delay and noise control is discussed further in Chapter 11.

3.4.5 Power Consumption

Power consumption usually impacts system performance through its indirect impact on size and weight. For example, in a notebook computer, weight considerations dictate the battery size and energy. Thus, to make a computer that lasts four hours on one battery means carefully controlling the power consumption. Similarly, in aerospace systems, increasing power also means increasing the weight of the power supply. In telecommunications systems, power consumption must be controlled so that the batteries required to power the system in the event of a power failure are not too large. Controlling power consumption also reduces power dissipation (the two have the same value), thus reducing the need for complex and large heat removal structures.

Whenever a 0-1-0 transition occurs on an interconnection, the amount of energy consumed in charging and discharging the interconnection capacitance is given by:

$$\text{Energy} = CV^2 \qquad (3\text{-}12)$$

where C is the total capacitance of the interconnection and V is the voltage swing. The power consumed is this energy per transition times the number of

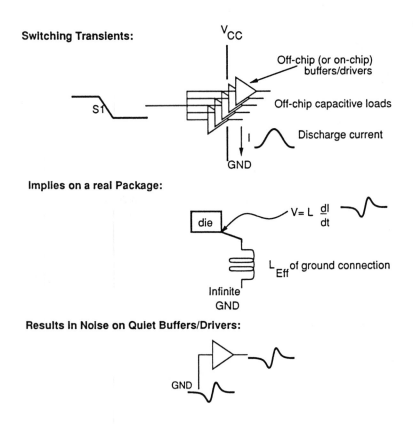

Figure 3-12   Simultaneous switching noise in digital systems.

transitions per second. Therefore, in any CMOS circuit where capacitive energy is the main form of power consumption, the power consumption increase is directly proportional to frequency of operation.

As the capacitance, C, is proportional to the length of the interconnection and the size of the chip attach, one advantage of MCM technology is reduced power consumption. For example, let us say that 10% of the system power consumption is due to the interconnect when mounted on a PWB and that migration of the product to an MCM reduces the power consumption due to the interconnect by a factor of five. Then the MCM product either would consume

8% less power than the PWB-based product or could be allowed to run at a faster rate of transitions per second (frequency) with no increase in power consumption.

3.4.6   Heat Dissipation

Consumed power is converted directly to dissipated heat. As the clock frequency and the number of transistors per chip increase, the heat produced by the chip increases. This heat causes the temperature of the chips to rise. With CMOS chips, higher temperatures affect system performance directly by slowing the transistors. In any system, high temperatures decrease the reliability. To maintain reliability, other performance factors might need to be compromised to improve heat removal. The structures used to remove heat might add considerably to the manufacturing cost. Thus, thermal issues are very important in MCM design because the heat density is much higher than in single chip packages, making cooling more difficult.

Heat is removed mainly by conducting it away from the chips and allowing it to convect into a circulating coolant. There are generally two choices for the heat conduction path: through-the-substrate or directly off the back of the chip. Also there are generally two choices for the coolant: air, often forced by a fan(s), or water (or another other liquid such as a fluorocarbon), often forced by a pump. Some possible combinations are shown in Figure 3-13. Unfortunately all of these alternatives involve some compromise in either another performance or cost factor.

Generally it is considered that through-the-substrate forced air cooling is the cheapest alternative. (Hence its very aggressive use by DEC, as described in Chapter 17.) An air cooled system has a number of advantages over a water cooled system. It reduces the need for expensive plumbing and seals. An air cooled system also is more likely to survive a fan failure than a water cooled system would a pump failure. By making the parts more accessible, maintenance and field repair costs are reduced. Through-the-substrate cooling tends to require less precise mechanical engineering than chip backside cooling and further helps in making the parts more accessible for repair.

The use of through-the-substrate forced air cooling might require some performance compromises. First, its use requires a highly conductive thermal path through the substrate. Unfortunately, the materials with the lowest dielectric constant such as glass-ceramic or polyimide tend also to have the lowest thermal conductivity (Tables 7.1 and 12.1, for example). This can be overcome by using either copper slugs, commonly called thermal vias, beneath the chip or sinking the chip into the substrate (Figure 3-14). This substantially reduces the capacity beneath the chip, particularly for laminate MCMs where the slug consumes all

Through-the-substrate, Forced Air:

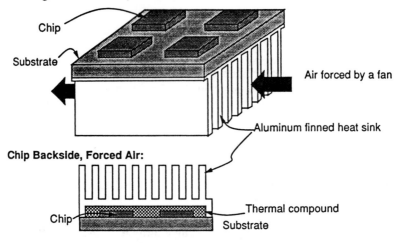

Chip

Substrate

Air forced by a fan

Aluminum finned heat sink

Chip Backside, Forced Air:

Chip

Thermal compound

Substrate

Through-the-substrate, Water:

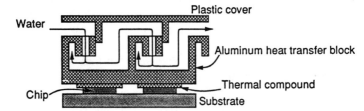

Forced Water

Chip Backside, Water:

Plastic cover

Water

Aluminum heat transfer block

Chip

Thermal compound

Substrate

Chip Backside, Circulating Fluorocarbon:

MCM immersed directly
in inert fluorcarbon liquid

Figure 3-13 Some of the alternatives that can be used to cool an MCM.

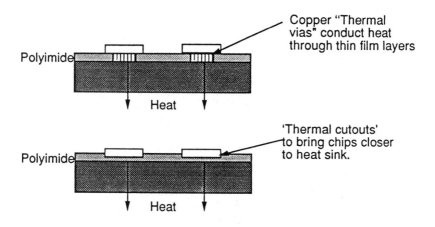

Figure 3-14 Thermal vias and thermal cutouts.

of the area beneath the chip. (Alternatively, an array of through-hole vias can be used, but their efficiency is limited.) The resulting lack of interconnect capacity might force an increase in chip pitch, P, and thus MCM size. Second, use of through-the-substrate cooling usually means that only edge connectors can be used to the next level of packaging, potentially reducing connection capacity. Third, even with the use of thermal vias, the heat density might be too high for the desired cooling mechanism. It is then necessary either to reduce heat density by forcing the chips further apart or to consider a more aggressive cooling approach.

When individually packaged, high power chips generally must be housed in ceramic or metal packages, or in plastic packages with direct ceramic or metal heat paths to the heatsink. The thermal tradeoffs between single chip and multichip packaging can not be easily summarized. While heat density is higher in the latter, one (larger) heatsink can be shared by multiple chips. The advantages of the multichip packaging relative to single chip packaging depends on the details.

3.4.7 Performance Tradeoffs

A number of performance to performance tradeoffs have been identified above. Increased interconnection density is the main driver to increased system performance. Increased density allows the chips to be spaced closer together, thus decreasing size, weight and delay. It also allows more chips to be placed in the same area, avoiding the bottleneck posed by the board to backplane connectors. The increased density offered by MCMs primarily benefits systems containing chips with hundreds of I/Os. It is erroneous to think that the main advantage of MCMs is only that they eliminate single chip packages. Repackaging a system with high I/O count chips as bare die on the same PWB results only in a small size reduction due to interconnection density considerations.

Sometimes it is necessary to sacrifice wiring density to satisfy thermal requirements if thermal vias become necessary. This should be weighed against the alternatives of using backside cooling or spreading the chips apart.

3.5 PACKAGING COST FACTORS

In this section, the different factors that contribute to system cost are described and explained. Details about how to model cost are provided in Chapter 4. In this section the intent is to explore their relationship to system level decision making by describing some of the more important cost performance tradeoffs. It must be noted that the unpackaged chip costs also must be included in the production cost. For any one board or module, the chip costs often exceeds the package cost if large, leading edge chips are used. The opposite is true if simple chips are used. When a multiple board system is considered, total packaging cost often exceeds total chip cost (see Chapter 18).

3.5.1   Production Cost

Production cost is the cost involved in getting the product out the factory door to the purchaser. It has two elements, manufacturing cost and the manufacturability cost.

Manufacturing Cost
The manufacturing cost is the cost of materials and process steps associated with the production of each part in the system and their assembly. This includes the chips, packages, connectors, heat removal mechanisms, power distribution features and the final packaging (casing, etc.). It includes the recurrent engineering (RE) cost elements (a cost that recurs for every part made) such as the purchase of the required materials and the labor and energy required to produce each part. It also includes the cost of purchasing the manufacturing equipment, a nonrecurrent engineering (NRE) cost element (a cost that is spread over a number of parts). A technical cost model for these elements is described in Chapter 4.

Table 3-2 Manufacturing Costs for Packaging and Interconnection Elements.

PACKAGE TYPE	TYPICAL COST
Plastic package Ceramic PGA < 144 pins Ceramic PGA > 144 pins	5¢ - $5.00 $5,000 NRE + 10¢ per pin ($14 or more for 144 pins) $25,000 NRE = 10¢ per pin ($50 or more for 500 pins)
Cofired ceramic MCM Thin film MCM Laminate MCM PWB	$3 per sq. inch per layer ($30 per sq. inch for 10 layers) $60 per sq. inch (expected to decrease to ~ $20) $3 - $5 per sq. inch < $1 per sq. inch
CMOS chip wafers	$25 - $150 per sq. inch

High volume production allows the cost of purchasing the manufacturing equipment to be distributed over a greater sales volume, minimizing the NRE cost apportioned to each part. The required manufacturing and assembly steps should be as simple as possible and should lend themselves to automation through the use of standard looking parts. For example, the fact that small MCMs often can be mounted in conventional packages (QFPs, PGAs, etc.) gives them an additional cost advantage over large MCMs that tend to require custom packages unusable by standard automatic assembly equipment [5].

Typical vendor prices for some lower level interconnect structures are listed in Table 3-2. It can be seen that high pin count PGAs, high layer count ceramic MCMs and thin film MCMs command a substantial price premium over plastic packages, PWBs and laminate MCMs. Not shown are the cost of connectors. High pin count connectors tend to be costly (see Chapter 18). The higher cost of using high density MCMs at the lower levels of the packaging hierarchy might be compensated for if the need for high pin count connectors at higher levels of the hierarchy were reduced.

Manufacturability Cost

Not all of the parts manufactured function correctly. Parts must be tested, repaired if possible, and retested after repair. The cost of test and repair and the impact of failures is referred to herein as manufacturability cost because they depend, in part, on how well the part was designed for manufacturability.

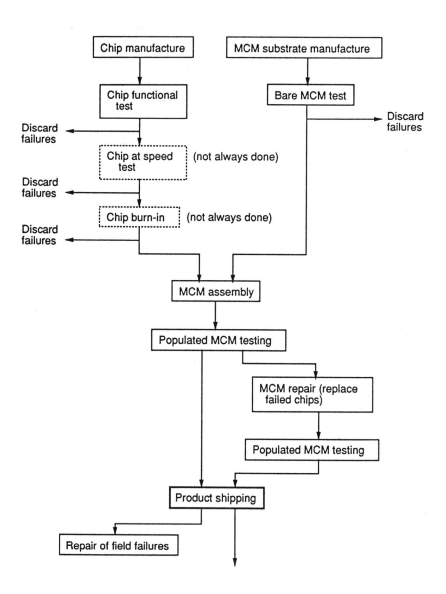

Figure 3-15 A simplified view of the MCM manufacturing process.

Assuming that this test and repair step is done only once in the manufacturing process (as shown in Figure 3-15), the final unit cost becomes:

Final unit cost = (Manufacturing cost per part + Test cost per part +
(1 - Initial yield) × (Repair cost per part +
Cost of retest per part)) / Final yield

where each cost component is averaged over all parts, good, bad or repaired.

Yield is defined as that percentage of parts that pass a test phase. Final yield is greater than initial yield if some of the failed parts can be repaired. For a chip, MCM or PWB, yield decreases with size. Thus, the cost premium of increasing a part size by a factor of two might be factor of eight if doubling the size doubles the manufacturing cost and quarters the yield. However, the greater effective interconnection capacity of the larger MCM often compensates for the poorer yield.

For example, consider die yield. The number of dies (chips) manufactured per wafer is given by:

$$\text{dies per wafer} \approx \frac{\pi \times (\text{wafer diameter}/2)^2}{\text{die area}} - \frac{\pi \times \text{wafer diameter}}{\sqrt{2} \times \text{die area}}$$
$$- \text{\# die sites used for process control}$$

$$(3\text{-}13)$$

and the die yield is given by:

$$\text{die yield} = \text{wafer yield} \frac{1}{(1 + \text{defect density} \times \text{die area}/\alpha)^{\alpha}} \quad (3\text{-}14)$$

where wafer yield is the percentage of wafers not containing a gross fault affecting every chip on them, and α is the defect clustering parameter, which tends to take on a value between 1 and 3 [6]. The defect density might run anywhere between 0.5 - 2 per cm^2. Defect density improves with process maturity. For a fixed wafer manufacturing cost (typically $500 to $2000), the die cost increases rapidly with chip area. If a large complex chip has a yield of only 10%, then its final cost is at least ten times its manufacturing cost. This is one reason why large chips are not always the best alternative to a small MCM. The presence of defects effectively limits the maximum size of chip that is manufacturable without some method of tolerating the faults produced. As chips usually cannot be repaired, initial yield and final yield are the same.

Testing and rework costs have been identified as critical MCM related cost issues. Many sources attribute one-quarter to one-half the cost of the final MCM to these subcosts. In the normal manufacturing flow (Figure 3-15) individual parts are tested before assembly and the assembled MCM also is tested. If the assembled board fails, then it is necessary to either scrap the board or rework it, that is find and replace the part that failed. Since most MCMs contain at least one high value chip reworking usually is preferred. However, the cost of reworking an MCM is generally high. The best way to minimize rework is to ensure that chips are fully tested before being mounted on the MCM. Currently, this requires that the die be TAB mounted for reasons explained below. TAB mounting itself is expensive and also consumes area on the MCM.

Thus a balance must be struck that optimizes the combined cost of test and the likelihood of rework. This balance is determined by the minimum of the total out the door cost. A simplified expression for this cost (based on the manufacturing process in Figure 3-15) is:

$$\text{Final cost} = \frac{\text{MC} + \text{TC} + (\text{P(TE)} + \text{P(AF)}) \times (\text{RC} + \text{ATC})}{\text{Final yield}} \quad (3\text{-}15)$$

where MC is the total IC and MCM manufacturing cost, TC is the total IC and bare and populated MCM test cost, P(TE) is the test escape probability, P(AF) is the probability of an assembly fault, RC is the rework cost and ATC is the cost of retesting the assembled substrate.

Test escape refers to an IC that passes its initial test but fails after assembly. This is a potentially significant problem for MCMs. The hardware used to probe the very small chip pads uses long non-controlled impedance leads, making it difficult to pass noise free, fast edge signals to the chip. Good design for test techniques, as discussed in Chapter 13, are needed. At the moment, the easiest way to test a chip at full speed is to mount it first in a package such as a PGA, QFP or TAB. If the chip is not tested at full speed, the test escape probability can be as high as 30% if the IC is a leading edge CMOS chip, but also can be very small if the IC is from a mature line. It also is higher for CMOS parts than for bipolar parts. This is discussed further in Chapters 13 and 18. Depending on the details of this combined cost, there are a number of options that can be considered:

- If the MCM contains no high value parts or has only one high value part likely to fail, then there is no need to fully test the ICs before assembly since the MCM can be scrapped at little expense, if it fails.

- If the MCM has only one high value part, that part can be placed by itself on the MCM where it is easier to test than in bare die form. If the part passes, then the MCM assembly is complete. If the part fails, then the MCM can be scrapped or the single die replaced. This approach reduces the risk that other parts are damaged during rework.

- If the MCM contains a number of high value dies (chips) likely to have high test escape probabilities, then full consideration should be given to properly testing the dies before assembly, perhaps by using TAB chip attach and testing the tape form.

- Sometimes parts can only be tested properly after assembly. For example, a complex microprocessor may be easier to test completely when attached to its memories rather than as a single IC. In that case, the process should be optimized toward inexpensive and easy rework such as flip attach techniques.

3.5.2   Post Production Costs

Post production costs are incurred once the system is in use. They include maintenance, field upgrades and the repair and replacement of failed parts. The sum of production and post production costs is referred to often as the life cycle cost. The goal of a designer is to minimize life cycle cost, of which production cost might only be a fraction.

Reliability depends on the elimination of possible failure mechanisms and the operation of the parts at a sufficiently low temperature. This is discussed throughout the chapters in **Part B — The Basics**.

Repairability relates, in part, to the size of the field replaceable unit. For example, if an MCM is soldered onto a board, the field replaceable unit is the board. If the MCM is socketed, and the field diagnostics located the failed MCM, then the MCM is the field replaceable unit. If that MCM is sealed in epoxy, it must be scrapped. However, if the MCM has a resealable lid, it might be repairable and have further value as a used part. A life cycle cost analysis points at the correct solution (though MCMs are considered to be so reliable that repair is not usually needed).

The down time required to locate the fault and replace the unit also often is important. For example, in military systems, the ability of a system to survive a mission and to be quickly and easily checked and repaired are very important. If liquid cooling is used, then replacing a failed unit takes much longer than when air cooling is used. Also, the greater compactness of an MCM system might make fault location more difficult if the design is not carefully thought out. It is difficult to probe a signal on an MCM in the field.

Later chapters pay particular attention to reliability exposures that come about through the use of specific MCM technologies and reliability results from actual MCM use (see Chapter 17). Particular attention should be given to establishing a quality engineering process so that reliability can be maximized.

3.5.3 Design and Prototyping Costs

Design and prototyping costs include engineer training, engineer time spent in design, purchase of CAE and CAD tools to help in the design, construction and testing of the prototype, followed by design changes that arise from testing. Particularly for small to medium production runs, design and prototyping costs can be a significant part of the final system cost. The following should be considered when estimating the impact of this factor:

- The cost of building and diagnosing faults in an MCM prototype is higher than for a PWB prototype. (In the PWB prototyping, signal lines can be probed easily to test the prototype.) Extra emphasis needs to be placed on using a design approach that results in first pass success of the prototype. This requires extra investment in computer design tools and engineer training. Note, however, this investment is not that much different than the investment required for achieving first pass success for PWB designs operating at similar speeds and power dissipation. Consideration should also be given to using rapid prototyping technology [7].

- Effort should be spent on learning the technology. One technique is to go through the entire design and prototyping cycle with a non-production part. One reason that laminates are currently a popular form of MCM is because of their similarity to the already familiar PWB technology.

- Existing designs and hardware should be reused as much as possible and reasonable. This involves using off the shelf parts (if available and suitable), using programmable or semi-custom logic (if suitable) and reusing portions of existing designs, if applicable, rather than creating new designs. This is very important for low volume commercial parts.

- The decision making process described in this chapter requires that models be built, data obtained and evaluations conducted. This is a design activity. The investment required to carry out this activity should be balanced against the likelihood that more detailed models, data, etc. lead to better decisions being made.

One possible impact of using existing designs is that, for low volume parts, it might be preferable to gain a performance improvement through repackaging a set of chips as an MCM, rather than redesigning them on one VLSI chip. (In any case, the one chip alternative is not a viable option if the chip set contains a mix of technologies.)

3.5.4   Time-to-Market

Reducing time-to-market often is more important than controlling design, prototype and production costs. One survey showed that being six months late to market resulted in an average 33% profit loss for that product, while a 9% production cost overrun resulted in a 21% loss, and a 50% design development cost overrun resulted in only a 3% profit loss. In a recent survey, engineering managers stated that they would rather have a 100% overrun in design and prototyping costs than be just three months late to market with a product. There are a number of reasons for this. If your competitors beat you to the market with a comparable product, they gain considerable market share and brand name recognition. This is difficult to quantify. One example is the stronger market presence of the Nintendo games over the other newer electronic games simply because it was introduced first. An earlier market entry has more opportunity to improve yield, thereby improving profits. Also, the technology is usually locked into place early in the design process. The design and production must be completed quickly from this point to prevent competitors from introducing a similar or more advanced product at the same time. Finally, design and prototyping costs are an up front investment that must produce a return. The longer investors must wait for their return, the higher that total return must be. (Who would you invest in? The company that could double your investment in two years or the one that would take one year?)

There are many reasons why time-to-market might be longer than necessary. One reason products might be late to market could be organizational, that is, excessive delay in making key decisions and an over reliance on complex design methods. Another possible reason is inexperience. Learning a new technology introduces delays into the design and manufacturing cycles. Remember that you are also inexperienced in pushing the older technology to new limits. It is very difficult to get a conventional 250 pin QFP to run reliably at 150 MHz.

Another potential reason for a long time-to-market, particularly critical for some MCM technologies right now, is the lack of infrastructure, as discussed in

Chapter 1. For example, you might need a source of tested, qualified solder bumped die and access to a high volume thin film MCM manufacturing line for success of your product. If both of these are difficult to obtain at the time of manufacture, product release must be delayed while you wait for their availability.

3.5.5 Cost Tradeoffs

The relative weight given to production cost, post production cost, design and prototyping cost and the impact of time-to-market depend on the details of the type of part being produced. For example, a part being produced in high volumes for use in applications where reliability is not critical, emphasizes production cost.

3.6 PACKAGING DECISIONS AND THE SYSTEM DESIGN PROCESS

During the system design process, the design engineering team needs to determine the organization of the system, the packaging hierarchy to be used, the packaging mixture to be used within each level of the hierarchy and the partitioning of the system functions between the chips and packages that comprise the system. The term "organization" means the block diagram for the system - what functional blocks make up the system and how they are connected. The term "packaging style" refers to the details of the package selected, including the type of package (QFP or thin film MCM), the size and layer count. It also includes the details of the connectors used between the packaging levels. The term "partitioning" refers to what functions are assigned to which chips and how the chips are assigned to different packages within the system. As part of the partitioning process, it might be necessary to determine a floor plan describing the approximate placement of components, chips or packages, with respect to each other.

Packaging and partitioning decisions are made by evaluating a set of alternatives. The performance and cost of each alternative is estimated and compared against the requirements and goals of the system. To do this, the requirements and goals of the system must be expressed in terms of the performance and cost factors described above and summarized in Table 3-3. If the performance and cost of each alternative is similarly expressed, then the alternative that best matches the goals of the system is the preferred option. This

Table 3-3 Performance and Cost Factors.

PERFORMANCE FACTORS
1. Size and weight 2. Interconnection capacity within each interconnection level 3. Connection capacity between interconnection levels 4. Electrical delay and noise 5. Power consumption 6. Heat dissipation
COST FACTORS

1. Production cost:
   a. Manufacturing cost
      (setting up the manufacturing line and buying the required materials)
   b. Manufacturability costs
      (running the manufacturing line, mainly the impact of test and yield)
2. Post production costs:
   (mainly replacing failed units—the effect of field reliability)
3. Design and prototyping costs
4. Time-to-market

process can be broken down into steps, some of which are carried out simultaneously:

1. Determine requirements and goals of the system.
2. Express requirements and goals in terms of performance and cost factors.
3. Determine partitioning and packaging alternatives.
4. Evaluate performance and cost of design alternatives in terms of performance and cost factors.
5. Decide on alternative that best meets the system's aims.

These are discussed and an example given in the next section.

3.7 DETERMINING SYSTEM REQUIREMENTS AND GOALS

The requirements of the system are the "must haves," those aspects of performance and price that the system must achieve. These become constraints on the performance and cost factors. The goals of the system are the "want to haves," those aspects of performance and price where maximization is highly desirable. Satisfaction of the goals must be judged in terms of tradeoffs between the performance and cost factors. One of the first steps in system design is to determine the requirements and goals and to determine the relative weight or importance of each one. These requirements and goals should closely match those of the end customer.

Six categories of electronic systems were identified in Chapter 1: consumer, aerospace and military, computers, biomedical and telecommunications and instrumentation. Each of these system types has different requirements and goals related to packaging performance and cost.

For example, most consumer products have a requirement that they be passively air cooled (no fan). Their goals are heavily weighed towards minimizing production cost, followed by maximizing customer satisfaction (minimizing post production costs) and minimizing size and weight. Rarely, is minimizing electrical delay an important goal.

Aerospace and military products tend to emphasize performance and post production cost factors in their goals. However, they often are given minimum requirements in terms of these factors. For example, a radar signal processor might be required to resolve a target with a 1 meter squared radar cross section at 100 miles, fit into one electronics bay, and the technician must be able to locate and repair a fault within half an hour. They usually must be air cooled. A recent trend has been to pay increased importance to minimizing production cost.

The requirements and goals of a computer system depend on the application. For example, a desktop workstation might have requirements that it be air cooled with a quiet fan, be software compatible with previous models and meet federal EMI standards. The goals of the system are to maximize computation performance while controlling cost. Marketing studies determine suitable minimum targets for these goals. It is shown in Section 3-10 how workstation performance depends on interconnect capacity and delay. On the other hand, a notebook computer must be passively air cooled and must be able to use a certain chip set. Ergonomics and price drive the goals of the system. Total system size, weight and user friendliness, together with battery life (power consumption) and selling price, all are given similar weight.

At the other end of the scale, supercomputer and mainframe applications require high performance, while cost is secondary. These applications require high interconnection capacities throughout the entire system. Designers need to avoid the low interconnection capacities usually associated with the higher levels of the packaging hierarchy. Thus, they tend to make extensive use of MCM technology. An unusual performance goal, sometimes appearing in large systems such as a supercomputer, is the need for scalability-the ability to make the same system in several different sizes without redesigning the system for each size. An example is being able to expand a system with 512 processors to one with 1024 processors, without having to slow down the system to compensate for the larger size.

The aims in biomedical systems vary widely according to application. However, they have tended to emphasize performance and post production cost factors over production cost.

In main telecommunications switches, international standards dictate many requirements for performance. The need to minimize system down time for maintenance and failures often is translated into a set of post production requirements and goals. For example, it is a common requirement that the switches be air cooled to simplify maintenance and repair. It also must be possible to run the system on batteries if main power fails. The goals of the system then are to maximize bandwidth (a function of interconnect capacity and electrical delay) and minimize life cycle cost.

High end instrumentation systems, such as test equipment, often must operate at higher speeds than the systems they are designed to test. The pin counts of the high speed bipolar and gallium arsenide parts used usually have been low. Thus, hybrid packaging has been a common solution. Recently, the sizes and pin counts of chips made in these technologies have increased dramatically. As the pin counts climb, it becomes necessary to use packaging technologies that combine high interconnect capacity and high speed, such as an MCM technology.

3.8 DETERMINING AND EVALUATING
PACKAGING ALTERNATIVES

Currently, the process of determining a set of suitable packaging alternatives requires that the designer have some insight into the broad range of alternatives available and how their relationship to the requirements and goals of the system. No general methodology for generating these alternatives exists today. Often, the alternatives are discovered during the evaluation process, so the current emphasis is on making this evaluation process as efficient as possible. Partitioning alternatives usually are generated on the basis that functions should be grouped into chips and packages to minimize the need for high cost connections between the different levels of the packaging hierarchy.

There are two levels of evaluation possible. First order evaluations use simple performance and cost measures and models, such as substrate efficiency and time of flight delay, to quickly evaluate a large number of alternatives [1], [8] and [9]. Computer aided engineering (CAE) tools, such as SPEC [10] and PEPPER [11], are becoming more available to help with this task. These first order evaluations often are useful in weeding out unsuitable alternatives. It often is necessary to conduct detailed evaluations to make the final decision and to determine the system parameters (such as the clock frequency). Detailed evaluations might be necessary when the possible options could violate a specific systems requirement (if the heat density is so high that air cooling might be difficult). CAE tools are helpful in this task (MetaSim [12] for electrical delay and the routability estimators included in commercial tools). SPEC and PEPPER have some of these detailed evaluations built in.

3.9 IMPACT OF SEMICONDUCTOR TECHNOLOGY

The designer must decide which semiconductor technology to use: CMOS, bipolar, BiCMOS (bipolar and CMOS mixed together) or gallium arsenide, and which chip design style to use. The design style has two elements: circuit style (TTL or ECL logic families) and implementation style (gate array or full custom design). The choice affects the decision making process as follows:

1. Determines the delay within each chip, both due to the circuits within the chip and the on-chip wiring.
2. Determines how many functions can be provided within one chip. This is highest with CMOS technology. Large CMOS chips also tend to have high pin counts.
3. Determines the on-chip interconnection capacity. This is generally high.
4. Determines the power dissipated by the chip. This is high now, even for CMOS chips. There is a speed/power tradeoff to be considered.
5. Determines semiconductor technology to use. Bipolar chips have lower test escape probabilities than CMOS chips.
6. Determines cost. Large, complex chips have a cost usually greater than the first level package cost.

3.10 EXAMPLE OF THE SYSTEM DESIGN PROCESS

In many computer products, particularly workstations, packaging technology has a main influence on system performance through its impact on memory access bandwidth. The memory is structured as a hierarchy (Figure 3-16), usually with

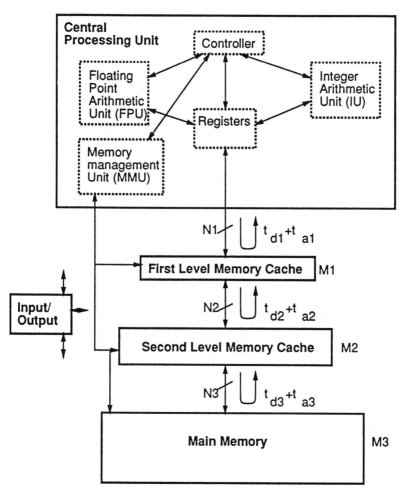

N_i = Number of bits fetched each access

$t_{di} + t_{ai}$ = Total round trip delay to fetch bits

M_i = Number of bytes at each level

Figure 3-16 Block diagram of the main elements in a single CPU computer. Packaging affects workstation performance through the widths and speed of the data paths between each level of the memory hierarchy.

a small amount of fast memory (first level cache) at the top of the hierarchy and a large amount of slow memory (main memory) at the bottom. This is part of the system organization. It is possible to write expressions that relate the computation performance of the system to the number of bits communicating between each level of the memory hierarchy (also called the fetch size, N1, N2 and N3, as shown in Figure 3-16), the interconnect delay between each level (t_{d1}, t_{d2}, t_{d3}), the memory access time of each level (how long it takes to fetch a memory location for each memory chip t_{a1}, t_{a2} and t_{a3}) and the number of memory locations at each level (M1, M2 and M3). The performance is expressed as a function or model:

$$\text{Memory Performance} = f(N1, N2 \ N3, t_{d1} + t_{a1}, t_{d2} + t_{a1}, t_{d3} + t_{a3}, M1, M2, M3) \quad (3\text{-}16)$$

The elements in this expression relate directly to packaging performance factors [13], [14] and [15]. This function tends to be most sensitive to the attributes $t_{d1} + t_{a1}$, M1 and N2 (N1 is usually fixed), and reasonably sensitive to M2 and $t_{d2} + t_{a2}$.

The choice of partitioning and packaging style has a large influence on the fetch sizes and interconnect delays. For example, the first level cache often is packaged with the CPU on one chip, allowing a minimum $t_{d1} + t_{a1}$. However, in 1992 technologies, this limits M1 to 8 - 16 kBytes, which really is too small. This is the case assumed here.

The progress of the IC technology (and the ability to integrate a fast, large, first level cache onto the CPU chip with good yields) must not be neglected. In particular, a company may decide not to pursue developing an MCM technology for its computer chips if an anticipated small chip MCM product would only have a two year performance lead on the equivalent function packaged entirely within one chip [16].

With the first level cache placed on the CPU chip, packaging determines performance mainly through its effect on N2, M2 and t_{d2}. Three options are presented and discussed here: a single chip package/PWB option, a laminate MCM option and a flip chip thin film MCM option.

If the CPU is packaged in a PGA, N2 is typically limited to 32 bits to 128 bits, to limit the PGA pin count, and t_{d2} might be as large as 23.1 ns. (Delays used here are based on the evaluations provided in reference [17].) This option provides the worst (though not bad) performance at the least cost.

The simplest MCM alternative is to package the CPU chip(s) as bare die on a fine pitch laminate MCM, either with TAB or wire bonded chip attach, and then connect the memories as single chip packages. This reduces the fanout and CPU footprint, decreasing t_{d2} and allowing for a modest increase in N1. The memories are left in their single chip packages because their footprint is barely larger than that of the bare die (they only require 20 or so pins). This also simplifies memory testing. t_{d2} is decreased from 23.1 ns to 17.9 ns, a 5.2 ns decrease. With the increased interconnect density it is possible to increase N1 to 128 bits or more. Based on assumptions beyond the scope of this text, a computation performance increase of around 11% is gained over the single chip package option.

With this laminate, the impact of thermal considerations is likely to be low. Thermal vias are used beneath the CPU but, if the chip is designed so that few signals run underneath this chip, then the impact on interconnection capacity is small. Manufacturing costs are low, possibly lower than the cost of using ceramic PGAs on a PWB, and the manufacturability cost also is low. In particular, the memory chips are easy to rework. It also might be possible to avoid reworking the CPU if it is a single die. The CPU can be tested after mounting it on the laminate, before mounting the memories, and scrapped with the low cost laminate if it fails. Other cost impacts also can be kept low. Sealing the bare chips in epoxy delivers sufficient environmental sealing at a low cost, and adequate mounting to the next level of packaging is through an standard edge connector. Overall, the price premium over the conventional option is small.

The most aggressive MCM alternative, in terms of performance, is to use flip chip solder bumps on a thin film MCM. With this approach, N2 can be very high and the shortest possible interconnection delay, short of using some three-dimensional technology, can be obtained. Then t_{d2} is decreased further 3.1 ns delay over the laminate approach when the memories are packed in short lead TAB frames and mounted on a thin film substrate [17]. The fetch size, N2, can be substantial, possibly even 512 bits or more. The overall computation performance increase over the single chip package option is about 20%.

The price premium is several hundred dollars however, and the risk of delayed time-to-market moderate unless close relationships had been established with parts suppliers and a carefully worked out test plan is used. If a silicon substrate is used, the cost of the package (typically a large ceramic PGA) can be higher than the substrate itself, possibly even $200 to $300. Manufacturability costs are higher as all the difficulties of working with bare die must be incorporated. An offsetting factor is that solder bump technology is the easiest bare die attach technology with which to do rework. If there is insufficient experience with flip chip thin film MCMs in the company, the risk and time-to-market impacts might be high. These considerations should be balanced against the performance gains.

3.11   SUMMARY

System design may be driven primarily by performance, by cost or by the desire to maximize the ratio of performance to cost. Due to continuing rapid advances in chip technology, as well as advances in customer's needs, the lowest cost packaging alternative often does not return satisfactory performance. This has led to a growing need for advanced, customized packaging. MCM technology represents the high performance end of advanced packaging technology.

The use of MCMs leads to improved system performance through the ability to pack the chips close together. This ability arises from the tight line and via pitches possible with MCM technology. These tight pitches reduce the impact of chip fanout on chip footprint, and results in sufficient interconnect capacity being available in a small area. This ability results in a reduction in interchip delay as compared with the conventional package alternatives. The reduction in delay is largest if the chips being packaged have high pin counts, as such high pin counts normally require a large fanout package and create a strong demand for wiring. For example, the size difference between a multiple 500 pin PGA mounted microprocessor array and the equivalent MCM mounted array is large, while the size difference between a surface mount packaged memory array and the equivalent MCM mounted array is small. In the former case, if system performance is very dependent on reducing interchip delays, the performance advantages of using MCMs are large. A further gain in performance comes about because the MCM-based solution fits onto fewer PWBs, reducing the need for higher levels of packaging in the packaging hierarchy. This allows the parts of the system to be more richly interconnected. The advantage is greatest with thin film MCMs, least with laminate MCMs. In all cases, however, the use of MCMs reduces size, weight and interconnect delays, and increases the total number of connections available.

The above discussion applies to the use of both large and small (< 10 chip) MCMs. Sometimes an alternative to a small MCM is to fabricate a single large ASIC chip. This is reasonable if the chip can be produced with sufficiently high yield. However, it is unreasonable if the production volume is too small to justify the extra design cost, if the time-to-market requirements are too short to justify the extra design time or if different semiconductor technologies have to be integrated.

There are a number of potential disadvantages to using an MCM solution. The heat density is higher, possibly requiring a more expensive heat dissipation solution. The manufacturing cost is likely to be higher, more so for thin film and high layer count cofired MCMs than for laminate MCMs. The test cost almost certainly is higher when bare die are packaged on MCMs rather than in easily tested, single chip packages. At the time of writing the risk of increased time-to-market is also greater with a bare die MCM solution due mostly to some infrastructure deficiencies. However, the infrastructure situation is improving rapidly.

Thus, the engineer must select the most appropriate technology mix for each system. This must be done by evaluating different packaging and partitioning alternatives against the system cost and performance goals. This is best done through the use of clearly identified cost and performance factors. For an advanced system, a large number of alternative courses of action should be generated, evaluated and compared using suitably detailed models and simulations. Only then can it be determined if the extra cost of advanced packaging can be justified by the system's needs. This process also enables the designer to determine which functional blocks in the system benefits most from the use of advanced packaging. As a general rule, the most highly connected functional blocks tend to benefit more from advanced packaging than lowly connected functional blocks, such as memories.

A somewhat idealized approach has been presented in which the system's cost and performance aims are clear and can be modeled numerically, as can the performance and cost of the different options. Unfortunately, this is not always the case. The customer's requirements might be vague and ill formed. In this case, scalability and flexibility of the solution becomes an important factor. Also, all of the models needed to do the evaluation might not exist. Quantifying time-to-market and test costs might be difficult. Nevertheless, by qualitatively understanding how the different performance and cost factors relate packaging alternatives to the systems aims, it is still possible to arrive at a sensible solution.

Acknowledgments

First, the author would like to take this opportunity to thank his co-editor, Dr. Daryl Ann Doane for her contributions to this book. In particular, I thank her for technical, organizational and editorial contributions to this chapter and to all of the other chapters in this book (all of which we edited as a team). Many of the concepts original to this book came about through long and detailed discussions between the two of us. It appears to be a result of extremely good fortune, rather than good planning, that our respective technical backgrounds and skills complemented each other to the point where this book is more than two times better than it would have been if either of us had done it alone.

With regard to this chapter, the author also thanks Bob Evans, Tom Gray, David LaPotin, Steven Lipa, Sharad Mehrotra, Slobodan Simovich, Douglas Thomae and the anonymous reviewers for their comments.

References

1   H. B. Bakoglu, *Circuits, Interconnections, and Packaging for VLSI*, New York: Addison Wesley, 1990.

2   D. Balderes, M. L. White, "Large General Purpose and Super-computing Packaging," R. R. Tummala, E. J. Rymaszewski, eds., *Microelectronics Packaging Handbook*, New York: Van Nostrand Reinhold, 1989, Chap. 16.

3   W. H. Knausenberger, L. W. Schaper, "Interconnection Costs of Various Substrates—The Myth of Cheap Wire," *IEEE Trans CHMT*, vol. CHMT-7, pp. 261-267, Sept. 1984.

4   W. R. Heller, W. F Mikhail, "Packaging Wiring and Terminals," R. R. Tummala, E. J. Rymaszewski, eds., *Microelectronics Packaging Handbook*, New York: Van Nostrand Reinhold, 1989, Chap. 2.

5   C. E. Bancroft, "Design for Assembly and Manufacture," *Electronics Materials Handbook, Vol. 1: Packaging*, Materials Park OH: ASM International, 1990, pp. 119-126.

6   J. L. Hennessy, "Trends in Processor and System Design and the Interaction with Advanced Packaging," *Proc. IEEE-MCMC Multichip Module Conf.*, (Santa Cruz, CA) pp. 1-3, March 1992.

7   R. Miracky, *et al.*, "Rapid Prototyping of Multichip Modules," *Proc. IEEE-MCMC Multichip Module Conf.*, (Santa Cruz, CA), pp. 163-167, March 1992.

8   L. L. Moresco, "Electronic System Packaging: The Search for Manufacturing the Optimum in a Sea of Constraints," *IEEE Trans CHMT*, vol. CHMT-13, no. 3, pp. 494-508, Sept. 1990.

9   J. P. Krusius, System Interconnection of High Density Multichip Modules," *Proc. Internat. Symp. on Advances in Interconnections and Packaging*, (Boston, MA), SPIE vol. 1390, pp. 202-213, Nov. 1990.

10 P. Sandborn, "A Tool for Evaluating Performance and Technology Tradeoffs in Integrated Circuit Packaging," *Proc. NEPCON-East*, (Boston, MA) pp. 569-576, June 1991.

11 D. P. LaPotin, "Early Analysis of Multichip Modules," *Proc 1990 IEPS Internat. Electr. Packaging Symp.*, (Marlboro, MA), pp. 557-563, Sept. 1990.

12 P. D. Franzon, *et al.*, "Tools to Aid in Wiring Rule Generation for High Speed Interconnects," *Proc IEEE and ACM Design Automation Conf.*, (Anaheim, CA), pp. 446-471, June 1992.

13 J. L. Hennessy, D. A. Patterson, *Computer Architecture, A Quantitative Approach*, San Mateo CA: Morgan Kauffman, 1990.

14 S. Przybylski, *Cache and Memory Hierarchy Design*, San Mateo CA: Morgan Kauffman, 1990.

15 J. D. Roberts, W. M. Dai, "Early System Analysis of Cache Performance for RISC systems," *Proc. IEEE-MCMC Multichip Module Conf.*, (Santa Cruz, CA), pp. 130-133, Mar. 1992.

16 W. M. Su, "MCM and Monolithic VLSI Perspectives Ion Dependencies, Integration, Performance and Economics," *Proc. IEEE Multichip Module Conf.*, (Santa Cruz,CA), pp. 4-7, Mar 1992.

17 J. Shiao, D, Nguyen, "Performance Modeling of a Cache System with Three

Interconnect Technologies: Cyanate Ester PCB, Chip-on-Board and CU/PI MCM," *Proc. IEEE Multichip Module Conf.*, (Santa Cruz, CA), pp. 134-137, Mar. 1992.

18   G. Messner, "Cost-density Analysis of Interconnections," *IEEE Trans. CHMT*, vol CHMT-10, no. 2, pp. 143-151, June 1987.

4

MCM PACKAGE SELECTION: COST ISSUES

Lee Hong Ng

4.1 INTRODUCTION

In the design of electronics packaging systems, there is rarely a single "best" solution; the final design is usually a tradeoff between different performance attributes (system speed, thermal constraints or size) and cost. In many cases, tradeoffs between cost and performance are the most important. Unfortunately, the analysis of cost and performance tradeoffs is a very complex task. On the performance end, the vast variety of design options available today to the packaging engineer precludes an exhaustive analysis of all viable alternatives. On the cost end, the treatment usually is even more cursory because of the complexity and uncertainty of cost before actual production.

While many may argue that performance analysis is more important in the design phase, one must be aware that cost is very dependent on the product design. In fact, up to 80% of the cost of a product is determined in the design phase [1]. If a design decision is based primarily on performance, one may discover, after the decision is made, that a slight modification might have lowered the cost drastically without an appreciable degradation in performance.

Obviously, both cost and performance issues must be addressed at the design phase to minimize sub-optimization. The designer must be able to think across the boundaries of different packaging approaches to optimize cost and performance for a specific design. In the previous chapter, cost was divided into four factors: production cost, post production cost (reliability, repair and maintainability), design and prototyping cost and time to market cost impacts. In this chapter, details about how to model production cost and examples of the cost considerations of all of these cost factors are given. Sections 4.2 to 4.5 present technical cost modeling (TCM) and its application to making cost-based decisions for MCMs and PWBs. The cost modeling approach presented in these sections is a process-based cost model. Section 4.6 discusses an alternative form of cost modeling used by a design bureau when it does not have access to manufacturing process information but does have access to vendor pricing information. This model is referred to as a design activity-based cost model.

4.1.1 The Importance of Cost

Product costing is important for strategic decisions, cost and performance evaluation, product pricing and product design, as well as to improve and manage existing operations. For example, product cost information is used by management to decide which products the company should drop to be more profitable, or it may be used to formulate a strategy for the company based upon its cost advantages in certain products. If the product cost information is inaccurate or biased, insensible decisions may result.

Product costing also is important in material and process selection, as well as to target areas for cost improvement and optimization. A consistent product costing framework also can be used to assess the cost position of suppliers, competitors and customers, and to evaluate "make versus buy" decisions.

Accurate product costing is especially important at the design phase where more than 80% of the cost of a product is determined. To be useful, cost evaluation at the design phase must incorporate the effects of design and manufacturing processes. This ensures that the design offering the optimum combination of cost and performance is selected. In many cases, process selection decisions also are made at the same time using product costing information. Thus, errors in product costing at this stage results in non-optimum designs being selected.

In many applications, the direct manufacturing cost is only a small part of the total system costs. The total system costs include operational cost such as fuel and power consumption, cost of cooling, prototype design and testing, repair and higher level connections. Again, all of these must be included in the cost analysis to minimize sub-optimization at the system level.

Table 4-1 Example of Equations for Traditional Cost Estimation.

$$Cost = Materials + Labor\ Cost \cdot (1 + \textbf{BURDEN}) + Tooling$$

$$\text{Material Cost} = \frac{\text{Part Weight} \cdot \text{Material Price}}{(1 - \text{Scrap})}$$

Labor Cost = CycleTime · Labor Wage

BURDEN = Other Costs (Depreciation, Energy, Indirect Costs)

Tooling = Cost of Tooling and Setup

4.2 TECHNIQUES FOR COST ANALYSIS

There are many methods available today to analyze the cost of products. They are broadly categorized into three methods: traditional cost analysis, activity-based cost analysis and technical cost modeling.

4.2.1 Traditional Cost Analysis

Traditionally, the task of cost analysis has been delegated to accountants, who are more familiar with the financial rather than the manufacturing aspect of a product. Traditional cost accounting systems, invented in the early 1900s, typically calculate the cost of a product based upon the labor content required for the product. Burden or overhead then is added to the product as a percentage of direct labor (or touch labor). Table 4-1 shows an example of the equations used in traditional cost estimation, and Table 4-2 shows an example of how burden or overhead rate is estimated [2].

In multi-step operations, the cost of a product is the sum of all the unit operations, each of which is estimated from the labor hours and materials required, adjusted by yield. Table 4-3 shows an example of the traditional cost analysis as applied to the assembly of a MCM [3]. In Table 4-3, the total material cost per module is $4,214.84 and total labor hour is 10.66 hours. Assuming that the overhead is 500% of direct labor cost of $12/hr, the total variable cost per module is $4,854 ($4,214.84 + 10.66 × $12 × 500%).

Traditional cost analysis works well when labor cost is the most influential cost driver. As manufacturing becomes more complex and automated, labor cost

Table 4-2 Variable Burden Calculation.

#	Item	Date	Materials	Labor	Energy ...			
	General Ledger					**Monthly Costs**	($000)	(%)
1	Memory Chips	1/1/90 ‖	$200,000			**Direct Costs**		
2	National Electric	1/1/90 ‖			$5,000	Materials	$320	60.2%
3	Resistors	1/5/90 ‖	$30,000			Direct Labor	$56	10.5%
4	Circuit Boards	1/10/90 ‖	$90,000			Energy	$11	2.1%
5	National Electric	1/15/90 ‖			$6,000	**Depreciation Costs**		
6	Monthly Salaries	1/31/90 ‖		$50,000		Equipment	$108	20.3%
7	FICA Payment	1/31/90 ‖		$6,000		Building s	37	7.0%
.			$320,000	$56,000	$11,000		$532	100.0%

Variable Burden = "Other Costs" ($11+$108+$37) / Labor Cost ($56) = 279%

Table 4-3 Traditional Cost Analysis for Assembly of MCMs.

Assumptions			
		Wafer Cost	$4,000
		Substrate Cost	$313
Glue Logic Die /Module	29	Packaged/Lid Cost	$234
VLSI Die/Module	12	Passive Cost	$0.85
Sub. Test Socket Cost	300	Cost/Logic Die	$1
Good Die /VLSI Wafer	40	Passives/Module	25

Task	Lab. Hrs./ Module	Mat'ls/ Module	Yield	Yielded Hrs/MOD	Yielded Mat'l/MOD
Attach Glue Logic	0.50	$342.00	99%	0.56	$386.20
Wirebond Glue logic	1.00		99%	1.12	$0.00
Logic Test/Repair	0.50	$1.90	96%	0.55	$2.10
Passives Attach	0.50	$21.00	100%	0.53	$22.31
Attach VLSI Devices	0.12	$1200.00	99%	0.13	$1,274.98
Wirebond VLSI Devices	0.60		99%	0.63	$0.00
VLSI Test/Repair	3	$1110.00	97%	3.12	$1,155.89
Scrap	3	$1137.00	100%	3.00	$1,137.00
Substrate-Pkg Attach	0.50	$234.00	100%	0.51	$236.36
Wirebond	0.20		100%	0.20	$0.00
Bond Monitor	0.10		100%	0.10	$0.00
To/From Test			99%	0.00	$0.00
Package Seal	0.20		100%	0.20	$0.00
Environmental Test	0.00		100%	0.00	$0.00
To Test			99%	0.00	$0.00
Subtotal				10.66	$4,214.84

becomes a smaller and smaller percentage of the total cost, while overhead becomes a larger and larger percentage. In fact, it has been reported that direct labor cost may be as low as 10% of product cost today [3], and yet more than 75% of the cost reduction effort reported was directed towards reducing labor cost. Since overhead typically is spread across the company or department arbitrarily, the cost calculated using such a method can be misleading [4]-[7].

Consider the case of a product manager trying to reduce the cost of a product using traditional cost accounting. Since the cost of the product is tied to direct labor, his or her first instinct is to reduce labor, which is equivalent to increasing throughput. By purchasing more automated equipment, the direct labor, and hence product cost, can be reduced. But a few months later, the overhead for the department increased because of the investment in the new equipment.

While the example is simplistic, it does illustrate how traditional accounting methods can mislead management decisions. Other shortcomings of traditional cost accounting methods include an insensitivity to production volume, tooling cost and other important manufacturing parameters. More important, it is difficult to justify approaches, such as Just-In-Time (JIT) manufacturing where inventory is kept to a minimum, Computer Integrated Manufacturing (CIM), manufacturing flexibility or automation using a cost estimation method based upon labor [8].

4.2.2 Activity-Based Cost Analysis

One way to overcome the problem is through the use of activity-based cost accounting (ABC) which traces costs to products according to the activities performed on them. The basic steps in establishing an activity-based cost system are listed below [9]:

1. Relevant activities (receiving, production, etc.) are identified.
2. Activities are organized by activity center.
3. Costs of the activities are determined.
4. A "cost driver" or allocation basis that relates the consumption of activities by product is identified.

Table 4-4 compares the allocation bases for the traditional and ABC systems [1]. In Table 4-4, the cost of purchasing is allocated to a product based upon the number of purchase orders (POs) issued for the product instead of the usual labor hours allocation. In this way, a product requiring multiple parts, and hence multiple POs, will cost more than one with fewer parts. Similarly, the cost of production setup is allocated based upon production changeovers rather than labor hours to better reflect cost.

Table 4-4 Allocation Bases for Traditional and ABC Systems.

INDIRECT COST	TRADITIONAL	ABC
Production Control	Labor hours	Parts planned
Inspection	Labor hours	Inspection
Warehousing	Labor hours	Store receipts and issues
Purchasing	Labor hours	Purchase orders
Receiving	Labor hours	Dock receipts
Production Steps	Labor hours	Production change overs

The key feature of ABC is that it indicates what products, customers, processes and product attributes create overhead cost. Since only traceable costs are controllable, the biggest benefit of using ABC is overhead control. ABC also can be used for cost justification for automation, total quality management (TQM) and just-in-time (JIT) manufacturing.

While ABC clearly improves the accuracy of product cost estimates, it has been argued that ABC may not be appropriate for strategic planning because it is an accounting system designed primarily for external financial reporting purposes [10]. In strategic planning and product design, it is important for managers to understand how manufacturing processes and technologies affect product design and cost. One way to do this is through the use of Technical Cost Modeling (TCM), a concept developed at M.I.T. [11]-[12].

4.2.3 Technical Cost Modeling

Unlike ABC, which is an accounting system, TCM is a process-based model which simulates the manufacturing operations to estimate cost. Thus, it is an *a priori* cost model based upon manufacturing simulation, rather than an *ex post* cost model based upon historical accounting data. As such, it can be used to estimate cost for hypothetical processes and product before actual production, and in designing the product for minimum cost (design-for-manufacturability).

Technical cost models can be probabilistic (stochastic) or deterministic. A probabilistic model requires the key input variables to be specified as probability distribution. Since the distributions are rarely vigorously measured, they are frequently assumed to correspond to common analytical functions, for example, uniform, Gaussian, Poisson, Weibull. The simulated cost in this case also is a distribution that results from the interaction of the random input variables. A deterministic model replaces all the distribution of the input variables with the expected value and calculates the expected value of the output variables. This results in a computationally more efficient analytic model.

4.3 TECHNICAL COST MODELING

TCM, also known as process-based cost modeling, is an extension of engineering process modeling with particular emphasis on capturing the cost implications of process variables and economic parameters. It approaches cost estimation by considering the individual elements that contribute to total cost. These individual estimates are derived from basic engineering principles, from the physics of manufacturing process and from clearly defined and verifiable economic assumptions. Since the cost estimates are grounded in engineering knowledge, critical assumptions, such as processing rates and materials consumption, interact in a consistent and logical manner to provide an accurate framework for economic analysis. Technical cost modeling can be tailored to a wide range of operating conditions, thus enabling new processing options to be investigated without extensive expenditures of capital and time.

A technical cost model does not make any assumption about the overall size of the operations. It assumes the existence of a facility capable of producing a specified number of products. In that regard, the cost model is *a priori* model which allows cost estimation before the product is actually made; it does not attempt to model any existing facility, although it can be modified to do so when necessary.

One advantage of TCM over simpler cost estimation techniques is that it not only provides an estimate of the total cost and its breakdown, but it also supports sensitivity analyses, which allow the cost consequences of yield, process rates, multi-shift operations, equipment utilization, downtime and other process variables to be investigated. Technical cost models have been used to prioritize investment, optimize designs, evaluate alternative processes and direct efforts at cost reduction and yield improvements.

4.3.1 Principles of Technical Cost Modeling

A technical cost model can be developed on any spreadsheet program and its general layout is shown in Figure 4-1. The power and flexibility of using a computer spreadsheet facilitates rapid data storage, data manipulation and output

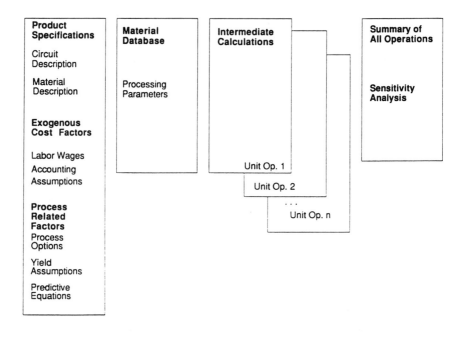

Figure 4-1 Layout of technical cost model.

recalculation. The models are composed of three distinct sections: inputs, intermediate calculations and cost summaries.

The input section of the TCM is divided into four segments: product specifications, exogenous cost factors, process-related factors and materials database. Product specifications vary depending upon the material and geometric configuration of the part being modeled. The exogenous cost factors reflect the economics of the work place and therefore vary with time and location. Process related factors are used to embody the mechanics of the process as it exists in industry. The final group of inputs, the materials database, contains currently available materials, their prices and properties and other material specific information.

The intermediate calculations sections display internal calculations for each unit operation in the process. The cost summary sections present a breakdown of cost into the variable and fixed cost elements. Variable cost elements include materials, utilities, direct labor and variable overhead, while fixed cost elements include capital equipment, tooling and fixed overhead. A detailed description of the estimation of fixed and variable cost and the methods for distributing them over the total number of components manufactured can be found in Busch [11]. The key principles of technical cost analysis are:

1. The total cost of a process is made up of many contributing elements that can be classified as either fixed or variable, depending upon whether or not they are affected by changes in the production volume.

2. Each cost element can be analyzed to establish the factors that affect its value. Depending on the process, the factors that affect cost may differ. For example, the firing time of a ceramic is dependent upon the material, while the screen printing time may not be affected by material.

3. Total cost can be estimated from the sum of the elements of cost for each contributing process. TCM essentially reduces the complex problem of cost analysis to a series of simpler estimating problems, and brings engineering expertise, rather than intuition, to bear on solving these problems.

Although a technical cost model embodies a number of simplifying engineering assumptions, the level of technical detail in the model far exceeds that of other more common cost estimation techniques. By improving the engineering analysis, one may improve the cost estimate further, but there are limits to the value of such modifications, as shown in Figure 4-2. Figure 4-2 shows that while a very accurate cost simulation model may produce minimum errors, its development and maintenance cost far outstrip its potential benefits.

While the ability to estimate production costs on the basis of a set of manufacturing and engineering assumptions is an attractive consequence of constructing a technical cost model, it is the framework for the calculation that provides the greatest benefit. By enforcing a discipline upon the cost estimating process, a consistent and easily justifiable cost estimate can be rapidly computed. Furthermore, the consequences of alternative processing and engineering assumptions can be evaluated on a consistent basis.

4.3.2 Applications of Technical Cost Modeling

Technical cost models can be applied to any manufacturing process. Because it is derived from manufacturing data, it is easy for engineers to understand and use. It has been successfully applied to the automotive [11]-[12], aerospace [13],

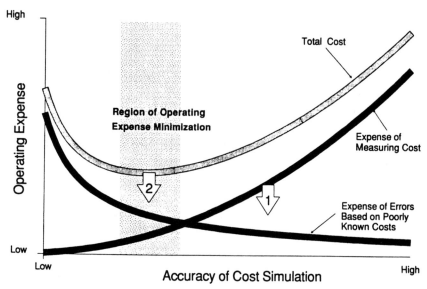

Figure 4-2 Optimization of cost simulation.

recycling [14] and electronics industries. TCM can be used to accomplish the following tasks:

- Estimate costs of products for guiding price quotation
- Establish direct comparisons between process alternatives
- Investigate effect of changes in manufacturing on overall cost
- Identify limiting process steps and parameters
- Determine the merits of specific process improvements
- Compare the merits of functionally equivalent designs
- Identify areas for future R&D

The next section presents some results on the estimation of cost for printed wiring boards, thick film and cofired substrates, thin film MCM and the cost of assembly and testing to illustrate the application of TCM to electronics packaging. *It must be emphasized that the results presented are intended to show the application of the models and the impact of materials and packaging technologies. The results encompass a large number of implied assumptions which are specific in the context they were analyzed and should not be taken as broad based cost projections.*

4.4 RESULTS OF TECHNICAL COST MODELING

4.4.1 Printed Wiring Boards

One of the first areas of application of TCM to electronics packaging was in the area of printed wiring boards (PWBs) [15]. In this section, only high density multilayer boards are discussed; the cost of double sided boards is reported elsewhere [16].

One of the issues facing the engineer is the tradeoff between the cost and the density of the board [17]. In the design of a PWB, it is possible to decrease the layer count by increasing the per layer interconnect density of the board. Suppose a board requires a total interconnect density of 120 in/sq. in. Using 7 mil lines, 2 lines per 100 mil grid, six signal layers are required. Alternatively, using 5 mil lines, 3 lines per 100 mil grid, only four signal layers are required. Assuming that one ground and one signal is required, the two alternative board configurations require an eight and six layer boards respectively.

Faced with these options, the design engineer needs to know the economic impact of the two designs. This can be accomplished by using the PWB technical cost model to simulate the production of the two boards. Figure 4-3 shows the cost of a finished board for both options as a function of both panel and inner layer yield, assuming a board size of 8.5" × 9.5". The top curve shows the cost of the eight layer board at an inner layer yield of 90% and the two shaded curves refer to that of the six layer board at different inner layer yields.

Figure 4-3 shows that the cost of the 7 mil line, eight layer board at 90% inner layer and overall yield is comparable to that of the 5 mil line, six layer board at 80% inner layer and 85% overall yield. It is expected that both the inner layer and panel yield for the two boards would be different in actual production, but the technical cost model allows the designer to find the breakeven point for each board. Alternatively, if yield as a function of the process capability of individual steps is available and coded into the technical cost model, the model will estimate the overall yield and the expected cost of the board for the designer.

Another common dilemma is the specification of materials. While epoxy glass is the cheapest option, its low glass transition temperature (T_g) limits the repairability of the board during assembly. In the case of boards populated with expensive components, repairability is very important. Several materials with higher T_g are available, but they are all more expensive. To investigate the cost penalty of using different materials, the technical cost model is used to simulate the manufacturing cost of the same board using different laminates, taking into account the differences in processing parameters such as drilling, oxide treatment, lamination time, and others for different materials.

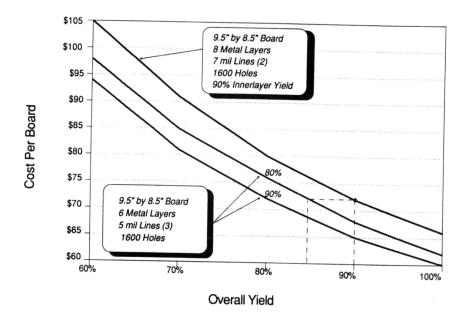

Figure 4-3 Cost tradeoffs between layer count and line width in PWBs.

Figure 4-4 shows the cost of the same eight layer board using different materials simulated using the technical cost model. At $2.50/sq. foot for the laminate, epoxy glass is the cheapest material. Using bis-maleimide triziane (BT) at $6.50/sq. foot, the same board would cost 25% more at the same yield. Finally, if polyimide (highest T_g) is used, the cost of the same board goes up by 60%, assuming an 80% overall yield due to the difficulty of processing polyimide. In all cases, although the raw material cost may be 2 to 4 times more than epoxy glass, the difference in the cost of the finished board is less than 100%. Thus, using raw material cost as the basis of material selection is very misleading.

In the analysis of the cost of the board using different materials, it was found that the cost penalty of going to a higher T_g material is not prohibitive. Therefore, in the case of expensive components mounted onto a circuit board, the use of a higher T_g material to ensure the success of rework may be well worth the additional board cost. Again, the PWB technical cost model allows these tradeoffs to be quantified. Of course, there are other performance variables, such

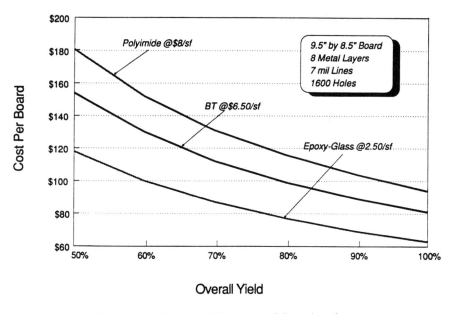

Figure 4-4 Effects of different materials on board cost.

as dielectric constant, CTE, dimensional stability, moisture adsorption, which must be considered in selecting an optimal board material.

Aside from density, another design option that has attracted much controversy is the use of buried vias. By providing an internal path to an adjacent signal plane, the use of buried vias can significantly improve the routability of a PWB design, thereby reducing the total number of signal layers required. There are, however, ramifications to buried via boards. Considerable additional plating and drilling time is required, since each signal layer must be drilled and plated, then printed and etched prior to lamination of the final board, which is again drilled and plated, printed and etched. Because these yields are cumulative, they can increase cost rapidly if not well controlled.

The PWB technical cost model has been used to assess the tradeoffs between through holes and buried vias [18]. In a case study using a board for a computer application (see Figure 4-5), the cost of a 22 layer PTH board is compared to an equivalent 18 layer board with 2,310 buried vias per laminate on three inner layers. Figure 4-5 shows that there is a broad range of yield for which the buried via construction might be cost competitive with the PTH board. Therefore, if the attainable yield of a buried via board in a specific facility falls

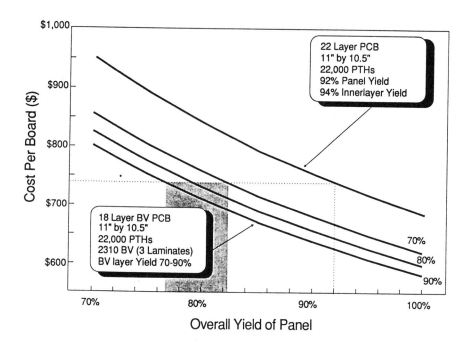

Figure 4-5 The cost impacts of buried vias in PWB.

within the shaded region, the designer can switch to the buried via construction without paying a cost penalty. Conversely, if the indicated yield is not achievable, the designer must decide if the improved performance is worth the additional cost.

4.4.2 Thick Film Substrates

Thick film technology was originally developed to interface and interconnect bare ICs on a hybrid circuit, which is essentially a multichip module. As chip density and complexity increase, the need for higher density and reliability in thick film boards has resulted in the development of better pastes and tighter process control. It now is possible to produce multilayer thick film substrates with 2 - 3 mil lines, although most thick film circuits produced today have line widths in the range of 6 - 10 mils.

Traditionally, thick film substrates are believed to be an expensive option particularly when compared to PWBs. However, studies have shown that they

Table 4-5 Cost Breakdown of SEM "E" Thick Film Substrate.

	COST	PERCENT (%)
Materials	$59	32
Labor	66	36
Overhead	20	11
Capital Equipment	13	7
Utilities	13	7
Tooling	13	7
Total	$184	100%

can be a cost effective option in some applications. Because of the small size of the thick film market and the fact that it is heavily military, the cost of multilayer thick film varies widely. In this section, some general cost results are discussed.

A six metal layer circuit 4" × 5" with 10 mil copper circuitry (SEM "D" format) is selected for detailed cost analysis. There are approximately 2,000 vias of 12 mil diameter per layer. The circuit has been fabricated in the industry and prices are available on these circuits. Using a thick film TCM and assuming an overall yield of 90%, the cost of the SEM "D" circuit with copper conductors is simulated and the cost breakdown is shown in Table 4-5.

Table 4-5 shows an estimated cost of $184 for the circuit, which translates into $1.53 /sq. in. per layer. Product materials comprise a hefty 32% of total cost, and labor makes up 36%. Note that the tooling in the model refers to the screen mesh and other tools; it does not include artwork generation or test fixtures for the circuit. For comparison, quotations obtained from the industry indicated a price range of $600 for a SEM "D" circuit. While this price is three times that estimated by the model, it is not surprising because of the levels of engineering support required for a military specification board and, of course, profit, risk premium and corporate overhead, all of which are not included in the TCM. The model estimates the production cost for the circuit, not the price at which it can be bought in the market.

The materials used for the thick film circuit affect the processing parameters. While different materials suppliers recommend different processing parameters for their materials, the main difference in processing is the type of atmosphere used for firing. The firing atmosphere is dictated by the type of conductor used: gold and other noble metals can be fired in air, but copper requires nitrogen firing.

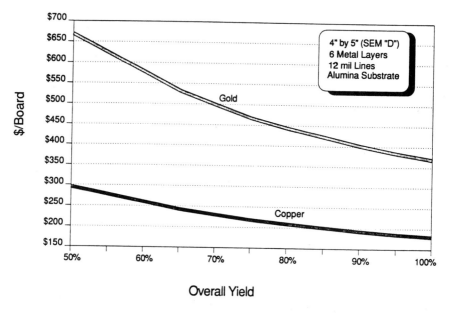

Figure 4-6 The cost impacts of conductors for thick film substrates.

Figure 4-6 shows the effects of conductor choice on the cost of the finished circuit as a function of yield. Since gold paste is 25 times more expensive than copper paste, raw material makes up about 68% of the total cost of the gold circuit, compared to only 26% for the copper circuit. At 90% yield, the gold board costs three times more to manufacture than the copper board.

4.4.3 Cofired Multilayer Substrates

Multilayer cofired ceramic technology has been used for microelectronics packaging since the 1950s. For MCM applications, two types of substrates have been evaluated: high temperature cofired ceramics (HTCC) and low temperature cofired ceramics (LTCC). In HTCC, the alumina green tape used has to be fired at above 1500°C, which allows only refractory metals such as tungsten and molybdenum to be used as conductors. This results in a circuit with high electrical resistance, which may not be suitable for high performance circuits. To overcome these shortcomings, newer materials that alleviate the requirement for refractory metals have been developed. These materials, called glass-ceramics, require a much lower firing temperature (800°C - 900°C), allowing

Table 4-6 Cost Breakdown of HTCC and LTCC.

	COST	HTCC (%)	COST	LTCC (%)
Materials	$16	41	$57	82
Labor	5	14	5	8
Capital Equipment	13	31	5	7
Overhead	3	8	1	2
Others	2	6	1	1
Total	$39	100%	$69	100%

noble metals such as gold, silver/palladium, and even copper, to be used in the circuit.

The incorporation of a new technology, however, often comes at a price. Currently, the cost of green tape for LTCCs is very high. Additionally, the yield of the finished circuit is highly uncertain at this time due to the small volumes in production today. Given the differences in processing parameters and the uncertainties in yield and material prices, it is difficult to assess the economic competitiveness of LTCC, but such an assessment has important implications for users and manufacturers of cofired substrates [19].

To analyze the economics of LTCC versus HTCC, a hypothetical substrate 2.5" × 2.5" with an eight metal layers for a MCM is used. It is assumed that the substrate has 8 mil lines and spacing, and there are an average of 800 8 mil vias per layer. It is assumed that both the finished HTCC and LTCC circuits have the same set of circuit attributes as listed in Table 4-6, and four circuits are produced on one 8" ceramic card (4-up). It is further assumed that the HTCC system uses alumina green tape with tungsten conductors and the LTCC system uses glass-ceramic tapes with silver/palladium conductors.

Using the circuit as the basis for cost simulation and making optimistic assumptions about production volume (10,000 per year) and conductor yield (95%), the cost of the HTCC circuit is estimated to be $39 ($0.80/sq. in. per layer). The cost breakdown (Table 4-6) shows three major elements: materials (42%), capital equipment (31%) and, to a lesser extent, labor (14%). Using the same circuit attributes, the cost analysis for LTCC circuits yielded a total cost of $69 ($1.40/sq. in. per layer). The overwhelming cost is in the product material (82%). From this cost breakdown, it is apparent that the material cost should be the first focus of any cost reduction efforts.

From Table 4-6, it is apparent that HTCC substrates are cheaper to produce at this time. However, it must be emphasized that the assumptions used for the

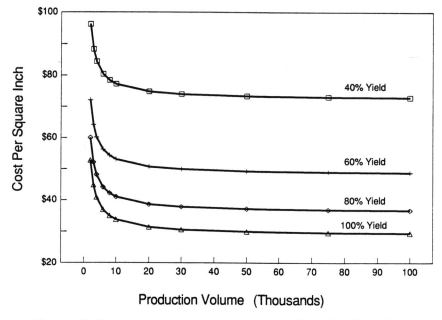

Figure 4-7 The impact of cost on the volume and yield for thin film MCMs.

analysis of LTCC represents industry averages for today. These assumptions, including material prices, yield and even the types of conductors used, are expected to change as the technology matures.

4.4.4 Thin Film MCMs

In the past few years, thin film MCM technology (MCM-D) has become the focal point of the electronics packaging industry. Currently, the cost of thin film MCM substrates is quite high, ranging from $40 to $70/sq. in., but has been predicted to fall to around $14/sq. in. in 1993 [20].

The biggest cost drivers for thin film MCMs are volume and yield [21]. Figure 4-7 shows the volume-yield relationship as generated using information collected from the manufacturers, material suppliers and equipment manufacturers. Figure 4-7 shows that the cost of fabricating thin film MCMs will level off at $30/sq. in. at high volume and reasonable yield for most manufacturers using current materials and processing technologies. However, that does not mean that the prices will not fall below $30/sq.in. in the future.

With improved materials and processing technologies, the cost should come down even further.

To better understand the potential for cost reduction, consider the cost breakdown of a high volume, high yield substrate [22]. Using a five metal layer substrate as the basis, the cost breakdown shows that capital equipment is the largest cost element, making up 45% of the cost of the circuit. This analysis assumes that all of the equipment is new and uses the capital recovery formula (similar to loan payment calculations) as the basis for distributing capital cost. Materials use makes up 21% and labor makes up 17% of total cost. In terms of processes, metallization makes up 46% of the total cost, while polyimide deposition and via formation takes up another 30%. From this cost breakdown, it is apparent that improving the metallization technology should be the first focus in any cost reduction efforts, with polyimide deposition and via generation coming a close second.

It has been reported that using photosensitive (PS) polyimide to generate vias can reduce the number of processing steps. However, the cost of PS polyimide is high and the yield may be low. Thus, the question as to whether the use of PS polyimide is economical depends on the tradeoff between material cost, processing parameters and yield, assuming that the performance of both dielectrics is acceptable. Figure 4-8 shows the cost of a finished circuit as a function of the cost of a polyimide, assuming an overall yield of 80% and high volume production [23]. The darkened lines show the current price range for photosensitive and non-photosensitive polyimides. Given the current price range, the use of PS polyimides does not result in any cost savings at the same yield. If, however, the prices of PS polyimides were to fall below $1.20/g, it could become cost competitive.

4.4.5 Cost of MCM Assembly

In the design and fabrication of the MCM, the chip-to-substrate connection technology is an important consideration. Unfortunately, there is no universal connection technology suitable for all applications. The connection choice depends upon a myriad of factors, ranging from the requirements of the application to the internal manufacturing and design capabilities. Currently, many of the high density MCMs produced are prototype modules, and wire bonding is favored in these applications. However, as these companies move from prototype to production, the choice of connection technology and its impact on the cost of the finished module will become critical to both the users and manufacturers of these MCMs.

In surveying the industry, little consensus was found in several critical parameters in chip assembly: chip yield, bond yield, repair and testing cycle time

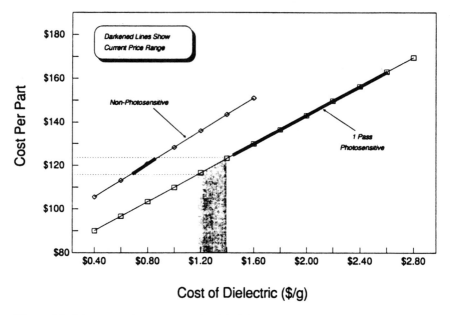

Figure 4-8 Substrate cost versus cost of polyimide (photosensitive and nonphotosensitive polyimides) for thin film MCMs.

and yield, and nonrecurring engineering (NRE) cost. This is an indication of the immaturity of the industry at large. Therefore, this section shows specific results and a sensitivity analysis for a given set of assumptions; the results are not intended to be representative of the industry.

The cost of assembly for any module, as presented in this analysis, is the sum of the cost of all the chips, the cost of bumping the chips and mounting them onto a TAB frame, if necessary, the cost of all the necessary materials (enclosures, adhesives, wires, TAB tapes, scrap), the cost of the assembly process (wire bonding, TAB and flip chip), the cost of electrical testing (functional structural test), the cost of repair and rework, and the cost of final inspection.

In this analysis, two cases of chip technologies are assumed: state-of-the-art and mature chip. When mature chips are used, the test escape rate (a defective chip not detected by testing) is low because the design and process are proven. When using state-of-the-art chips, however, the test escape rate can be very high. In this case, the use of TAB to pre-test and burn-in the chips can reduce the test escape rate and lower the cost of rework. The assumptions used for this analysis are discussed elsewhere [24].

Using a chip assembly technical cost model and the assumptions listed in Table 4-7 with mature chip technology, the cost breakdown for an MCM using wire bonding and TAB is tabulated in Table 4-8. As expected, the cost of the chip (at 85% of total cost) dominates the cost of the finished module. There is only a small percentage of difference in the connection costs between TAB and wire bonding when mature chip technology is used. Therefore, the connection choice in this case should not be driven by cost alone, but rather, the capability of the manufacturers and future requirements.

However, if state-of-the-art chips with a higher test escape rate are used, the use of TAB mounting reduces system cost substantially. Assuming that chip costs remain constant and the percent of known good die is improved from 60 - 95% for the ASICs and from 85 - 98% for memory devices, the cost of the same module is $223 lower when using TAB. The cost savings results from the need for less rework or repair and fewer scrapped chips.

The assembly cost module can be also used to investigate the effects of yield per bond on overall cost. The results of the cost breakdown for the different connection schemes as a function of yield per bond is shown in Figure 4-9. At 99.95% yield per bond for TAB, the cost of the system is estimated at $2,700. At the same yield, wire bonding is expensive. However, if the yield of wire bonding is 99.98%, the cost of TAB and wire bonding become comparable.

4.5 APPLICATIONS TO SUBSTRATE SELECTION

Thus far, this chapter has presented an extensive analysis on the cost of different substrates under a specific set of assumptions. For example, it is estimated that the cost of a thin film MCM will level off at $30 sq. in., for a given manufacturing line, while the current cost of a 22 layer PWB is $7/sq. in. While this information is useful, it does not provide a common metric for substrate technology selection; a design implemented on thin film MCM is likely to be smaller than the same system on PWB.

To compare different substrate technologies, it is necessary to reduce all the different packaging technologies to a common metric. Some common metrics suggested in the industry includes cost per interconnection density [25] and cost per substrate pad [20]. An example of the price/density plot is given in Figure 3-6 where the vertical axis is the cost/sq. in. and the horizontal axis is the available density in in./sq. in. The plot has been used in a number of tradeoff studies of packaging methods.

While these price/density metrics are very useful for general macroscopic comparison, they may not be a good selection tool at the microscopic level when actual design constraints are considered. TCM can be applied to substrate

Table 4-7 Assumptions for Module Assembly and Test Cost Estimation.

CHIP SET DESCRIPTION	NUMBER	PRICE/UNIT	MATURE CHIP (% GOOD)
ASIC Chips	2	$700	85
Memory	8	72	98
Substrate	1	175	99
Enclosure	1	100	100
Total I/Os	1000		
Process Assumption	Wire Bonding		TAB
Yield/Bond	99.99%		99.95
Rework Yield	90%		90
Production Volume	1000		1000

Table 4-8 Cost Breakdown of MCM Using Wire Bonding and TAB.

	WIRE BONDING		TAB	
	Price/Unit	% Good	Price/Unit	% Good
Mature Chip Technology				
Chips	$2,273	83	$2,258	83
Assembly	9	0	192	7
Testing	125	5	151	6
Rework	216	8	107	4
Final Assembly	101	4	101	4
Total	$2,723	100%	$2,809	100%
State-of-the-Art Chip Technology				
Chips	$2,576	84	$2,294	75
Assembly	10	0	193	6
Testing	142	5	153	5
Rework	244	8	109	4
Final Assembly	101	3	101	3
Total	$3,073	100%	$2,850	100%

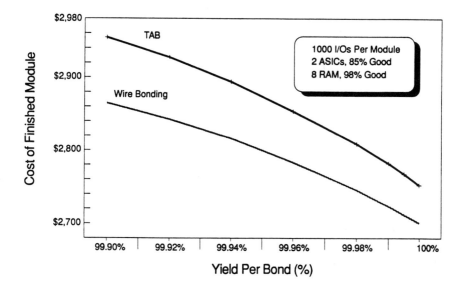

Figure 4-9 The impact of cost on the bond yield of MCMs for TAB and wire bonding.

technology selection when actual design constraints are available. Before the cost models can be used, however, it is necessary to have a procedure to translate a design from one scheme (such as DIPs on PWB) into another (bare chips on ceramic substrate), given a set of constraints and assumptions for each packaging scheme.

A simple way to do the translation is to first set the area so that it can accommodate the required chips, with a suitable space provided for the leads. The number of layers is then calculated so that the routing density is the same in each option [26]. These results then can be fed into the different substrate technical cost models for substrate cost estimation.

As an example, consider a system consisting of two 4.9" × 4.6" piggy-backed PWBs using through-hole packages on one side, and surface mounted discrete components on the other side. The two boards each use 10 mil lines and have 10 and 8 layers respectively. Table 4-9 estimates and tabulates the requirements of the same system using different packaging schemes and needing the same area for the die, the discrete components and the unpopulated area.

Table 4-9 shows that it is possible to reduce the two boards into one board of the same size by using surface mounted devices (SMDs) on both sides. For

Table 4-9 Attributes of Different Substrates for Cost Comparison.

	PWB	THICK FILM	THIN FILM
Length (in.)	4.9	4.9	3.3
Width (in.)	4.6	4.6	3.3
Line Width (mils)	7	8	1
Via Pitch (mils)	100	16	4
Pad Layers	2	1	1
Total Layers	10	5	4

this example, it is assumed that the original board size is preserved for compatibility reasons when using thick film substrate and PWBs. In the case of a thin film MCM, it is assumed that bare chips are used instead of SMDs. The minimum required size is 3.3" square, and only four metal layers are required because of the high connectivity of thin film MCMs.

Using the attributes listed in Table 4-9, the costs of the different substrates estimated by the technical cost models are plotted in Figure 4-10. Note that because yield varies greatly from one facility to another, it is used as a variable in the figure with the darkened sections indicating the expected yield ranges. In Figure 4-10, there are two curves for the PWB, one using polyimide (PI) and the other using epoxy glass (FR-4). The curve for the thick film substrate is for copper conductors.

Figure 4-10 shows that the lowest cost option is achieved by using conventional PWB technology. However, if CTE mismatch is a problem, then epoxy-Kevlar or copper-Invar-copper boards may have to be used. In that case, ceramic boards, at 2 to 3 times the cost of conventional boards, may be attractive. Although the bare chip on thin film option offers greater potential for improved performance, it is considerably more expensive given the yield and volume levels existing today. However, the thin film substrate can be cheaper than thick film with gold, if its yield is better than 65%. Conversely, thick film with gold is cheaper if its yield is better than 76%. As can be seen in this example, the competitive position of each substrate technology is very dependent on the design constraints and manufacturing, especially yield.

Once the cost of substrates has been estimated, the cost of assembly and backplanes also can be estimated to provide the cost at the system level. This approach not only provides a good framework for cost comparison of different packaging options at the design phase, but it also allows the designer to understand what drives the cost in each of the packaging options, leading to improved and more economical design.

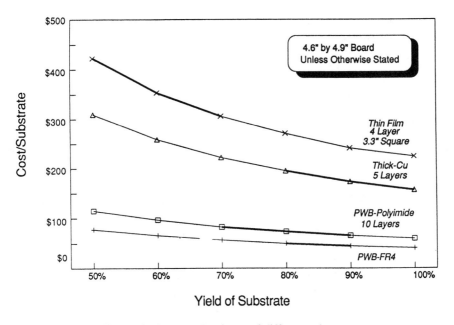

Figure 4-10 Example of cost of different substrates.

4.6 DESIGN ACTIVITY-BASED COST MODELING[1]

In Section 4.2.3 technical cost modeling was defined as a cost model based on a simulation of manufacturing operations.   Doing this requires detailed knowledge of those operations.   Many applications design companies do not manufacture the components themselves and, thus, do not have sufficiently detailed knowledge.   The manufacturing costs are determined by the prices set by the vendors who manufacture their parts for them.   Their total cost also includes the cost of supporting design, prototyping, vendor interfacing and customer support.   Such companies need a design-based, not a manufacturing-based cost model.

It is worth noting that an applications company might have valid use for a manufacturing activity-based TCM.   If such a model includes a set of good assumptions about cost, then it can be used to form a first estimate of the cost

[1]   Contributed by Paul D. Franzon

Table 4-10 Cost Elements that Comprise a Design Activity-Based Cost Model.

Recurring Manufacturing Costs	
R1	Substrate Cost = Panel Cost/Number of substrates per panel
R2	Cost per part of other components: heatsink, connectors, frame, box, etc.
R3	Chip cost (might be untested, partially tested or fully qualified)
R4	Cost of TAB frame
R5	Cost of assembly for each subsystem
R6	Cost of testing each subsystem
R7	Percentage of test escapes and assembly-related failures
R8	Average cost, per part, of rework
R9	Final subassembly yield
R10	Cost of components of higher levels of packaging (hermetic seals, MCM-PWB connectors, PWBs, backplanes, cabinets, etc.)
R11	Cost of assembly for system
R12	Cost of test for system
R13	Percentage of assembly-related failures
R14	Average cost, per system or rework
R15	Final system assembly yield
Nonrecurring Design and Prototyping Costs	
N1	Designer time for analysis and design
N2	Management and support staff time including time spent on vendor interfacing
N3	Cost of computers, CAE/CAD tools, other design equipment as apportioned to project
N4	Cost of offices, electricity, etc.
N5	Designer time for generating test plan
N6	Time to debug prototype and improve design for manufacturing
Recurring Post-Production Costs	
P1	Customer support services and repair
P2	Impact on customer satisfaction

for different alternatives without having to obtain detailed vendor quotes. It also might be useful when vendor cost information is uncertain. For example, current vendor pricing of thin film MCMs do not anticipate the future improvements discussed in Section 4.4 above, Also, current vendor pricing of bare chips is uncertain. Knowing the basis for these costs would allow a designer to anticipate future prices and also would provide bargaining power.

The inputs that are needed to construct a design activity-based cost model are given in Table 4-10. There are three types of inputs in this model: recurrent manufacturing costs, nonrecurrent design and prototyping costs, and recurrent post-production costs. Recurrent costs are incurred for each unit manufactured. Nonrecurrent costs are incurred only once for each design or sometimes for a set of designs.

Most of the recurring manufacturing costs (R1-R15) are obtained from vendor quotes and require the provision of preliminary design information (such as layer count, feature size, production volume). The most straightforward way to investigate alternatives is to obtain quotes from vendors for these alternatives. One approach to generating this preliminary design information for different substrate technology alternatives was given in Section 4.5. Often, technology decisions (such as selecting from substrate alternatives) can be made on the basis of rough cost estimates which are themselves based on rough design data.

Typical current high-volume quoted prices for different substrates are given in Table 4-11. Such quoted prices usually include the expected effect of yield. A proviso is needed, however, when working with this type of data. Despite the mainly per unit inch prices shown in Table 4-11, vendors actually work in terms of the cost per complete panels (or complete substrate), not the cost per part. For example, PWBs typically are made in 24" × 18" panels, usually of which 22" × 16" is usable usually. MCM-Ls are made in smaller panels. Cofired MCM-Cs are made in 9" squares (which shrink by 15 - 20% during firing). MCM-Ds usually are made in 5" or 6" diameter wafers or "rounds." It is up to the designer, in conjunction with the vendor, to maximize the number of individual components manufactured on each panel or wafer. If N components are made, the number of parts per panel often is referred to as "N-up." The prices given in Table 4-11 were generally rough panel per unit area prices. They do not include the effect of wastage if all of each panel cannot be used.

There are other components besides the substrates, including the chips, connectors, heatsinks, power supply, frames, boxes. It is straightforward (though sometimes time consuming) to obtain vendor quotes for all the physical component alternatives being considered.

Some of the recurring costs, such as test cost and rework need might be difficult for the vendor to estimate when presented with only preliminary design data. it might be necessary for the designer to estimate these cost factors and work with the vendor to estimate their impact. This can be done by using Equation 3-16 in Chapter 3, repeated here:

$$\text{Final Cost} = \frac{MC + TC + (P(TE) + P(AF)) \times (RC + ATC)}{\text{Final Yield}},$$

(4-1)

Table 4-11 Some Typical Vendor Substrate Prices.

TECHNOLOGY	TYPICAL COST ($ PER SQUARE INCH)	TYPICAL PANEL SIZE
PWB 5-6 mil lines 　2 layers 　Each additional layer pair	$0.1 $0.2	24" × 18"
MCM-L per layer 　3 mil lines 　12 mil PTHs	$1.50	12" × 18"
MCM-C (cofired) 4 mil layers 　6 layers 　10 layers 　30 layers	$15 - $20 $25 - $30 $50 - $75	9" × 9" 9" × 9" 9" × 9"
MCM-D 　5 layer	$400-$500 per round	6" diameter rounds

where MC is the total IC and MCM manufacturing cost, TC is the total IC and MCM test cost, P(TE) is the test escape probability, P(AF) is the probability of an assembly fault, RC is the rework cost and ATC is the cost of retesting the assembled substrate.

The cost model in Table 4-10 assumes a two step process: MCM assembly and test, followed by system assembly and test. Thus, in Equation 4-1, the manufacturing cost MC for the MCM assembly is the sum of R1 to R5, and the manufacturing cost for the system is the sum of R10 and R11. Many manufacturing processes would be expected to have more than two steps.

The design costs (N1 to N6) are nonrecurring costs. Designer time has to be estimated upfront in the design cycle. This is a difficult task, particularly given the propensity of some managers to underestimate how long it takes someone else to do something. An example of the impact of the design time (and incidently manufacturing cost) for different cooling approaches is given in Table 4-12.

Also contained in the model are two post-production cost elements: customer service costs and the cost impact of customer satisfaction. These were discussed in Chapter 3 in terms of maintenance, reliability and repair. To estimate these costs, reliability models and maintenance effort models are necessary. With the use of these models, it is possible to judge the life cycle cost impact of maintenance, repair and replacement.

Table 4-12 Impact of Different Cooling Techniques on Design Time and Production Cost per Part for a Signal Processing MCM. (Courtesy E-Systems, Melpar Division.)

TECHNOLOGY	ANALYSIS	DESIGN	PROCURE	MATERIAL
Natural convection with no finned heatsink	80 hours	0 hours	0 hours	$0
Natural convection with finned heatsink	120 hours	40 hours	40 hours	$25
Forced air convection with no finned heatsink	80 hours	20 hours	10 hours	$100
Forced air convection with finned heatsink	120 hours	70 hours	40 hours	$200
Liquid cooled	160 hours	100 hours	40 hours	$500

Often the designer has to weigh one cost factor against another and make a decision as to the route that will minimize total cost. For example, it is necessary to balance the cost of fully testing the chips against the cost of extra MCM rework because of chip failures (test escapes), or, alternatively, the cost of procuring TAB frames and testing the chips. This was discussed earlier in this chapter and in Chapter 3.

Another example is to consider the cost impact of putting extra features in the design so that faults on the assembled substrate can be located easily and then diagnosed. Though these features might increase size and manufacturing cost (MC) they might also reduce test cost, rework cost and reduce the time required to debug the prototype. This would involve considering boundary scan techniques (see Chapter 13) if the design team also is designing the chips. On the other hand, if the chip design is fixed, testability could be enhanced by bringing certain internal signals within the MCM to the edge connector just for test purposes. For one module, produced by E-Systems, this was done by adding multiplexer chips to the MCM for test signal injection and monitoring. Despite a resulting increase in substrate size of 20% and an increase in I/O of 10%, they considered this extra cost to be well worthwhile because of the resultant savings in prototyping time.

The designer also should consider tradeoffs that make the module easier to assemble with standard equipment. The vendor should be able to indicate what cost savings can be passed on to the designer by doing this. The vendor should also indicate what design features will maximize yield (and hopefully pass these savings on to the designer).

4.7 SUMMARY

As this is a book mainly about MCM components, this chapter has focused on MCM cost at the component level. Two types of model were presented. Both of the models required a fair level of detail in their inputs. For example, they require a good idea of the substrate size, number of layers etc. In the early phases of design for a large system (large being larger than a single board), this level of detail often is not available. Thus, the inputs to a cost model at the system level will look different than the inputs to the models above.

A system level model is important because it is easy for the system designer to be sidetracked by the fact that an MCM will be more expensive than a similar PWB populated by the same chips in packaged form. As an MCM-based system allows multiple racks of boards to be reduced in size to one rack, the system wide cost savings on the higher levels of packaging can be substantial. This is discussed further in Chapter 18.

Two approaches are commonly used to make decisions early in the system design phase. The first approach is to use a system level cost model, similar in concept to the models presented above, that can use summary and estimates, rather than detailed, information as its inputs. Typical inputs to such a model would include summary information for each MCM such as number of chips per substrate, substrate types, number of MCM signal I/Os and power density. The system level cost model would then estimate substrate cost, heat dissipation cost, test overhead cost etc. to arrive at a system cost estimate. By estimating these inputs for different system implementations, cost estimates can be generated.

The second approach is to compare different system implementations through the use of a set of system metrics that can be easily translated from one implementation to another. For example, if the required interconnection density for a single chip packaged version is determined first then the size and cost of a thin film version can be estimates by considering its interconnection density. Consider the case where the required PWB interconnection density was 51 cm/cm^2 and the interconnect density of an MCM alternative is 100 cm/cm^2. Then the substrate size is expected to be one-half the latter and the system is costed appropriately.

In this chapter, however, two component level cost models were presented. One model, called technical cost modeling (TCM), was manufacturing-based to make manufacturing-related decisions when the designer knows the manufacturing details (usually only the case when the manufacturing activities are in-house). The other cost model was a design activity-based model for use when the only manufacturing-related information available to the designer are the vendor's quoted prices.

Through the application of technical cost modeling to different substrate technologies, it was found that the cost of thin film technology will likely remain high unless the volume hurdle can be overcome. It also was found that though the use of photoimagable polyimide can save on the cost associated with extra via processing steps, it will remain more expensive overall until the cost of this polyimide comes down. Thus, the TCM can be used to judge tradeoffs for minimizing manufacturing costs.

The design activity-based cost model can be used to compare vendor alternatives and also to judge tradeoffs to minimize total cost. For example, a comparisons of the cost involved in adding chips to improve testability versus the cost and time saved in testing and debugging the MCM as a result of this shows that doing this is worthwhile.

Acknowledgments

The author would like to acknowledge both of the editors for their suggestions regarding approach, style and content. She would also like to the P. Franzon for contributing Section 4.6. He in turn would like to thank J. Giordano, M. Montesano and J. O'Malley of E-Systems (Melpar Division) for the contributions used, as acknowledged in Section 4.6, for their helpful discussions and for their shared insights.

References

1   M. C. O'Guin, "Activity-Based Costing: Unlocking Our Competitive Edge," *Manufacturing Systems*, pp. 35-39, Dec. 1990.
2   J. R. Dieffenbach, "Technical Cost Modeling as Cost Simulation Tool," *Soc. of Cost Estimating and Analysis Nat. Conf.*, (Boston MA), June 1991.
3   M. M. Salatino, R. C. Bracken, "Assembly Choices in Mulitchip Module Fabrication," *Proc. of Internat. Conf. on Multichip Modules*, (Denver CO), April 1992.
4   R. Cooper, R. S. Kaplan, "How Cost Accounting Distorts Product Cost," *Management/Accounting*, pp. 20-27, April 1988.
5   F. S. Worthy, "Accounting Bore You? Wake Up," *Fortune*, Oct. 12, 1987.
6   R. Cooper, "The Rise of Activity-Based Costing," *J. of Cost Management For the Manufac. Industry*, vol. 2, no. 2, pp. 4554, Summer 1988.
7   R. Cooper, R. S. Kaplan, "Measure Costs Right: Make the Right Decisions," *Harvard Business Review*, pp. 96-103, Sept. - Oct. 1988, pp 96-103.
8   "Why Costs Need to be Put in Step with the March of Automation," *Financial Times*, June 30, 1987.

9   R. B. Troxel, M. G. Weber, Jr., "The Evolution of Activity-Based Costing," *J. of Cost Management*, pp. 14-22, Spring 1990.

10. H. T. Johnson, "Beyond Product Costing: A Challenge to Cost Management's Conventional Wisdom," *J. of Cost Management*, pp. 1521, Fall 1990.

11 J. V. Busch, "Primary Fabrication Methods and Costs in Polymer Processing for Automotive Applications," Massachusetts Institute of Technology, SM Thesis, June 1983.

12 B. Poggiali, "Production Cost Modeling: A Spreadsheet Methodology," Massachusetts Institute of Technology, SM Thesis, August 1985.

13 N. V. Nallicheri, J. P. Clark, "Competition between Powder Metallurgy and Other Near Net Shape Processes: Case Studies in the Automotive and Aerospace Industries," *KONA Powder and Particle*, no. 8, pp. 105-118, 1990.

14 A. E. Mascarin, J. R. Dieffenbach, "Fender Material Systems: A Life Cycle Cost Comparison," *SAE International Congress & Exposition*, (Detroit MI), Feb. 1992. Published in *SAE Technical Paper Series* No. 920373.

15 L. H. Ng, F. R. Field, "Cost Modeling for Printed Circuit Board Fabrication," *Printed Circuit Fabrication*, vol. 12, no 2, Mar. 1989.

16 L. H. Ng, F. R. Field, "Technical Cost Modeling: A New Approach To Cost Analysis," *IPC Technical Conf.*, (Boston MA), April 1990.

17 L. H. Ng, "Design For Manufacturability: The Cost Connection," *Proc. of Surface Mount Internat.*, (San Jose CA), Aug. 1991.

18 L. H. Ng, Z. Emstad, "Buried Vias, Friend or Foe?," *Proc. of Internat. Electr. Packaging Conf.* , (Marlborough MA), Sept. 1990.

19 M. R. Cox, L. H. Ng, "An Economic Comparison of High and Low Temperature Cofired Ceramics," *Inside ISHM*, vol. 19, no. 1, pp. 31-33, Jan./Feb. 1992.

20 C. L. Lassen, "Integrating Multichip Modules into Electronic Equipment - The Technical and Commercial Tradeoffs," *Proc. of Internat. Electr. Packaging Conf.* , (Marlborough MA), Sept. 1990.

21 L. H. Ng, "Multichip Module Cost: The Volume-Yield Relationship," *Surface Mount Techn.*, March 1991 and *Inside ISHM*, March/April 1991.

22 L. H. Ng, "What Drives the Cost of Multichip Module?," *Proc. of Japan Internat. Electr. Manufac. Technology Symp.*, (Tokyo Japan), June 26-28, 1991.

23 L. H. Ng, "Economic Impact of Processing Technologies on Thin Film MCMs," *Proc. of 42nd ECTC Conf.*, (San Diego CA), May 1992.

24 E. Jan Vardaman, L. H. Ng, "A Cost/Performance Analysis of Interconnect Options," *Proc. of Internat. Symp. on Hybrid Microelectr.*, (Orlando FL), Oct. 1991.

25 G. Messner, "Cost-Density Analysis of Interconnections," *IEEE CHMT Transaction*, vol. CHMT-10, p. 143, June 1987.

26 L. H. Ng, "Economic Comparison of Alternative Packaging Systems 1: Packaged Chips," *Proc. of NEPCON West*, (Anaheim CA), Feb. 1992.

Part B—The Basics

> *The White Rabbit put on his spectacles,*
> *"Where shall I begin,*
> *please your majesty?" he asked.*
> *"Begin at the beginning," the King said,*
> *"and go on till you come to the end:*
> *then stop."*

Alice in Wonderland
by Lewis Carroll

Part A - The Framework was like preparing the reader for a race. We defined the course and alerted the reader to most of the important decisions. Now we are at the starting line!

The nine chapters in **Part B - The Basics** provide an understanding of the technical fundamentals in the design and fabrication of multichip modules. Two broad topical areas are covered. Chapters 5 to 10 describe the physical components of multichip modules and their fabrication. These include first and second level connections, for which Chapters 9 and 10, respectively, present the available alternatives. Chapters 11, 12 and 13 cover the design sciences (electrical, thermal and test, respectively) that must be understood in order to make a successful MCM product.

Chapters 5, 6 and 7, respectively, describe in detail the three basic MCM fabrication technologies: MCM-L (based on laminate structures), MCM-C (based on ceramic materials) and MCM-D (constructed with deposited films). Typical fabrication sequences are described, along with the available major technical alternatives and the strengths and weaknesses of the possible choices. These chapters seek to answer such questions as: *"What portions of the technology are the strongest and most mature (useful for reliable design)?"* and *"What are the limitations in current fabrication sequences, and how may they be overcome?"*

Chapters 11 to 13 concentrate on what makes an MCM different from a design science point of view. Chapter 13 also concentrates on important manufacturing aspects associated with testing.

The chapters in this Part can be read either to extend knowledge gained in another area into the area of MCMs, or to learn about an entirely new subject. For example, the reader who is already familiar with electrical design of printed wiring boards can extend his knowledge into the area of MCMs by reading Chapter 11. He can also gain from Chapter 5 an understanding of printed wiring board fabrication technologies and how they are used in MCM-L structures. Alternatively, the reader who is not familiar with these topics will gain a complete understanding from Chapters 11 and 5. All chapters are intended to be accessable to nonspecialists.

Thus, the aim of these chapters is to be tutorial in nature, as well as to provide specific information that will assist the reader in the transition to MCM technology. For example, the chapter on dielectrics (Chapter 8) provides a concise but complete picture of the important new area of dielectric materials for thin film MCMs (MCM-Ds). These chapters are not intended, however, to provide an exhaustive treatise on all aspects of their topics. Guidance is given to other sources as required.

Specific guidelines are given where appropriate. For example, the test chapter (Chapter 13) ends with specific guidelines on how to reduce test cost and effort. The practicing test engineer, for example, will be able to use these guidelines for help in designing MCM products. However, other engineers should not be deterred from reading this chapter, since it provides an excellent understandable tutorial on this important topic.

For the first time, this Part permits the reader to gain a clear understanding of the following:

- The factors that differentiate the MCM technologies and, thus, how to choose between them.
- The important issues related to dielectric selection.
- The relative merits of different chip connection and MCM connection alternatives.
- The importance of testing, and guidelines for controlling its cost.
- The basic issues and available alternatives in electrical and thermal design.

For the first time, this Part gives

- A complete presentation of the important MCM-L technologies and alternatives.
- A complete and consistent presentation of first level (die to MCM) and second level (MCM to PWB) connection alternatives.

5

LAMINATE-BASED TECHNOLOGIES FOR MULTICHIP MODULES

Leo M. Higgins III

5.1 INTRODUCTION

Subsystems based upon ICs wire bonded directly to printed writing bonds, have been important constituents of electronic products since the early 1970s. The density of these early systems was quite low and frequently did not connect many unpackaged die. Systems built with this process were said to be based upon chip-on-board (COB) technology. The term multichip module (MCM) has been widely used since the mid-1970s, but was not applied to modules based upon PWBs until the end of the 1980s. In the most common phraseology, the term MCM-L has come to imply an IC assembly comprised of multiple wire bonded die on a PWB. Other types of connection technologies, TAB and flip chip also are practiced in MCM-L systems, but COB assemblies have been the most common. COB has come to imply the use of wire bonding and epoxy glob-top encapsulation, which are usually the lowest cost connection and sealing methods. These systems have typically been tested and burned in at the module level, and defective units are disposable. Higher cost, higher performance die often must be tested and burned-in prior to assembly, and the system cost forces planning for MCM repair. These higher performance die often have higher I/O counts, driving a higher module interconnect density than with the lower cost, disposable modules. Thus, COB modules have developed into a subset of MCM-

L, wherein disposable modules, built with untested die, are often implied. The range of MCM-L technologies has widened considerably in the areas of the materials and methods of substrate fabrication, and in the types of die interconnection to the board. In all instances, the minimum set of MCM-L attributes is the connection of unpackaged die on a substrate, whose manufacture is based upon laminate process technology [1].

5.1.1 MCM-L Amid the Spectrum of MCM Substrate Technologies

While the definitions of MCM substrates are discussed in Chapter 1, a cursory review of MCM substrate technologies is useful in view of the broad range of technologies available with laminate structures.

PWBs are known as organic boards since the primary constituent of the board dielectric is an organic polymer. The dielectric layers are supported most often by a reinforcing fabric, usually based upon woven glass fibers. Usually, there is no substrate underlying, or supporting, the laminate structure, as there is with MCM-D substrates where the typical dielectric (organic or inorganic) is very thin and would be structurally inadequate if not formed upon more rigid substrates such as Si, ceramic, metal or an organic laminate (PWB/MCM-L).

All MCM substrate technologies offer a wide range of material and structural options. Conductor options also are available with MCM-D (Cu, Al), and with MCM-C (W, Mo, Cu, Ag, AgPd). MCM-L conductors primarily feature Cu, but in certain structures Al and polymer thick films are used. MCM-D has numerous organic dielectric possibilities, and while silicon dioxide is the prevalent inorganic dielectric, other possibilities such as silicon nitride and spin on glasses are plausible. MCM-C similarly offers a wide range of ceramic dielectric materials. MCM-L allows the use of many polymer dielectrics and reinforcing structures and materials (woven fabric, random matt, porous thermoplastics, particulates such as glass, graphite, fused silica).

Laminate boards can attain a high degree of rigidity after lamination due to the presence of the high modulus of elasticity fibers in the reinforcement phase, while particulate fillers are not as effective in imparting stiffness to laminates since they are not a continuous phase in the xy-plane. When dielectric layers use no fiber reinforcement, the resulting laminate can be quite flexible, leading to the term flexible ("flex") circuits. Combining both types of laminates in a single laminate product leads to structures called "rigid-flex." Thin film-type processing on MCM-L surfaces has been reported by several companies. Such processing includes fully additive and subtractive processing of conductors as well as the processing of dielectrics from liquid precursors or from supported or unsupported films. This creates a hybrid structure which takes advantage of the low cost of MCM-L and the enormous interconnect density provided by thin film technologies. Development work, and low volume manufacture of laminate structures formed on a more rigid base (metal and ceramic), is underway. This type of structure may be necessary in order to use the very thin dielectric layer film materials ($\leq$ 50 µm) which have been developed recently.

5.1.2 MCM-L Attributes

MCM-L has a broad range of desirable attributes, with the primary set including low cost for one and two conductor layer structures for interconnecting a few die, forming a *few chips module* (FCM) and multilayer PWBs with a high interconnect density for the interconnection of many die. The FCM is a type of MCM-L where the interconnect density requirements usually do not push the envelope of process technology. Rather, this type of MCM-L provides a very low cost means to increase interconnect density for a small set ($\sim\leq$ 4) of die. An FCM may be described as an MCM on which the number of chip to chip interconnections is less than the total number of MCM I/Os. This type of MCM-L may make novel use of vias, solder masks, encapsulation and finishing metals to achieve the primary objective of low cost.

Higher density MCM-Ls require an engineering effort to focus on the achievement of higher interconnect density and performance, which includes smaller drilled holes, buried vias, finer lines and spaces, thinner dielectrics, low loss dielectrics and thermal management. Other desirable features include two-sided assembly, large area substrates and the elimination of connectors through the use of flex or rigid-flex structures.

The primary shortcoming of MCM-L has been the limitation of interconnect density due to the use of plated through-holes (PTH) for layer to layer interconnections, and the relatively coarse line widths (compared to thin film processing) and spacings commonly practiced (typically $\geq$ 0.075 mm). New circuit definition technologies, the increasingly widespread use of blind and buried vias, improved drilling technology (use of 0.2 mm diameter bits is nearing production capability in some companies), the use of punching instead of drilling for small vias in thin dielectrics and the coupling of thin film surface layer processing with MCM-L structures, continue to permit the increase of MCM-L interconnect densities.

Another problem which MCM-L technology must address is the relatively high coefficient of thermal expansion (CTE) of the substrates. The use of large die on MCMs is forcing critical evaluation of the reliability of die to MCM interfaces, including both die attach and electrical connection. This reliability is critical, especially when the die connection is made by flip chip or short lead flip tape automated bonding (TAB) due to the low, thermally-induced stress compliance afforded by these techniques. Flip chip connection is seen as the ultimate MCM connection method, but it offers the least compliance. MCM-D and MCM-C offer the use of low thermal expansion substrates. MCM-D can be offered on Si and metal substrates for a near perfect CTE match to bonded die. MCM-C can be made from aluminum nitride, mullite, cordierite, lithium alumino-silicates and glass ceramics. These all offer near perfect CTE matches. Since the die temperature is usually more than the substrate in actual use environments, due to transient effects and the thermal resistance through the die to substrate interface, the ideal MCM substrate should have a CTE slightly higher than that of silicon. The development of new, low CTE dielectric systems for MCM-L and the emerging technology of dielectric layer lamination on low CTE base substrates offers the promise of MCM-L solutions for this thermomechanical strain problem.

Despite the delay in recognition of laminates as MCMs, especially conventional high density PWBs, demand for cost effective MCM solutions for emerging electronic systems has pressed MCM-L into the mainstream. Due to the low cost of MCM-L (relative to other MCM substrate technologies) and the wide diversity of materials, properties and manufacturing process technologies used with MCM-L, it has become a very popular type of substrate for high density, high performance MCMs. The broad vendor base, and the extensive use of PWBs in almost all current electronic systems, indicates the wide range of application functions and system frequencies supported by MCM-L.

This chapter begins by describing the standard construction process for an MCM-L or PWB. This is the process most widely used in the industry. Following that, the desirable properties of the materials used in this process are presented together with a discussion of their limitations. Because of these limitations, a number of alternatives are being pursued actively. These alternative materials and processes are presented. Bare chip mounting creates some unique requirements on MCM-Ls. These are discussed in the section on connection and repair. Finally, some examples of MCM-L configurations are given.

5.2 STANDARD MCM-L CONSTRUCTION PROCESS

Five basic process steps are used in the manufacture of MCM-Ls and PWBs. They are:

1. The preparation of individual copper foil clad dielectric layers.
2. The photolithographic patterning and etching of conductors on those layers.
3. The drilling of vias through individual layers or partial laminates to form blind and buried vias, and drilling through the total laminate

thickness to form plated through-holes (PTH).

4. The lamination of the individual layers and sublaminates onto each other form a multilayer MCM-L. Multiple lamination steps are needed to form blind and buried via structures. A single lamination step is usually used if the board has no blind or buried vias.

5. The plating of drilled holes in single layers or partial laminates, or through the entire board thickness, to create blind and buried vias, and PTHs, respectively. The plating of the surface metallurgy follows.

Each of these basic processes consists of several steps, which are summarized in Figure 5-1. For more information, the reader is referred to PWB handbooks by Coombs [2] and Clark [3].

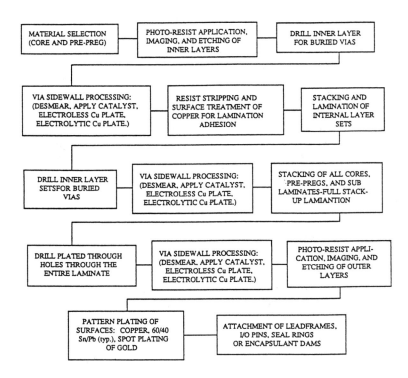

Figure 5-1 Typical substrate process flow.

5.2.1 Dielectric Material

Two types of dielectric layers are used in typical PWBs: cores and prepregs. In both cases, the dielectric reinforcement material or fabric, typically fiberglass with the E-glass composition, is run through a coating process to impregnate the material with the selected polymer. The fiberglass may be woven in many different weaves to allow the formation of layers with various thicknesses and glass to polymer ratios. Core material is fully cured, while the prepreg material is partially cured before the material is cut to size. Core and prepreg thicknesses are selected to meet mechanical and electrical performance criteria, for example, to provide the correct value of characteristic impedance. The core layers have copper conductors patterned on one or both sides. The prepreg layers are then placed between the core layers to cause adhesion of the various layers in the stackup during the lamination period, when temperatures are elevated and pressure is applied.

5.2.2 Copper Foil (Conductor) Processing

The primary PWB conductor usually is electro deposited (ED) or rolled annealed copper foil. The foil is treated on one surface (or both, for internal Cu planes) to enhance adhesion to the polymer impregnated core or prepreg layers. Often, this surface is cleaned with mechanical abrasion, chemically microetched to increase surface topography and area, and chemically oxidized to form a thin, passivating adherent layer of CuO, further increasing foil surface area. In many instances, copper is coated with a very thin layer of tin for passivation, and laminate adhesion. The immersion tin process is less sensitive to copper material cleanliness, and protects the copper in subsequent etch processing [4].

The copper foil is roll laminated (to the dielectric), after which the full, or B-stage, dielectric curing takes place. The dimensional stability of the core and prepreg layers can be affected by the manufacturing process and storage conditions (temperature and humidity). The prepreg layers are particularly sensitive, and are stored under refrigerated, atmospheric control conditions for best results. If refrigerated, the pre-pregs should be brought to room temperature in the absence of water, especially when hygroscopic materials, such as polyimide, are used.

5.2.3 Inner Layer Photolithographic Processing

The type and thickness of copper used is based upon the specific layer function and the fineness of the features to be etched (for example, 110 μm and 19 μm thick Cu may be used for ground planes and fine feature layers, respectively). The core and pre-pregs surfaces are processed as described in Section 5.2.2, and photoresists (PR) then are applied. PR films are laminated to the layers and liquid PRs are applied by roller coating, spraying, screen printing or curtain coating. Resists also may be applied with electrostatic spraying and by electrophoretic deposition techniques (which refer to electroplating of organic materials in this case). Liquid resists and advanced deposition techniques often provide finer etch resolution than the older PR film lamination process.

Proponents of film PRs claim the fixed film thickness permits a more readily controlled process for etching of fine lines than the use of liquid PRs, where the user's process must provide the resist thickness control. Pattern exposure, resist development and the copper foil etching follows. Copper etchants are usually cupric chloride or alkaline ammonical based systems. The photoresist is then removed chemically and the copper surface is processed to form the desired CuO layer needed for adhesion to the overlying layer during lamination [5].

The artwork used to expose the photoresist frequently is not an exact geometric replica of the desired etched pattern geometry. The layers are subject to xy-shrinkage or expansion, but in a well characterized process, the xy-dimensional stability of the core and prepreg layers during storage, processing and lamination is controlled and well quantified. The stability data are used to modify the artwork, expanding or contracting the dimensions of the layer pattern. This modification may be graded uniformly across the entire layer, or it may vary from dimensionally insensitive to sensitive regions.

5.2.4 Blind and Buried Via Formation

Standard PWBs utilize holes drilled through the entire multilayer board thickness to electrically connect the desired metal planes. After drilling, these holes are copper plated to affect the electrical connection, forming plated through-holes (PTH). Most high density MCM-Ls utilize layer to layer electrical connections which do not span the entire board thickness. Blind vias extend from the surface into the desired layer(s), while buried vias only interconnect internal planes. After the pattern etching process, the blind and buried vias are drilled through individual layers, or sub-laminate stackups formed by the lamination of the desired inner layers. The side walls of the drilled holes, and any exposed, etched copper features are then plated with copper. Then final board lamination is performed upon the stackup of the single layers and sublaminates which form the board. The drilling and plating of the total board thickness PTHs follows.

Blind and buried vias also may be formed by precisely drilling part way through the formed laminate or sublaminate. After drilling, the laminate structure is plated to form the interconnection from the underlying layer to the top layer. This process often is considered more difficult, but may be necessary for some structures.

The blind and buried via process is considered expensive in many circumstances, due to the difficulties in handling thin core and prepreg layers and the multiple drilling, plating and lamination operations required. This reality is mitigated somewhat by the ease with which the drilling and plating operations are performed. Small via drilling and plating of single layers and thin sublaminates is simple when compared to full laminates PTHs since via aspect ratios are smaller than those of PTHs. This greatly facilitates hole drilling, cleaning and plating. Also, the use of blind and buried vias in a high density MCM-L reduces significantly the number of PTHs needed. If the density of the MCM-L is very high, there is great design pressure to use small diameter PTHs. Since the PTHs are drilled through the entire board thickness, small diameters result in high aspect ratio PTHs. This also can have a major effect on production throughput and yield. If the use of these vias reduces the number of layers needed to achieve a circuit, there is a cost benefit of fewer layers to offset the added blind and buried via cost.

5.2.5 Lamination

The purpose of the lamination step is to "glue" all of the layers together. The pressures and temperatures which are used must be carefully controlled to drive out entrapped air, absorbed water, and solvents retained in the prepreg. The B-staged (partially cured) pre-pregs, which are placed between the cores, are the adhesive layers. During lamination, the partially cured polymer in the pre-pregs softens, flows and wets the adjoining surfaces, effecting the bond [6].

In the first step of the lamination process, the etched core and prepreg layers are stacked in precise registry using mechanical tooling and alignment pins. Laminating the type of high density PWBs used for MCM-L with vacuum presses permits the use of lower pressures, temperatures and times. The vacuum assists in the removal of entrapped air, retained solvents and water, permitting the use of less aggressive lamination parameters. These conditions also aid the control of the thickness of prepreg layers, improving the control of electrical properties such as capacitance, crosstalk, and characteristic impedance (Z_o). The optimum lamination time, temperature and pressure are interrelated. They are highly dependent upon the polymer system used in the prepreg and core, the ratio of polymer and reinforcing fabric volumes, the fabric weave, the thickness of the copper planes, the presence of blind and buried via layers or sublaminates and the dimensional control required.

5.2.6 Drilling

Computer aided design (CAD) data from the design software used to create the artwork also is used to create a drill list, which defines the sizes and locations of the holes to be drilled. This information is downloaded to high speed, multi-spindle drilling machines which determine the location of the next hole to be drilled by use of the mechanical data.

There are many sources of drilled hole location errors. To increase production throughput, it is common practice to stack the boards on the tooling under each drill spindle. Drilling through multiple boards with each drill stroke increases productivity, but increases the risk of hole position errors. New drill bits are available which utilize advanced materials and which incorporate well designed flutes in the drill. Improved drill bits are important, but not sufficient to solve stack drilling problems. The registration tooling, and the registration features drilled and routed in the boards prior to layer processing, must be accurate and consistent. Since each board may not have contracted or expanded identically in processing, the stack tooling must be able to accommodate these dimensional differences while providing suitable dimensional registry to the xy-program driving the positioning system on the drill machine [7].

Another major source of drill position error is wandering of the drill, which implies drill tip movement in the xy-plane during the z-axis excursion through the board. This can be caused by inaccuracies in the drill head (vibration, precession, worn bearings, nonorthogonal z-axis motion), but many errors are due to characteristics of the board stack. Typically, special cover and back up sheeting materials are placed over and under the stack. These entry and backup materials are intended to act as drill guides and to reduce drill breakage. Three layer laminates of special aluminum alloy foil with a cellulose core and aluminum alloy about a thin wood core are used widely as entry and backup materials, respectively [8].

In general, features which increase drill position errors and drill breakage are thick board stacks, high aspect ratio holes (~ > 3:1), high glass content, excessive drill rates, loosely stacked boards and dull drills. Core and prepreg layers measuring 100 μm thick are commonly used in the fabrication of high density, low thickness PWBs. High volume production use of 200 - 300 μm diameter drill bits for these structures is ongoing, while the most widely used drill diameters are as large as 500 μm.

Drill hole quality also is critical for the construction of the highest density surface layers. In these cases, the drilled hole is equal to, or smaller than the width of the surface feature it electrically connects to internal conductors. This reduces the surface area required to fanout the surface conductors from the die to vias, or PTHs. Drills as small as 50 μm diameter are being used in development work, while 100 - 150 μm drills are being used in prototype manufacture of MCM-Ls.

5.2.7 Plating of Drilled Holes

During drilling of vias and PTHs, the frictionally heated drill bit causes polymer residue to smear and coat the hole wall. This residue must be removed to allow plated copper to bond to the structurally sound bulk dielectric and to the cross section of the copper features through which the hole was drilled. This operation is referred to as desmear and is commonly performed using concentrated sulphuric acid, alkaline permanganate solutions or oxygen plasma cleaning. The solutions used also must make the hole walls hydrophilic to enhance wetting by electroless Cu plating solutions and to eliminate bubble entrapment in the holes. The desmear process difficulty increases with increasing hole aspect ratio and decreasing hole diameter [9].

After desmear, the polymer and glass surfaces of the drilled holes must be treated to permit plating and good adhesion. The units are immersed in a palladium-tin compound solution for surface activation, permitting deposition of both the palladium and tin. It is necessary to remove the tin in a subsequent post activation cleaning step. The board is then placed in an electroless Cu plating bath where the bath chemistry, temperature and mechanical agitation are critical to ensure uniform Cu deposition, good Cu adhesion, and proper Cu microstructure for good mechanical properties [10]. Electrolytic Cu plating follows to build the copper thickness to the desired levels. The copper surfaces of the inner layers are then photolithographically processed and etched to form the desired conductor patterns. The surface of the conductors are then cleaned, etched and oxidized to enhance adhesion to neighbor layers. A schematic cross section of an MCM-L/PWB structure is shown in Figure 5-2.

5.2.8 Processing of Surface Layers

The final plating of the MCM-L surface defines perhaps the greatest difference between conventional PWBs and MCM-Ls. Conventional PWBs are made for the surface mount, or through-hole, soldering of packaged devices which usually possess metal leads. Thus, the copper attach pads on the PWB surface are plated typically with Pb-Sn solder, suitable for the mass reflow of surface mount devices, or for the attachment of through-hole components with wave soldering. The surface metallurgy of MCM-L often requires multiple process steps to permit selective plating of gold on some sites and lead-tin solder on others. These steps involve the application of plating resists, imaging, developing, plating, resist stripping and a repetition of this sequence for each different metal finish to be applied. Soft gold plating may be desired for wire bond pads, while hard gold is needed for edge connectors, and 60/40 tin-lead solder may be required for soldering of TAB leads.

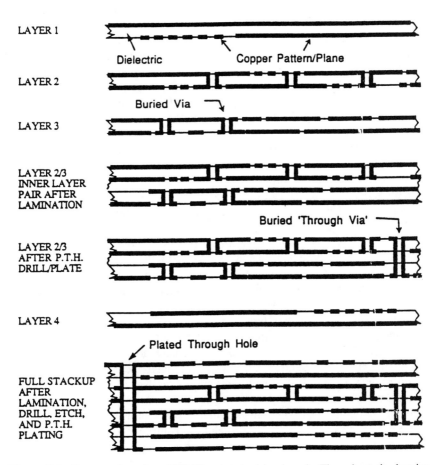

Figure 5-2 Cross section of an MCM-L printed wiring board. The schematic drawing shows construction of a typical MCM-L substrate, exhibiting blind and buried vias and a typical lamination sequence.

After the PWB is fully laminated, the PTHs are drilled through the entire board thickness, and plated using processes similar to those described above. A PR is applied to the copper foil surface, and is imaged and developed. The PTH sidewalls and the copper exposed on the surface is then plated with Cu/Ni/solder (typically 63/37 Sn/Pb solder). The PR then is stripped and the newly exposed

Cu is etched away, effectively using the solder as an etch resist. The surface is PR coated again and patterned to protect the regions where solder plating is desired. The exposed solder is etched away. At this point, the solder may be reflowed. The surface then is coated with a photoimageable solder mask, imaged and developed, exposing the solder plated pads. This procedure is called solder mask over bare copper (SMOBC) [11]. Alternative SMOBC process flows are possible. Areas requiring future Ni/Au plating or other non-solder treatment, are to be masked off through much of the SMOBC process. The wire bond pads typically are plated with 5 μm of Ni, and 0.5 - 1.0 μm of high purity (~ 99.99%) soft Au. The use of 0.2 - 0.3 μm of electroless gold for wire bond pads is increasing in Japan.

Additional applications of Prs and more photolithographic processing are often required to permit the final spot plating of the Ni/Au regions. Reflowing of the solder, before or after Ni/Au plating, improves resistance to oxide formation, but it causes the pad surface to dome due to the surface tension of the molten solder. This domed surface can cause lead alignment problems during TAB outer lead bonding (OLB).

Selective, or spot plating, is a critical process for MCM-L since it is very common to connect wire bonded and TABed die on the same MCM. In the near future, flip chip attachment on MCM-L will become common, so it may be possible that three distinct bare die connection methods (wire bond, TAB and flip chip) will be practiced on a single MCM. This is likely to require even more sophisticated selective plating procedures.

Electrolytic plating is commonly used for the deposition of Ni, Au and Pb/Sn solder finishes. This requires that the features to be plated are connected to a plating bus. The need for connections to fanout to the bus can interfere with circuit routing design. If the fanout nets are not on the surface, they will not be etched away after plating. High frequency systems will encounter signal integrity problems if plating bus connections remain after removal of the plating bus. These useless nets act like signal stubs and can contribute to circuit noise due to increases in driver loading, crosstalk, reflections, radiated noise and coupled radiation. Electroless plating does not require the use of plating bus connections, but it usually does not permit the deposition of the desired thicknesses of Au and solder.

5.3 MATERIAL CONSIDERATIONS

There are four types of materials used in the fabrication of a laminate:

1. **Core dielectric layers.** These layers are rigid materials, usually comprised of reinforcing fiberglass fabric and a fully cured epoxy

matrix. Some core dielectrics use no reinforcement.

2. **Prepreg dielectric layers.** These layers are flexible materials, usually composed of a fiberglass reinforcement fabric and a B-staged epoxy matrix. This material attains rigidity with full cure during the lamination and post lamination bake processes.

3. **Conductor materials.** The typical conductor material used for power distribution and signal interconnection is etched copper foil.

4. **Finishing materials.** Typically tin-lead solders or nickel-gold, are plated on the external surface copper bonding pads to permit die connection.

The epoxy-glass dielectric described above is referred to as FR-4 material. In the following sections, the ideal material properties are discussed and alternative materials are presented and compared.

5.3.1 Dielectric Layers

MCM-L dielectric layers are usually composite materials consisting of a reinforcing material and a continuous polymer matrix phase. Flexible circuits, and some advanced MCM-L substrates, use unreinforced polymer dielectric layers. In this section, the various polymers, reinforcement media, and the resulting dielectric layer products are discussed in terms of their physical and electrical properties, and their effects on MCM-L processing. Table 5-1 provides a listing of the physical and electrical properties of a wide range of materials used in MCM-L fabrication.

The laminate structure of the typical PWB is a ternary composite of phases: resin, reinforcement fiber or filler and copper conductor. The dielectric constitutes the major volume of the MCM-L. The copper conductor planes are commonly 9 - 105 μm thick, while the dielectric layers typically range from 25 - 500 μm in thickness. All three phases contribute to the electrical and mechanical properties of the final board. Polymer and reinforcement phase effects are discussed in the following section.

With standard types of laminates, both the fiberglass fabric and the matrix resin are continuous in the xy-plane, while only the resin is structurally continuous along the z-axis. Thus, laminate structures exhibit anisotropy, where the macroscopic properties of laminate structures are quite uniform in the xy-plane, but different in the z-axis. The differences in the basic material properties of the resin, reinforcement and the conductor lead to stresses in the laminate

Table 5-1 Properties of Materials Used in MCM-L Fabrication.

Materials	xy-Plane Thermal Expansion (ppm / °C)	z-Axis Thermal Expansion (ppm / °C) (<Tg)/(>Tg))	Tg; Glass Transition Temperature (°C)	Thermal Conductivity (W / m · K)
FR4 (epoxy-'E' glass)	16-18	60 / 320	125-140	0.16-0.4
Polyimide (PI)-E glass	13-15	40 /190	225-260	0.3-0.6
High Tg epoxy-E glass:				
Risho CS-3665	13-14	50-220	200	
Teflon-E glass	20		75	0.26
Epoxy-aramid (PPDETA)	6.5		172	0.18
PI-Kevlar 108	5.0-8.0	85	250	
Epoxy-fused silica	6.0-12.0		125	
PI-fused silica	6.0-12.0	30	250	
BT epoxy-Kevlar 120		73.7		
High Tg epoxy-fused silica #525	6.0-12.0	65		
'Gore-Ply' (cyanate ester-expanded PTFE)	55		190	
PI-unwoven Kevlar 'ROHSI 2800' (Rogers)	16	24	Thermoplas.	
PI film: Kapton H	20-25			
Upilex S				
Polyester film	25-30			
Epoxy resin (#5010)	55			
PTFE	224	224		
'E'-glass	5			
'S'-glass	2.3			
Aramid fabric	-2			
Unwoven Aramid / PI	1.35-2.25			
Fused silica fabric	0.54			
Copper (CDA 102)	17.3	17.3		393
Aluminum (elemental)		22.1		240
Aluminum (6061)		21.1		200
Molybdenum	5	5		146
Kovar		5.3		17
Cu/Invar/Cu: 20/60/20	5.5			169 (Z=23)
12.5/75/12.5	3.15			114 (Z=18)
Cu/Mo/Cu: 20/60/20	6.7 (20 C)			141(Z=113)
13/74/13	5.8 (20 C)			122(Z=98)

Table 5-1 Properties of Materials Used in MCM-L Fabrication (continued).

Materials	Tensile Modulus (10E6 psi)	Tensile Strength (10E3 psi)	Dielectric Constant (1 MHz, 25°C)	Volume Resistivity (Ω-cm.)	Dissipation Factor (%)
FR4 (epoxy-'E' glass)	2.5	40	4.0-5.5	4.00E+14	2.2
Polyimide (PI)-E glass	2.8	50	4.0-5.0	4.00E+14	1.3
High Tg epoxy-E glass: Risho CS-3665			4.3	3.00E+14	1.3
Teflon-E glass	0.2		2.3-2.6	1.00E+10	0.2
Epoxy-aramid (PPDETA)	4.4		3.7		2.6
PI-Kevlar 108	4	30	3.95	1.00E+12	1.7
Epoxy-fused silica					
PI-fused silica				1.00E+09	
BT epoxy-Kevlar 120			3.51		1.1
High Tg epoxy-fused silica #525					1.3
'Gore-Ply' (cyanate ester-expanded PTFE)			2.6	> 10E+07	0.3
PI-unwoven Kevlar 'ROHSI 2800' (Rogers)	0.12		2.8		0.3
PI film: Kapton H	0.4	25	3.5	1.00E+12	0.25
Upilex S	1.3	57	3.5	1.00E+11	0.13
Polyester film		20-40	2.8-3.2		0.3-1.6
Epoxy resin (#5010)	0.39		3.8		
PTFE	0.05		2.2		
'E'-glass			6.3		
'S'-glass			5.3		
Aramid fabric	18.5	440	2.3		0.7
Unwoven Aramid / PI					
Fused silica fabric					
Copper (CDA 102)		32-55		1.67E-06	
Aluminum (elemental)		45		2.66E-06	
Aluminum (6061)		35		4.30E-06	
Molybdenum		95			
Kovar		75		4.70E-05	
Cu/Invar/Cu: 20/60/20	19	60			
12.5/75/12.5	19	60			
Cu/Mo/Cu: 20/60/20	20	100		3.40E-06	
13/74/13	39			3.80E-06	

structure as the individual phases attempt to behave independently during environmental and electrical stressing. One of the major reliability problems associated with PWBs is barrel-cracking, or the cracking of the plated through-hole copper during temperature cycling. This is due to the fact that the z-axis thermal expansion is much higher than the xy-thermal expansion of typical boards. The thermal expansion of copper is ~18 ppm/°C, from 0°C - 125°C, which closely matches the range of xy-thermal expansion (~13 - 18) offered by typical fiberglass-based laminates. The z-axis thermal expansion can be 3 to 5 times the thermal expansion of copper, at temperatures below the board glass transition temperature (T_g) and 10 - 20 times higher above the board T_g.

Desirable dielectric properties shown are in Table 5-2. With most materials, the dielectric layers typically take two forms, the core layers (fully cured in advance), and the prepreg layers (partially cured, or B-staged, in advance). The compositional makeup of the two types of layers can vary in the board, but designers strive to achieve a balanced structure, where the copper and dielectric content are fairly uniform throughout the structure. This balanced structure reduces bowing, or camber, after lamination due to variations in the thermal expansion and cure shrinkage of individual layers on opposing sides of the board's "neutral plane," which extends through the center of the laminate. This is why MCM-Ls are usually symmetric through the board center in terms of signal and reference layers. If the board is thin and flexible, solder masks also are used in equal proportion on both surfaces, even if it is really only needed on one surface. The solder masks usually have higher thermal expansions than the board and induce board camber at room temperature, or while the board heats up in an assembly operation or in an electronic system.

Varying the ratio of reinforcement to polymer and the type of fabric reinforcement and polymer, permits moderate variations in the properties of the final board, but it may induce significant variations in fabrication process control. For example, increasing the resin content in the core and prepreg may reduce the dielectric constant of the board. On other hand, a higher polymer content may cause greater lateral displacement during lamination. This can cause problems in dimension control or cause difficulty in hitting the internal pads during hole drilling. Increasing the glass fabric content, or changing the fabric weave, may improve layer thickness control after lamination, but also may induce lamination voiding and cause increased drill wear and breakage. Difficulties in process control and subsequent reliability and assembly problems, (the result of the low T_g for FR-4 resins and the resultant high CTE at typical assembly temperatures), have been the driving forces behind the development of new polymers and dielectric layers.

Table 5-2 Desirable MCM-L Substrate Properties.

ELECTRICAL	
Low dielectric constant	1.0 - 3.0
Homogeneous dielectric	Isotropic properties
Low dissipation factor	$\leq 0.1\%$ for desired frequencies
PHYSICAL	
Low moisture absorption at saturation	< 0.01%
Low rate of moisture absorption	Slower than FR-4 (< 0.05% sat.)
High glass transition temperature	> 250°C
Low CTE (xy-plane)	Close to Si (3.3 ppm/°C, 0-200°C)
Low CTE (z-axis)	Match Cu (18 ppm/°C, 0-200°C)
High Cu adhesion	> FR-4
Low lateral deformation during lamination	< FR-4
High modulus of elasticity and high strength	Resist encapsulant induced camber

5.3.2 Polymers for Dielectric Layers

A variety of high T_g polymers are available for board manufacturers. FR-4 grade PWBs are made from a relatively low glass transition temperature (T_g = ~ 125°C) epoxy (commonly supplied by Shell Chemical) with an internal E-glass fabric reinforcement. Other widely available polymer materials include tetra- and multifunctional epoxies (T_g range of ~150°C - 200°C), cyanate esters (T_g = ~180°C - 200°C), bismaleimide triazine (BT) (T_g = 175°C - 190°C), and polyimides (T_g = ~240°C - 270°C). Triazine is a cyanate ester, and often is blended with bismaleimide (one of the precursor constituents of polyimide) in a range of customer specified ratios to achieve the desired properties. Mitsubishi is the primary source of the BT resins.

Microwave applications often demand the use of very low loss materials made from fluorinated thermoplastic glass fabric laminates. New materials are constantly being developed and evaluated. Recently Hercules Corp. announced a silicon-carbon thermoset polymer (SYCAR) which is being evaluated for PWBs in conjunction with PolyClad Corp. This material has a T_g of ~180°C, low dielectric constant, and extremely low moisture absorption characteristics [12]. Low moisture absorption aids processing, reduces risk of delamination during high temperature assembly operations and reduces risk of conductor corrosion and polymer microcracking. Polymers typically adsorb and desorb water as a function of the local relative humidity, and processing environment. This dependency upon the relative humidity of the environment causes variations in the dielectric constant and in the loss characteristics (both increase with increasing water absorption) as polymer water content adjusts dynamically to the ambient. A material which shows low moisture sensitivity provide a more stable transmission line environment with much less variation of line capacitance and characteristic impedance than other polymers. General Electric [13] recently has come out with a high T_g epoxy (olefin-modified) which is cost competitive with conventional low T_g FR-4 epoxies.

5.3.3 Reinforcements for Dielectric Layers

FR-4 grade dielectric materials use woven fiberglass fabrics for reinforcement. The most commonly used glass is E glass, the designation applied to the specific silicate glass composition with a dielectric constant of 6.3 and a thermal expansion of 5 ppm/°C. Other glasses are available, but are not commonly used due to higher cost. An example of an alternative glass fabric is S-glass, which has a dielectric constant of 5.3 and a CTE of 2.3 ppm/°C [14].

Reinforcing fabrics are available in a wide number of weaves, in which the following parameters are varied: fiber diameter, fibers per thread, number of threads per inch, tightness of the weave and type of weave. The ratio of polymer resin to fiber volume in the dielectric layers and the fabric characteristics determine the properties of the core and prepreg layers. The fabric reinforced dielectric is a nonhomogeneous material. During lamination, the prepreg resin is displaced as it flows to accommodate the raised copper pattern to which it is bonding. This results in resin-rich and glass-rich regions. The resin flow and the nonhomogeneity of the dielectric actually create microregions of variable dielectric constant in the dielectric. The knuckles (xy-crossovers) of the fabric create high spots which actually may press against the copper pattern regions after lamination. At very high frequency, the local variations in the dielectric medium surrounding the conductor transmission line contribute to pulse dispersion and scattering. This effect usually is not too significant until very high frequencies, where the signal wavelengths approach the dimensions of the fabric features.

5.3.4 Copper Conductors

Copper foils used for PCBs all have high copper purities (> 99%). The foils are made usually by rolling, but also can be made by electrodeposition (ED). The microstructural morphology of rolled copper is affected by the noncopper additives, the heat treatment of the original large cross section copper feedstock, the rolling conditions and the annealing process. The microstructure of the ED copper is controlled by the chemistry of the plating bath and the plating conditions (temperature, current density, etc.). The microstructure and copper chemistry affect preprocessing steps prior to lamination to prepreg and core layers, and subsequent micro etching processes prior to board lamination. The etching behavior of the copper is also greatly dependent upon the microstructure and foil chemistry.

Copper deposited in PTHs and on the surfaces of the PWB, develops properties highly dependent upon the plating bath chemistry and plating conditions. Probably the most critical concern is the thermomechanical properties of the PTH copper. Due to the high z-axis thermal expansion mismatch of most PWBs and PTH copper, it is desirable to use high elongation, high ductility and high tensile strength copper. Proper plating bath chemistry (usually proprietary to vendors) and good process controls are needed to attain the desired set of properties. The strain on the copper increases with board thickness, so this type of copper is critical for thick boards, high aspect ratio PTHs and when materials with very high z-axis expansions are being used. The reduction in the quantity of PTHs in aboard through the use of very low aspect ratio blind and buried vias reduces the risk of copper failure.

5.4 FLEXIBLE ("FLEX") CIRCUITS

Flex circuit MCM-Ls are constructed with dielectric and conductor planes which permit the finished circuit to retain mechanical flexibility after circuit finishing. The typical dielectrics are unreinforced polymer films, most commonly polyimide or polyester. The conductors typically are copper, aluminum or polymer thick films. The conductor-dielectric layers are formed usually by bonding the metal foil to the dielectric with an adhesive, or by electroplating copper to an adhesion layer deposited on a dielectric surface. The degree of flexibility is a function of the physical properties of the dielectric, copper and adhesive (if used), the dielectric and conductor plane thicknesses, and the number of circuit layers.

Multilayer structures can be formed using adhesive layers between conductor planes. Vias, PTHs and laminates are fabricated using processes similar to those used for rigid PWBs. SHELDAHL Corp. uses a special film to laminate layers together. The film has an adhesive matrix with properly sized and spaced solder balls dispersed throughout. During lamination, the solder particles melt as the adhesive is softening. The solder contacts and wets any capture pads about PTHs that were formed previously and about plated blind and buried vias, effecting plane to plane conductivity. This eliminates the need for drilling and plating holes for inner layer sublaminates, providing a savings in processing costs [15]. It is likely that unreinforced, flexible dielectric layers, or multilayer structures, will begin to be laminated into the interior of rigid laminate structures to exploit the lower dielectric constant and the << 100 µm thicknesses commonly available with these materials. When flex circuits are laminated to the surface of rigid boards, leaving flex circuit sections extending beyond the edge of the rigid component, the resulting module is commonly called rigid flex. Rigid flex segments may be needed for attachment of massive components and connectors.

Very low cost, low circuit density flex circuits are made with polyester films. These types of films are used commonly with polymer thick film conductors (carbon-loaded, or carbon/silver-loaded epoxy) to make keyboard, phone circuitry and low cost consumer electronics. CASIO Corp., for example, uses polyester films as the basis of many of their circuits. Calculator circuits often are fabricated from polyester films, with aluminum foil laminated to a surface. This surface is etched to form the circuitry; the opposing side is patterned with conductive epoxy. Side to side connections are made by filling vias with the conductive epoxy. Additional conductive layer patterns are formed on top of the aluminum or base conductive polymer layer by patterning a dielectric layer between the conductor planes. Components are attached typically using anisotropic conductive adhesives (z-axis conductive only) because standard polyethylene terephthalate (polyester) films begin to decompose at ~180°C, which is less than the eutectic temperature of Sn/Pb solder. For calculator circuits, TABed LCD driver and microprocessor die are attached to form the keyboard circuitry. A laminated, multilayer version of this type of MCM-L forms a circuit which represents the lowest cost for a low interconnect density MCM.

5.4.1 Flex Circuit and Connector Integration

PWB and MCM assemblies often require the use of connector components to connect between the circuit and the remainder of the system. The use of a connector involves forming two electrical connection interfaces. The PWB or MCM must connect with the connector, and the connector must couple to the rest of the system. If the original circuit was a flex circuit, the functionality of the connector can be integrated, in many instances, with the circuit itself. This may reduce part cost, increase ease of assembly and improve reliability due to the elimination of a connection interface. Since a transmission line experiences some level of an impedance discontinuity at the interface with the connector, elimination of this interface provides an enhanced level of signal integrity in the flex. The flex circuit also allows the incorporation of an integral cable(s) into the MCM-L. This cable can twist, turn and bend, permitting system design options not available with other types of substrates.

5.5 ADVANCED MCM-L MATERIAL AND PROCESS TECHNOLOGY

The standard materials discussed in previous sections may present some limitations when attempts are made to use these materials to build advanced high performance MCM-L systems. These issues are summarized below:

1. Standard photolithographic patterning and copper etching limits the minimum line widths and spacings.

2. The low glass transition temperature (T_g) of FR-4 limits high temperature assembly processes and results in high thermal expansion values both below and above the T_g.

3. High density MCMs may utilize large die with low stress compliance interconnections (straight TAB or flip chip). In these cases, the relatively high thermal expansion of FR-4 may not be acceptable.

4. The dielectric constant of FR-4 may not be suitable for high frequency, low power systems.

5. The minimum thickness of standard FR-4 dielectric (~ 100 μm) may not permit attainment of the desired characteristic impedance value if very fine line technology is used. The minimum thickness also may impede attaining thin, low mass modules for portable electronic systems.

6. The need to drill PTHs greatly reduces available signal routing channels in a module. The large diameter of even the smallest drilled vias still negatively impacts routing.

Advanced MCM-L technologies which address these, and other, problems are discussed in this section.

5.5.1 Integral Termination Resistor Technology

High frequency transmission line interconnects often require terminations to reduce reflections due to characteristic impedance mismatches. Resistors are most commonly used for line termination, although active termination with diodes has been used. If the interconnect is highly lossy (DC resistance is high), it may be possible to use the net itself for series damping of reflections. These lossy nets cause significant rise time degradation and signal attenuation, introducing RC time delay effects. None of these characteristics are beneficial to interconnect performance. In most instances, resistors are used in series or parallel termination schemes. It is possible to integrate resistors into laminate structures using two or three processes.

Nickel alloy films can be formed on core layers used for standard PWBs. The film is then photo patterned and etched to form the desired resistive element. This layer is laminated into the interior of the PWB. The normal processing of drilling and hole plating interconnects the resistor to the desired net and terminating voltage plane. This process is licensed from OHMEGA Corp. and is utilized by a large number of PWB vendors. Since ECL series termination uses a high value pull down resistor and a resistor in series with the driver and the load, this structure is more difficult to construct than a parallel termination structure where a single resistor is close to and in parallel with the receiver.

Polymer thick film resistors also are being used to form integral resistors in PWBs. In this case, internal layers are screen printed with resistor paste, and laminated into the stackup. This provides another method that is potentially lower cost [16].

In prototype structures, thin film resistors have been formed on the surface of high temperature PWBs, permitting high density termination resistor arrays on the MCM-L surface. The composition of the possible resistors may vary from titanium nitride, to nichrome, etc. The surface real estate is very valuable, and is frequently better used for device attachment, so the internal resistor schemes often have the most appeal.

5.5.2 Additive Conductor Processing

The conductor processing described above is referred to as subtractive processing since the circuits are defined by etching, or removing, copper material (as discussed in Chapters 2 and 7). Another method of circuit definition, additive processing (AP), is becoming more popular as the size of circuit features is decreased and the mechanical requirements of conductors are getting more severe, due to greater board thicknesses, and board bending requirements. With additive processing, the surface of the PWB is activated with a plating catalyst, commonly palladium, coated with a photoresist developed to expose the areas where plating is desired, and then immersed in the plating solution for the controlled deposition of copper. Copper thickness control is easier than standard electrolytic Cu plating, and the plating in high aspect ratio PTHs is more uniform. The properties of AP copper are dependent upon the plating bath chemistry and deposition process controls, but the copper usually possesses higher ductility, elongation and tensile strength than circuits fabricated from laminated copper foils, using standard subtractive processing. The full additive copper, deposited on PTHs, possesses greater resistance to barrel cracking than electrolytically deposited copper (due to the higher ductility provided by AP copper), but AP copper usually shows lower adhesion strength than subtractive copper. This can be a problem with surface bond pads, which are exposed to considerable thermomechanical abuse during the device bonding and attachment process and during repair cycles.

One of the major attributes of AP copper is the generally accepted ability for this process to produce fine lines more readily than subtractive processing. The subtractive process uses photolithographic and etching processes which remove the copper exposed through the photoresist, so there is a risk of uneven etching of the resulting conductor line. The enchant tends to etch under the protective resist as it etches through the copper foil (undercutting). With a narrow conductor line width, the undercut from both sides of the conductor can approach the width of the line. The use of thin copper foils (for example one-half ounce copper, 18 μm thick) reduces the undercutting problem. Unless the undercutting is well controlled, the etching of 50 μm lines, may force the use of a line spacing of 75 - 100 μm. AP conductors tends to be square or rectangular in cross section, since AP circuits are defined by plating up through photoimaged resists, which typically possess straight sidewalls. Since there is no undercutting, 50 μm lines may be processed with finer spacings than typically possible with subtractive processing, offering the potential for AP lines at < 50 μm. Since the patterned PR is controlling the line width, it also is possible to plate close to the full thickness of the PR to achieve 50 μm lines, which are thicker than those typically possible with subtractive processing. With standard AP, the PR usually is removed after plating the conductors.

The use of a similar process allows the construction of a multilayer additive circuit. In this case, the initial PR layer is not removed after plating the initial AP circuit layer. A secondary PR layer is applied, covering the initial AP copper pattern. This PR layer is imaged to permit via etching, and the via post then is plated. The secondary PR is activated with a plating catalyst. A ternary PR layer is applied, imaged and developed, exposing the underlying activated, secondary polymer layer, in the regions in which the deposition of the AP circuitry is desired. AP of the Cu circuitry follows. The use of this type of additive processing is being developed by a number of companies, including IBM, Ibiden and Litronics.

The unfilled liquid polymers typically have been photosensitive acrylate-modified epoxies, which are not as environmentally stable as unmodified epoxies (lower solvent resistance, moisture resistance and glass transition temperature). The dielectric and PR films usually are applied in liquid form, but unfilled polymer films also are being used. Unmodified epoxies (liquids or B-staged films) are being evaluated for this type of surface circuit fabrication. Photoimaged and developed via or circuit formation is not normally possible with these unmodified epoxy films.

These additive surface layers provide very high routing density. One of the chief factors limiting the extension of classical MCM-L to very high performance, high density circuits is the inability to achieve the fanout from a dense array of interconnections without greatly increasing the effective die connection footprint. The ability to form conductor and dielectric features with this "near thin film" process technology offers the greatest extendability to future systems for MCM-L. This near thin film technology includes AP, use of unreinforced dielectrics (liquids or films) and actual thin film patterning of buried conductor planes, constructed upon an MCM-L base. This process offers the lowest cost option for MCM-D type of circuitry, due to the use of the MCM-L substrate base and the unique processes used to create the surface pattern layer(s). This type of substrate offers the promise of great extendability for MCM-L into very high performance systems, and represents the future of MCM-L.

5.5.3 Advanced Reinforcement, Rigidifying Materials

The use of standard E-glass fabric as the reinforcement material in FR-4 boards has a number of limitations. The dielectric constant and thermal expansion are higher than desired. The use of a woven fabric creates a core layer surface texture which may impede the formation of very fine conductor lines and spaces. The use of novel fabric structures and materials, particulate filler replacement of fabrics, and the use of porous polymer film reinforcements provide significant improvements. In another approach, laminates are built upon preformed rigid substrates. Organic laminates constructed on cofired multilayer ceramic or MCM-L bases also are possibilities. This progression of improved materials and processes is leading toward the merger of laminate (MCM-L) and thin film (MCM-D) technologies.

Advanced Dielectric Filler Materials and Structures

Compositech Corp. has developed nonwoven, glass reinforced dielectric materials with various polymer matrices. The elimination of fabric knuckles in boards using these dielectrics for all layers, or only surface layers, can provide very smooth surfaces, aiding in the formation of fine geometry surface conductor features and the assembly of very fine pitch surface mount devices [17].

Gore Associates Inc. manufactures a highly porous, expanded poly-tetra-fluoroethylene (PTFE) film for use as the reinforcement for dielectric layers. This material is a high temperature thermoplastic possessing very low dielectric constants, ε_r, of 2.2 and 1.2, respectively, in bulk and expanded states. It can be coated or partially impregnated with a variety of polymers to form dielectric layers that can be laminated. A wide range of thermomechanical and electrical properties results. The film can be manufactured in thin layers, permitting construction of 25 - 50 μm dielectric layers. When processed with a low dielectric constant (2.9), and high glass transition temperature, cyanate ester polymer, a high performance material designated Gore-Ply, is formed. The Gore-Ply™ (ε_r = 2.7) layers are used as pre-pregs to laminate Gore-Clad layers. These are comprised of a polyimide film core, with copper foil laminated to the polyimide using a material similar to very thin Gore-Ply layers. The Gore-Clad layers serve a role similar to the fully cured core layers in conventional PWBs. Entire boards can be constructed with this material, or it may be used only in critical planes [18].

Rogers Corp. makes a material designated RO-2800, which is a PTFE material incorporating filler materials that are a blend of particulate crystalline and glassy ceramic materials. A dielectric with low, nearly isotropic thermal expansion, and a low dielectric constant (ε_r = 2.8) is the result. Since this material is a thermoplastic, the designation of prepreg and core layers has no meaning. Copper-clad layers are processed in typical fashion, and laminated at temperatures higher than those used for thermoset PWB materials. Since the dielectric reinforcement is not continuous in the xy-plane, processes which guarantee that excessive lateral flow does not occur must be used [19].

Fused silica (silicon dioxide, SiO_2) fabrics also have been used when its low thermal expansion (~ 0.5 ppm/°C, 0°C - 100°C) and lower dielectric constant (ε_r = 3.8) are desirable. The resulting dielectrics have CTEs of 6 - 12 ppm/°C, with low to normal resin contents. The materials with a CTE of 6 - 7 ppm/°C are not readily laminated due to the low resin content. The hardness of fused silica creates dielectric layers which are harder to drill, causing increased drill wear and breakage. Fused silica fabric has limited availability and is more difficult to manufacture due to the very high temperatures needed to form the fibers, and the very high viscosity of fused silica during the fiber drawing process [20], [14].

Organic aramid fiber fabrics also are used for MCM-C. The fabric is similar to those used in the fabrication of tires and bullet proof vests. These fibers are exceptionally strong, and tough, but also contribute to high drill wear and breakage. Adhesion of the polymer matrix material to the aramid fibers has been a problem. Adhesion has been improved in recent years, but may still pose a concern with some material systems and applications due to the high moisture absorption shown by aramid fibers. Poor adhesion in any laminate can lead to separation at the interface of the polymer and the fibers, and polymer microcracking, due to thermal or mechanical stress cycling. The low density, and negative thermal expansion of aramid allows the formation of low weight, low CTE PWBs. PWBs containing Aramid also are considerably more expensive than standard FR-4 boards but in a pinch, they may stop a bullet!

Non-woven, random orientation, aramid fiber fabric for PWB reinforcement also is available. This material also provides smooth surfaces, and essentially provides isotropic properties in the xy-plane. The non-periodic nature of the fiber overlap and orientation allows smooth surfaces after lamination. Of course, as with the woven aramid fabric, a low xy-thermal expansion results.

Graphite, or carbon fibers, also are used to construct reinforcing fabrics for use in the construction of very stiff and thin boards, possessing a low thermal expansion and improved thermal conductivity. Uses of this type of fabric have been limited to aerospace or military applications.

Rigid MCM-L Bases and Cores

Ibiden Corp. developed CERACOM™ substrate technology, which uses a porous ceramic substrate as the base of a single sided substrate, or as the core of a two sided substrate. The preferred ceramic is based upon cordierite, a magnesium-alumino-silicate crystalline phase, possessing a low thermal expansion, a low dielectric constant, a high thermal conductivity (relative to standard laminate structures) and a high modulus of elasticity (relative to standard laminates). Laminate structures for the CERACOM technology are prepared by lamination of prepreg layers on the epoxy impregnated porous cordierite. Conductor layers can be processed with standard subtractive and additive procedures. Structures that interface with flip chip solder bump arrays on 250 μm pitch are readily fabricated with fine pitch additive plating of conductors. PTHs are drilled with the same equipment used with standard boards. The hardness of the ceramic material reduces drill life. The thermal expansion of the resulting CERACOM module increases with the number of laminate layers, and expansions of 5 - 7 ppm/°C are typical for structures with three conductor layers on each side of the substrate. This structure is likely to have a conductor layer formed on each surface of the cordierite, and two additional conductor planes formed over the base conductor plane on each side of the core. The middle plane on each side can be a power or ground plane, allowing the construction of embedded microstrip transmission lines on all signal interconnect layers. This type of substrate has been used for flip chip connections on a 250 μm chip array I/O pitch. The low thermal expansion of this MCM-L substrate offers great promise for flip chip modules, particularly when large die are used [21]-[22].

The use of a base substrate for the construction of a laminate, as exemplified by CERACOM, may represent an emerging generation of MCM-L substrates. In the case of CERACOM, the cordierite ceramic core possesses no interconnect functionality, except as provided by the PTHs. There is no reason why a low CTE multilayer ceramic base could not be used in the construction of an MCM-L substrate. In this case, the base would possess a number of interconnect planes which interconnect to the organic dielectric and copper layers laminated to the surface of the bases. The interconnection would be affected by use of PTHs (in CERACOM™), adhesive contacting pads on the ceramic surface or by the use of thin film metallurgy to connect from the surface pad to the laminate via sidewall to the surface contact pad. A low CTE metal base also could be used for the construction of an MCM-L laminate stackup.

In another approach to reducing the thermal expansion of the MCM-L substrate, low CTE metal foil layers are laminated into the PWB stackup. Materials typically used are copper clad Invar (Cu/Invar/Cu), and copper clad molybdenum (Cu/Mo/Cu). The CTE of the metal foils is a function of the cross sectional ratio of Cu to Invar, and of Cu to Mo. Boards with a CTE of approximately 6 - 7 ppm/°C are attainable using these metals. During processing, oversize drills are used to drill some of the holes in the metal plane. These holes are filled with an insulative epoxy plug to prevent PTH sidewall metal from shorting to the heavy metal planes. Subsequently, the laminated substrate is drilled through the via plugs and in locations not previously drilled where electrical connections to the plane are to be made. Typical hole processing and plating follows. This technology was developed for surface mount board technology for military programs, where low CTE boards were needed to permit the use of leadless ceramic chip carriers. The low CTE requirement extends to the use of large flip chip and flip TAB die, where little or no stress compliance is built into the connection.

5.6 IC CONNECTION AND REPAIR: LAMINATE IMPLICATIONS

IC connection techniques have great influence in the selection process of the laminate technology. The common connection methods are wire bonding (WB), tape automated bonding (TAB) and flip chip (solder bumped die connections). Each of the bonding processes requires a set of joining process specifications and board specifications. The chip joining processes are described in detail in Chapter 9. The joining process specifications designate the temperature, time, pressure, atmosphere, and chemical exposure to which the die and the MCM-L substrate are exposed during and after the joining process. The board specifications must meet the requirements of the joining process, including the minimum line width and spacing for the surface contact pads, the glass transition temperature of the board (or more specifically, the board properties at the joining temperature), the maximum temperature exposure the board can tolerate, chemical durability and metallurgical compatibility with the joining materials. Many MCM-L applications require that the module system be repairable, which involves separation of the die's electrical connections, die removal, die bond pad preparation, die attach site preparation, die reattachment and the re-bond of the

I/O connections. The joining process and rework procedures subject the module to considerable thermal, mechanical and chemical abuse. The substrate and the surface contact pads must withstand this treatment without structural degradation or without reduction of surface metal adhesion.

5.6.1 Wire Bonding to MCM-L Substrates

Both aluminum and gold wires are used to connect die to MCM-L substrates. Gold wires are used in most consumer electronic chip-on-board MCM-L applications, where nonhermetic, glob-top encapsulation is widely practiced. In both cases, the deposition of a barrier layer of 5 μm of low stress Ni over the copper conductors is recommended. The final plating is typically a soft 99.99+% purity electrolytic gold film, measuring 0.5 - 0.75 μm in thickness. Some wire bonding to electroless Au plating with a thickness of approximately 0.25 μm is underway in Japan.

The pressure to reduce the cost of electronics is causing an increase in the use of lower cost FR-4 boards for Al wire bonded MCM-L applications. FR-4 materials are well suited to Al wedge bonding since the required bonding temperatures are << 100°C. The level of leachable ionics in glob-top materials has been greatly reduced in the past few years, largely eliminating the major cause of reliability concerns with the use of aluminum wires with glop-topping. Gold wire bonding requires a temperature of approximately 150°C, which is higher than the T_g of FR-4 (~125°C); therefore, reliability concerns often preclude the use of FR-4 for Au wire bonded die. If the wire bond count is high, the substrate needs to be held at the wire bond temperature for several minutes (assuming a bond rate of 2 to 8 wires per minute). At temperatures greater than the board T_g, the polymer matrix softens and the board undergoes an increase in thermal expansion. In this softened state, the board material can transfer the bond pressure from the bond point into the surrounding material via viscoelastic deformation of the underlying material. This can degrade the adhesion of the bond pad to the MCM-L dielectric material. The ultrasonic energy applied to the bond point also is attenuated quite readily above the T_g. This is due to the significant reduction in elastic modulus which occurs above the T_g. The likely result is inadequate bond strength and degradation of the bond pad to board adhesion. Therefore, a substrate material with a T_g greater than the necessary Au bonding temperature is desirable. Wire bonding processes are covered in detail in Section 9.3.

Module rework is another major consideration. The substrate, assembly process and materials must be selected to permit the removal of defective die. The use of a reworkable die attach adhesive is required. If a module test indicates a defective die, the wires are pulled or cut by hand under a microscope, and the die is removed at temperatures high enough to soften the adhesive. The reattachment and rebonding of die require that the substrate structure be strong enough to withstand the necessary processes. The bond pad adhesion must not be degraded by the repair processes. The bond pads also must be designed large enough to permit one or more reworks, since it may be impossible to achieve a good bond on the same site from which a wire was removed. Some very low T_g epoxies and thermoplastic materials do not need much site rework since they soften so well with applied heat.

5.6.2 TAB Bonding on MCM-L Substrates

Tape bonded die can be attached to a substrate using either thermodes (solder reflow or conductive adhesives) or single point bonding. Au to Au single point bonding is most common, but single point solder reflow can be performed at a much slower bond rate. A challenge for MCM-L is the etching of the fine bond pads often required, especially for in-line TAB (no fanout from the die to the bond pad). MCM-L boards with TAB bond pad pitches of 100 μm are available. TAB bonding is discussed in detail in Section 9.4.

Other MCM-L challenges involve the stresses the TAB outer lead bonding (OLB) process exerts on the bond pads. In this process, the substrate is heated to 100° - 200°C, the hot thermode (> 250°C) presses on each lead. Also, potentially corrosive fluxes are used, the residues of which usually are cleaned away. Improper board materials and OLB processes can induce mechanical and thermal stresses which can char the board, induce internal delamination or weaken bond pad adhesion. The rework of a TAB site also is a relatively complex issue. If the die back (or the die face in the case of flip TAB connections) is bonded adhesively to the substrate, this adhesive must be softened simultaneously with solder at the OLB sites. At this point, the die and the TAB leadframe are pulled from the PWB. The bond sites often are cleaned of residual solder, especially if the bond pad surfaces are rough or if Au content in the solder exceeds 3 - 4%. It also may be required to add solder to the pads before a replacement die is bonded. The die attach site is prepared as previously described.

5.6.3 Flip Chip Bonding on MCM-L Substrates

Flip chip bonding classically has taken the form defined by IBM, wherein a refractory lead-tin solder (historically 95/5, but other metallurgies are used) is used to bump the die. The bumped die are connected to ceramic substrates at temperatures exceeding 300°C. With MCM-L, this connection process is not possible because of the high temperatures. An alternative process is required, which permits joining at a temperature suited to organic boards. The general flip chip process is discussed in detail in Sections 9.5 and 9.6.

The process being pursued by Motorola and IBM involves surface preparation of the MCM-L which allows a lower temperature Pb/Sn solder join between the substrate and the die bumps, allowing the use of FR-4 modules. To reduce the incidence of temperature cycling-induced failures in a flip chip connection, the gap between the die face and the substrate is filled with a specially formulated polymeric material. This material becomes the major load bearing member, leading to a significant enhancement in temperature cycling life. Rework and die replacement can be done only prior to performing the underfill. This process is referred to as direct chip attach [23]. There are numerous other means to bump die and to achieve an connection to MCM-L. Sections 9.5 and 9.6 provide more information.

5.7 VERSIONS OF ASSEMBLED MCM-L SYSTEMS

MCM-L assemblies take many forms. In previous sections, the wide variety of substrates was discussed. In this section, some of the possible MCM-L configurations are presented. Several of these forms are generically unique to MCM-L technology, while others are formed with any MCM technology. The forms contrast mainly in the details of I/O configuration (edge versus area), the die sealing and lidding approach, type(s) of die connection technologies used and single or dual sided assembly.

5.7.1 Substrate I/O Configurations

Perimeter I/O
Perimeter lead configurations are discussed in Chapter 9. MCM-L uses several methods of perimeter lead configurations. In its simplest form, the lead frame can be soldered to surface pads, similar to brazed MCM-C assemblies. If the MCM-L substrate perimeter is defined by a cut through a row of PTHs, which connect to the bond pads, added lead adhesion strength can be achieved by solder fillet wetting of both the underside of the lead and the sidewall of the sectioned PTH. This is common practice for leaded, fine pitch, cofired ceramic packages.

Figure 5-3 is a prototype MCM-L from Motorola exemplifying this type of lead frame attachment. It consists of a 32 mm sq. BT glass substrate with 100 I/Os on 0.8 mm pitch, a back side bonded lead frame heat spreader, one digital signal processor (DSP) die, three fast static random access memories (FSRAMs), a number of resistors and capacitors, black glob-top encapsulant and a white

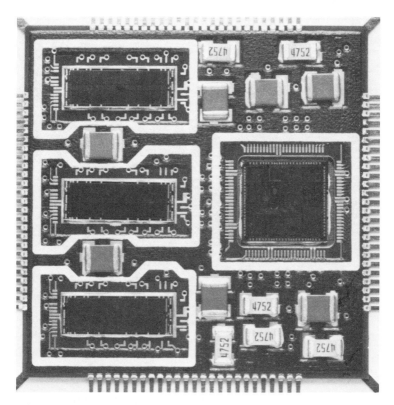

Figure 5-3 Prototype DSP/FSRAM module. Features: four conductor planes, BT glass dielectric, $Z_0 = 60 \ \Omega$, 40 MHz operation, 100 I.O, 0.8 mm pitch, glob-topped. (Courtesy of Motorola.)

epoxy flow dam. Glob-top encapsulation provides good mechanical and environmental protection for the die, but the specific requirements are not well defined. The thermal expansion, modulus of elasticity, moisture permeability, extractable ion content and wettability and adhesion of the encapsulant to all wetted materials must be carefully evaluated for each material system.

The square DSP die is attached through an opening in the substrate, directly onto the copper heat spreader. The FSRAMs are attached to the substrate surface, on top of PTHs functioning as thermal vias to the heat spreader. Au

Figure 5-4 Module with molded perimeter. Features: Standard PQFP footprint, silicone elastomer encapsulation, high temperature lead-to-PWB connections, 0.65 mm pitch. (Courtesy of Matsushita.)

plated leads are excised and formed into a gull-wing shape, permitting die down assembly to a board and heat sinking away from the board.

Lead adhesion may not be adequate with this type of lead frame attachment. Matsushita Corp. molds a frame around the perimeter of a similar substrate, effectively encapsulating the board perimeter and the bond pad region (see Figure 5-4). Due to the temperatures involved in transfer molding, eutectic Pb/Sn solder joining is not used, and a joining operation with higher temperature stability is practiced. Hestia Technology, Inc. uses an adhesive to bond a preformed frame to the substrate with perimeter soldered leadframes, embedding the leads between the substrate and the frame. In other instances, a high rheology adhesive material is dispensed along the edge of the substrate, covering the ends of the leads and the bond pads, effectively encapsulating the lead tips. A version of this method was used on the Motorola MCM-L shown in Figure 5-3. In all cases, these molded, preformed and dispensed frames also act as dams to constrain the flow of material used to coat or pot the substrate components, if the frames are formed on the die side of the substrate. These frames also can act as a base against which lids to the MCM-L are attached. The attached die may be sealed by several methods, including potting with silicone gel (as done with the Matsushita module in Figure 5-4), using epoxy or silicone elastomer attachment of a lid, glob-topping with an epoxy compound, lid sealing without an encapsulant, transfer molding of the die side surface (overmolding) or full transfer molding of the entire substrate.

Ibiden Corp. uses a configuration in which the leadframe is laminated directly into the MCM-L substrate. Ibiden calls this substrate technology PACKTHOL™. Die attach cavities are formed in the top and/or bottom portions of the substrate, exposing the copper leadframe plane as the cavity floor. These cavity floor segments can be electrically connected and share common connections with other cavity floor segments. In other instances, the cavity floor is contiguous with several 150 - 200 µm leads, acting as heat pipes, which connect the die to the substrate solder joints. In Figure 5-3, the die attach floor is the opposing surface of the back side copper heat spreader. The presence of the cavities significantly reduces the circuit routing capacity of the substrate. This configuration is not likely to provide a high interconnect density. The die are sealed typically with glob-topping or transfer molding of the substrate. Transfer molding of the substrate provides level of reliability similar to a standard molded quad flat pack (PQFP).

The I/O provided by perimeter leadframes does not provide adequate I/O for small, high I/O density modules. In these cases, very fine pitch flex circuits are used to provide connection between the MCM-L substrate and the board. IBM uses similar flex circuits with their experimental RS-6000 workstation and PS-2 computer modules, which are MCM-D based [24]. This type of flex connection also is used in the Digital Equipment Corp. VAX-9000 MCM-D module.

MCC developed an MCM utilizing both TAB and wire bonding on a single substrate with approximately 1700 internal interconnection nets. The TAB assembly pads are on pitches as fine as 100 µm. The TAB pads are solder plated and the wire bond sites are gold plated. The substrate is designed to be connected to the next level through the use of a perimeter pad connector. The substrate is built with both MCM-D and MCM-L technologies. Both substrate types achieve the same xy-dimensions. The MCM-D substrates are built with five metal layers. The MCM-L substrates eight metal layers and built with 3 mil line and space design rules. The MCM-L substrates are made by Assist Inc. and Diceon Inc., and were priced at levels substantially less than their MCM-D counterparts [25]. This MCM-L is shown in Figure 5-5.

Another very high performance MCM-L, made by Litronics Corp., is shown in Figure 5-6. This module uses TAB for all the die connections. The substrate

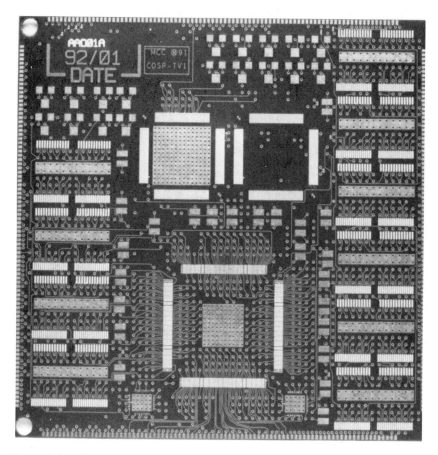

Figure 5-5 Microelectronics and Computer Technology Corporation (MCC) designed MCM for evaluation of a two processor RISC chip set. Features: 2.5" × 2.5", eight copper planes, 10 mil vias on 40 mil grid, 3 mil lines and spaces internal, 2 mil lines and spaces on bond surface, Z_o = 50 Ω, > 65 MHz operation, selective plated Au and Pb/Sn surface, TAB reflow and Au wire bond assembly. (Courtesy of MCC.)

is designed to interconnect with a stack of other similar modules to create a super computer functionality in less than a one liter volume. The stack utilizes Cinch Corp. fuzz button connectors (as discussed in Chapter 10) to connect the substrates.

The lead frame or flex circuits, described above, add significant cost to the MCM-L substrate. The use of a socket also adds considerable cost to the system. For the lowest cost applications, it is desirable to eliminate the lead

Figure 5-6 Single board element of a massively parallel computer. Features: 10 copper planes, polyimide/glass, approximately 3.5" diameter board, three sets of dual stripline signal pairs, 3 mil lines and spaces, buried 8 mil vias, 12 mil plated through-holes, approximately 4000 layer holes, over 1600 nets, all single point Au/Au TAB bonding. (Courtesy of Litronic Industries Incorporated.)

frame or flex. A form of MCM-L which uses perimeter pads, much like those of leadless ceramic chip carriers, is gaining widespread popularity in Japan. This configuration is appropriate for relatively small MCM-Ls, with few die ($\sim\leq 6$) and a low interconnect density, where a two sided substrate (two metal layers), or multilayer board, can be employed for the interconnection. The I/O pads usually are arranged in a single row at the perimeter of the substrate area array I/O pads. Forming a land grid array is also quite feasible. They are connected directly to bottom or top side interconnect nets, with PTHs or with perimeter castellations. The castellations are PTHs bisected by the action of cutting the substrate free from a multi-up panel of parts. The electrical integrity from the top side net to the bottom side pad is maintained after the cutout process. This type of surface mount technology (SMT) MCM-L is soldered, or attached with conductive adhesion, directly to the motherboard. Since the MCM-L substrate and the motherboard usually are made of the same material (FR-4, BT resin/glass etc.), the thermal expansion mismatch between the substrate and the motherboard is minimal. There is a thermal resistance imposed by the air gap between the MCM-L substrate and the motherboard, but for low power modules the temperature difference is small. If a thermally conductive underfill adhesive is applied between the substrate and the motherboard, the thermal stress between the two assembled elements is reduced further and the under fill acts the primary stress bearing member, reducing the stress on the solder joints. This type of substrate is likely to see great use as a vehicle for a few-chips-module (FCM), where the interconnect density is likely to be low enough for the use of a two sided, or a 3 - 4 layer substrate. The FCM arbitrarily is defined as a substrate interconnecting a small number of die (~≤ 6), where the number of die to die connections is less than the number of module I/O [26].

Area Array I/O
MCM-L offers the option of area array I/O (discussed in Chapter 9), which take the form of pins, pads or solder bumps. These I/O contacts usually are dispersed in an array about the bottom side of the substrate. If it is desired to mount the die on the same side of the substrate as the I/Os, then the bond pads must be allocated into rows and columns to leave space for component attach. Figure 5-7 shows a substrate, made by Eastern Electronic Components used in Epson portable computers for graphics control. The substrate uses a connector soldered on two edges of the substrate for connection to the next level. The connector headers are comprised of two rows of pins in a molded plastic bar, forming a type of a perimeter pin grid array (PGA) I/O [27].

Figure 5-8 shows an MCM-L substrate, made by Ibiden Corp. incorporating mixed connection technologies. Three sites for the attachment of wire bonded die are in the central portion of the substrates. The two centermost sites have the PWB material routed out to allow direct die attach to heatsinks attached with adhesive to the back of the module. The third wire bond site permits die attach directly to the PWB surface, where four thermal vias (PTHs plated with thick copper) are formed to reduce thermal resistance through the   board. The presence of the routed openings and the thermal vias reduces board routing efficiency. In cases where routing requirements are very dense, this reduced efficiency forces the addition of more layers to the stackup. The presence of routed openings and dense fields of thermal vias also can cause the average

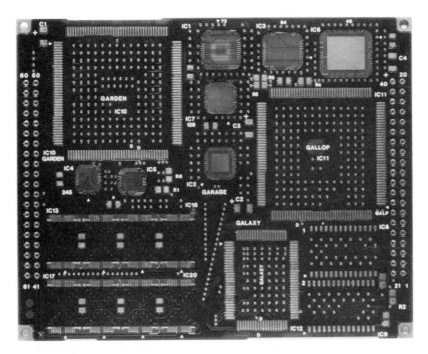

Figure 5-7 Graphics controller MCM from a Seiko/Epson portable computer. Features: selective plated Au and Pb/Sn for TAB and wire bond assembly, glob-top encapsulation, perimeter pinned headers used for I/O. (Courtesy of Eastern Electronic Components Ltd.)

length of the MCM-L nets to increase. This increases signal parasitics and propagation delay, possibly reducing high speed system performance.

The remaining components are packaged surface mount devices. Footprints for three PLCCs can be seen across the top, and in the lower third of the substrate footprints for two SOJ packages also are shown. The opposite side of the substrate supports six other SMT devices. Finally, the module I/O is provided by an array of PTHs along the bottom edge, permitting the insertion of a through-hole connector.

The I/O array in Figure 5-8 populates only one edge of the package. Figure 5-9 shows a Hestia Technologies, Inc. PGA MCM-L made with a polyimide/glass substrate. This particular substrate has no I/O pins in the center of the bottom side, but some other PGAs have fully populated arrays. The I/O arrays also can take the form of pads or solder spheres. I/O arrays can

Figure 5-8 RISC microprocessor system module. Features: routed windows for back side heatsink attachment, wire bonding and SMT, perimeter pinner connector for I/O. (Courtesy of Ibiden Inc.)

be connected to the next level using solder, adhesives or connectors. Connection to motherboards is discussed in detail in Chapter 10.

Motorola Inc. has used solder ball I/O on near die size packages for several years. They are developing a new type of solder ball I/O package called an OMPAC (overmolded pad array carrier). This type of solder ball package connection is called Controlled Collapse Chip Carrier Connection (C5). OMPACs have been discussed as single and multichip packages. The OMPAC is made from a BT glass substrate, with bottom side solder balls attached

Figure 5-9 Pin grid array MCM. Features: Polyimide/glass structure, bottom side chip capacitor attachment orientation, use of a surface ring for encapsulant containment, metal lid. (Courtesy of Hestia Technologies, Inc.)

typically on a 0.040" - 0.060" pitch. After device attach, the substrate is overmolded (transfer molding only on the die side of the package), but alternative sealing processes, such as glob-topping or lidding can be used [28].

Overmolding has been used for other types of MCM-L. Figure 5-10 shows an overmolded module, from Citizen Corp. which interconnects four static random access memories (SRAMs). The mold compound encapsulates the die-side surface, while maintaining interconnect access to the back surface. In this case, the module I/O takes the form of a row of surface mount contact pads along a single edge. The memory module is based upon a very thin, two sided board, providing a very low profile and mass.

Dual Sided MCM-L Assembly
One of the advantages of MCM-L is the ability to use both substrate surfaces for

Figure 5-10 Overmolded SRAM module: Use of two-sided board, SIMM-type I/O format of a row of pads on one edge for use with surface attachment, low profile molding. (Courtesy of Citizen, Inc.)

device connections. There have been instances of PWB and ceramic substrate assembly where both surfaces are utilized, but in almost all cases, the devices are packaged. The MCM design must be configured very carefully to allow use of both sides of the substrate, since the substrate under the site of a die connection may need to be supported mechanically during assembly or provide for heat dissipation while in use. Similar access will be needed to effect repairs to the module in most cases. This dual surface assembly is not usually an option with MCM-D. Silicon substrates with through vias allowing circuit buildup on both surfaces have not been reported to date. In theory, a cofired multilayer ceramic base, containing internal circuitry, could permit the development of thin film interconnections on one surface, while permitting the connection of parts to gold or solder plated pads on the opposite ceramic module surface. If a high density laminate is used as the base for final construction of thin film interconnection circuitry on the top surface, it might be possible to connect bare die on the bottom of the laminate base. The use of both module surfaces implies a perimeter module I/O configuration. Area array I/O, from a field of die which require cooling, may be very difficult to achieve in a product.

Thermal management of any dual sided assembly module may be more complicated. Unless excellent lateral heat spreading is built into the substrate by virtue of the materials used, it may be impossible to cool the bottom side components for a planar mounting on a motherboard. Provisions for heatsinking into the motherboard, or for the insertion of heat pipes between the die and the motherboard, can be implemented, but the series thermal resistance, the complexity, the board area demand, and the cost may preclude this possibility.

If it is not possible to dissipate module heat from the MCM substrate, then heat must be removed directly from the die surface. This is facilitated most readily when the connection method permits die back contact (flip chip or flip TAB), but it also may be accomplished by contact to the junction surface of the wire bonded die. In all three modes of connection, the lid and heatsink(s) must make adhesive and mechanical contact with the back side, or face, of the die. Dual surface MCM-L may be applied in designs where modules are stacked in the z-direction, permitting air to be ducted through the stack, cooling both surfaces simultaneously. Another design where dual surface modules have promise is the SIMM (Single In-line Memory Module) configuration. (See Chapter 10 for a description of the SIMM connection design.) The distinguishing characteristic of SIMM is that all of the module I/O are on one edge, on one or both sides of the module. This type of MCM mounts perpendicular to a motherboard, typically with a socket. A similar ducted and directed air thermal management solution is appropriate here. A schematic of a dual surface MCM-L with flip chip assembly is shown in Figure 5-11.

5.8 MCM-L SYSTEM EXTENDABILITY AND COST ISSUES

The cost of an MCM substrate, for any material technology, is controlled by the yield of good functional product. Due to the complexity of manufacture, a 99% yield from a 40 to 50 step process results in low product yield, increasing costs (as discussed in Chapter 3). The manufacturing base for PWBs is very broad and competition is high, forcing competitive pricing. The base of manufacturers willing to build PWBs with the characteristics necessary for MCM-L is fairly limited at this time, but is growing. Therefore, the pricing for MCM-L substrates continues to be quite reasonable when compared with other MCM technologies.

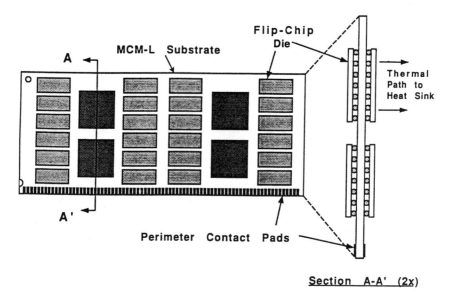

Figure 5-11 Schematic of dual surface MCM-L assembly. Features: SIMM I/O, dual surface assembly, flip chip assembly.

The choice of dielectric material has the greatest effect on the material cost of the substrates. If the MCM-L substrate is made from standard FR-4 material, it is much less expensive than if it were made of polyimide/aramid fabric. However, as the cost of the substrate usually is much less than the cost of mounted chips often it does not make sense to cut corners on substrate material in an effort to minimize expense. The higher price materials may offer higher reliability due to their higher glass transition temperatures (lower CTE in the use temperature range, less copper feature adhesion degradation during high temperature assembly operations etc.). Conversely, it is folly to use a much higher performance board than is needed for an application.

The cost drivers for MCM-L substrates [29] are:

- High layer count
- Small drill diameter
- Tight characteristic impedance tolerance specifications
- Narrow etched feature widths and spacings
- High number of drilled holes

- Nonstandard material selection (composition, thicknesses)
- Plating complexity: more than one metal finish on the surfaces
- Tight thickness specifications
- High plated through hole aspect ratio
- Use of blind and buried vias
- The incorporation of leadframes

High layer count requirements result in a considerable reduction in yield. One approach to solve this problem is the testing of inner layers and inner layer sublaminates prior to committing them to final lamination. Of course, the cost of testing must be balanced against the yield benefit. DEC utilized a version of this approach with the MCM-D used in the VAX-9000 (see Chapter 17).

In some analyses comparing PWB and MCM-D technologies it often is assumed that the former remains static while the latter will grow. Thus, there is some crossover point where MCM-D becomes cost competitive. While it appears that thin film interconnections offer the greatest density (≤ 25 μm feature sizes are now available), it is unclear where the cost crossover point with MCM-L actually occurs. It also is possible that thin film conductor processing may become common on dielectric layers prior to lamination (inner layer construction), or on the surface of a completed base laminate for fanout purposes. It is unclear which arbitrary MCM technology designation would be applied to this type of structure. MCM-L fabricators will continue to improve their technologies and experience economies of scale as the MCM industry accelerates. Therefore, the costs of both technologies should continue to fall in the near future. Since many high volume MCM applications do not require the interconnection density of MCM-D or MCM-C or the hermetic characteristics of MCM-C, it is likely that the demand for classical and emerging MCM-L substrates and assemblies will be quite high over the next several years. This may offer the MCM-L fabricators the opportunity to ride the economy of scale curve before the MCM-D vendors. Initially, this may widen the price disparity between the two technologies.

It also appears that the blending of MCM-L and MCM-D technologies will create a set of substrate technologies which satisfies the very high end of MCM-L applications and the low-to-high end of projected MCM-D applications. This may create a cost arena not currently anticipated by market forecasts, where this hybrid MCM-L/D will even satisfy very high end applications.

References

1   S. Crum, "30 Years of Electronics Technology," *Electr. Packaging & Production*, vol. 31, no. 7, pp. 42-46, July 1991.

2   C. F. Coombs, *Handbook of Printed Circuits*, 3rd edition, New York: McGraw Hill, 1988.

3   R. Clark, *Handbook of Printed Circuit Manufacturing*, New York: Van Nostrand Reinhold, 1985.

4   L. Smith-Vargo, "Give New Strength to Multilayer Bonding," *Electr. Packaging & Production*, vol. 27, no. 2, pp. 52-55, Feb. 1987.

5.   G. Ginsberg, "New Technology for Imaging Circuits on Multilayer Boards," *Electr. Packaging & Production*, supplement: *Fabricating Advanced Printed Circuit Boards*, vol. 30, no. 8, pp. 32-37, May 1990.

6   C. Guiles, "The Importance of Thermomechanical Properties of High Performance PCB Laminate Materials," *Electronics Manufacturing*, vol. 34, no. 10, pp. 41-44, Oct. 1988.

7   R. Schaefer, R. Ellett, "Higher Productivity from Higher-Stack Drilling," *PC FAB*, vol. 14, no. 2, pp. 62-74, Feb. 1991.

8   J. Murray, "Small-Hole Technology: It Takes More than a Drilling Machine," *PC FAB*, vol. 14, no. 2, pp. 52-59, Feb. 1991.

9   K. Nargi, "Using Subtractive Methods For Multilayer Board Fabrication," *Electronics Manufacturing*, vol. 33, no. 11, pp. 9-11, Nov. 1987.

10   "Wet Processing, Drilling and Inspection For MLBs," *Electr. Packaging & Production*, supplement: *Fabricating Advanced Printed Circuit Boards*, vol. 30, no. 8, pp. 41-45, May 1990.

11   D. Jacobus, K. Ferris, "Surface Mount Boards With Blind and Buried Vias," *Surface Mount Techn.*, vol. 3, no. 2, pp. 44-49, Feb. 1989.

12   Private communication.

13   Private communication.

14   C. Guiles, "High Performance Materials For Printed Wiring Boards," *Proc. NEPCON West*, (Anaheim CA), pp. 349-359, 1989.

15   K. Gilleo, "A New Multilayer Circuit Board Based on Anisotropicity," *Proc. NEPCON West*, (Anaheim CA), pp. 8-31, 1990.

16   C. Martin, "Buried Planar Resistor Technology," *Electr. Packaging & Production*, vol. 31, no. 5, p. 81, May 1991.

17   Private communication.

18   Gore-Ply™, Gore-Clad™ product literature, *W. L. Gore Associates Incorporated*.

19   J. Olenick, *et al.*, "Fluoropolymer Composite Dielectric Substrates and Invisicon," *Proc. ISHM/IEPS Internat. Conf. on Multichip Modules*, (Denver CO), pp. 470-481, 1992.

20   L. Gates, W. Reimmann, "Quartz Fiber in PCBs Improves Temperature Stability," *Electr. Packaging & Production*, vol. 23, no. 5, p. 68-73, May 1983.

21   M. Kato, "Development and Application of MCM-L: Practical Solution for Commercial Type MCM," *Proc. Internat. Conf. on Multichip Modules*, (Denver CO), pp. 485-489, 1992.

22   CERACOM™ product literature, *Ibiden Corporation*.

23   Private communication.

24   W. Pence, D. McQueeney, J. Mosley, "Design of a Silicon-on-Silicon Multi-Chip Module for a High-Performance PS/2 Workstation," *Proc. Multi-Chip Module Conf.*, IEEE, (Santa Cruz CA), pp. 110-113, 1992.

25   Private communication.

26   L. Higgins III, "Perspectives on Multichip Modules: Substrate Alternatives," *Proc. IEEE-MCMC Multi-Chip Module Conf.*, (Santa Cruz CA), pp. 12-15, 1992.

27   K. Sakurai, M. Masuda, "Multichip Modules Assembled Using TAB Technology," *Proc. Internat. Tape Automated Bonding Conf.*, (San Jose CA), pp. 105-115, 1991.

28   B. Freyman, B. Miles, "A DSP-Based Multichip Modules Employing Multilayer Printed Circuit Board Technology," *Proc. Internat. Conf. on Multichip Modules*, (Denver CO), p. 490, 1992.

29   P. Fischer, "A Parallel Approach to High Density," *PC FAB*, vol. 14, no. 11, pp. 34-45 Nov. 1991.

30   G. Geschwind, R. M. Clary, "Multichip Modules: An Overview," *PC FAB*, vol. 14, no. 11, pp. 28-38, Nov. 1990.

6

THICK FILM AND CERAMIC TECHNOLOGIES FOR HYBRID MULTICHIP MODULES

William A. Vitriol

6.1 INTRODUCTION

Ceramic based multichip modules (MCMs) have been used in the electronics industry for over 20 years; however, the term MCM-C has become popular only in the last few years. For the purpose of this chapter, ceramic MCMs are considered as sophisticated extensions of simple hybrid circuits in which bare chips are mounted on substrates and interconnected using screen printed conductor materials. Three ceramic based technologies are used to make the various microelectronic circuits that are considered MCM-C structures. They are thick film multilayer (TFM), high temperature cofired ceramic (HTCC), and low temperature cofired ceramic (LTCC) MCM-C technologies. In fact, standard TFM hybrid circuits may be the oldest type of MCM-C.

6.1.1 Definitions of MCM-C Technologies

Thick Film Technology
Thick film is a technology in which specially developed pastes (also referred to as inks) are deposited and patterned by screen printing onto a ceramic substrate. The paste can be formulated to produce any number of different passive electrical components such as conductors, resistors, inductors and capacitors.

The pastes are applied by a screen printing process. After printing, the pastes are dried at temperatures of 85°C - 150°C to remove low boiling point solvents and then fired at temperatures ranging from 400°C - 1000°C. The substrate in the thick film-based technology is a passive inorganic material used to provide a mechanical platform for the thick film circuit. Typically substrates are made from 96% alumina. For MCM applications multilayer circuits are generally used instead of single layer circuits because of greater density requirements. Figure 6-1 shows an exploded view of a fabrication sequence for a single layer of a TFM circuit. The first conductor layer is printed, dried and fired on the substrate. Next, the dielectric (or insulating) layers are printed, dried and fired. Depending on the materials and complexity of the circuit as many as four dielectric layers may be processed before the next conductor layer. After the dielectric is fired the vias (vertical conductors that provide electrical connection between dielectric layers) are filled, dried and fired. This sequence of operations is repeated until the complete multilayer circuit is produced.

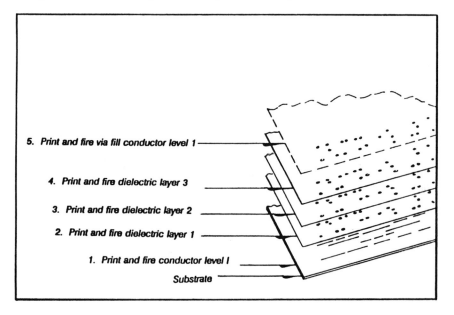

Figure 6-1 Exploded view of a fabrication sequence for single level of a thick film multilayer.

High Temperature Cofired Ceramic (HTCC) Technology

Cofired multilayer ceramic technologies have been developed to improve packaging densities by increasing circuit density, reducing the number of interconnects and by shortening conductor lengths. The building blocks for cofired ceramic structures are green (unfired) sheets of dielectric tape formed from a slurry of alumina powder, small amounts of glass and various organic components using a fabrication process called doctor blading. Doctor blading is a method of forming the slurry into a thin sheet by the use of a knife blade placed over a moving carrier to control slurry thickness. The dried, "green tapes" formed from the slurry, are usually 0.010" thick and have the consistency of fresh unchewed chewing gum!

Vias, or holes, then are formed in each individual sheet of tape. These vias are filled with a specially formulated conductive material to make electrical connection between metal layers. Each sheet then is patterned with a conductor material using screen printing techniques. After all the required sheets of alumina dielectric have been processed, they are stacked, aligned and laminated together using temperature and pressure. Figure 6-2 shows a drawing of the major steps involved in making a cofired part (both LTCC and HTCC). The finished product is formed by firing the laminated module at a high temperature, approximately 1600°C. At high temperatures only refractory metals such as tungsten (W) and molymanganese (Mo-Mn) can be used as conductors and they must be fired in a reducing (hydrogen) atmosphere to keep the conductors from oxidizing. The entire assembly shrinks approximately 17% in the x- and y-directions and 20% in the z-direction. Figure 6-3 shows an exploded view of the individual sheets of tape and is representative of either cofired technology.

Low Temperature Cofired Ceramic (LTCC) Technology

The conductor materials used to make high temperature cofired structures are limited by their low conductivity. Developments have sought to combine the advantages afforded by the cofired process (dielectric tape formation) with the advantages of standard thick film materials (use of metals with higher conductivity). For example, lowering the firing temperature of the dielectric tape to 850°C would make it possible to use thick film metals such as gold, silver and copper. These metals have a much higher conductivity than tungsten or molymanganese, allowing hybrid circuits to be made with faster processing speeds. The processing sequences using HTCC and LTCC technology are shown in Figures 6-4 and 6-5. The initial development in LTCC technology used a commercially available tape system from Dupont called Green Tape™.

Tape Transfer (TTRAN) is a variation of the LTCC process. It combines the serial processing of thick film technology with the use of dielectric tape. A single layer of cast tape replaces multiple screen printings of the thick film

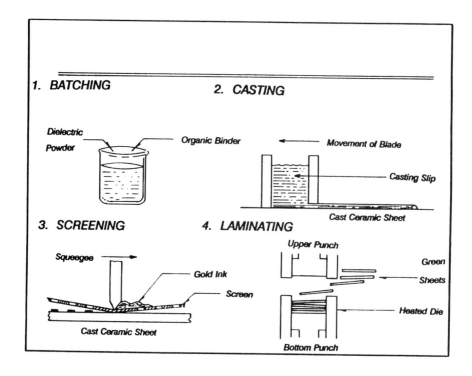

Figure 6-2 Cofired ceramic process steps.

dielectric layers in the TTRAN process [1]. Sheets of dielectric tape, with preformed vias, are registered and laminated to fired 96% alumina substrate at a fixed temperature and pressure. Since the tape adheres to the substrate it does not shrink in the x- or y- direction, but does shrink in the z-direction. Figure 6-6 compares the TTRAN process sequences to thick film processing for a single level of a multilayer circuit. Notice that fewer process steps are required when tape processing is used relative to thick film processing. It takes three or four individually printed and fired layers of thick film dielectric (15 μm - 25 μm thick per layer) to replace a single sheet of 3 mil (75 μm) thick tape laminated to a substrate and fired. The TTRAN process is lower cost and has higher yields relative to conventional TFM circuits [2]. Since it is fired on an alumina substrate, it exhibits superior mechanical and dimensional control compared to LTCC.

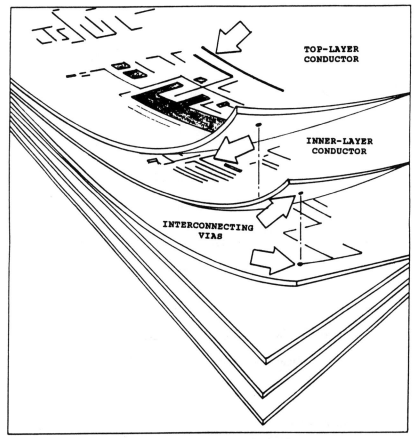

Figure 6-3 Exploded view of cofired ceramic structure.

6.1.2 General Comparison of MCM-C Technologies

The major difference between TFM MCMs and HTCC or LTCC MCMs, other than the fact that a substrate support platform is required for TFM circuits, is that thick film parts are processed serially, layer by layer. Each conductor level of a multilayer circuit may require as many as six printing steps and firing operations per level. This consists of three (sometimes four) individually printed and fired layers of dielectric, two via fills and the conductor. A similar structure

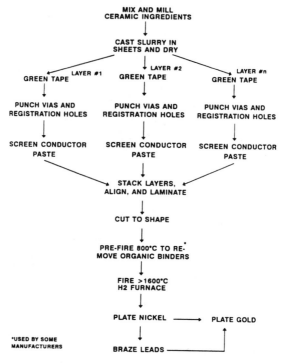

Figure 6-4 Process sequence for HTCCs.

using the cofired technology requires only a single step to fire all the laminated layers and then possibly three or four post firing process steps. The three technologies are described in Table 6-1.

In the cofired ceramic technologies, the substrate and the package form a single monolithic structure. The HTCC or LTCC package is usually sealed hermetically by attaching a metal lid to a metal seal ring which was previously soldered or brazed to the substrate. In contrast, surface mounted devices on TFM substrates are not hermetically sealed, so the substrate must be mounted into a package or the devices must be passivated in some other manner. Standard thick film sealing techniques can be used to attach the lid. Another possibility is to make an all ceramic package by forming a ceramic sidewall, integral to the package, during fabrication of the structure. Again, a metal lid can be attached to the ceramic sidewall to provide hermeticity. For high density applications, cavities can be manufactured in both the high and low temperature

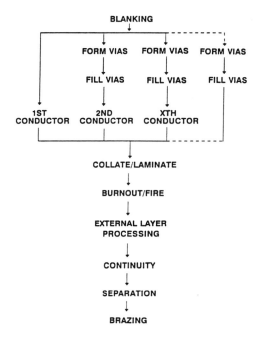

Figure 6-5 Process sequence for LTCCs.

ceramic structures to accommodate active and passive devices. When cavities are used, multiple bond shelves or ledges are used to fanout fine pitch bond fingers (pads) so that wire or tape automated bonds (TAB) can be made easily and reliably. Figure 6-7 shows a cofired package demonstrating multiple bond shelves. Tables 6-2a and 6-2b compare the three MCM-C technologies in terms of the technology options and the mechanical and electrical characteristics.

These tables show that LTCC MCM-Cs have a number of advantages over HTCC modules. One advantage is the use of high conductivity metals such as gold, silver or copper for the LTCC technology in place of the tungsten metal required for HTCC. Gold and silver do not oxidize to any great degree so they do not have to be plated to protect the metal. This is not the case when copper is used. Another big advantage of LTCC is that the basic ceramic substrate can be modified to provide any number of combinations of dielectric materials with different electrical and physical properties. For example, the dielectric constant can be varied between 4 and 400 and the coefficient of thermal expansion (CTE) can be designed to match silicon, gallium arsenide, or alumina. Finally, the same

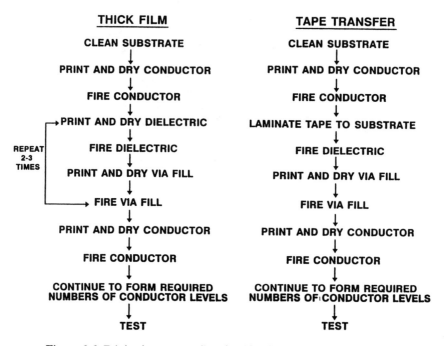

THICK FILM

CLEAN SUBSTRATE
↓
PRINT AND DRY CONDUCTOR
↓
FIRE CONDUCTOR
↓
PRINT AND DRY DIELECTRIC
↓
REPEAT 2-3 TIMES — FIRE DIELECTRIC
↓
PRINT AND DRY VIA FILL
↓
FIRE VIA FILL
↓
PRINT AND DRY CONDUCTOR
↓
FIRE CONDUCTOR
↓
CONTINUE TO FORM REQUIRED NUMBERS OF CONDUCTOR LEVELS
↓
TEST

TAPE TRANSFER

CLEAN SUBSTRATE
↓
PRINT AND DRY CONDUCTOR
↓
FIRE CONDUCTOR
↓
LAMINATE TAPE TO SUBSTRATE
↓
FIRE DIELECTRIC
↓
PRINT AND DRY VIA FILL
↓
FIRE VIA FILL
↓
PRINT AND DRY CONDUCTOR
↓
FIRE CONDUCTOR
↓
CONTINUE TO FORM REQUIRED NUMBERS OF CONDUCTOR LEVELS
↓
TEST

Figure 6-6 Fabrication process flow for thick film multilayer circuit: Comparison between thick film and tape transfer.

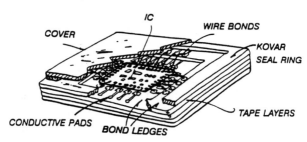

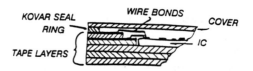

Figure 6-7 MCM-C package demonstrating the use of multiple bonding ledges.

Table 6-1 Comparison of Characteristic Differences of MCM-C Technologies.

MULTILAYER THICK FILM	LTCC	HTCC
• Requires a mechanical support base	• Monolithic structure	
• Serial processing	• Parallel processing	
• 6 prints per layer (3 dielectric, 2 via fill, 1 conductor)	• 2 prints per layer (1 via fill, 1 conductor)	
• 60 sequential firing steps for up to 10 layers	• 1 cofire step plus three postfire steps	
• Must protect devices	• Hermetic package	
	• Fires at 850°C	• Fires at 1600°C
	• Uses standard thick film conductors—gold and silver	• Requires refractory metals—tungsten and molymanganese
	• Fires in air	• Fires in hydrogen
	• Standard thick film processing time	• Long processing time to remove organics
	• Tailorable dielectrics	• Uses alumina dielectric

thick film resistor and capacitor materials developed for standard thick film circuits can be adapted to LTCC circuits. Both buried and surface resistors and capacitors have been designed into LTCC parts [3].

Tables 6-2a and 6-2b also show some advantages in the use of high temperature cofired MCMs. HTCC is a mature technology; its materials and processes are better understood. Also, alumina has a much higher mechanical strength than LTCC dielectric materials, making packages stronger and more durable. Finally, the thermal conductivity of alumina is almost 20 times higher than that of the basic LTCC dielectric materials.

Table 6-2a   Feature Benefit Matrix for Multilayer Thick Film, HTCC and LTCC: Technology Characteristics. (Ranking order; 1 = best.)

Feature	Thick Film	HTCC	LTCC	Benefits
1. Mature materials and processes	1	1	3	Higher yields
2. Shorter proof of design cycles	3	3	1	Faster response to design changes
3. Lower nonrecurring expense (NRE)	1	3	1	Lower prototype costs
4. High print resolution	3	2	1	High frequency application
5. Unlimited number of layers	3	1	1	Highest possible circuit density
6. Postfiring process	1	3	1	Conventional thick film material
7. Wide processing options	2	3	1	Greatest design flexibility
8. Brazing	N/A	1	2**	Network and package processes
9. Tight process controls	2	3	1	Improved overall yield
10. Accommodates chip wire and surface mount assembly	1	1	1	Design flexibility
11. Cavities	3	1	1	Multiple bonding shelves

**   Low temperature braze.

Table 6-2b Feature Benefit Matrix for Multilayer Thick Film, HTCC and LTCC: Mechanical and Electrical Characteristics. (Ranking order; 1 = best.)

Feature	Thick Film	HTCC	LTCC	Benefit
1. Camber	2	2	1	Improved wire bond assembly yield
2. Surface roughness	3	2	1	Better high frequency performance
3. Top layer dimensional stability	1	3	2	Improved wire bond, assembly yield stability
4. CTE matched to alumina or silicon	2**	3	1	Capability of assembly
5. Thermal conductivity	2	1	3***	Good thermal characteristics
6. Hermeticity	2	1	1	Development of packages
7. High conductivity metallization	1	3	1	Smaller line and space designs
8. Low K values	1	3	1	Improved high frequency performance
9. Excellent dielectric control	3	1	1	More consistent electrical performance
10. High mechanical strength	2	1	3	More rugged package

**   Varies with substrate material.
*** Thermal via designs enhance thermal conductivity.

6.2 MATERIAL CONSIDERATIONS

6.2.1   Thick Film Technology

Substrates
A substrate is the base onto which all thick film circuits are printed. It provides mechanical support, electrical insulation and assists in dissipating heat. Substrates can be obtained in various shapes, sizes and thicknesses. The shape

Table 6-3 Common Substrate Materials by Application.

SUBSTRATE TYPE	APPLICATION

Alumina
  99+%
 	
  96% (glass)
 	
  85% - 94% (glass)
 	
  40% - 60% (glass)
	Thick/thin film circuits
	Microwave integrated circuits
	
	Thick film single/multilayer circuits
	
	Refractory thick film/cofired circuits
	Thick film/LTCC

Beryllia	Thick/thin film power circuits
Aluminum Nitride	Thick/thin film power circuits Refractory thick film circuits

can be pressed and fired to meet a variety of configurations or laser scribed to size specification. Sizes may range from 1" × 0.33" up to 6" × 13" and from 0.010" - 0.060" thick. For special applications thicknesses as great as 0.125" can be made. Because of the variation in available sizes, circuits can be processed in a multiple-up format, scribed and singulated (separated) in subsequent operations to improve handling efficiencies. Substrates may be laser drilled to permit use of the other side of the substrate, with connections made using through-hole technology. Substrates and their applications typically are defined by the amount of glass contained in the ceramic body. For example, a 99+% alumina substrate is used for thin film circuits and for high frequency applications, while HTCC bodies range from 85 - 94% alumina, and LTCC formulations from 40 - 60% alumina. The majority of thick film circuits are printed on 96% alumina because it provides the best compromise of cost, strength, resistor stability and conductor adhesion. Table 6-3 summarizes commonly used substrate materials by their principal application. Table 6-4 lists the advantages and disadvantages of these same substrate types.

Most substrates used in the electronics industry are alumina; however several other ceramic materials have found limited application based on their specific material properties. Beryllia (BeO), one such material, has one of the highest thermal conductivities of any ceramic material and is used most widely for high power handling applications. A major consideration in using BeO is its toxicity. Special handling is required for its use in certain processing steps. Also, thick film materials specifically formulated to match the properties of the ceramic must be used, since adhesion promoters used with alumina do not work with BeO.

Table 6-4 Advantages and Disadvantages of Substrates for Thick Film Circuits.

SUBSTRATE TYPE	ADVANTAGES	DISADVANTAGES
99+% Alumina	Low microwave loss High strength Good stability for resistors Good thermal conductivity No alkali migration—low loss Good surface finish	High cost Low conductor adhesion
96% Alumina	Lower cost High strength Good stability for resistors Good conductor adhesion	Higher microwave loss
Beryllia	Highest thermal conductivity	High cost Requires special thick film materials Toxic
Aluminum Nitride	High thermal conductivity Low CTE	High cost Requires special thick film materials

More recently another ceramic material, aluminum nitride (AlN), has come into wide use. AlN has some very unique properties that make its use highly desirable [4]. AlN, like BeO, has a very high thermal conductivity. Since the thermal conductivity of AlN increases with temperature, unlike BeO which decreases with temperature, AlN is a better material for high power applications. Additionally, AlN has none of the toxicity problems associated with BeO. Another desirable property of AlN is its close coefficient of thermal expansion (CTE) match to silicon. This means that it is more compatible with the large silicon devices that are being designed into MCMs. A great deal of research and development time has been spent to attach AlN to both MCM-D (deposited) substrates and LTCC MCM-C structures thus taking advantage of the thermal benefits of AlN [5]. Table 6-5 shows some of the electrical and mechanical properties of substrate materials, including AlN, used in hybrid applications.

Conductors
Thick film conductors are made from mixtures of metal powders, small amounts of vitreous or oxide powders and an organic phase that determines the printing

Table 6-5 Properties of Commonly Used Ceramic Substrate Materials.

PROPERTIES	MATERIALS				
	AlN	BeO	Al$_2$O$_3$	Al$_2$O$_3$	
Purity	98 - 99.8%	99.5%	96%	99.5%	
Color	Dark gray to translucent	White	White	White	
Density (g/cc)	3.255	3.01	3.70	3.87	
Specific Heat (cal/g°K)	0.177				
Thermal Conductivity (W/m-k) at 25°C	80 - 260	250 - 260	20 - 35	20 - 35	
CTE (10^{-6}/°C) (25°C - 400°C)	4.4	9.0	7.1	7.6	
Dielectric Constant at 1 MHz	8.8 - 8.9	6.5	9.5	9.9	
Dielectric Loss Tangent at 1 MHz	0.0007 - 0.0020	0.0004	0.0004	0.0001	
Resistivity (Ω-cm)	> 10^{13}	10^{15}	10^{14}	10^{14}	
Dielectric Strength (kV/mm)	10 - 14	9.5	26	24	
Flexural Strength (N/mm^2)	280 - 320	170 - 240	250	400	

Note: CTE (25°C-400°C) for Si = 3.35 x 10^{-6}/°C, for GaAs = 5.73 x 10^{-6}/°C.

characteristics of the ink or the paste. In most cases the conductor inks are designed to be fired between 850°C and 950°C. The glass or oxide constituents of the ink are developed so that they react with the substrate and form a glass or oxide bond. This reaction provides the adhesion mechanism critical to the performance and reliability of the electrical circuit. The main function of a thick film conductor is to provide the electrical connections between different points in the circuit. Conductors also are used as resistor terminations to provide low contact resistance, as capacitor electrodes, as attachment pads for active and passive devices and as low value resistors.

Most commonly used thick film conductors fall into two categories: noble metals that can be fired in air, and non-noble metals that must be fired in an inert or reducing atmosphere. The noble metals include platinum, palladium, silver and gold, as well as binary and ternary mixtures of these metals. The non-noble metals include copper, aluminum and nickel. Of the non-noble metals only copper plays a major role in thick film technology.

Critical parameters characterizing a conductor are its conductivity, solder wettability, solder leach resistance (the ability of a conductor to not dissolve in the solder), adhesion and wire bondability. Good adhesion to the substrate is the one property required of all thick film conductors. A great deal of time and effort has gone into improving adhesion.

There are three basic mechanisms of adhesion between the substrate and the metal: glass bonding (fritted), oxide bonding (fritless) and mixed bonding. Fritted conductor systems add small amounts of low melting point glass powder to the metal. The glass powder, typically a lead borosilicate, melts during the firing process and wets both the particles of metal and the substrate. If too much glass is added, excess glass can float to the surface and cause problems with soldering and wire bonding. If not enough glass is used, adhesion is reduced.

Fritless adhesion mechanisms were developed originally for use with air fired gold formulations but were quickly adapted for use in nitrogen atmospheres and with copper conductors. In fritless conductors the glass is replaced by small amounts of copper oxide. During firing the copper oxide reacts with the alumina substrate to form a spinel (copper aluminate) structure that creates a chemical bond between the substrate and the metal. Too much oxide in the formulation will result in an oxide layer on the surface of the copper, reducing solderability and wire bondability. Too little copper oxide reduces adhesion. In many cases manufacturers plate the top copper layer with gold to eliminate problems associated with oxide formation.

A mixed bonded system produces the best of both worlds. By using both oxide and glass, the amount of each can be reduced, minimizing the deleterious effect of too much of either additive. Typically, better adhesion and improved wire bond strength are obtained with mixed bonded conductors.

Table 6-6 Comparison of Key Conductor Metals for Thick Film Circuits.

PROPERTIES	SILVER	COPPER	GOLD
Conductivity	Excellent	Excellent	Excellent
Resistor Compatibility	Good	Good	Excellent
Print Definition	Good	Moderate	Excellent
Solder Leach Resistance	Poor	Excellent	Poor
Wire Bondability	Good*	Good*	Excellent
Processing	Very good	Moderate	Excellent
Cost	Very low	Low	High

* Aluminum wire

Three metals (gold, silver and copper) have found widespread use in thick film conductor inks used to make hybrid circuits. In addition, metals such as platinum and palladium have been added to the gold and silver, in either binary or ternary combinations, to enhance specific properties such as solderability and solder leach resistance. The properties of these materials are compared in Table 6-6.

Gold is useful because it is an inert metal and does not oxidize during subsequent processing steps. It is also highly conductive, and easily wire bondable. Active devices can be attached readily and reliably to gold pads on the substrate. The biggest drawback to the use of gold is its cost. Gold is approximately 10 to 20 times more expensive than silver or copper. Additionally, gold leaches (dissolves) readily with commonly used solder and is not typically used where soldering is required. If soldering is necessary, gold/tin solder can be used. Gold conductors find their greatest application where high reliability is an issue, particularly for military, medical and space related requirements. The addition of varying amounts of platinum and palladium to gold will improve the adhesion and the solder leach resistance at the expense of conductivity and wire bondability.

Silver is used in place of gold specifically where cost is an issue. Silver has a higher conductivity than gold and has lower losses at microwave frequencies. The biggest drawback to the use of pure silver in hybrid circuits is that it will migrate in the presence of voltage, humidity and ionic contaminants if not passivated properly. This electromigration of silver has precluded its use in many military and high reliability applications. Silver over time will tarnish making for more difficult soldering and bonding. While pure silver conductors leach readily, they have a significantly higher leach resistance to commonly used solders such as 62 Sn/36 Pb/2 Ag than do gold conductors. Components can be solder attached reliably to silver conductors for many applications. The addition of platinum or palladium improves solder leach resistance and inhibits migration, but lowers conductivity and increases cost. Silver bearing thick film conductors are used mostly for commercial circuits and as capacitor and resistor terminations. If silver is buried within a hermetic material, so that moisture or ionic contaminants are not present, no migration can take place [6-7].

A copper thick film technology was developed because of the high cost of gold and the reliability related issues of silver. Even though copper will oxidize, it is migration resistant and, therefore, acceptable for military and high reliability applications. Some advantages of copper are its high electrical conductivity, its low loss at microwave frequencies and its low cost. Values for all of these properties fall between those of gold and silver. The overall fabrication cost of copper circuits includes the additional cost of nitrogen gas. As mentioned previously, copper will oxidize, so the surface needs to be protected, usually by plating with gold. The plating process also adds labor costs to the product.

Dielectrics

A dielectric material is a material that does not conduct electricity. In the strictest sense of the word, a dielectric refers to a material used in a capacitor. An insulator is a material that provides electrical isolation between conductive circuit elements. In the microelectronics industry the two terms have become synonymous. Thick film dielectric inks are mixtures of dielectric powders and an organic vehicle. The dielectric portion of the ink may be a low melting point glass, a mixture of glass and ceramic powders or a crystallizable material. As with thick film conductors, the organic portion of the ink defines the rheological characteristics and determines how well the ink prints. There are five major applications or uses for dielectric inks in a hybrid circuit: crossovers, insulating layers in multilayer circuits, capacitors, encapsulation, and hermetic seals. The critical properties that define whether a dielectric material can be used in a given application are the temperature and frequency and voltage response of its dielectric constant, dissipation factor, insulation resistance and dielectric breakdown voltage.

Crossover Dielectrics. The crossover dielectric is a low dielectric constant insulating material used in single layer thick film hybrid circuits to separate two conductor traces or lines. A crossover configuration allows the production of the simplest type of three dimensional circuit. Crossover dielectrics are designed to separate conductors occupying the same location on the substrate, effectively increasing the available circuit area. The first crossover dielectrics were made from low melting point glasses such as lead aluminum borosilicates and were designed to be compatible with conductors fired around 850°C. Additionally these materials had to match the CTE of the substrate so that stresses were minimized during the processing of the hybrid circuit. It is important to be aware of some of the problems that can occur when using glass crossovers in a hybrid circuit. For example, depending on the number of times the glass is fired, reactions between the glass and the conductor can result in short circuits. Excessive flow of the glass can cause swimming (movement) of the conductor, possibly distorting the printed pattern.

Because of these problems, glass-ceramic formulations were developed to replace glass dielectrics. Glass-ceramic materials are made as a glass, printed and then crystallized during the firing process. The crystalline phase that forms is dispersed uniformly throughout the glassy matrix so that the resultant material has very uniform properties. The crystalline material has a higher softening point than the parent glass and is stable under multiple refirings as long as it is not fired higher than the original processing temperature. Another major advantage is that glass-ceramic crossovers have less tendency to form pinholes, minimizing short circuits. The glass-ceramic materials do not react with conductor materials to the same extent as their glassy counterparts. Other terms synonymous with glass-ceramics are crystallizable and devitrifying glasses.

Multilayer Dielectrics Once a hybrid circuit with crossovers has been designed and built there is no more area available on the substrate to increase the circuit density. Circuit density may be increased by increasing the size of the substrate, by printing on two sides of the substrate or by building a multilayer circuit. Since the latter has the fewest limitations, circuit designers have long been designing their hybrid circuits with multiple layers. By definition a TFM circuit is one containing multiple internal conductor traces separated by insulating dielectric layers.

Electrical connections between layers are made through vias in the dielectric filled with conductor. The same technical issues (of using glass-ceramic materials with the crossover dielectrics) presented even more serious challenges in making multilayer circuits. Material requirements for multilayer dielectrics are demanding due to the multiple firing steps required in processing. Glass-ceramics make the best multilayer dielectrics. As mentioned in the previous section, these stable materials limit problems such as blistering, conductor reactions and shorting. One common problem with devitrifying multilayer dielectrics is the incompletion of the crystallization process during the first firing step. Crystallization will continue through additional firings and the CTE of the layers may change making it difficult to control the CTE of the composite.

A glass-ceramic material developed for a multilayer circuit application must have certain properties that are not required when used as a crossover material. For example, because the overall thickness of the multilayer dielectric can be quite large and screen printed to cover the whole substrate, the dielectric material must have a CTE closely matched to the alumina substrate. Mismatch produces excessive substrate camber, making it difficult to print additional dielectric layers, to attach devices to the surface and to mount the substrate to a heatsink or to a package base. The via holes must remain open to allow electrical connections between the layers. Finally, the fired dielectric material should have a very good surface finish so that fine line patterning of the conductors is possible.

Capacitor Dielectrics As mentioned above, dielectrics also can be used as capacitors, but not so much in MCM applications. Typical materials are combinations of barium titanate ($BaTiO_3$) and glass to lower the firing temperature. Thick film capacitors for resistor capacitor (RC) networks or as blocking or bypass capacitors have dielectric constant values ranging from 20 - 2000. Typically they are porous and have poor mechanical integrity, but serve the purpose for which they were developed. Low firing (850°C - 1000°C) crystallizable capacitor formulations, called relaxor dielectrics, have been developed, which have very high dielectric constants (> 20,000).

Resistors Thick film resistors have been used in the microelectronics industry for over 40 years in applications ranging from simple resistor networks to printed resistors as part of very complex multilayer circuits. Most of the modern thick film resistor inks are combinations of heavy metal (lead or bismuth) ruthenate compositions or ruthenium dioxide dispersed in a glassy matrix. As with other thick film inks, these active ingredients are mixed with an organic vehicle that defines the printing characteristics of the ink. Resistor inks are usually formulated in a series of decade values ranging from 1 $\Omega/\square$ to 1 M$\Omega/\square$. The individual decade value inks, called end members, are blended in varying proportions to obtain resistivity values between those specified by the end members. The actual resistivity range is determined by the type of resistive element in the ink and by the ratio of this material to the glassy phase. When fired, (typically to 850°C) the glass softens and wets the resistive material and substrate. Typical electrical parameters of resistor inks are given in Table 6-7.

Resistor values, for design purposes, are determined by the equation shown in Figure 6-8. The key terms in the equation are sheet resistivity ($\Omega/\square$) and number of squares (defined as the resistor length divided by the resistor width). The resistance can be determined from this equation if the geometry of the resistor and the value of the ink are known. Thickness of the printed resistor is

Table 6-7 Typical Fired Resistor Properties[1].

Sheet Resistivity[2] ($\Omega/\square$)	TCR[2] (ppm/°C)	Short Term Overload Voltage[3] (V/mm)	Standard Working Voltage[4] (V/mm)	Maximum Rated Power Dissipation[5] (mW/mm²)	Blendable Series
10 ±20%	0 ±100	45	1.8	320	A
100 ±20%	0 ±100	22	9	810	A
1k ±20%	0 ±100	62	25	625	A
3k ±20%	0 ±100	77	30	300	A
3k ±20%	0 ±100	90	36	430	B
10k ±20%	0 ±100	145	58	340	B
100k ±20%	0 ±100	245	98	96	B
1M ±20%	0 ±100	310	124	15	B

1   Typical resistor properties based on laboratory tests using recommended processing conditions; termination - DuPont Pd/Ag Conductor 8134 prefired over DuPont Dielectric 5704 at 850°C; substrate - 96% alumina; printing - 200 mesh stainless steel screen (8 µm - 12 µm; firing 30 minute cycle to peak temperature of 850°C for 10 minutes.

2   Shipping specifications. Resistor geometry - 1.5 mm × 1.5 mm. Temperature coefficient of resistance - 55°C to +25°C to 125°C.

3   Short term overload voltage - tested under military specifications of MIL-R-83401D, Paragraph 3.15.

4   Standard working voltage - 0.4 × short term overload voltage.

5   Maximum rated power dissipation. $\dfrac{(\text{Standard Working Voltage})^2}{\text{Resistance}}$

another important factor affecting sheet resistivity. Values can be increased or decreased by varying the thickness of the printed resistor.

6.2.2   High Temperature Cofired Ceramic Technology

Dielectric Materials

The common dielectric material used to fabricate HTCC substrates or circuits is alumina with varying amounts of glass. Depending on the application, the percentage of alumina may vary from 88 - 96% but typically falls between 92 - 96%. Because of the maturity of this technology, the properties of both the raw materials and the fired ceramic have been well characterized. Table 6-8 shows

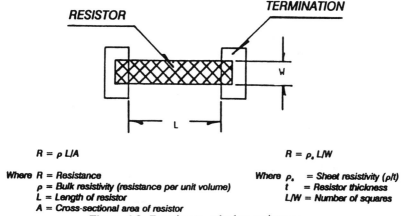

RESISTOR

TERMINATION

$$R = \rho\, L/A$$

$$R = \rho_s\, L/W$$

Where R = Resistance
   ρ = Bulk resistivity (resistance per unit volume)
   L = Length of resistor
   A = Cross-sectional area of resistor

Where ρ_s = Sheet resistivity (ρ/t)
   t   = Resistor thickness
   L/W = Number of squares

Figure 6-8 Equation to calculate resistance.

the key physical and electrical properties of fired alumina ceramic material. The critical parameters to note are flexural strength (which relates to the ability of the ceramic to withstand bending forces) and thermal conductivity. These properties differentiate HTCC dielectric materials from LTCC dielectric materials.

Conductors

All of the processing parameters for HTCC circuits are established by the fact that the alumina dielectric is a refractory material and must be sintered at approximately 1600°C. This limits the selection of conductors to high temperature metals such as tungsten and molymanganese. The thick film inks used for conductor traces and as via fills have been modified to match the sintering temperature and shrinkage rate of the alumina body. Since tungsten and molymanganese both oxidize readily in air, any exposed metal must be plated to protect the conductor surface. Typically two different metals are plated sequentially to achieve the best results. Nickel is plated first for solderability and brazability. Gold is then applied to keep the nickel from oxidizing so that reliable gold wire bonds and soldered leads can be formed. Both electrolytic and electroless nickel plating is used, but gold is usually plated electrolessly. Plating is a process that is not required by the other two MCM-C technologies. Also, the relatively high resistivity of tungsten and molymanganese metals is a limitation on their use for high speed, high frequency applications.

6.2.3   Low Temperature Cofired Ceramic Technology

Dielectric Tape Materials

The driving force in developing a low temperature cofired process was the use

Table 6-8 Physical and Electrical Properties of Ceramic Material for HTCC Multichip Modules.

Composition	92% - 96% alumina
Color Available	White or black
Flexural Strength	55,000 psi (380 N/mm^2)
CTE 25°C - 200°C	6.3 x 10^{-6}/°C
Thermal Conductivity at 100°C	15 W/m-K
Volume Resistivity at 25°C	10^{14} ohm-cm
Dielectric Constant at 25°C	8.9
Hermeticity	1 x 10^{-8} cc/sec

of the high conductivity metals and thick film processing. The alumina dielectric used in the HTCC process was replaced with dielectric materials that fire below 1000°C, specifically around 850°C. Such materials were available from TFM circuit technologies. Standard thick film dielectric formulations in tape form became the basic building block for developing the LTCC technology. The low temperature dielectric allowed standard thick film materials and processes to be used when manufacturing LTCC MCMs.

Two basic types of dielectric materials are used in LTCC fabrication today. These are alumina filled glasses and crystallizable ceramics [8]. In the alumina-filled glass, glass softens and wets the alumina powder during firing, providing a dense hermetic structure. Most of the critical parameters of the fired composition, such as firing temperature, CTE, mechanical strength and thermal conductivity, are dictated by the nature and properties of the glass. Most typically the fired properties are designed to match the CTE of standard alumina ceramics. They are also designed to have dielectric constants from 6 to 9. Another benefit of using alumina filled glasses is that the glass softens during firing, allowing the dielectric to conform to the setter on which it is fired. The finished part can be extremely flat. Alumina filled glass dielectric tape, Green Tape®, is available from Dupont. Other thick film suppliers, such as ESL and EMCA, also manufacture and sell tapes.

Another dielectric formulation is a glass-ceramic or crystallizable material, a magnesium aluminum silicate, called cordierite [8]. As discussed in Section 6.1.2, glass-ceramic materials exhibit many of the same advantages over alumina filled glass type formulations. For example, LTCC modules made from glass-ceramic materials are much less sensitive to the adverse effects of multiple firings. This is advantageous when parts are subject to additional postfiring steps. Also, the ceramic does not soften when fired as much as alumina filled glass parts soften. Glass-ceramic dielectric tapes are available from Ferro and Dupont. Different variations of commonly used LTCC dielectric materials and their electrical and physical properties are shown in Table 6-9.

The most significant feature of the cordierite glass-ceramic system is its stability when fired in nitrogen for use with copper conductors. Another process [9] is the firing of the LTCC/copper laminate in a steam-hydrogen gas mixture instead of in a nitrogen-oxygen gas mixture. The steam-hydrogen firing atmosphere ensures that a proper range of oxygen potentials exists, at the temperatures of interest, to remove completely carbon residue and to allow complete densification of the glass-ceramic without oxidizing the copper.

Conductors
The use of thick film dielectric formulations for the tape allows the use of the high conductivity conductors such as gold, silver and copper to make the LTCC MCMs. Most of the work has been with gold conductors.

Cofired top conductors, inner layer and via fill golds are available which eliminate registration problems associated with nonuniform fired shrinkage. Shrinkage variations are eliminated by printing the top conductor either on an unlaminated sheet of tape or on the laminate prior to firing. The number of processing steps, the process time, and the cost of the finished part are all reduced by cofiring all of the top layer metallizations.

Gold conductors available from commercial suppliers exhibit physical properties that are comparable to their thick film counterparts. Gold materials have been tested and have been shown to exceed Mil Spec requirements for 1.0, 1.5 and 2.0 mil gold wire bond adhesion. These same conductors also have been tested and passed requirements for TAB applications. See the discussion in Section 9.4.

LTCC modules must also interface to other modules. A low temperature brazing capability (the joining of two metals together using a filler material with a lower melting temperature than either of the base metals) has recently been developed using either gold/tin solder or gold/indium solder [10]. Brazing allows reliable attachment of leads and seal rings giving the LTCC technology a packaging capability. Figure 6-9 shows an example of an LTCC module with brazed leads (25 mil pitch) and a brazed Kovar seal ring demonstrating the capability of manufacturing packages that can be sealed hermetically. The part is 2.8" × 2.8", made from 10 layers of 6.5 mil tape. There are over 10,000 6 mil electrical vias, 20 mil thermal vias, 5 mil lines and spaces and ground and power planes.

238 THICK FILM & CERAMIC TECHNOLOGIES FOR HYBRID MCMs

Table 6-9 Physical and Electrical Properties of Common LTCC Dielectric Materials.

Matrix	Fillers	Firing Temp. (°C)	Firing Atm	Dielectric Constant	Tan δ	TCE (ppm)	Shrinkage (%)	Shrinkage Tolerance (%)
BaO-Alumino-Borosilicate	Al_2O_3 Forsterite	850-900	air	5.0-6.5	0.0008-0.002	3.8 - 6.8		
Alumino-Borosilicate	Proprietary	850	air	7.8	0.002	7.9	12	±0.2
Borosilicate	Al_2O_3	900	reducing	5.6		4.0	16	
Pb-Alumino-Borosilicate	Al_2O_3 $CaZrO_3$	850	air	9 - 12	0.001 - 0.003		16	
Cordierite	None	925 - 1050	air/neutral	5.3 - 5.7		2.4 - 5.5		
Spodumene	None	850 - 990	air/neutral	5.0 - 6.5		2.0 - 8.3		
MgO-CaO-Alumino-Borosilicate	Al_2O_3	1000	reducing	7.1	0.0025			
BaO, SiO_2, Al_2O_3, CaO, B_2O_3	None	950 - 1000	reducing	6.1	0.0007	8	13.5	
CaO-Alumino-Borosilicate	Al_2O_3	880	air	7.7	0.0003	5.5	16	
Pb-Borosilicate	1) Al_2O_3 2) Cordierite 3) SiO_2	900 900 900	air air air	7.8 5.0 3.9	0.003 0.005 0.003	4.2 7.9 1.9	13 13.7 13.9	< 0.3 < 0.3 < 0.3

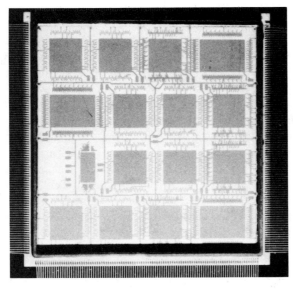

Figure 6-9 LTCC module demonstrating brazed leads and Kovar seal ring.

The full cost advantage of the LTCC technology requires that the expensive gold conductor be replaced with a reliable, lower cost metal. Both copper and silver conductor systems offer possibilities. The silver conductor system is easier to implement, since it can be fired in an air ambient. Formulations of silver and silver/palladium mixtures for inner layers, via fills and top layer conductor formulations compatible with commercially available dielectric tape formulations exist. These systems are used currently to build LTCC modules.

Silver and silver/palladium conductors also can be used on top of the LTCC structure to satisfy wire bonding and soldering needs for commercial applications where humidity is not a major issue. The metal system is not acceptable for military and commercial circuits requiring high reliability. Silver based conductors do not meet the environmental, wire bonding or soldering requirements demanded for those kinds of circuits. Gold conductor must be used where gold wire bonding is needed. For solderability requirements, the best results have been obtained using gold/platinum conductors. A low cost option to gold/platinum is a low firing (< 600°C) copper conductor. This type of conductor will meet all of the soldering and adhesion requirements of high reliability circuits without degrading the properties of air fired conductors and dielectrics when fired in nitrogen [11].

Resistors

Resistors currently play a limited role in LTCC designs. The same standard thick film resistor inks developed to use on top of thick film dielectric can be used with the LTCC dielectric layers because the ceramic has already been fired. One problem is that the irregular fired surface makes it difficult to print the correct thickness of resistor. This problem is magnified when very small resistors (< 0.030" × 0.030") are needed. This nonflat or contoured surface occurs after firing because of differences in compressibility between the printed conductor and the dielectric tape.

A big advantage of the LTCC technology is that buried resistors can be printed and cofired with the rest of the part. This feature allows the MCM manufacturer to reserve more of the surface area of LTCC modules for active devices rather than for passive components. The major drawback to burying resistors is that resistor materials have not been developed fully for this purpose. Reaction with the dielectric tape can occur making it difficult to achieve correct design values. Buried resistors cannot be trimmed easily, so that resistor values must be designed with wide tolerances.

6.3 PROCESSING

6.3.1 Thick Films

Screen Printing

Thick film circuits are manufactured with a series of screen printing passes. A flow diagram for processing a TFM circuit is shown in Figure 6-6. The process starts with the development of a thick film layout from the circuit schematic. Layouts are generated typically on a computer aided design (CAD) system. Each print must have its own artwork. From a CAD system, a photoplotter is used to generate the mylar film or glass mask necessary for each screen, even in multiple-up configurations. The artwork positive (the pattern to be printed is dark) is placed against a photosensitive emulsion coating on the proper mesh size screen and exposed, defining the desired pattern. Once the screen is made, it is placed in a screen printing machine and covered with ink. The pattern is aligned to the substrate or to a previous pattern. A squeegee forces the thick film material through the open areas of the screen onto the substrate. The part is then dried and fired. This printing cycle is repeated for each print pass required.

The printing process requires control of a number of key parameters: print head speed, durometer (hardness) of the squeegee, sharpness of the squeegee, angle of attack and snapoff distance. Other parameters needing control are the screen mesh type, the wire diameter, the number of openings per inch, the emulsion buildup, the emulsion type and the ink viscosity. Screen printing machines may be fed automatically or by hand. Automatic systems may include the use of vision systems to reduce set up time and to optimize printer throughput.

Drying and Firing

After each pattern is printed the ink must be dried to remove organic solvents that give the ink its desired rheological properties. A typical drying temperature is 85°C - 150°C for five to ten minutes. Drying can be done in either a box oven or a continuous belt dryer. The drying process plays a critical role in obtaining the fine line definition required for the circuit. After each printed layer of the multilayer is dried, the part must be fired. Firing is a well defined process and can be accomplished using conventional furnace equipment. A typical thick film furnace profile is shown in Figure 6-10. Most materials are fired to 850°C with a 10 minute dwell time at peak temperature in clean, dry air. Other critical parts of the firing profile are the heating and cooling rates. As the part is heated, the organic binder portion of the ink is burned off, usually by 500°C. From

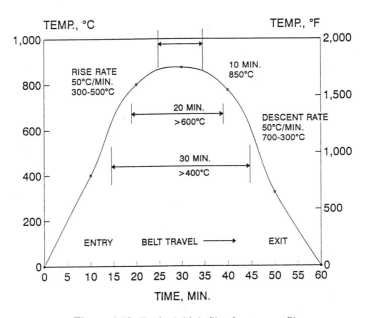

Figure 6-10 Typical thick film furnace profile.

500°C - 850°C any glass or oxide in the paste will soften and react with other elements of the ink or the material on which it is printed (the alumina substrate or a dielectric layer). It is during this part of the profile that sintering and densification occur. The rate of heating is important because it controls how rapidly and how completely the organics are removed. The cooling rate is critical because thermal stresses are set up during cooling. If parts are very large, if the substrate has been laser machined or if there are laser drilled holes, cracking may occur during cooling.

6.3.2 High Temperature Cofired Ceramics

Process Flow

The process sequence for a typical HTCC process is given in Figure 6-4. It begins with the dielectric tape. Initially, each lot of cast tape is inspected for defects such as pits, thin spots and contamination and is characterized prior to use. Tape is cast in thicknesses from 0.004" - 0.025" depending on the electrical and mechanical requirements of the parts being fabricated.

Once the tape is accepted it is blanked (cut into individual sheets) to a standard size using either a blanking die or a computer controlled cutting machine. Each sheet of the tape is punched to form vias, cavities and/or castellations (grooves formed in the edges of the ceramic to aid in solder attach and subsequent inspection). These features can be made using either hard tool punching dies or with computer aided punching machines. After all of the tape machining steps have been completed, the vias are filled with conductor ink and the conductor is printed, both usually with tungsten metal. Typical conductor thicknesses vary from 0.0003" - 0.0005". As with thick film processing each conductor print must be dried prior to performing any additional processing. Next the individual sheets are stacked onto each other and laminated for fixed times, temperatures, and pressures. Laminating pressures can vary from 2000 psi to 4000 psi; laminating temperature varies from room temperature to 100°C depending upon the tape formulation used by each manufacturer. In most cases the completed laminate is scribed and/or cut into individual parts or to the final shape of an individual part.

The individual laminated parts are fired to approximately 1600°C in a reducing atmosphere to sinter the structure and to prevent oxidation of the refractory metals. Firing can be done either in an automatically controlled high temperature box furnace or in large continuous pusher type furnaces. The overall burnout and firing cycle is quite long, 24 to 48 hours. Typical HTCC parts when fired shrink approximately 16% ±0.5 in the x- and y-directions and 20% ±2 in the z-direction.

6.3.3 Low Temperature Cofired Ceramics

Process Flow
Extensive hands-on labor is required for making LTCC circuits, because of the newness of the technology and the lack of high volume applications. Most manufacturers of LTCC products buy the tape formulated and cast by materials suppliers, which shifts the burden of defining good material to the user. Tape is cast on mylar and stored in this fashion. The first step in the manufacturing process is to inspect the tape for defects, such as pits, holes and contamination. Areas that are not usable are marked and removed from the roll. Then the tape is stripped from the mylar and blanked using steel ruled dies or some type of cutter to a specific standard size such as 5" × 5". Because of problems associated with tape relaxation (shrinkage) after removal of the mylar, the tape goes through a low temperature preconditioning bake to stabilize it. Preconditioning allows the tape to be stored without further shrinkage.

The process sequence for LTCC parts is shown in Figure 6-5. It is very similar to the process flow for HTCC parts. Each sheet is rotated 90° to eliminate any possibility of nonuniform firing shrinkage resulting from tape casting direction. Registration holes and vias are formed in the individual sheets of tape. The vias are filled using specially formulated via fill conductor inks designed to shrink at the same rate as the dielectric material when fired. Conductors are printed using standard screen printing equipment and processes. After printing all conductors, any cavities in the tape are cut or nibbled using an automatic punch. Finished sheets are stacked on a laminating plate over tooling pins. The laminating fixture is placed in a conventional uniaxial laminator or in an isostatic laminator. The typical laminating cycle is 3000 psi at 70°C for ten minutes, but pressures can vary from 2000 - 4000 psi and temperatures from 60°C - 80°C. Isostatic lamination is more advantageous than uniaxial lamination because the same uniform pressure is applied to the whole part, and because parts of varying sizes can be laminated simultaneously.

The part can be fired after lamination using either of two slightly different procedures. The first procedure requires a separate furnace to burnout the organics. After burn out the part is fired in a conventional thick film belt furnace using a profile similar to the one shown in Figure 6-10. The second firing process uses a modified thick film furnace so that burnout and firing can be done as a single step. For this firing process there are two separate temperature holds. The first, typically at 500°C, is the binder burnout step. The hold, typically at 850°C, allows the ceramic material to densify. The single step furnace profile is faster than the two step furnace process, but may not be as desirable for large complex modules. Fired parts typically shrink 12% ±0.2 in the x- and y-directions and 17% ±2 in the z-direction. The tighter shrinkage

Table 6-10 Thick Film Multilayer Design Guidelines.

PARAMETERS	STANDARD	CUSTOM
Substrate Size	< 7" x 7"	> 7" x 7"
Substrate Thickness	> 0.025"	< 0.025"; 0.050"
Maximum Number of Layers	< 8	> 8
Dielectric Thickness	> 0.045 μm	> 0.045 μm
Minimum Line Width	0.008"	0.005"
Minimum line Spacing	0.008"	0.005"
Minimum Via Diameter	0.008"	0.005"

tolerance is a major advantage over HTCC technology. The fired dielectric material, whether an alumina filled glass or a glass-ceramic, is very dense and has proven to be hermetic. Hermeticity is the key material property necessary for developing a low cost silver LTCC process. Moisture will not penetrate the package and buried silver will not migrate since the dielectric is hermetic.

6.4 DESIGN RULES

Well defined design rules exist for both TFM and HTCC technologies since these are both mature technologies. Typical design guidelines for these two technologies are given in Tables 6-10 and 6-11. LTCC is a relatively new technology and is evolving so that design rules are still changing. A typical set of LTCC design rules for circuits being built today is listed in Table 6-12.

When comparing the design rules for LTCC circuits with thick film and HTCC, the number of dielectric layers and the pitch of the printed conductors are the areas of difference. Typically the number of layers is greater and the pitch is lower with LTCC designs. LTCC circuits generally require some additional means to handle thermal dissipation. Since the thermal conductivity of the LTCC dielectric material is much lower than the alumina used for thick film and HTCC, thermal vias or slots are designed into these circuits for power management. The addition of thermal vias increases the thermal conductivity of the LTCC module so that it becomes a thermal equivalent to 96% alumina. An example of a part with thermal arrays is shown in Figure 6-11. This module is

Table 6-11 HTCC Design Guidelines.

PARAMETERS	STANDARD	CUSTOM
Size (max.area)	6"× 6"	6" × 6"
Number of Layers	≤ 15	≥ 15
Layer Thickness	0.010" - 0.025"	0.005" - 0.010"
Tolerances Length and Width Thickness Camber	±0.8%, NLT ±0.004 ±10%, NLT ±0.003 0.003"/', NLT ±0.002	±0.5%, NLT ±0.002 ±5%, NLT ±0.001 0.002"/', NLT ±0.001
Conductor Width	0.007" minimum	0.004" minimum
Conductor Spacing	0.007 minimum	0.004" minimum
Via Diameter	0.007 nom.	0.005" nom.
Via Cover Pad	0.014"8	0.006"
Via Center to Center	0.024"	0.010"
Via Center to Center (with one conductor)	0.034"	0.020"
Sheet Resistivity	0.012 Ω/□	0.008 Ω/□
Via Hole Resistance	0.010 Ω/□	0.005 Ω/□

a MIL STD 883 Group D test coupon, built by CTS Microelectronics Division. It is made from nine layers of 6.5 mil tape and has arrays of 20 mil thermal vias stacked from top to bottom beneath the two large die pads for thermal management. When fully assembled there is 60% silicon coverage of the top surface. This test coupon also demonstrates the ability to print 3 mil lines and spaces.

6.5 APPLICATIONS

Current and future applications for MCM-C can be divided into three major categories: military/aerospace, medical, and commercial.

Table 6-12 LTCC Design and Producibility Guidelines.

Parameters	Preferred Manufacturing	Prototype Manufacturing	Future Development
Maximum Substrate Size (fired)	1.5" × 1.5"	3.5" × 3.5"	12" × 12"
Minimum Line Width (cofired)	10 mil	5 mil	2 mil
Minimum Line Space 8(cofired)	10 mil	5 mil	2 mil
Minimum Line Width (postfired)	10 mil	5 mil	2 mil
Minimum Line Space (postfired)	10 mil	5 mil	2 mil
Minimum Via Diameter	8	6	4
Maximum Layers of Stacked Vias	3	4	6
Maximum Thermal Via Diameter: 4.5 mil thick tape ≥4.5 mil thick tape	15 mil 20 mil	20 mil 25 mil	25 mil/slots 25 mil/slots
Maximum Number of Metal Layers	10	15	25+

The initial development of LTCC MCMs, using commercially available material, was a result of government funded programs. This development is discussed further in Chapter 15 of this book as a case study. The first applications were for memory circuits. LTCC modules were fabricated with over 1 megabyte of memory packaged in a ceramic block less than an inch square.

There is still interest in using MCM-C technology in military electronics, particularly in the areas of microwave and millimeter wave packages. The technology driver here is improved performance and lower cost. LTCC could become the dominant military packaging technology as dielectric materials are modified to perform better at higher frequencies and as the less expensive metals become acceptable.

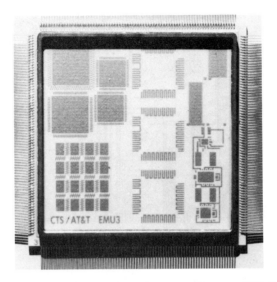

Figure 6-11 LTCC module with thermal arrays. (Courtesy CTS Corporation.)

The medical industry is another industry that has adopted the use of MCM-Cs, in particular, LTCC circuits. These applications require high reliability, reduced weight and smaller packages at lower cost. Large numbers of medical pacemakers, hearing aids and defibrillators are being manufactured using LTCC technology. Figure 6-12 shows a heart pacemaker manufactured for a supplier of medical hybrids. It uses seven layers of 6.5 mil tape and has 900 8 mil electrical vias, 5 mil lines and spaces and is double sided. The hearing aid in Figure 6-13 is manufactured with five layers of 7.5 mil tape and has 10 MΩ, 20% resistors postfired on the surface. This part also is double sided and has a hole punched in it after lamination.

The overall growth in MCM-C usage will be determined by high volume commercial applications, such as occur in the computer, transportation and telecommunications industries. The need for high performance CPUs, microprocessors and memory circuits, combined with the growing popularity of workstations, personal and laptop computers makes the computer industry a critical area of focus. The drive to reduce the size and weight of packages combined with the need for increased electrical performance makes MCM-C packaging technologies ideal for these applications. The electronics requirements of the transportation industry continue to grow becoming a major component in

Figure 6-12 LTCC heart pacemaker. (Courtesy Pacesetter Systems Inc.)

the cost of new cars. Some applications include on board computers and collision avoidance radar, as well as electronic sensors to determine brake wear and fuel levels. The part shown in Figure 6-14 is an aircraft antilock brake hybrid circuit made with eight layers of 6.5 mil tape, 8 mil vias and a stepped through cavity. It contains three decade value 1% postfired resistors. All these applications require low cost, small, rugged electronic packages. MCM-C hybrid circuits meet these requirements. MCM-Cs will play a major role in telecommunications, particularly home satellite receivers and their related electrical circuitry, where the requirements are also for low cost, light weight hybrid circuits such as MCM-Cs.

6.6 FUTURE TRENDS

At a recent conference on MCM technology [12] a great deal of discussion took place on the merits and future of each individual MCM technology (MCM-L, MCM-D and MCM-C). Manufacturers of MCM-L made it quite clear that they have the low cost alternative. They claimed that MCM-L would be the technology leader within the next ten years, if advances are made in fine line processing techniques and in methods to handle the thermal dissipation.

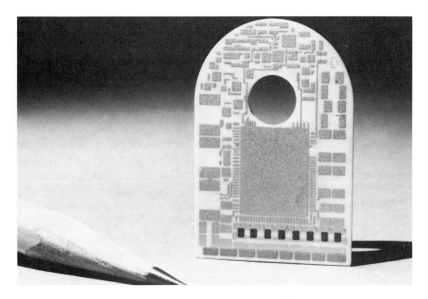

Figure 6-13 LTCC hearing aid. (Courtesy MiniMed Technologies.)

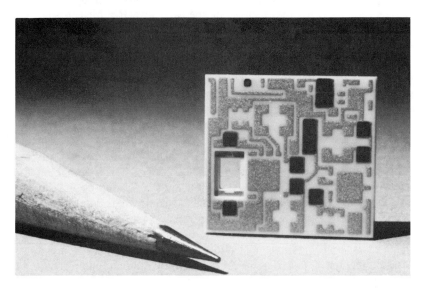

Figure 6-14 LTCC antilock brake substrate. (Courtesy Allied Signal Bendix Engine Controls.)

Engineers and scientists involved with MCM-D argued their case with equal fervor. They pointed to future requirements for high speed, high density circuits and indicated that no other technology could do what MCM-D could do.

Surprisingly no one vouched for the future of MCM-Cs. The consensus, however, was that it is the technology of the present! Specifically, we now consider high frequency applications, new ceramic materials, buried passive components, fine pitch interconnections and thermal management techniques as areas for future development.

Input from military systems' houses and newly funded government programs indicate that MCMs will need to be capable of performing at very high frequencies (in the 40 - 60 GHz range) in future years. LTCC materials are expected to meet this capability. Some new low firing ceramic dielectric formulations already operate at these frequencies with low transmission loss. Silver will be used as an interconnect metal for MCMs. The combined benefits of reduced transmission loss and the lower cost of silver conductor will encourage the replacement of gold. Using lower dielectric constant tape materials will produce a positive benefit for high frequency and high speed requirements. Dielectric materials with dielectric constants as low as 4, and developmental materials having dielectric constants as low as 2.5, will be available for both thick film and LTCC MCMs. Similarly, new high temperature materials for HTCC MCMs, such as mullite, with a CTE similar to silicon and a dielectric constant lower than alumina (6.8), will open application areas for HTCC [13].

Fine line technology is another future trend that will influence the use of MCM-Cs. A great deal of time and money is being spent to reduce the pitch of conductor traces on thick film and cofired ceramic modules. While 5 mil lines and spaces are common now, future goals include 2 mil lines and spaces. One approach is the use of standard screen printing techniques with new fine print conductor pastes and advanced wire materials and emulsions for making the screens. Wires are now available as fine as 0.7 mils in 325 and 400 mesh screens. Screen mesh open areas greater than 50% allow denser lines to be printed. These advanced wires also are much stronger and do not stretch as easily and quickly as standard wire materials. Finally, new emulsion materials allow much sharper line definition than is possible with the older emulsions.

New techniques for attaining top layer fine lines are being investigated and, in some cases, are being implemented into advanced designs. Technologies such as photolithography, sputtering, laser delineation of solid ground planes and ink writing are all techniques for reducing the conductor pitch. The ultimate goal may be to couple thin film MCM-D technology with multilayer thick film and cofired MCM-C technology (MCM-C/D). See the discussion in Chapter 7.

Power management is becoming more important and more difficult to accomplish. It is not high power circuits alone where heat management is critical. As more and more low power devices are packed onto ceramic modules, the overall heat dissipation requirement of these packages becomes an issue. Several future trends in this area are becoming apparent. AlN is an important substrate material for TFM circuits and HTCCs. Development of compatible thick film materials (gold, silver and tungsten) is continuing. The increased availability and increasingly lower cost of AlN has renewed interest in its use. The tape transfer process can be used on AlN substrates. Tapes compatible with AlN material also have been developed. Combining the use of AlN substrate/package base with the ability to make cavities in the tape, so that high power devices can be mounted directly on the substrate, makes this future combination of new technologies quite interesting. AlN can also be coupled to LTCC structures using pressure assisted sintering techniques. The same benefits as with TTRAN on AlN are realized. A more standard approach to power management for LTCC modules has been the use of arrays of thermal vias beneath the high power devices. Such configurations are limited in capability. A better approach is to attach heatsinks (such as AlN, W/Cu and any number of the new thermally conductive composite materials) to the fired module. The heatsink can be soldered or brazed to the bottom of the package and the devices mounted directly on them.

Another possible trend is the use of buried components within the LTCC body. Buried resistors are examples here, but there may be an even greater demand for buried capacitors and inductors.

6.7 ENGINEERING CHOICES

The first decision a circuit designer makes is which of the three major MCM technologies should be chosen (MCM-L, MCM-D or MCM-C). This decision is based generally on three specific areas: familiarity with the technology, technology innovations or drivers and cost.

In many cases the selection is based on familiarity with and availability of a specific technology. For example, people comfortable with or already using organic printed wiring boards (PWBs) would tend towards MCM-L technology since this is what they understand best. Similarly circuit designers working with and knowledgeable about thin film or semiconductor technology would most likely pick MCM-D technology as their first choice. Finally, it is apparent also that practitioners of thick film technology look upon MCM-C technology as being the closest, in materials and processing, to what they do best.

Looking at technology drivers and cost as critical decision makers the choice becomes much more complex. Suppliers of MCM-L claim that their product is less costly to manufacture and is therefore more attractive to use [14].

Unfortunately there are a number of applications where MCM-Ls cannot be used until technological issues such as high density, power management and hermeticity are fully resolved.

The choice between using MCM-D or MCM-C technology is not clear cut since both technologies require different uses of ceramic materials, either as a platform or as the package itself. The determination of the selection is made usually because of circuit density and cost requirements.

Most manufacturers of MCM-D circuits point to three specific advantages of their technology. The first advantage is the ability to achieve finer conductor geometries with thin film processing. The second advantage is the lower dielectric constant of the polyimide dielectric. The final advantage is the higher thermal conductivity of the silicon, alumina or AlN substrate on which the multilayer thin film circuit is processed. The major shortcoming of this process is that it is essentially a serial, batch process; therefore, process times and manufacturing costs are quite high. Unfortunately MCM-D technology does not readily lend itself to high volume production. This may limit the area of potential applications.

Manufacturers of MCM-C circuits will highlight their ability to process larger parts with many more layers than can be made with MCM-D technology. While line density and spacing may not be as fine as that attainable with thin film technology, there are other technology tradeoffs: lower cost, higher volume manufacturing, faster prototype turn around time, tailorable ceramic dielectric properties, the capability to add integral passive components and to seal the package hermetically.

The technology innovations and process changes being made in both the materials and manufacturing processes for MCM-Cs (specifically thick film and LTCC) and MCM-Ds will make it more difficult in the future to determine which of these technologies to choose. Manufacturers of MCM-Ds could build circuits that have tighter pitches (approaching 2 mil lines and spaces rather than micron size). The defect density of the substrate would become less of an issue. Yields go up and costs come down. Manufacturers of thick film and LTCC circuits are reducing the print geometry of their conductor traces. New technology in the area of screen emulsions and wires and the development of fine print conductors allow lines and spaces approaching 2 mils, albeit not yet for high volume manufacturing. Finally, new dielectric materials have been developed with dielectric constants as low as 3.5, making inorganic systems equivalent to their organic counterparts. What this all means is, as the performance gap between MCM-D and thick film and LTCC MCM-C technologies continues to diminish, the deciding factor in choosing one over the other will come down to cost, yield and volume throughput.

What is not so well defined is where LTCC technology should be used instead of the other two MCM-C technologies. Simple, low layer count circuits will be less expensive to build with thick film than with either of the other ceramic technologies. However, the industry appears to be moving away from this concept toward denser MCMs with increasing silicon area demands. As this happens, LTCC circuits with their potentially lower cost and quicker manufacturing turnaround time should dominate, especially as this technology moves up the learning curve.

LTCC circuits, since they use higher conductivity metals, also have a major advantage over HTCC MCMs. There is an increasing interest in MCMs that operate at millimeter wave frequencies (20 - 40 GHz) and higher. HTCC circuits using tungsten metal cannot be used for this application. LTCC technology also has the advantage of being much more flexible than HTCC technology. The potential for dielectric materials with varying dielectric constants and CTE makes it even more attractive.

In general the decision on which one of the three MCM-C technologies to use is still an open issue. Both TFM and HTCC circuits, because of their maturity, appear to be the technology of today. However, the future may belong to LTCC technology. As materials improve, processes become better controlled and the technology matures, the natural advantages of LTCC could predominate and make it the MCM-C of choice. Most likely all of the technologies will continue to be used depending on the particular application required.

Acknowledgments

The author wishes to thank his many colleagues at CTS Microelectronics and CTS Corporate Headquarters for their assistance with figures and tables and Jon Krause (Zenith Microcircuits Corporation) for his many helpful suggestions during a review of this chapter.

References

1   W. A. Vitriol, *et al.*, "Development of a New Tape Dielectric Technology for Thick Film Multilayer Applications," *Proc. Internat. Symp. on Microelect.*, (Dallas TX), pp. 487-495, Oct. 1986.

2   P. Danner, "High Density Ceramic Modules," *Proc. Internat. Conf. on Multichip Modules*, (Denver, CO), pp. 325-328, April 1992.

3   R. G. Pond, *et al.*, "Processing and Reliability of Resistors Incorporated within Low Temperature Cofired Ceramic Structures," *Proc. Internat. Symp. on Microelect.*, (Dallas TX), pp. 461-472, Oct. 1986.

4   D. D. Marchant, T. E. Nemecek, "Aluminum Nitride: Preparation, Processing, and Properties," *Ceramic Substrates and Packages for Electronic Applications, Advances*

in Ceramics, vol. 26, M. F. Yan, *et al.*, eds., American Ceramic Society Publication, pp. 52-81, 1989.

5   K. R. Mikeska, R. H. Jensen, "Pressure Assisted Sintering of Multilayer Packages," *Ceramic Transactions, Materials and Processes for Microelectronic Systems*, vol. 15, K. M. Nair, *et al.*, eds., American Ceramic Society Publication, pp. 29-35, 1990.

6   R. R. Sutherland, I. D. E. Videlo, "A Comparison of the Reliability of Copper and Palladium-silver Thick Film Crossovers," *IEEE Trans. on Components, Hybrids, and Manuf. Tech.*, vol. CHMT-12, no. 4, pp. 676-682, Dec. 1987.

7   S. N. Mesher, "A Comparison of the Reliability of Silver Compatible Multilayer Dielectric Materials," *Proc. Internat. Symp. on Microelect.*, (Minneapolis MN), pp. 1-7, Sept. 1987.

8   J. H. Alexander, *et al.*, "A Low Temperature Cofiring Tape System Based on a Crystallizing Glass," *Proc. Internat. Symp. on Microelect.*, (Orlando FL), pp. 414-425, Oct. 1991.

9   A. H. Kumar, R. R. Tummala, "State-of-the-art, Glass-ceramic/copper, Multilayer Substrate for High Performance Computers", *Internat. J. for Hybrid Microelect.*, vol. 14, no. 4, pp. 137-150, Dec. 1991.

10   T. R. Bloom, "Parameters for Au/In Brazing to Thick Film Au on Green Tape," *Proc. Internat. Symp. on Microelect.*, (Orlando FL), pp. 513-516, Oct. 1991.

11   S. Nishigaki, *et al.*, "LFC-III: A New Low Temperature Multilayered Ceramic Substrate with Au(top)-Ag(internal)-Ti/Mo/Cu(bottom) Conductor System," *Proc. Internat. Symp. on Microelect.*, (Minneapolis MN), pp. 400-407, Sept. 1987.

12   *Proc. Internat. Conf. on Multichip Modules*, (Denver CO), pp. 37, April 1-3, 1992.

13   R. E. Sigliano, K. Gaughan, "Ceramic Material Options for MCMs," *Proc. Internat. Conf. on Multichip Modules*, (Denver CO), pp. 291-299, April 1992.

14   B. Freyman, B. Miles, "A DSP-based MCM Employing Advanced Multilayer Printed Circuit Board Technology," *Proc. Internat. Conf. on Multichip Modules*, (Denver, CO), pp. 490-498, April 1992.

7

THIN FILM MULTILAYER INTERCONNECTION TECHNOLOGIES FOR MULTICHIP MODULES

Ronald J. Jensen

7.1 INTRODUCTION

Historically there has been a wide gap between the feature geometries on integrated circuits (ICs) (typically ~1 μm) and those of IC packages and printed wiring boards (PWBs) (typically 50 - 100 μm). To achieve the ultimate density and speed in electronic systems, this gap must be closed, and it is inevitable that the thin film processes used to fabricate ICs will be required for IC packaging. In recent years, there has been active development of thin film multilayer (TFML) structures to provide high density interconnections between ICs in multichip modules (MCMs). The materials, processes, and designs for TFML interconnections are the focus of this chapter.

The term "thin film" does not refer to a specific range of film thicknesses, but implies the use of IC processes to achieve high density patterns in conductor and dielectric layers, roughly 2 - 25 μm thick. This chapter focuses on multilayer structures to distinguish the technology from single layer thin film metallizations (such as thin film resistors or bonding pads) that have been used on cofired ceramic or hybrid substrates for many years. The TFML interconnections generally use high conductivity conductor materials (such as copper, aluminum or gold). The conductor layers are separated by deposited dielectric layers, usually polymers with low dielectric constants. Within this broad definition there are many options for selecting substrate, conductor and dielectric materials. Many different processes are available for depositing and patterning the multilayer structures. Also, TFML interconnections may be implemented in a variety of package designs.

The development of TFML interconnections has been a major trend in electronics packaging over the last 5 to 10 years, coinciding with the emergence of MCM technologies. The term "MCM-D" was introduced to describe a wide variety of MCMs using thin film interconnections with deposited dielectrics. MCM-D generally is recognized as the most important and extendable MCM technology for future systems. However, the growth of the MCM-D market has been slower than expected, largely due to high costs, limited availability of substrates and advances in competing technologies such as ceramic-based MCMs (MCM-C) or MCMs based on laminated PWB technology (MCM-L). It is important to examine critically the strengths and weaknesses of thin film interconnections relative to other packaging technologies.

This chapter gives a general description of TFML interconnections, their characteristics and their benefits relative to alternative interconnection technologies. It examines, in detail, the options for substrate, conductor and dielectric materials, followed by options for processing multilayer structures. Design strategies for implementing thin film interconnections in MCMs are examined and illustrated with examples from industry. Finally, the tradeoffs involved in selecting an interconnect technology are summarized and the potential for the growth and extension of thin film interconnections is explored.

7.2 CHARACTERISTICS, BENEFITS OF TFML INTERCONNECTIONS

The primary emphasis in multichip integration is to achieve high silicon density in the MCM. This results in a smaller package, lower system cost and higher speed due to the shorter interconnection length between chips. However, it creates a need for a multichip substrate with high interconnection density. MCM-D approaches meet this need by using IC processes (such as photolithography, vacuum deposition, wet and dry etching) to pattern multiple layers of interconnections in deposited thin films. High conductivity metals, primarily copper (Cu), gold (Au) or aluminum (Al), are used for the conductor layers. The most widely used dielectric materials are polymers, particularly polyimides, which have high thermal, chemical and mechanical stability and low dielectric constants. The high conductivity conductor and low dielectric constant dielectric produce interconnections with low resistance and capacitance that can transmit high speed signals with little degradation. The characteristics of TFML interconnections and their benefits relative to other interconnection technologies are described below.

7.2.1 Packaging Structures Using TFML Interconnections

There are a variety of ways to incorporate TFML interconnections into multichip packaging structures. Figure 7-1 shows three basic approaches. The thin films may be patterned on a blank substrate mounted in a second level package such as a perimeter leaded flat pack (middle approach in Figure 7-1). Electrical connections (such as wire bonds) are made between the substrate and the package, and the package then may be hermetically sealed. Alternatively, the TFML interconnections may be patterned on a multilayer ceramic substrate that contains pins or leads, power and ground distribution layers and sealing structures. In this approach the substrate serves as the package body. In the third option, the thin films are patterned on a large substrate used like a PWB for interconnecting bare or packaged ICs. This is essentially an extension of chip on board technology to much higher interconnection densities.

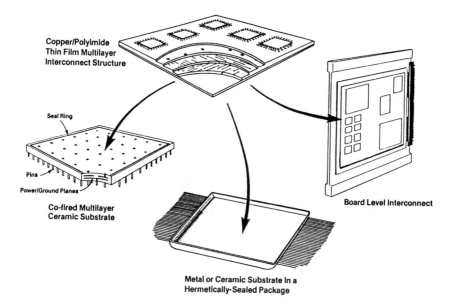

Figure 7-1 Implementation of TFML interconnections in IC packaging structures.

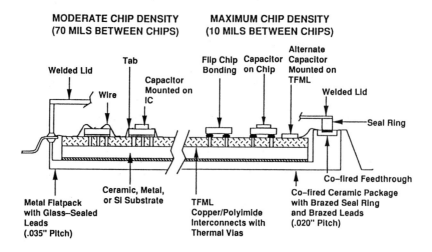

Figure 7-2 Cross section of an MCM containing a TFML interconnecting substrate, showing several options for IC bonding and package materials.

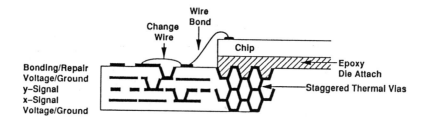

Figure 7-3 Detailed cross section of thin film layers on a TFML interconnect substrate.

Figure 7-2 is a cross section representing several different options for bonding chips and mounting TFML substrates into a second level package. The chips may be attached and bonded to the TFML substrate using a variety of technologies (refer to Chapter 9): Au or Al wire bonding, tape automated bonding (TAB) with face up or face down chips or flip chip solder bonding. Flip chip bonding provides the closest spacing between chips.

Figure 7-3 shows a detailed cross section of the thin film layers. A dense array of vias (similar to electrical vias) may be provided beneath the die attach pad to promote the conduction of heat to the substrate (discussed in Section 7.5). For flip chip bonded ICs, most of the heat is removed from the backside of the ICs, although a significant amount of heat can be removed through the solder bumps and electrical vias to which they are attached.

7.2.2 Signal Line Characteristics

Layer Configuration

The signal interconnections are usually routed on two or more layers of orthogonally oriented conductor lines (x and y layers) as shown in Figure 7-3. Additional metal layers may be used to distribute power and ground voltages; these are essentially planes (solid or meshed) with low resistance and inductance. For high speed applications (generally > 100 MHz clock frequency or < 1 ns signal rise time), the interconnections are fabricated as transmission lines with controlled characteristic impedance, typically 40 - 60 Ω. Figure 7-4 shows the most common transmission line structures for TFML interconnections:

- **Microstrip** - with the signal line on top of the dielectric over a reference plane
- **Stripline** - with the signal line sandwiched between reference planes
- **Offset stripline** - with two signal layers offset symmetrically about the center line between reference planes

The offset stripline is used in most high speed applications because it eliminates one reference plane, provides uniform impedance and protects against crosstalk (noise coupling between adjacent signal lines). A common structure for lower speed applications is the buried microstrip, which has two layers of signal lines above one or more adjacent power planes. This structure is somewhat easier to fabricate, and the capacitance between adjacent power planes provides some noise decoupling.

Electrical Characteristics

Characteristic Impedance. The electrical characteristics of the signal interconnections are related to their geometry and material properties. If we assume the interconnections are lossless (have negligible resistance), the characteristic impedance (Z_o) of the stripline [1] shown in Figure 7-4 is given by Equation 7-1:

$$Z_o = \frac{30\pi}{\sqrt{\varepsilon_r}} \frac{(1 - \frac{t}{b})}{\frac{w}{b} + \frac{1}{\pi}\left\{ 2 \ln\left[\frac{1}{1 - t/b} + 1\right] - \frac{t}{b} \ln\left[\frac{1}{(1 - t/b)^2} - 1\right] \right\}}$$

(7-1)

where ε_r is the relative dielectric constant of the dielectric material, and b, t and w are the dielectric thickness, conductor thickness and conductor linewidth, respectively. Equation 7-1 is valid for $w/(b-t) < 0.35$ and $t/b < 0.25$. In reality, TFML interconnections are usually somewhat lossy and Z_o is frequency dependent. (Refer to Chapter 11 for a more detailed discussion of electrical characteristics.) However, Equation 7-1 is still useful for seeing the effects of geometrical parameters on Z_o. A high Z_o requires a low dielectric constant, a large dielectric thickness, and a small conductor cross section (small t/b and w/b). In general, these geometries require high aspect ratio structures (high thickness/width). A high Z_o (> 40 Ω) is usually required to maintain low power dissipation and delay in driver circuits, low switching noise and high noise tolerance [2]; conventional values of Z_o are 50 - 60 Ω.

  Resistance The DC resistance of an interconnection line is given by Equation 7-2:

$$R = \frac{\rho L}{w t} = R_s \frac{L}{w}$$

(7-2)

where ρ is the conductor resistivity, R_s is the sheet resistance and L is the length of the interconnection. For low resistive losses, the conductor should have low resistivity and the signal lines should be short with a large cross section (wt). Since the linewidth must be narrow for high wiring density, low resistance interconnects require a large conductor thickness and thus a large aspect ratio (t/w).

  At high frequencies, the current is forced out of the center of the conductor toward the surfaces adjacent to the nearest ground plane, a phenomenon known as skin effect. The depth at which the current has decreased to 1/e of its density at the surface is called the skin depth, defined as $\delta = \sqrt{(\rho/\pi f \mu)}$, where f is the frequency and μ is the conductor permeability (1.26×10^{-6} H/m for

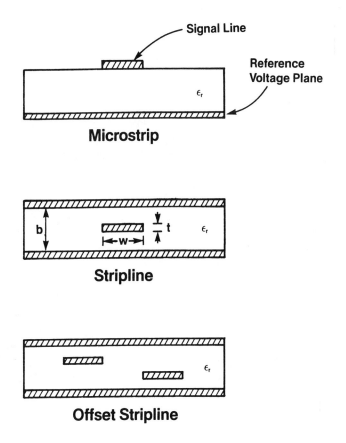

Figure 7-4 Transmission line structures for TMFL interconnections.

nonmagnetic conductors). For Cu, the skin depth is 2 μm at 1 GHz. When the skin depth is less than the conductor thickness, the AC resistance of the line is greater than the DC resistance, resulting in dispersion and degradation of short rise time signals. The skin effect may be a concern also if an interface metal (discussed in Section 7.3.2) with high resistivity is used around the primary conductor. If the thickness of the interface metal is a significant fraction of the skin depth (more than a few hundred nm), most of the current will be conducted in the interface metal and the line will be more lossy than if the primary conductor were used alone.

Capacitance and Propagation Delay. The capacitance per unit length of a lossless interconnect is given by Equation 7-3:

$$C = \frac{\sqrt{\varepsilon_r}}{c Z_o} \qquad (7\text{-}3)$$

where c is the speed of light in vacuum. Thus the geometries that produce a high characteristic impedance produce low interconnection capacitance. Low capacitance is desirable to minimize the RC delay time and the power required to charge an interconnection.

A final important characteristic is the propagation delay time of a lossless interconnection. The propagation delay per unit length (τ_{pd}) is the inverse of the phase velocity (v_p) of electromagnetic waves in the dielectric, as shown in Equation 7-4:

$$\tau_{pd} = \frac{1}{v_p} = \frac{\sqrt{\varepsilon_r}}{c} = 33\sqrt{\varepsilon_r} \quad ps/cm. \qquad (7\text{-}4)$$

Propagation delay is minimized by keeping lines short and by using low dielectric constant materials (discussed in Section 7.3.1). The propagation delay is the minimum delay for a lossless line. RC charging time and resistive losses in the interconnection cause additional delay.

7.2.3 Interconnect Design Rules

TFML interconnections encompass a wide range of designs, making it difficult to define a typical set of design rules. Table 7-1 compares the materials, geometries and electrical characteristics of four TFML designs:

1. An interconnect technology developed at Honeywell for GaAs MCMs that has been characterized at frequencies up to 10 GHz [3],
2. An MCM-D substrate offered by Alcoa [4], based on the Advanced VLSI Package (AVP) developed at AT&T,
3. The technology used by DEC in the High Density Signal Carrier for the VAX-9000, described in Chapter 17, and
4. An MCM-D technology developed by IBM called VHSIC (VCOS chips on silicon technology) [5].

Table 7-1 TFML Design Rules.

	Honeywell	Alcoa	DEC	IBM-VCOS
Substrate				
Material	99.5% Alumina	Silicon	Aluminum	Silicon
Size	2.5" - 5" square	6" round	6" round	100 mm round
Layer Construction	offset stripline	buried microstrip	offset stripline	offset stripline
# metal layers	5	5	5*	4
Conductor				
Material	Copper	Copper	Copper	Aluminum
Signal line geometries:				
Thickness (μm)	5	4	10	3
Linewidth (μm)	25	19-23	18	15
Line Pitch (μm)	100-125	50-75	75	25
Power plane thickness (μm)	5	2	18	3
Dielectric				
Material	Polyimide	Polyimide	Polyimide	Polyimide
Dielectric constant	3.5	3.4	3.5	3.5
Thickness (μm) §	15-25	6-12	25	2.5
Via diameter (μm)	25	20		8x8
Electrical Characteristics				
(Signal Lines)				
Sheet resistance (mΩ/sq)	3.5	4.5	1.7†	8.8†
Resistance (Ω/cm)	1.35	2.0, 2.4	1	5.9†
Capacitance (pf/cm)	1.2	1.0, 1.2	1.2	3.4†
Characteristic Impedance (Ω)	50	50,58	60	not controlled
Propagation delay (ps/cm) ‡	62	61	62	62
Via resistance (mΩ/via)	2.5	<5		
Insertion Loss (dB/cm)				
@2 GHz	-0.3			
@9 GHz	-0.9			

* signal core only
† calculated values from design geometries
§ signal line to nearest reference plane
‡ propagation delay for lossless line

Table 7-1 illustrates the wide range of materials and designs possible with TFML technology. Substrates include alumina, silicon and Al in both round and square shapes. Conductor materials include Cu and Al. There is a wide range in layer thickness, signal linewidth and pitch and electrical characteristics.

One of the most significant differences between TFML designs is the conductor line cross section and resulting electrical performance. For low speed systems (< 100 MHz), narrow linewidths of 10 - 15 μm and conductor thicknesses of 2 - 3 μm may be used. The VCOS technology is typical of this type of design. The small cross section of the signal lines (15 μm wide × 3 μm thick) and the Al conductor produce a relatively high DC resistance of 6 Ω/cm, which will significantly degrade high speed signals. The advantage of the narrow linewidth is that it permits high routing density (signal line pitch of 25 μm). The narrow lines also permit the dielectric layers to be relatively thin without producing high capacitance. The thin dielectric layers permit higher via density (although the VCOS technology has unusually thin polyimide layers).

For high speed systems (> 100 MHz), larger conductor cross sections are needed (5 - 10 μm thick × 18 - 25 μm wide) to reduce loss at high frequencies. A thicker dielectric is required to maintain low capacitance/high impedance. The result is a lower routing density of 50 - 125 μm, still adequate for most designs. The Honeywell and DEC designs are representative of this type of interconnection. Typical interconnect resistance for these geometries is 1 - 2 Ω/cm. The Honeywell design for striplines produced an insertion loss of 0.3 dB/cm at 2 GHz and 0.9 dB/cm at 9 GHz [3]. Since signal amplitude losses should be kept to less than ~1 dB, the interconnects for multi-GHz systems must be kept very short (< 1 cm), or long interconnects must have larger cross sections [6].

The Alcoa design rules achieve a compromise between relatively high speed and high interconnection density. These geometries will satisfy most MCM-D applications and are typical of most TFML designs. The Alcoa design uses a buried microstrip with two voltage planes below two signal layers. This produces a slightly different characteristic impedance on each signal layer, although the conductor linewidths can be adjusted to equalize the impedance on the two layers. A capacitance of > 0.7 nF/cm^2 is provided by using a dual silicon nitride/oxide dielectric between the power plane and conductive silicon substrate.

7.2.4 Comparison With Alternative Interconnection Technologies

Advantages of TFML Interconnections
Table 7-2 compares the characteristics of TFML interconnections with those of the primary alternative technologies: thick film screen printed multilayer structures (discussed in Chapter 6), high temperature cofired ceramic (HTCC - see Chapter 6) and laminated interconnections based on PWB technology (MCM-L - see Chapter 5). The advantages of TFML technology are apparent. First,

Table 7-2 Comparison of MCM Interconnection Technologies.

	THIN FILM	THICK FILM	COFIRED CERAMIC[2]	LAMINATE
Conductor Material[1]	Cu (Al,Au)	Cu (Au)	W (Mo)	Cu
Thickness (μm)	5	15	15	25
Line width (μm)	10 - 25	100 - 150	100 - 125	75-125
Line pitch (μm)	50 - 125	250 - 350	250 - 625	150-250
Bond pad pitch (μm)	100	250 - 350	200 - 300	200
Max. # of layers	4 - 10	5 - 10+	50+	40+
Dielectric Material	Polyimide	Glass-ceramic	Alumina	Epoxy/glass
Dielectric Constant	3.5	6 - 9	9.5	4.8
Thickness/layer (μm)	25	35 - 65	100 - 750	120
Min. via diameter (μm)	25	200	100 - 200	300
Electrical Characteristics				
Propagation Delay (ps/cm)	62	90	102	72
Sheet Resistance (mΩ/□)	3.4	3	10	0.7
Line Resistance[3] (Ω/cm)	1.3 - 3.4	0.2 - 0.3	0.8 - 1	0.06-0.09
Stripline Capacitance[4] (pF/cm)	1.25	4.3	2.1	1.46

[1] Primary conductor material; alternative materials in parentheses.
[2] High temperature cofired ceramic
[3] For range of line widths shown
[4] For 50 Ω characteristic impedance line (minimum capacitance for thick film).

thin film structures provide higher interconnection density and a higher bond pad density. This is due to the thin film patterning processes, which can achieve much higher resolution and higher aspect ratios than the screen printing and hole punching processes used for thick film and cofired ceramic interconnections, or the through hole drilling and plating processes used in MCM-L. The laminated approaches (cofired ceramic and MCM-L) achieve high interconnection density by stacking a large number of layers. Eventually this process becomes more costly than the thin film approach. It also may be prohibited by height or weight limitations. The thin film technology is the only one that can match the finest bond pad pitch used on ICs (~100 μm). Other technologies require a fanout or double row of bond pads on the substrate, reducing the chip packing density.

The TFML interconnections achieve the best high speed performance due to the material properties and high aspect ratio patterning. Low temperature deposition processes, such as sputtering and plating, permit the use of high conductivity metals (such as Cu and Al) as opposed to the low conductivity refractory metals (W and Mo) used in HTCC. For typical conductor thicknesses of 3 - 8 μm, the high conductivity metals produce a low sheet resistance of 2 - 6 mΩ/square, versus 10 mΩ/square in cofired ceramic. This results in lower voltage drops and less switching noise on power distribution planes. The high conductivity metals also produce low resistivity for narrow signal lines (Equation 7-2). (Low resistivity is obtained in MCM-C and MCM-L by using wider lines at the expense of routing density.) Lines with low resistivity cause less degradation in signal amplitude and rise time. The polymer dielectrics used in TFML interconnections have a significantly lower dielectric constant than typical HTCC or thick film glass-ceramics. The low dielectric constant enhances high speed performance by reducing propagation delay (Equation 7-4), interconnect capacitance (Equation 7-3) and crosstalk between closely spaced signal lines.

The thin film approach is more versatile than the other interconnection technologies because the multilayer structures can be fabricated on a wide variety of substrates and incorporated into a variety of package designs. High thermal conductivity substrates such as metals, aluminum nitride (AlN), or even diamond may be used for applications requiring high thermal dissipation. Substrates that match the coefficient of thermal expansion (CTE) of silicon (including silicon itself) may be used to improve the reliability of flip chip bonded ICs. Silicon substrates with active driver circuits or passive resistor or capacitor structures also may be used.

Finally, TFML technology is extendable to even higher interconnection density and speed. A number of thin film interconnection systems are being evaluated for multi-GHz digital applications [5]. The material system is also compatible with some planar optical waveguide approaches and may even be extended to thin film superconductor interconnections (discussed briefly at the end of Chapter 2 and this Chapter).

Issues

In spite of numerous advantages that are widely recognized, thin film interconnections are being implemented more slowly than initially expected. This can be attributed to a variety of factors: high costs, an unstable vendor base, a lack of standardized designs and materials, an inadequate infrastructure for MCM design and test, and continual improvements in competing MCM technologies. Due to the use of polymer dielectrics, with insufficient life cycle and field performance data, the thin film systems are generally seen as risky and less robust toward assembly processes than more established technologies such as cofired ceramic. These factors must also be considered in selecting an MCM technology. The tradeoffs are discussed in more detail at the end of this chapter.

7.3 MATERIALS FOR THIN FILM INTERCONNECTION SYSTEMS

7.3.1 Substrate Materials

Substrate Requirements

TFML interconnections can be fabricated on a wide variety of substrates, including metals, ceramics and silicon wafers. The selection of a substrate material is dictated by a combination of thermal, mechanical and processing considerations and is also dependent on the second level package design. However, there are some universal requirements or desirable characteristics for all substrates.

The substrate should have high thermal conductivity to dissipate heat generated by ICs. The substrate should also have a high flexural strength to avoid fracture. A lightweight (low density) substrate is obviously desirable for MCM applications such as airborne or space-based systems or medical implants.

The CTE of the substrate is dictated by several considerations. If the substrate is mounted in a second level package, its CTE should match that of the package material. It may also be important for the substrate CTE to match that of silicon (2.3 - 3.5 ppm/°C) to avoid excessive stress in the die attach material or flip chip bonds for large die. Finally, the substrate CTE should match that of the dielectric, especially for polymer dielectric materials. As the polymer film cools from the cure temperature or glass transition temperature, the CTE mismatch causes stress (usually tensile) in the polymer film, resulting in bowing of the substrate. The bowing can complicate photolithographic patterning and the bonding of the substrate into a second level package. It can be shown (refer to equations in Chapter 16) that the camber warpage of the substrate, B, is given by Equation 7-5:

$$B = \left[\frac{d^2}{4}\right] \left[\frac{3\,t_f}{t_s^2}\right] \left[\frac{E_f}{E_s}\right] \left[\frac{1 - v_s}{1 - v_f}\right] (\alpha_f - \alpha_s)(T_c - T_r)$$

$$(7\text{-}5)$$

where d is the substrate diameter or diagonal, t_f and t_s are the thickness of the dielectric film and substrate, respectively, E_f and E_s are the Young's modulus of the film and substrate, v_f and v_s are Poisson ratios, α_f and α_s are CTEs of the film and substrate, T_c is the cure temperature (or glass transition temperature if lower than the cure temperature) and T_r is the room or point of use temperature.

The camber increases linearly with dielectric film thickness and depends on the square of substrate diameter/thickness. A substrate with a high modulus (E_s) and a CTE matched to the dielectric will experience less bowing.

From a fabrication standpoint, the substrate must be inert to all of the process chemicals (including both wet etchants and plasma gases) and temperatures (typically up to 450°C) used in TFML processing. For photolithographic processes the substrate must be flat (typically less than 50 - 100 µm total camber) and the surface must be smooth (typically less than 0.1 - 0.2 µm average roughness) and free of defects. Lapping or grinding is often required to achieve sufficient flatness in ceramic or metal substrates. Lapping removes material by rubbing the substrate against a slurry of relatively large abrasive particles; this produces a rough surface that may require polishing with smaller particles to achieve an acceptable surface roughness. The substrate must be machineable, since it is often cut to the final size or diced into multiple circuits. Finally, an inexpensive substrate available from multiple vendors is considered ideal. Table 7-3 lists some candidate substrate materials with their thermal and mechanical properties.

Ceramic Substrates

Ceramics, particularly alumina, are among the most frequently used substrate materials. This is partly an outgrowth of their use in thick film and single layer thin film hybrid circuits (refer to Chapter 6). The hybrid industry has provided an infrastructure for fabricating and machining (lapping, polishing, laser scribing etc.) high quality ceramic substrates. Most ceramics also have the advantage of being electrical insulators. This permits metal layers to be patterned directly on the substrate without an insulating film. The insulating substrate also can have internal metal layers or feedthroughs to the backside of the substrate. The main drawback to ceramics is the warped, rough surface of as-fired substrates. Ceramics can be lapped to an acceptable flatness, but lapping increases the surface roughness and exposes voids in the grain structure that can cause defects in thin film patterns. Ceramics with larger grains, generally, will have larger voids. After lapping, the surface can be polished to a near mirror smoothness (~10 nm average roughness). However, voids will still be present after polishing; their size and density must be minimized. A layer of polyimide is sometimes used on the ceramic substrate to fill the voids and produce a smooth surface for metal patterning.

Ceramic substrates can be fabricated by tape casting, hot pressing or injection molding. Multilayer structures are fabricated by the lamination and cofiring processes described in Chapter 6. Internal metal layers in the multilayer ceramic can be used to distribute power or to provide interconnection to external leads. In addition, metal features such as seal rings, perimeter leads or pins can

Table 7-3 Properties of Substrates for TFML Interconnections.

MATERIAL	DENSITY (g/cm^3)	CTE (ppm/K)	YOUNG'S MODULUS (GPa)	THERMAL CONDUCTIVITY (W/m-K)
Polyimide	1.4	40	2.5	0.15
Si	2.3	2.6	113	148
Al$_2$O$_3$ (99.6%)	3.9	6.3-6.7	360	20-35
Al$_2$O$_3$ (92% cofired)	3.6	6.7	275	17-20
BeO (99.5%)	3.0	6.9	350	251
AlN	3.3	4	340	160-190
SiC	3.1	3.7	400	270
Mo	10.2	4.9	324	138
Cu	8.9	16.8	110	398
Al	2.7	2.5	62	237
Au	18.9	14.3		318
Steel (AISI 1010)	7.9	11.3	192	64
Kovar (Fe/Ni/Co)	8.4	6.1	138	16
Cu/Invar/Cu (20/60/20)	8.4	6.4	134	170
Cu/Mo/Cu (20/60/20)	9.7	7	248	208
Cu/Mo/Cu (13/74/13)	9.9	5.7	269	242
CuW (20/80)	17	8	283	186
Natural Diamond	3.5	1.1		2000
CVD Diamond	3.5	1.5-2.0	890-970	400-1600

Notes: • All properties at 25°C or 300 °K.
                • Thermal conductivity is for lateral conduction (for clad materials).

be brazed to metal pads on the ceramic.

The most widely used ceramic is alumina (Al_2O_3). Al_2O_3 has a high modulus, a high strength-to-weight ratio, is chemically and thermally stable and is a good insulator. The main drawback to Al_2O_3 is its relatively low thermal conductivity. The highest quality Al_2O_3 substrates are fabricated by tape casting, which can produce a 99.5 - 99.9% Al_2O_3 material with a small grain size (1 - 3 μm). Tape casting typically produces an average surface roughness of 75 - 200 nm and a camber of about 10 - 30 μm/cm after firing. Tape cast Al_2O_3 can be lapped to a flatness of < 10 μm/cm and polished to a surface roughness of 10 nm. These techniques produce acceptable void sizes and an excellent surface for subsequent thin film processing.

Cofired Al_2O_3 is typically 92% Al_2O_3, resulting in a lower thermal conductivity and modulus than tape cast Al_2O_3. The cofired material has a larger grain size and therefore greater surface roughness and larger voids than tape cast Al_2O_3. After firing, the camber is typically 30 μm/cm, so lapping is usually required. The voids produced by lapping and polishing can be 10 μm or greater in diameter and may limit the feature sizes that can be patterned directly on the substrate. A smoothing layer of polyimide is required for critical feature dimensions less than about 20 μm.

Alternative ceramics with significantly higher thermal conductivity than Al_2O_3 have been developed. The most prominent alternatives are beryllia (BeO), aluminum nitride (AlN) [7] and silicon carbide (SiC). AlN and SiC have a low CTE that is closely matched to silicon. The main drawback to AlN is the difficulty in achieving good metal adhesion; it cannot form the metal oxide bonds that are usually involved in metal adhesion. AlN also has a larger grain size than tape cast alumina and therefore larger voids in the surface. SiC has a very high modulus of elasticity and excellent thermal conductivity. Furthermore, SiC can be polished to an excellent surface finish. However, it has a high dielectric constant which may preclude the patterning of signal lines directly on the substrate. A significant drawback to both AlN and SiC is their limited availability and the lack of industry experience with the materials. The main drawback to BeO is the health risk posed by airborne particles, which complicates machining processes.

BeO and AlN can be fabricated into multilayer structures, although the multilayer system for AlN is still in a developmental stage. Alternative multilayer ceramics such as glass-ceramic mixtures [8] and mullite [9] (refer also to Chapter 6) have been used as substrates for thin film interconnections. Their main advantages are a close CTE match to silicon and good dielectric properties (low ε_r) for internal interconnect layers. Their main drawback is low thermal conductivity.

Metal Substrates

A variety of metals may be used as substrates for thin film interconnections. In general, the advantages of metal substrates are high thermal conductivity and relatively low cost. Because the metal substrate is electrically conductive, an insulator is required beneath the first patterned metal layer. (Note that a conductive substrate may provide a good ground plane.) Also, since metal substrates are reactive toward many of the metal etchants used in TFML processing, protective coatings may be required on the backside of the substrate. The flatness and surface roughness of metal substrates varies greatly and depends on the method of fabrication. However, most metals can be heat flattened, ground, lapped and/or polished to produce an excellent surface for thin film patterning.

Al and Cu substrates have excellent thermal conductivity. They are also inexpensive and are easily machineable. Al has the additional advantage of light weight. A disadvantage of Al and Cu is that they are quite reactive toward common metal etchants. These metals also have a much lower modulus than the ceramics and a high CTE that is poorly matched to silicon. DEC has used Al as a temporary substrate that is etched away after fabricating copper/polyimide TFML interconnection structures or power planes in their VAX-9000 Multichip Unit (refer to Chapter 17).

Molybdenum (Mo) and tungsten (W) are attractive metal substrates because of their high modulus of elasticity and low CTE. However, these metals are quite heavy and difficult to machine. Molybdenum is frequently used as a heat spreader and as a CTE buffer in high power devices.

Desirable properties may be produced in metal substrates by alloying or cladding a combination of materials. Cu clad and nickel clad Mo, produced by hot roll pressing, combine the high modulus and low CTE of Mo with the high thermal conductivity of Cu or the solderability of nickel [10]. Cu/W alloys (typically 10 - 20% Cu), produced by powder metallurgy, also achieve a high modulus, low CTE and high thermal conductivity. However, they are heavier and more difficult to machine than clad Mo. The CTE of these materials may be tailored by varying the Cu composition (for Cu/W) or the relative thickness of the Mo and cladding layers.

Silicon and Diamond Substrates

Silicon is an attractive substrate because it has a relatively high thermal conductivity and a perfect CTE match to silicon die. More importantly, silicon wafers are widely available in standardized sizes that are adaptable to IC process equipment. They also have a defect free polished surface that is excellent for thin film processing. For these reasons, silicon has been used widely in MCM-D systems, as discussed in Chapter 16. An additional advantage of silicon is the possibility for directly fabricating active driver circuits or passive devices on the substrate. The passive devices, such as implanted resistors or thin film capacitors, are relatively straightforward to fabricate. For silicon substrates with active devices, it is argued that thermal performance is enhanced by fabricating high power driver circuits in the substrate and that different circuit technologies can be used in the substrate and ICs. However, complex processes (including dopant implantations and isolation structures) are required to fabricate the active devices. It is debatable whether this is an economical use of processed silicon area.

The main drawback to silicon is its low modulus and low CTE, resulting in significant warpage when conventional high CTE polyimides are used as a dielectric material. 50 μm of high CTE polyimide (35 ppm/°C) deposited on a Si wafer 100 mm diameter × 0.5 mm thick (standard thickness) will cause over 500 μm of bowing; the wafer must be 1.4 mm thick to maintain an acceptable camber of < 75 μm. This phenomenon has been a major driver in the development of low CTE polyimides that match the CTE of silicon.

Recently, synthetic diamond has been investigated as a substrate material for thin film interconnect systems because of its extremely high thermal conductivity, as high as 1600 W/m-K (four times that of Cu) [11]. Diamond also has the advantages of a very high modulus and a good CTE match to silicon. There is a high level of activity in fabricating polycrystalline synthetic diamond, primarily through plasma-enhanced chemical vapor deposition processes. Substrates as large as 150 mm in diameter and 1 mm thick are being produced with large crystal sizes and excellent optical and thermal properties. However, these substrates are very expensive; to be marketable the diamond growth rate must increase and the production cost must decrease.

7.3.2 Conductor Materials

The conductor material system for TFML interconnections consists of the primary current-carrying conductor, various interface metals for adhesion and/or diffusion barriers, and top layer metals that are compatible with the assembly process. The requirements and options for each of these materials is different.

Primary Conductor
While many conductor materials may be used for signal interconnection and power distribution, TFML technologies use the highest conductivity metals that are stable and processable, namely, Cu, Al and Au. Copper has the highest electrical conductivity (0.596×10^6 S/cm) and high thermal conductivity. Cu is

also inexpensive. It can be deposited by sputtering or by low cost electroplating processes, and is readily wetted by lead/tin solders. However, Cu has poor adhesion to polyimide. Furthermore, when polyamic acid (a precursor to polyimide) is deposited on Cu, the Cu is oxidized and forms precipitates that migrate several microns into the polyimide during curing [12]. For this reason, it is necessary to have adhesion/barrier metal layers between the Cu and certain polymer dielectrics.

Au has the advantages of high electrical conductivity (0.452×10^6 S/cm), excellent corrosion resistance and good compatibility with assembly processes such as wire bonding. It requires adhesion metals due to its weak interaction with polymer dielectrics. However, Au is prohibitively expensive for most MCM applications. It is more commonly used as a top layer metal (discussed later).

Al has been widely used as a conductor material because the semiconductor industry has extensive experience in depositing and patterning Al (or Al-4% Cu alloys) for IC interconnections. Also, Al has relatively good adhesion to polymer dielectrics. This simplifies the processing by eliminating adhesion metals. The biggest drawback to Al is its low electrical conductivity (0.377×10^6 S/cm, about 60% of the conductivity of Cu), which limits the high speed performance of interconnections.

To summarize the choices in selecting a conductor material: Al is preferred for low cost and low speed applications; Cu is preferred for the highest speed applications, and Au may be used occasionally where the highest reliability is required. Because of its extendability to future high speed applications, Cu is the conductor material most widely used in the industry.

Interface Metals
Interface metals are used to achieve good adhesion between the primary conductor (particularly Cu) and the polymeric dielectric, or to prevent chemical attack of the metal or interdiffusion of metals. The most frequently used interface or barrier metals include chromium (Cr), titanium (Ti), 10% Ti/90% W, nickel (Ni) and Mo. Most of the barrier metals are deposited by sputtering to a thickness of 20 - 200 nm. The polymer interface studied most widely is polyimide.

Cr has excellent adhesion to polyimide, even after temperature/humidity aging. There is extensive reliability data for the Cr/Cu/Cr material system used by IBM and others in thin film interconnections. However, because Cr interacts so strongly with polyimide, it may leave a conductive residue after being etched. (The correct etch process eliminates this residue.) Ti also has excellent adhesion to polyimide and is easily etched. However, its conductivity is somewhat lower than Cr. TiW has good adhesion and is a common diffusion barrier for gold metallization. However, TiW is prone to being deposited with high internal stress. The stressed TiW films can flake from the walls of sputtering systems and produce particulates. Ni is another common diffusion barrier for Au, but it has only a moderate interaction with polyimide and thus, moderate adhesion. The most important advantage of Ni is that it can be electrolytically or electroless plated. Electroless plating is useful in protecting the sidewalls of Cu conductor lines from attack by polyamic acid.

Top Layer Metallization

The top metal layer in a TFML interconnection system serves a different function and has different requirements than the internal layers. It must be resistant to corrosion, strongly adherent to the dielectric and compatible with the assembly processes (wire bonding, soldering or die attachment). Au is used most commonly as a top layer metal because of its excellent corrosion resistance and its compatibility with die attachment and wire bonding processes. Au requires an adhesion/barrier metal between itself and the dielectric or underlying conductor. The most common interface metals for Au are Ni, Cr or TiW. Cr or TiW are sputtered, Ni is usually electroplated and Au may be sputtered, electroplated or electroless plated. For wire bonding or die attach, the Au is usually 1 - 3 μm thick. For soldering, excessive Au embrittles the solder, so a thin flash (less than 0.1 μm) of Au is used over the solderable metal, either Ni or Cu. For flip chip solder bonding, a metallization sequence such as sputtered Cr/Ni-Cu plated with Ni and Au [8] or Cr/Cu with a flash of Au [12] is used.

7.3.3 Dielectric Materials

There are many options for the dielectric material in a thin film interconnection system. The dielectric must have good thermal and chemical stability over the range of conditions encountered in thin film processing, assembly and end use. The CTE of the dielectric should be well matched to the substrate to minimize stress and warping of the substrate. Processes must be capable of depositing the dielectric in thicknesses up to 25 μm with low stress and no cracks or pinholes. Ideally the dielectric should smooth or planarize the large topography created by underlying conductor patterns. The most important electrical property for the dielectric material is a low relative dielectric constant (ε_r) as discussed in Section 7.2. The dielectric material also should have a reasonably low dissipation factor to avoid significant losses in the dielectric at high frequencies.

Polymer Dielectrics

High thermal stability polymers are the most frequently used dielectric materials for TFML interconnections, although silicon oxides have also been used.

Polymers are attractive in general because they have low dielectric constants and can be deposited from solution using spin or spray coating processes. Polymer dielectrics are discussed in detail in Chapter 8.

The most common polymer dielectrics are polyimides, a broad class of polymers characterized by aromatic groups and an imide ring in the repeat structure. Polyimides are deposited as a soluble precursor, usually polyamic acid, or as a soluble ester or acetylene terminated polyimide that is already imidized. The solution viscosity or coating parameters can be varied to achieve a wide range in film thickness. The polymer solution flows on the substrate, thus planarizing the conductor and via topography. Curing the solution at temperatures up to 450°C removes the solvents and converts the precursor to a stable, insoluble polyimide.

Fully-cured polyimide films are generally stable to 500°C, mechanically tough and flexible, and inert to process solvents and chemicals. Polyimides also have a low dielectric constant (2.0 - 3.5), a relatively low dissipation factor and high breakdown voltage. Polyimides that were developed first and are still widely used, such as PMDA-ODA and BTDA-based chemistries (refer to Chapter 8), have two significant drawbacks: (1) high moisture absorption resulting in a large variation in dielectric constant (30% variation in ε_r over the full range of relative humidity) [14], and (2) a large CTE (20 - 50 ppm/°C) that causes warping of low-CTE substrates (refer to Equation 5). To alleviate these short comings, low thermal expansion polyimides with a rigid backbone structure have been introduced [15]. The low CTE polyimides also have less moisture absorption, a lower dielectric constant (2.9 - 3.0) and higher tensile strength than the earlier polyimides.

A number of alternative polymers with certain advantages over polyimides also have been introduced. Polyphenylquinoxaline (PPQ) has a low dielectric constant (2.9), low moisture absorption, good mechanical properties and good adhesion. Polyquinolines have a low dielectric constant (2.6), very low moisture absorption, low stress when deposited on silicon, good planarization, long shelf life at room temperature and do not cause corrosion of Cu [16]. Polybenzocyclobutenes (PBCBs) have several desirable properties including excellent planarization, a low cure temperature (250°C), a low dielectric constant (2.7), low dissipation factor, low moisture absorption (0.3%) and good adhesion to Cu without metal adhesion layers. However, the PBCBs have a very high CTE (50 ppm/°C) and low elongation at break, and are susceptible to oxidation during curing or temperature aging. The oxidation problem has been solved by curing under a nitrogen atmosphere and adding an antioxidant to the formulation. Fluorinated poly(aryl ethers) have very low moisture absorption and an extremely stable dielectric constant (2.7) with temperature aging. The development of specialty polymers for thin film interconnections will continue to be an active area for development as the MCM-D market grows.

Silicon Dioxide Dielectric

Silicon dioxide (SiO_2 is used as a dielectric medium by nCHIP, a company that manufactures "Silicon Circuit Boards" consisting of thin film interconnections on silicon substrates. The SiO_2 dielectric layers are deposited by a plasma chemical vapor deposition (CVD) process that produces films up to 20 μm thick under moderate compressive stress. nCHIP cites several advantages of the SiO_2 dielectric:

1. It requires fewer processing steps than polymer dielectrics.
2. The compressive stress in the film cancels the intrinsic tensile stress of metal films and produces a flat substrate.
3. SiO_2 has a reasonably high thermal conductivity (1.5 W/m°C, ten times greater than most polymers), eliminating the need for thermal vias to conduct heat through the interconnect layers.
4. SiO_2 does not have the high moisture absorption of polymers.
5. The dielectric is rugged, wire bondable and more reliable under thermal cycling, shock or nonhermetic environments.

Potential disadvantages of this dielectric are its slightly higher dielectric constant ($\varepsilon_r = 4.0$), its lack of planarization, susceptibility to pinholes and the slow deposition rates and high cost of the process gases, particularly silane. Although the CVD process requires fewer steps than polymer deposition and curing, it is unclear which dielectric will be more cost effective in a manufacturing environment. Additional vendors must offer SiO_2 dielectric and comparative studies of polymer dielectrics versus SiO_2 must be conducted before this dielectric will find widespread use in TFML interconnections.

7.4 THIN FILM MULTILAYER PROCESSING

TFML interconnections are fabricated using a repetitive sequence of unit processes for depositing and patterning conductor and dielectric layers. This section begins by describing the unit processes that are used most frequently. Then two basic approaches for patterning conductor features, additive and subtractive processing, are described and compared. Finally, some of the unique challenges imposed by TFML designs are discussed.

7.4.1 Conductor Deposition and Patterning Processes

Conductor Deposition

Conductor materials can be deposited by vacuum-based processes such as evaporation, sputtering, ion plating or by wet processes such as electrolytic or electroless plating. Several textbooks provide detailed descriptions of thin film deposition processes [17]-[18]. A recent paper compares the film properties (resistivity, stress, microstructure and adhesion) of Cu, Al and Au deposited by sputtering, evaporation, enhanced ion plating and electroplating [19]. The metal deposition processes that are used most frequently are described below.

Sputtering. Sputtering is a vacuum-based process in which the ions (usually Ar^+) in a plasma are accelerated toward a target material, ejecting or "sputtering" the target atoms by momentum transfer. The sputtered atoms deposit on surfaces exposed to the plasma, including substrates. The plasma can be sustained by either an RF or DC discharge. Magnets can be used to concentrate the plasma and enhance the sputtering rates without overheating the target. The substrate is heated by condensation, and both the target and substrate are heated by ion bombardment. The ability to cool the substrate or target can determine the maximum deposition rate.

Sputtering is the most widely used vacuum deposition process because it offers a number of advantages:

1. Almost any material can be deposited, including insulators, adhesion metals, thin film resistors (such as tantalum nitride or Ni-Cr) or the primary conductor.
2. Excellent adhesion to the substrate or dielectric material can be achieved by backsputtering or bombarding the surface with argon ions prior to depositing the metal.
3. Films can be deposited with excellent control of thickness, stress and morphology through variations of process parameters such as pressure, power and electrical bias of the substrates.
4. Sputtered thin films have low resistivity due to the high packing density of the atoms.
5. The films achieve conformal coverage of via holes and other topography.
6. Automated equipment is available for sputtering because of its widespread use in the IC industry.

The main drawbacks to sputtering are its relatively low throughput rate (because of slow deposition rates and limited batch size) and the high cost of equipment.

Evaporation. To deposit metals by evaporation, the metal is heated to a vapor using an electrical resistance heater or electron beam, and the vapor condenses on the substrates. Because the process is carried out at much lower pressures than sputtering (10^{-5} - 10^{-6} torr versus 10^{-2} - 10^{-3} torr for sputtering), the mean free path of the vaporized metal atoms is tens of meters, and deposition occurs only on surfaces that are directly exposed by "line of sight" to the deposition source. Therefore vertical surfaces such as via sidewalls are not coated as thickly as horizontal surfaces and overhanging surfaces can completely shadow underlying surfaces from deposition. The line of sight deposition can be used to advantage in liftoff processes (discussed later). In general, evaporated films have poorer adhesion than sputtered films because the atoms arrive at the substrate with lower energy. There also is less control of film stress and microstructure with evaporated films, since there is no controllable substrate bias to affect ion bombardment of the growing film.

Alternative Vacuum Deposition Processes. A variety of alternative vacuum deposition processes may be considered for thin film interconnections. Examples are ion cluster evaporation, ion plating (or enhanced ion plating), cathodic arc deposition, or ion beam sputtering. In ion cluster evaporation and ion plating, a thermally evaporated material is ionized and the metal ions are accelerated toward a biased substrate. These processes achieve good adhesion and packing density due to the high energy of arriving atoms. Film morphology and properties can be controlled by varying the evaporation rate and substrate bias. Cathodic arc deposition and ion beam sputtering are alternative processes for sputtering a target material. Cathodic arc deposition can achieve very high deposition rates. The primary limitation to all of these alternative processes is the relative lack of industry experience and automated equipment for high throughput processing.

Electroplating. Metals can be deposited from ions in a liquid solution by electroplating or electroless plating. Electroplating requires the application of an electrical potential to a metal seed layer on the substrate, while electroless plating can deposit metal through chemical reactions without an applied potential. Electroless plating is limited to thin coatings and therefore is useful mainly for interface or barrier metals. Electroless plating of Ni also has been used to fill vias [20].

For the thick conductor layers required in most TFML interconnections, electrolytic plating is used. Electroplating is well understood and widely used for top metal Ni/Au metallization. Electrolytic plating first requires deposition of a seed layer, usually by sputtering. An electrical contact is made to the seed layer and a current is applied. The substrate acts as the anode in an electrolytic cell and the metal ions are reduced and deposited on the substrate. The plated metal can be deposited as a blanket coating or in open areas defined in a photoresist film (the additive approach described later).

The main advantage of electroplating is the high deposition rate, typically on the order of μm per minute. Therefore, electroplating, often is used for thick power/ground planes or thick signal lines (> 5 μm). A drawback of plating is the difficulty in controlling the process and film properties. The plating deposition rate, uniformity, film stress and morphology are dependent on a variety of process parameters, such as bath composition, temperature, agitation, electrode design and the plated area. The bath composition must be continually monitored and adjusted. Impurities in the plating bath can cause poor adhesion, defects or variation in the properties of plated films. In general, electroplated films have lower density and higher resistivity than vacuum deposited films. Finally, electroplating cannot deposit some metals and allloys that can be deposited by sputtering.

Conductor Patterning
Conductor layers can be patterned by either subtractive or additive processes, as discussed in Section 7.4.3. The unit processes discussed below are primarily used for subtractive patterning (etching) or direct writing of conductor patterns.

Wet Etching. Wet etching removes conductor material at a controlled rate by oxidizing the metal and converting it to a soluble product in an acid solution. It is the patterning process used most widely for TFML structures because it involves inexpensive equipment and can achieve high etch rates. The ideal etchant will selectively etch the desired metal without attacking the substrate or interface metals. The main limitation of wet etching is the resolution or aspect ratio achieved in conductor features. Because the chemical reactions proceed at an equal rate in all directions, wet etching undercuts the photoresist and produces a sloped sidewall. Figure 7-5 shows the undercut and sloped sidewall in a sputtered copper film 8 μm thick that has been spray etched. The undercut limits wet etching to aspect ratios below about 0.5, sufficient for most thin film interconnect designs.

Wet etching may be done by immersion or by spraying the etchant. Spray etching produces more vertical sidewalls and achieves better etch uniformity and higher etch rates by removing depleted reactants. Accurate control of linewidth requires a method for detecting the endpoint of etch. Bath parameters such as temperature, oxidation potential, pH and additives such as surfactants must also be controlled.

Dry Etching Dry vacuum processes for metal etching include ion beam etching, reactive ion beam etching (RIBE) and reactive ion etching (RIE). These techniques can produce much higher aspect ratios because the processes are activated by charged species accelerated perpendicularly to the metal surface.

Figure 7-5 SEM micrograph of the sidewall in a spray etched copper film (8 μm thick), showing undercut beneath the photoresist.

Ion beam etching (or ion milling) uses a beam of inert gas ions (such as Ar$^+$) accelerated perpendicularly or at an angle to the substrate to dislodge metal atoms (as in sputtering). Ion milling can etch very high aspect ratio features in any material, however, it etches at a slow rate, has poor selectivity (many materials etch at similar rates) and causes heating of the substrate and photoresist. RIBE is a similar process, but uses a reactive gas to achieve higher rates or better selectivity. RIE uses a plasma, usually sustained by an rf discharge between parallel plates, as a source of reactive ions and neutral species that etch the conductor material. RIE achieves higher etch rates and better selectivity than ion milling or RIBE. RIE of Al is a widely practiced process in the IC industry, whereas the RIE of Cu requires high electrode temperatures to volatilize the reaction products, and is not a practical option at this time. The biggest drawback to all of the dry etch processes is the high cost of equipment.

Laser Patterning Some of the most promising alternative processes for patterning conductors are based on the use of lasers to write conductor lines directly. The two basic approaches for laser direct writing are: (1) using a laser to decompose liquid or gaseous organometallic compounds, and (2) decomposing catalytic metals such as palladium (Pd), followed by selective electroless plating of Cu or Au. The direct writing processes eliminate the need for masks or photolithographic processes and can be computer driven, but are limited in writing speed and thickness of the deposited conductor. Thus, they are most attractive for quick turnaround, low volume prototype fabrication, or for the local repair of conductor faults. Figure 7-6 shows links in a conductor pattern made by localized laser irradiation of a Cu formate film [21].

Lasers also can be used to etch conductor material, either by direct ablation in an inert atmosphere or by initiating reactions in a reactive gas or liquid. Laser ablation has been used for many years to trim thin film resistors. The main limitation to laser etching is the long time required to clear large areas. Thus,

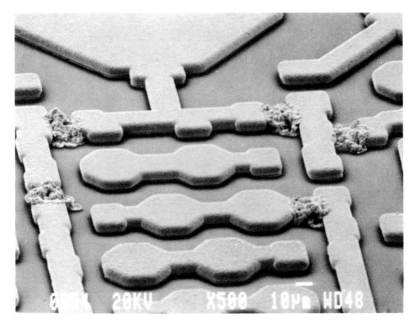

Figure 7-6 Links in a copper conductor pattern filled in by localized laser irradiation of copper formate solution. (Courtesy of R. Miracky, Microelectronics and Computer Technology Corp.)

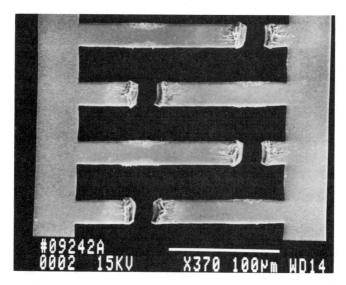

Figure 7-7 Openings in gold-nickel conductor lines (5 μm thick × 15 μm wide), cut by a frequency-doubled Nd:YAG laser. (Courtesy of R. Miracky, Microelectronics and Computer Technology Corp.)

laser etching processes are most useful for defect repair. Figure 7-7 shows Au-Ni conductor lines 5 μm thick × 15 μm wide on polyimide, cut by a frequency-doubled Nd:YAG laser [21]. Note the minimal damage to the polyimide.

7.4.2 Dielectric Deposition and Patterning Processes

Because the predominant dielectric materials are polymers deposited from solution, the following discussion focuses on polymer deposition and patterning processes, particularly those used for polyimide.

Polymer Deposition Processes
Polymer films can be deposited from solution by a variety of techniques including spin coating, spraying, dipping, screen printing, roll coating and extrusion. In all of these processes, the low viscosity of the deposited solution

allows it to flow and planarize underlying topography. The thickness, uniformity and planarization of the film depend on the viscosity and solids content of the solution. Surface tension in the coating causes edge bead, a build up of polymer near the edge of the substrate that can interfere with the patterning of features near the edge.

After deposition, the films are cured by heating at a controlled rate in an oven, hot plate or tube furnace to evaporate solvents and complete such reactions as the conversion of polyamic acid to polyimide. As the film cures, its viscosity increases until it is not able to flow; further volume reduction causes the film to partially conform to underlying topography. The final properties and adhesion of the polymer films depend on curing rate and final curing temperature, ranging from 250°C - 450°C. The atmosphere of the curing chamber is also important. In general, a nonoxidizing atmosphere is preferred to avoid oxidation of underlying conductor layers or the polymer itself.

Multiple coats may be used to achieve the layer thickness required for high impedance interconnections. The planarization improves with each coat. An important consideration in multiple coatings is the polymer-to-polymer adhesion. Good adhesion requires polymer interdiffusion at this interface. Interdiffusion depends on the degree of cure and the T_g of the underlying film. In general, a lower cure temperature or a lower T_g for the underlying film will improve polymer to polymer adhesion. Polymers that have poor self adhesion are sometimes treated with a plasma to roughen the surface prior to subsequent coatings.

An organosilane-based adhesion promoter is often used to improve the adhesion of the first polyimide coat to silicon or ceramic substrates [22], [23]. The adhesion promoter is applied as a very thin film, usually by spin coating. In recent years, polyimide formulations that are self priming or that contain adhesion promoters have become available.

Spin Coating. The most common process for depositing polymer films is spin coating. This well characterized process is used frequently for depositing photoresists in IC fabrication. Highly automated equipment, known as track systems, can load substrates from a cassette, dispense the polymer solution, spin the substrate at a controlled speed and acceleration, remove the edge bead by applying a solvent to the outer edge of the spinning substrate and perform an initial cure on hot plates. Adhesion promoters also can be applied on the track system. The film thickness can be controlled accurately by varying the spin speed and time and very uniform thicknesses can be obtained. The main drawback to spin coating is its relatively low throughput rate and difficulty in spin coating large, square substrates. Spin coating also wastes 90 - 95% of the dispensed polymer.

Spray Coating. An attractive alternative to spin coating is spray coating. In systems designed for microelectronics processing, substrates are loaded on a conveyor that moves beneath a reciprocating arm that dispenses the solution at a controlled rate through an atomizing nozzle. The thickness can be controlled accurately by varying the flow rate and the conveyor speed. The diluting solvent, solution viscosity, solution concentration, nozzle diameter and nozzle distance from the substrate are critical to the film uniformity. Spray coating is an ideal production process because of its high throughput capability and its ability to coat a wide range of film thicknesses on a wide range of substrate shapes and sizes, including large, square substrates. However, spray coating produces a larger edge bead and more local thickness variation than spin coating. Also, the process is sensitive to particulates, which can cause local dewetting and pinholes in sprayed films. The correct solvent and solution concentration will permit enough flow on the substrate to prevent pinholes, yet not enough to cause pooling and large thickness variation.

Extrusion. FAS Technologies recently introduced an extrusion process for coating polymer films [24]. The polymer solution is dispensed by a positive displacement, diaphragm driven pump with a master cylinder for filtering the solution at low pressure and a slave cylinder for accumulating and dispensing the solution. The metered solution is forced through a linear orifice in an extrusion head that is ~0.25 mm above the surface of the substrates that are moving beneath the head. This casts a polymer film across the full width of the substrate. The extrusion process can deposit films 1 - 100 μm thick with 2 - 5%, uniformity across the substrate and from substrate to substrate.

The extrusion process has several important advantages:

1. The large material waste of spinning or spraying is eliminated, since most of the polymer stays on the substrate.
2. Films can be deposited on large square substrates.
3. Edge bead can be eliminated and a clear zone can be left around the perimeter of the substrate.
4. A wide range of thicknesses can be deposited in a single pass.
5. High throughput can be achieved.

Possible concerns include the effect of substrate warpage on coating uniformity, and the effects of particulates, vias and topography on film defects. Further characterization of this process is required, but it has the potential for significantly reducing the material cost in high volume production.

Patterning of Polyimide Films
Polyimide films may be patterned by a variety of processes including wet or dry etching through a photolithographically defined mask, direct photopatterning of

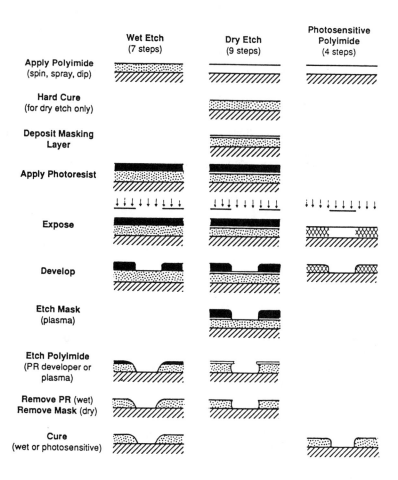

Figure 7-8 Process steps for patterning polyimide films by three different methods: wet etching, reactive ion etching and photopatterning of photosensitive polyimide.

photosensitive polyimides or laser ablation. The process steps involved in wet etching, dry etching and photopatterning are shown in Figure 7-8.

Wet Etching. Polyimide films can be wet etched by partially curing the film, patterning a photoresist coating, and dissolving the exposed polymer in an aqueous base solution. Many positive photoresist developers etch the partially cured polyimide, permitting development and wet etching in a single step. Wet etching is a simple and inexpensive process, but produces low aspect ratio features because of the isotropic etch. The etch rate and geometries are highly dependent on the extent of partial curing, which is difficult to control. After patterning, the film must be fully cured. This results in further shrinkage and loss of resolution. More importantly, residues left after etching can cause high via resistance. In spite of these problems, the economic incentives for a wet etch process are significant, and there has been considerable activity in developing polyimides or other polymers that can be wet etched. Polyimide formulations that are designed for a two step patterning process (separate photoresist developer and polyimide etchant) have recently been demonstrated to give steep sidewall angles and very smooth features. Figure 7-9 shows a 25 μm wide opening etched in a 10 μm thick film by such a two step process [25].

Dry Etching. Dry etching processes, such as plasma etching, RIE or RIBE, can produce much higher aspect ratio features in polymer films by using charged species in a plasma or ion beam to initiate etching in a direction perpendicular to the substrate surface. The most common dry etching process is RIE. Substrates are placed on the powered electrode sustaining an RF plasma in a gas mixture of oxygen and a fluorine containing gas such as CF_4, SF_6 or CHF_3. Because the polymer dielectrics etch at nearly the same rate as photoresists, the RIE of thick films requires that a masking layer of a slow etching material, such as a metal, silicon dioxide or spin-on glass, is deposited and patterned on top of the polymer. The full process sequence is shown in the middle column of Figure 7-8.

Dry etch processes can be used on fully cured polymer films and are less sensitive than wet etching to the specific chemistry of the polymer. Very low via resistance can be achieved, since the plasma etches any polymer residue at the bottom of vias (although metal oxidation or redeposition of materials from the RIE mask or chamber must be minimized). RIE processes can pattern high aspect ratio vias with nearly vertical sidewalls. More often, process parameters are varied to produce a controlled sidewall angle of 70 - 80 degrees from horizonal for better metal step coverage. Figure 7-10 shows a via etched by RIE in a 25 μm thick polyimide film, with a sidewall angle of 70 degrees. The main drawbacks to the dry etch processes are the high equipment cost and limited throughput rates.

Photosensitive Polyimide. The number of process steps required to pattern polyimide films is greatly reduced by using formulations that are photosensitive. The chemistry of photosensitive polyimides (PSPls) is discussed in Chapter 8. Most PSPls are negative acting so that areas exposed to ultraviolet (UV)

0682   5KV   X2,200   10µm WD48

Figure 7-9 Ultradel 4212® polyimide patterned by spray etching with a specially formulated wet etchant. The coating will be 10 µm thick when fully cured, and the photoresist opening is 25 µm wide. (Courtesy of H. Neuhaus, Amoco Chemcial Company.)

radiation become cross linked and insoluble. The unexposed material is dissolved in a developer solution and the patterned film is cured to complete the imidization reactions and drive off solvents and crosslinking agents.

The photopatterning process involves considerably fewer steps, as shown in Figure 7-8 and eliminates the need for expensive plasma equipment. However, the resolution and aspect ratio of patterned features, particularly via holes, are limited by several factors:

- Developer solutions cause swelling of the crosslinked polymer, which can cause a small opening to close or form webs,
- There is considerable shrinkage during the final cure (typically 50% thickness loss), causing a loss in via aspect ratio, and

Figure 7-10 Via hole 25 μm in diameter etched in a polymide film 25 μm thick by reactive ion etching.

Polyimide and photoinitiators are highly absorbing at UV wavelengths, which limits the depth of crosslinking in the film.

During development, the uncrosslinked polymer beneath the crosslinked surface layer is dissolved, resulting in an undercut sidewall. This problem may be solved by filtering the highly absorbed wavelengths. Thick PSPI films also may be patterned by multiple coating and development of thin films. A final concern with PSPIs is the poor mechanical properties of the cured films, a result of their lower molecular weight. In spite of these shortcomings, the potential cost advantage offered by PSPI is so great that significant development efforts continue. Many Japanese companies manufacturing thin film interconnections are using PSPIs.

Laser Etching. A promising future technique for patterning polymer films is etching by laser ablation. Polyimides and other polymers can be thermally or photochemically decomposed by a variety of lasers and wavelengths, including CO_2, Nd-YAG, Ar$^+$ and excimer lasers. Pulsed excimer lasers operating at wavelengths of 193 - 351 nm have produced the best results with cleanly etched via holes and minimal thermal decomposition or residues. The mechanism is believed to be a photochemical process in which chemical bonds in the polymer are broken and the products are explosively ejected. Because of the high absorption of UV radiation and the poor thermal conductivity of polyimides and the short pulse duration (typically 10 - 20 ns), the etched depth per pulse is about 0.1 - 0.5 μm with little thermal diffusion. Typical fluences for etching polymers are 100 - 600 J/cm^2, which is below the damage threshold for most metals (typically > 1 - 2 J/cm^2) [26]. Thus, metals are a good mask or etch stop for laser etching.

There are three methods for patterning polymers by laser ablation: (1) individual vias may be written directly with an aperture-controlled or focused beam that may be rastered, (2) complex patterns may be ablated by projecting a broad laser beam through a special lens and mask, or (3) a broad beam may be scanned over a contact mask deposited and patterned on the polymer. Direct writing of individual vias offers the advantages of maskless processing: direct computer control, short turnaround time, low tooling costs (no masks) and high yield. However, the writing speed for vias is limited to a few vias per second, which may not be an acceptable throughput rate for manufacturing TFML substrates with thousands of vias per layer.

A large number of vias can be etched simultaneously by depositing and patterning a masking material (such as Cu or Al) on the polymer and scanning a broad beam across the surface [26]. In this process, the laser essentially replaces a reactive ion etcher. Figure 7-11 shows a via etched in a 12 μm thick polyimide film by exposing a broad beam XeCl (308 nm) excimer laser through a deposited Al mask 2 μm thick. The mask is still on the polyimide. Note the lack of undercut that typically occurs in RIE of vias. Broad beam projection patterning through a special projection mask is very attractive if the optics and specialized masks can be developed. Projection patterning eliminates all of the process steps, equipment and yield defects associated with photolithography, mask deposition and polymer etching. Projection laser etching is used to pattern polyimide layers on the latest IBM Thermal Conduction Module (TCM) for the ES/9000 mainframe computers [27].

7.4.3 Basic Process Approaches: Additive and Subtractive

In PWB manufacturing, there are two basic approaches for patterning conductor features: additive and subtractive processing. In the additive approach, a negative image of the conductor pattern is defined in a photoresist film, and conductor is deposited in the open areas. In the subtractive approach, a blanket coating of conductor is deposited, a positive image of the conductor pattern is defined in photoresist and exposed metal is etched away. These process approaches are described in more detail and compared below.

Figure 7-11 Via hole etched in a 12 μm thick polyimide film by broad beam laser ablation through a deposited aluminum mask 2 μm thick. The mask is still on the polyimide. (Courtesy of T. Tessier, Motorola Corporate Manufacturing Research Center.)

Additive Approach

The most common additive process, based on selective electroplating, is shown in Figure 7-12. First, a thin blanket coating of conductor (usually an adhesion metal followed by Cu) is deposited on the substrate to act as an electrical contact. A negative image of the conductor pattern is defined in a photoresist film thicker than the desired conductor thickness, and the conductor lines are electroplated in the open areas. The photoresist is stripped and another photoresist film is applied and patterned to define the via posts for interconnection between layers. The vias posts are electroplated and the photoresist and preplate layers are stripped. At this point, the exposed Cu may be coated with an electroless Ni to prevent interaction with the dielectric. Next, the dielectric (polyimide) is coated and the tops of the via posts are exposed by mechanical polishing [28] or by etching via holes in the dielectric (as in the DEC process, Chapter 17). The process sequence is repeated to build up additional layers.

An alternative additive process is based on liftoff of conductor material. A negative image of the conductor pattern is defined in photoresist and a blanket

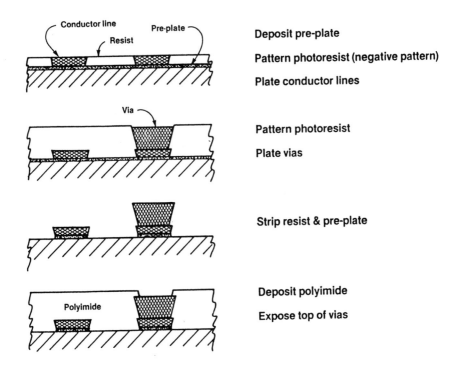

Deposit pre-plate

Pattern photoresist (negative pattern)

Plate conductor lines

Pattern photoresist

Plate vias

Strip resist & pre-plate

Deposit polyimide

Expose top of vias

Figure 7-12 Simplified process sequence for patterning conductor lines and via posts by selective plating, an additive approach.

coating of conductor is deposited, usually by evaporation. By dissolving the photoresist, the metal deposited on top of the resist is removed and the patterned conductor is left behind. In an alternative liftoff process known as dielectric assisted liftoff, the dielectric is deposited first, followed by a soluble release layer, such as polysulfone, and a hard masking layer such as silicon nitride or oxide. All three layers are patterned (usually by RIE), leaving an overhanging ledge in the hard mask. The conductor is evaporated, and the release layer is dissolved to remove the conductor that is not in the etched areas. This leaves a self planarized conductor pattern.

Subtractive Approach
A simplified subtractive process is shown in Figure 7-13. The conductor is deposited as a blanket coating by a method such as sputtering, possibly followed

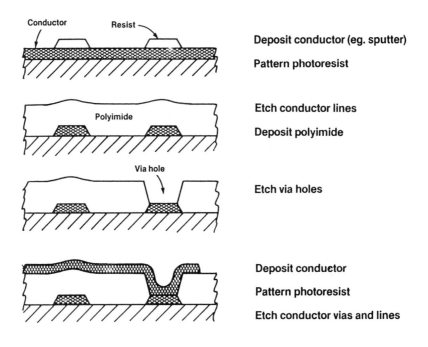

Figure 7-13 Simplified process sequence for patterning conductor lines and filling vias by subtractive patterning.

by electroplating to achieve the desired thickness. A positive image of the conductor pattern is defined in a photoresist film that may be thinner than the conductor film. The exposed conductor is removed by wet or dry etching and the photoresist film is stripped. The dielectric is coated in a blanket layer and an RIE mask may be deposited on the dielectric. A photoresist film is patterned with openings for vias. The vias are etched by a wet or dry process and the mask or resist is stripped. Another blanket layer of conductor is deposited with the conductor conformally coating the sidewalls of the via hole. The vias and next layer of conductor lines are then patterned in a single photolithography and etching step. This sequence is repeated for additional layers. An important difference from the additive approach is that vias cannot be vertically stacked, but must be staggered or staircased through multiple dielectric layers. The cross section in Figure 7-14 shows conformal staggered vias through three dielectric layers.

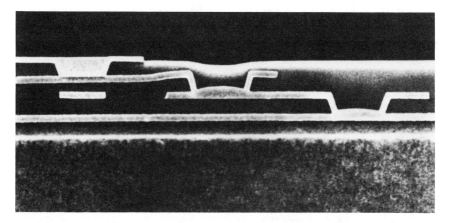

Figure 7-14 Cross section of a four metal layer TFML interconnection, with staircased vias through three dielectric layers. All metal layers are 5 μm thick. Notice the planarization of the via topography by the polyimide. (Courtesy Advanced Packaging Systems.)

Comparison of Additive and Subtractive Approaches

The additive approach has the advantage of being able to pattern very high aspect ratio features, because it is easier to define vertical wall features in photoresist than in etched metals. The additive approach can also form vertically stacked vias, leading to higher routing densities and possibly better heat transfer for thermal vias. However, there is some concern with stress and CTE mismatch in vertically stacked vias. The additional metal interface at the top of the via post (not present in the conformal via) may cause reliability problems or high via resistance. The mechanical polishing process that planarizes the dielectric down to the via posts is difficult to control. Etching to expose the vias involves additional process steps and equipment. Electroplating baths require tight controls (as discussed earlier) and plating rates and uniformity depend on the plated area and feature sizes. Finally, electroplating produces a higher resistivity conductor than sputtering or evaporation and cannot be used for Al interconnects.

The subtractive approach generally produces lower aspect ratios than selective plating because of undercut in the conductor etching process. The routing density also is lower because of staggered vias that require more area. However, subtractive patterning has a number of advantages. It requires fewer processing steps than either selective plating or dielectric assisted liftoff. Any conductor material can be used, and there are more process options for patterning both the conductor and dielectric. In general, the subtractive approach is similar to the processes used for fabricating multilevel interconnections on ICs.

Most of the TFML processes reported in the literature have used subtractive approaches. Cu conductor layers are usually sputtered and then electroplated to the final thickness, while Al is sputtered. IBM uses a combination of subtractive etching and liftoff approaches for the TCM in the ES/9000 computer [27]. Several TFML substrate foundries are using selective plating approaches, including Alcoa (producing the Advanced VLSI Package (AVP) developed by AT&T) and DEC (see Chapter 17 for a process description). The most notable example of an additive approach with mechanically polished dielectric layers is the process developed by the Microelectronics and Computer Technology Corp. (MCC) [28].

7.4.4 Unique Process Requirements

The TFML processes employ many of the techniques and equipment used in IC fabrication. However, the substrates and geometries required for TFML interconnections present a number of unique challenges for conventional IC processing. The following section describes the unique process requirements imposed by the relatively thick films, high aspect ratio features and large substrates used for TFML interconnections.

Thick Film, High Aspect Ratio
Because of the requirement for low resistive losses and low interconnect capacitance, the conductor and dielectric layers must be relatively thick, on the order of 2 - 5 μm for conductor layers and 5 - 25 μm for dielectric layers. These thick films require long processing times, especially for dry (plasma-based) deposition and etching processes. Thick films also produce high strain energy that leads to substrate warping (Equation 7-5) and creates the potential for cracking or spontaneous delamination of films.

In order to achieve high interconnect density, conductor linewidths and dielectric vias must be closely spaced. The combination of small, lateral spacing and thick films means that features with high aspect ratios (thickness/width) must be patterned. In general, this requires anisotropic patterning processes, such as RIE, that etch faster vertically than horizontally. The high aspect ratio features create large topography as layers are built up, and this topography must be planarized for accurate photolithographic patterning.

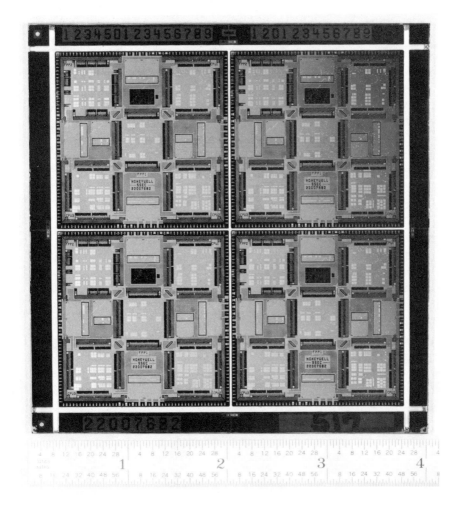

Figure 7-15 4" square TFML substrate containing four interconnection substrates separated by saw streets. The substrate is alumina with four metal layers of copper/polyimide interconnections. The individual substrates are assembled with chips and mounted in the package in Figure 7-16. (Courtesy of Honeywell Solid State Electronics Center.)

Large Substrates

MCMs and substrates vary greatly in size; MCMs up to 125 mm square are common and currently in production [8]. For smaller MCMs, it is usually most cost effective to pattern several interconnection substrates on a larger starting substrate (the term "multiple up" is frequently used in the PWB industry to describe multiple circuits repeated on a substrate). Figure 7-15 shows four interconnection substrates patterned on a 4" × 4" × 0.050" thick ceramic substrate with four layers of TFML interconnections. The individual substrates (approximately 1.8" square) are separated by sawing, assembled with ICs and mounted into the MCM shown in Figure 7-16.

The large area and square shape of the substrate presents some unique challenges. First, the warpage induced by CTE mismatch increases linearly with substrate area or with the square of the substrate side length or diameter (Equation 7-5). Large substrates must be thicker to prevent this warpage; however, thick substrates increase the volume and weight of the MCM.

Choosing a square or round starting substrate depends on a number of factors. Round substrates in standard sizes of 100, 125, 150 or 200 mm are handled conveniently by automated wafer handling equipment. However, the final MCM is usually square or rectangular (as dictated by the shape of PWB or electronic chassis), and square starting substrates provide more efficient area utilization for single or multiple up substrates. Large square substrates are more difficult to handle with automated equipment designed for round wafers. It also is more difficult to spin coat polymers and remove the edge bead on square substrates. Some of the techniques and equipment developed for the fabrication of photomasks or flat panel displays are well adapted to the processing of large square MCM substrates.

Yield

The most significant impact of the large substrate area is its effect on process yield. This is illustrated by a simple yield model that assumes that defects are distributed randomly on a surface without clustering. (This is not usually observed in practice and is a worst case assumption.) The distribution of defects is described by a Poisson distribution. The yield, or probability of having no fault producing defects, is given by Equation 7-6:

$$Y = \exp(-A_c D) \qquad (7\text{-}6)$$

where A_c is the critical area in which a defect will cause a fault and D is the area density of defects larger than a critical size. For particles in the size range of interest, D is inversely proportional to the square of particle size. If we also assume that the critical area A_c is proportional to the total substrate area $A = L^2$,

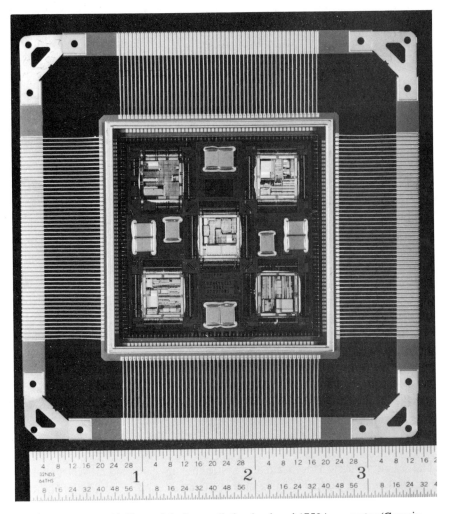

Figure 7-16 Multichip module for a radiation hardened 1750A computer (Generic VHSIC Spaceborne Computer), containing five chips on a TFML substrate in a 2.1" square cofired ceramic package with 200 leads. (Courtesy of Honeywell Solid State Electronics Center.)

where L is the side length of a square substrate, then:

$$Y = \exp\left[-k\ (L/x_c)^2\right] \qquad (7\text{-}7)$$

where x_c is the minimum critical defect size [29]. For faults in features such as conductor lines, x_c is some fraction (typically 50%) of the linewidth. Although critical features in thin film interconnection substrates are roughly an order of magnitude larger than IC feature sizes (10 μm versus 1 μm), the side length is also an order of magnitude larger (10 cm versus 1 cm) and the yield, as predicted by the proportionality L/x_c, is equivalent to IC process yields. As the substrate size, L, or total interconnect length increases, the predicted yield will decrease exponentially with L^2, as shown by Equation 7-7.

Another way of viewing the yield challenge is to consider that a silicon wafer with a number of critical defects will still yield some good die after testing and dicing, whereas a thin film interconnection substrate that is comparable in size to the wafer must have no critical defects to yield a functional product. Because of this severe yield constraint, many substrate designs incorporate features that permit repair of a limited number of interconnects after testing. Repair methods can involve wire bonding (Figure 7-3) or direct patterning of top surface conductor lines (Figures 7-6, 7-7) [21].

7.5 DESIGN STRATEGIES FOR THIN FILM INTERCONNECTIONS

The following section discusses general strategies for achieving efficient MCM designs using TFML technology. Then it describes several implementations of TFML technology, with examples of MCM-Ds offered by various companies.

7.5.1. General Design Strategies

The design of an MCM is dictated by system requirements and the IC technology balanced against the capabilities of the interconnection technology. The total number of chips in an MCM is determined by system partitioning. It is usually desirable to capture a function within the MCM in order to reduce the number of off package connections. Once the number of chips is determined, the maximum packing density of chips within the MCM (and the MCM size) is determined by one of the following limiting factors:

1. The footprints of the chips and other components
2. The number and density of off package connections
3. The interconnect routing density
4. The thermal density

It is important to recognize the limiting factor so that technology advancements can be directed toward that factor. For example, in an MCM containing memory die, the packing density is probably limited by the component footprints rather than the interconnection density, and the emphasis in design and technology selection should be placed on fine pitch bonding or three dimensional packaging rather than on high routing density. For an MCM with microprocessor chips, the routing density or the off package connections probably determines the minimum package size.

The TFML technology offers reduced component footprints as well as high routing density. For wire bonding or tape automated bonding, it is desirable for the substrate to match the bond pad pitch on the IC, which can be as small as 100 μm in current technology. This is not difficult for thin film patterning processes, whereas MCM-C technologies normally require a double row of bond pads to achieve this density. For even higher packing density, flip chip bonding is desirable. The thin film system is well suited for flip chip bonding because it can be processed on a CTE matched substrate (such as Si or AlN) and can meet the flatness and density requirements for flip chip bonding. Another way to achieve high chip packing density is to pattern the interconnection over the face of the die, as is the High Density Interconnect (HDI) overlay approach discussed in the next section.

The routing density of TFML interconnections is limited by either the linewidth and spacing or, more often, by the via pitch. Photolithographic capabilities and yield place a lower limit on feature sizes. In general, linewidths and via diameters can be reduced by decreasing the conductor and dielectric thickness. However, this increases the interconnect resistance and capacitance, and thereby reduces the high speed performance of the interconnections. Thus the design of TFML interconnections is determined by a tradeoff between large conductor cross sections for high speed performance and fine lines/thin layers for high interconnect density. Electrical simulation of interconnects is important for performing this tradeoff.

There are several ways of improving the thermal performance of a thin film interconnection system. Since polymer dielectrics are poor thermal conductors, there is a high thermal impedance through the interconnect layers. The specific thermal impedance for vertical conduction through 40 μm of polyimide with no spreading would be approximately $2.5°C \ cm^2/W$. The most common method for reducing thermal impedance is to pattern a dense array of metallized vias through all the dielectric layers beneath the chip attach pads (Figure 7-3; thermal vias can also be seen on the die attach sites of the substrate in Figure 7-15). It has been shown that Cu vias 35 μm in diameter on a 100 μm triangular pitch can reduce the thermal impedance through four 10 μm thick polyimide layers to less than 2.6°C/W for the heat generated by a 0.5 cm square chip (specific thermal impedance of 0.1°C cm^2/W) [30]. The thermal vias permit partial routing through the via array beneath the chips. For even lower thermal impedance, an opening can be patterned through the interconnection layers, placing the chip in direct contact with the substrate. However, this greatly reduces the area available for interconnection routing.

The optimum way to achieve efficient heat removal plus high interconnect density is to have the interconnections on the active side of the chips and heat removal from the backside. This is achieved in flip chip bonding; however, compliant materials (such as conductive pastes or solders) or special designs (such as the IBM Thermal Conduction Module or liquid immersion cooling) are required to remove heat from the backside of the chips [31]. Another alternative is the HDI approach, which places the chips in direct contact with the substrate and fabricates the interconnection on an overlay laminated to the face of the chips (discussed more in the next section). In this technology, the thermal density is limited by the ability to remove heat from the substrate, rather than from the ICs.

7.5.2 Implementations of Thin Film Interconnections

Thin Film Substrate in a Package

There are many ways to incorporate TFML interconnections into high density packaging structures, as shown in Figure 7-1. The most widely used approach is to fabricate TFML interconnections on a substrate that is populated with chips and mounted into a second level package, similar to bonding a chip into a single chip package. This permits the use of a wide variety of substrate and package materials. The multichip substrate can be procured as a component by a hybrid or IC packaging facility that manufactures MCMs. Thin film interconnection substrates have been available (albeit in small quantities) from a number of vendors, including Alcoa (shown in Table 7-1), nChip, Polycon and Advanced Packaging Systems.

Options for the second level package include metal flatpacks with glass insulated leads, cofired ceramic packages of Al_2O_3 or AlN [32], and plastic packages. The metal and ceramic packages can be hermetically sealed after thoroughly baking out the polymer dielectric material. For this reason, they are the most widely used package for military qualified MCM-Ds. Figure 7-16 shows an MCM-D for a radiation hardened MIL-STD-1750A computer

(Honeywell's Generic VHSIC Spaceborne Computer) containing five chips on a TFML substrate in a 2.1" square cofired ceramic package with 200 leads. The substrate (shown in Figure 7-15) has four metal layers and 1400 cm of interconnect wiring. There are 1100 gold wire bonds between the chips and substrate and 200 wire bonds between the substrate and package. In an earlier design, the same substrate was mounted in a Kovar flat pack with glass feedthrough leads [33]. A complementary memory module with nine radiation hardened 8k × 8 SRAM memory die and a memory line driver chip in a 2.6" × 1.6" cofired ceramic package is shown in Section 9.3. An IEEE task force has recently developed a recommendation to JEDEC for a series of standard package sizes for MCM-D substrates. This will reduce the tooling expense for packages and permit standardization of testing and handling fixtures.

Plastic packaging may be done in two ways: (1) the assembled multichip substrate is soldered to a premolded plastic leadframe, an encapsulant is applied for mechanical and environmental protection and a cover and optional heat sink is attached, or (2) the multichip substrate is soldered to a leadframe and then molded into an epoxy-based material. The most significant implementation of plastic packaged MCM-D is AT&T's Polyhic technology [34]. They are producing several standard sizes of plastic leaded chip carriers, quad flat packs and dual in line packages containing MCM-D substrates.

TFML on MCM-C
Patterning the TFML interconnections directly on a package base such as cofired ceramic eliminates the substrate to package bonds, a drawback of the first approach. Power and ground distribution layers can also be incorporated into the substrate. This approach has been termed MCM-D/C, for MCM-D on ceramic. MCM-D/C has been widely implemented in large mainframe computer systems. The outstanding examples are: the NEC SX series of supercomputers which use copper/polyimide interconnections on a multilayer Al_2O_3 cofired ceramic substrate as an interconnect for individually packaged flip TAB carriers [35], and latest version of the IBM thermal conduction module for the ES/9000 computer, which uses 63 layers of cofired glass-ceramic with Cu metallization to handle the signal interconnection and power/ground distribution and a thin film layer of Cu on polyimide for partial redistribution of the flip chip bonding pads to engineering change pads [8], [27]. Motorola, Honeywell and the Microelectronics Center of North Carolina have also reported on the development of this approach.

While the multilayer ceramic substrate is attractive because it serves as the package housing and second level interconnection, it presents a number of technical problems:

1. It is difficult to handle large, pinned substrates in IC process equipment.
2. Lapping and polishing are normally required to achieve a sufficiently flat and defect free surface for thin film patterning.
3. The thin film patterns must mate with the vias in the cofired ceramic, which have a large uncertainty in their location.
4. No seal rings can be brazed on the top surface, since a flat surface is required for photopatterning processes.
5. Substrates are expensive, and thus the TFML structures must have high yield or be reworkable.

The most likely users of MCM-D/C will continue to be vertically integrated manufacturers that control all aspects of MCM fabrication and assembly.

Chip-on-Board

In conventional chip-on-board packaging, unpackaged ICs are bonded directly to an interconnecting substrate (PWB or flexible tape) and protected with a glob top encapsulant. This approach has been used for a number of years in low cost, low density applications such as watches. Thin film interconnections offer the possibility of extending this approach to much higher levels of integration with higher interconnect density. A ceramic, metal or multilayer ceramic substrate equivalent in size to a PWB is patterned with TFML interconnections.

This approach is distinguished from conventional MCMs by the relatively large size of the substrate and the removal of sealing structures such as seal rings and lids. This requires a protective coating for the chips to prevent corrosion or mechanical damage. There has been active development of reliability without hermeticity in recent years and a number of promising coatings have been introduced, including glob top encapsulants [36] and vapor deposited coatings. The chip on substrate approach offers advantages in both cost and performance since it eliminates several components and assembly steps as well as the thermal and electrical impedance of additional packaging interfaces. However, it requires high confidence in functional die before assembly, and sophisticated testing methods are needed for the assembled board.

High Density Interconnect (HDI) Overlay Approach

Some thin film interconnection systems have unique features that fall outside the standard approaches discussed above. One such system is the General Electric high density interconnection (HDI) represented in Figure 7-17 [37]. In this approach, the chips are mounted in cavities in a substrate. A Kapton (polyimide) film is laminated to the face of the chips and a laser is used to etch vias holes for contact to the chip bonding pads. A TFML interconnect structure is built on

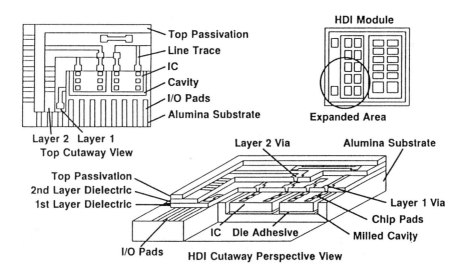

Figure 7-17 Representation of the General Electric High Density Interconnect overlay approach. (Courtesy of R. Fillion, GE.)

the Kapton overlay, using the laser to etch via holes in the polymer dielectric and to pattern the conductor photoresist. Cu conductor is deposited by sputtering and plating and patterned by wet etching. The completed substrate is mounted into a second level package.

The advantage of this approach is that die can be tightly packed, essentially side by side, and heat can be removed efficiently since the die are in direct contact with the substrate. High routing density can be achieved since no thermal vias are required through the interconnect layers. Chip bonding operations (such as wire bonding) are eliminated by the metallurgical via contact to the chips. Rework is the biggest concern with this approach. The overlay can be removed by heating and peeling, but removal of defective die requires a more complicated procedure. There is also a concern with the cost and throughput of the laser-based processes. Finally, there are limited sources for this type of interconnect.

DEC High Density Signal Carrier
A thin film interconnection system with several unique features is used by DEC in the High Density Signal Carrier (HDSC) for their VAX-9000 series of mainframe computers. In this technology, copper/polyimide thin film

interconnections are built on a temporary Al substrate and then detached, leaving a free standing film. One film contains four power distribution layers (the power core); the other contains two interconnect layers surrounded by reference planes, plus the top bonding layer (the signal core). These two cores are laminated together and the power core is electrically connected to the top surface by drilling and plating through holes as in PWB processing. Cutouts are excised for the chips, and the structure is laminated to a Cr-Cu substrate. This approach offers excellent power distribution and heat removal for very high power density ECL ICs. However, since the chips are in direct contact with the metal substrate, there is no wiring under the chips, limiting the interconnection density. The DEC technology is described in greater detail in Chapter 17.

MCC Quick Turnaround Interconnect

Another unique concept for using TFML interconnections is the quick turnaround interconnection (QTAI) proposed by MCC. The QTAI is designed to reduce the turnaround time and tooling costs of new interconnect designs. This is done using a generic substrate with power and ground distribution layers and reconfigurable interconnection layers consisting of short x- and y- segments with vias to the top surface (analogous to a gate array for IC design). The substrate can be personalized with one metal layer which connects the required segments to achieve the interconnections. The top layer also provides the chip bonding pads. The QTAI has been demonstrated in a crossbar switch containing 16 identical ICs (0.300" × 0.360" chips with 103 leads/chip) and 336 external connections, on a substrate approximately 2" square [38].

7.6 APPLICATIONS, GROWTH OF THIN FILM INTERCONNECTIONS

7.6.1 Interconnection Technology Selection

The thin film interconnection system offers many fundamental performance advantages, as discussed in Section 7.2, namely: high interconnection density, high speed signal transmission, low impedance power distribution and high bonding density. It also offers flexibility in terms of substrate materials, thermal designs and package implementation. However, the selection of an interconnection technology involves a number of additional considerations.

The interconnection technology must be compatible and robust for a variety of assembly processes. IC assembly operations have more experience with die attachment and bonding to ceramic substrates. In general, the harder ceramic surface permits the use of higher force, temperature, and energy in thermosonic and ultrasonic wire bonding. The bonding and die attach pads on ceramic substrates are also more robust toward rework processes than thin film metallization on polymer dielectrics. Finally, there is less concern with absorbed moisture in hermetically sealed MCM-C packages. All of the assembly and sealing operations have been demonstrated for polymer-based TFML systems, but there is less experience and a smaller data base for the technology.

Because of the relative immaturity of TFML technology, there is less life cycle and field performance data compared to older technologies (although IBM, AT&T and DEC have generated extensive life cycle data internally). There is a general concern in the military community with the reliability of packages containing polymers. Some of these concerns arise from early failures attributed to impurities in epoxy die attach materials. In general, ceramic interconnect systems are considered more reliable at this time. More accelerated life cycle testing, failure analysis and field testing of TFML systems is required.

The most important barrier to the use of MCM-D is its high cost and limited availability. Initial applications have been in customized MCMs for large mainframe or supercomputers (as in the NEC, IBM and DEC examples discussed earlier) or in prototype military systems. None of these applications has generated the high volume of standardized modules and substrates necessary to reduce costs and provide a broad manufacturing base. Most of the high volume manufacturing facilities are for captive products. A vendor infrastructure has not developed for providing thin film substrates or assembled MCM-Ds as a component, as exists for single chip and multichip ceramic packages. The development of this infrastructure is the most important requirement for the widespread application of thin film interconnection systems.

7.6.2 Evolution of TFML Applications

Since thin film interconnections cannot compete with alternative interconnect technologies on a cost basis at this time, initial applications of the technology must be performance driven. Thin film interconnections must be required for small size and/or weight, as in space-based electronics, military avionics, or medical implants. Alternatively, TFML interconnects must be demanded by the requirements of fast clock speeds and high interconnect density, as in supercomputers, mainframes, or high speed digital processors using GaAs ICs.

The emergence of TFML technology in these specialized applications will be followed by higher volume products where size is still a discriminator, such as laptop or notebook computers, commercial avionics, or possibly telecommunications electronics. As the range of applications and production volumes increase, the required vendor infrastructure will develop, and costs will be driven down to make MCM-D competitive with MCM-C and MCM-L technologies. At the same time, assembly experience and reliability data will grow, increasing confidence in the reliability of the technology. An unresolved issue is whether thin film substrates will be provided as a component by traditional package manufacturers (cofired ceramic vendors) or specialized thin film substrate foundries, or whether assembled MCM-Ds will be manufactured by captive systems houses, IC foundries, hybrid manufacturers or full service MCM foundries.

7.6.3 Future Applications

Thin film interconnection systems provide a technology base for even higher performance packaging technologies that are currently under development. Some examples are mentioned briefly:

1. Thin film interconnections are needed for hybrid wafer scale integration (HWSI), in which a wafer is tested and diced, and the good die are reconstructed into an array with essentially the same density as the original wafer. This approach requires flip chip bonding to a CTE matched substrate (possibly with active or passive devices, as discussed in Section 7.3), or an HDI-type of overlay interconnection. HWSI achieves the high circuit density of monolithic wafer scale integration, but avoids the need for redundant circuitry.

2. Stacking of die, especially memory ICs, is being developed to achieve very high volumetric density with additional interconnection in the vertical direction. Thin film interconnections are sometimes patterned on the face of these die cubes. TFML substrates are often used to interconnect the die stacks [39] because of the bonding density, flatness, and special metallization needed.

3. Integrated optical waveguides for chip-to-chip interconnections have been built using thin film structures with the same materials system as for electrical interconnects, using polyimides for the waveguide media [40]. This offers the possibility of a combination of optical and electrical interconnections in the same substrate.

4. Superconductors have been proposed for overcoming the speed/density limitations of traditional conductor materials [41]. Very thin, fine line superconductor interconnections will be lossless, as long as the temperature and current are below the critical values. This will permit almost unlimited interconnection density, lossless signal and power distribution and very thin multilayer structures.

References

1   I. J. Bahl and R. Garg, "A Designer's Guide to Stripline Circuits," *Microwaves*, vol. 17, pp. 90-95, Jan. 1978.
2   E.E. Davidson, "Electrical Design of a High Speed Computer Packaging System," *IEEE Trans. CHMT.*, vol. CHMT-6, no. 3, pp. 272-282, 1982.
3   K. Jayaraj, *et al.*, "Performance of Low Loss, High Speed Interconnects for Multi-GHz Digital System," *Proc. IEEE Nat. Aerospace and Elect. Conf.*, (Dayton, OH), pp. 1674-1681, May 1989.
4   "Alcoa Thin Film Design Guide," Rev. 92-04, *Alcoa Electronic Packaging Inc.*, San Diego, CA, 1992.
5   F. D. Austin, D. C. Green, M. A. Robbins, "VCOS (VHSIC Chips-on-Silicon): Packaging, Performance and Applications," *Digest of Papers, Government Microcircuit Appl. Conf.*, (Las Vegas, NV), pp. 577-581, Nov. 1990.
6   B. K. Gilbert, G. W. Pan, "Packaging of GaAs Signal Processors on Multichip Modules," *IEEE Trans. CHMT.*, vol. CHMT-15, no. 1, pp. 15-28, 1992.
7   B.C. Foster, *et al.*, "Advanced Ceramic Substrate for Multichip Modules with Multilevel Thin Film Interconnects," *IEEE Trans. CHMT.*, vol. CHMT-14, no. 4, pp. 784-789, 1991.
8   E. E. Davidson, *et al.*, "The Design of the ES/9000 Module," *IEEE Trans. CHMT.*, vol. CHMT-14, no. 4, pp. 744-748, 1991.
9   T. Inuoe, *et al.*, "Microcarrier for LSI Chip Used in the HITAC M-880 Processor Group," *IEEE Trans. CHMT.*, vol. CHMT-15, no. 1, pp. 7-14, 1992.
10 R. D. Nicholson, R. S. Fusco, "Copper Clad Molybdenum for High Performance Electronics," *Proc. ASM 3rd Conf. on Electronic Packaging*, (Minneapolis, MN), pp. 87-90, April 1987.
11 R. C. Eden, "Applicability of Diamond Substrates to Multi-chip Modules," *Proc. Internat. Symp. Microelect.*, (Orlando, FL), pp. 363-367, Oct. 1991.
12 Y. H. Kim, *et al.*, "Adhesion and Interface Investigation of Polyimide on Metals," *J. Adhesion Sci. Tech.*, vol. 2, no. 2, pp. 95-105, 1988.
13 N. G. Koopman, T. C. Reiley, P. A. Totta, "Chip-to-Package Interconnections," *Microelectronics Packaging Handbook*, R. R. Tummala, E. J. Rymaszewski, eds., New York: Van Nostrand Reinhold, 1989, Chapter 6.
14 R. J. Jensen, J. P. Cummings, H. Vora, "Copper/Polyimde Materials System for High Performance Packaging," *IEEE Trans. Components Hybrids Manuf. Techn.*, vol. 7, no. 4, pp. 384-393, 1984.
15 B. T. Merriman, *et al.*, "New Low Coefficient of Thermal Expansion Polyimide for Inorganic Substrates," *Proc. 39th Electronic Components Conf.*, (Houston TX), pp. 155-159, May 1989.
16 N. H. Hendricks, "Thermoplastic Polyquinolines: New Organic Dielectrics For Highly Demanding Packaging Applications," *Proc. Internat. Symp. Microelectr.*, (Orlando FL), pp 105-109, Oct. 1991.
17 S. M. Sze, ed. *VLSI Technology*, 2nd edition, New York,: McGraw-Hill, 1988.
18 J. L. Vossen, W. Kern, eds, *Thin Film Processes*, New York: Academic, 1978.

19 D. Darrow, S. Vilmer-Bagen, "A Comparative Analysis of Thin Film Metallization Methodologies for High Density Multilayer Hybrids," *Proc. Internat. Conf. Multichip Module*, (Denver, CO), pp. 56-70, April 1992.

20 C. J. Bartlett, J. M. Segelken, N. A. Teneketges, "Multichip Packaging Design for VLSI-based Systems," *IEEE Trans. CHMT.*, vol. CHMT-12, no. 4, pp. 6479-653, 1987.

21 R. Miracky, *et al.*, "Laser Customization of Multichip Modules," *Digest of Papers, Government Microcircuit Appl. Conf.*, (Orlando, FL), pp. 235-238, Nov. 1991.

22 D. S. Soane, Z. Martynenko, *Polymers in Microelectronics: Fundamentals and Applications*, Amsterdam: Elsevier, 1989, Chapter 4.

23 C. Speerschneider, *et al.*, "Honeywell's VHSIC Phase 2 Packaging Technology," *Proc. VHSIC Packagingin Conf.*, (Houston TX), pp. 131-143, April 1987.

24 T. Snodgrass, G. Blackwell, "Advanced Dispensing and Coating Technologies for Polyimide Films," *Proc. Internat. Conf. Multichip Modules*, (Denver, CO), pp. 428-435, April 1992.

25 H. J. Neuhaus, "A High Resolution, Anisotropic Wet Patterning Process Technology for MCM Production," *Proc. Internat. Conf. Multichip Modules*, (Denver, CO), pp. 256-263, April 1992.

26 T. Tessier, "Compatibility of Common MCM-D Dielectrics with Scanning Laser Ablation Via Generation Processes," *Proc. 42nd Elect. Components and Tech. Conf.*, (San Diego, CA), pp. 763-769, May 1992.

27 T. F. Redmond, C. Prasad, G. A. Walker, "Polyimide Copper Thin Film Redistribution on Glass-ceramic/Copper Multilevel Substrates," *Proc. 41st Elect. Components and Tech. Conf.*, (Atlanta, GA), pp. 689-692, May 1991.

28 J. T. Pan, S. Poon, B. Nelson, "A Planar Approach to High Density Copper-Polyimide Interconnect Fabrication," *Proc 8th Internat. Elect. Packaging Conf.* (Dallas, TX), pp. 174-189, Nov. 1988.

29 C. W. Ho, *VLSI Electronics: Microstructure Science*, N. G. Einspruch, ed., New York: Academic, 1982, vol. 5, chapter 3.

30 A. Keely, C. Ryan, "Achieving a Balance of Thermal Performance and Routing Density in a Multichip Module Substrate,' *Proc. NEPCON East '91*, (Boston, MA), pp. 551-561, June, 1991.

31 R. Darveaux, I. Turlik, "Backside Cooling of Flip Chip Devices in Multichip Modules," *Proc. Internat. Conf. Multichip Modules*, (Denver, CO), pp. 230-241, April 1992.

32 J. K. Hagge, "Ultra-reliable Packaging for Silicon-on-Silicon WSI," *IEEE Trans. CHMT.*, vol. CHMT-12, no. 2, pp 170-179, 1989.

33 T. J. Moravec, *et al.*, "Multichip Modules for Today's VLSI Circuits," *Elect. Packaging and Prod.*, vol. 30, no. 11, pp. 48-53, Nov. 1990.

34 R. E. Maurer, "The AT&T MCM Packaging Program," *Proc. Internat. Conf. Multichip Modules*, (Denver, CO). pp. 28-33, April 1992.

35 T. Watari, H. Murano, "Packaging Technology for the NEC SX Supercomputer," *IEEE Trans CHMT.*, vol CHMT-8, no. 4, pp. 462-467, 1985.

36 B. E. Goblish, *et al.*, "The Reliability of EP/TAB Integrated Circuits," *Proc. Internat. Elect. Packaging Conf.*, (Marlborough, MA), pp. 858-874, Sept. 1990.

37 L. M. Levinson, *et al.*, "High Density Interconnects Using Laser Lithography," *Proc. Internat. Symp. Hybrid Microelect.*, (Seattle, WA), pp. 301-306, Oct. 1988.

38 D. C. Carey, L. Paradisio, "A Collaborative VHSIC Multichip Module Design Using Programmable Copper/Polyimide Interconnect," *Digest of Papers, Government Microcircuit Appl. Conf.*, (Las Vegas, NV), pp. 533-535, Nov. 1990.

39 R. Bruns, W. Chase, D. Frew, "Utilizing Three Dimensional Memory Packaging and Silicon-on-Silicon Technology for Next Generation Recording Devices," *Proc. Internat. Conf. Multichip Modules*, (Denver, CO), pp. 34-40, April 1992.

40 J. P. G. Bristow, *et al.*, "Polymer Waveguide-Based Optical Backplane for Fine-Grained Computing," *Optical Interconnects in the COputer Environment*, Bellingham WA: SPIE - Internat. Soc. for Optical Engineering, 1990, pp. 102-114.

41 R. C. Frye, *Elect. Packaging Materials Sci. III*, R. Jaccodine, K. A. Sindahl, eds., Pittsburgh, PA: Materials Research Society, vol. 108, pp. 27-38, 1988.

8

Selection Criteria for Multichip Module Dielectrics

Claudius Feger and Christine Feger

8.1 INTRODUCTION

In the last decade, polymeric dielectrics have been widely accepted as materials of choice for interlayer dielectrics in MCMs [1]-[2]. The driving forces for this development are:

1. The low dielectric constant, usually exhibited by organic polymers, allows higher packaging densities in comparison to ceramic packages that have a higher dielectric constant. Furthermore, a lower dielectric constant dielectric material leads to faster transmission speeds and to lower power consumption.

2. Polymers are relatively easy to process with procedures developed for semiconductor thin films and therefore are widely understood. Although the equipment and materials can be expensive in comparison to thick film technology, most companies already have the knowledge and some equipment in place and can start development work without a large investment in new technology.

3. Many interesting combinations of properties can be tailored by making changes in the chemical composition of these polymers. The search for

the ultimate dielectric polymer, coupled with the emergence of a lucrative market for such specialty polymers, has increased tremendously the number of chemical companies offering polymers to the electronics packaging industry. Thus engineers today have the luxury and responsibility to select from an ever increasing market offering the best material for their specific application.

To apply polymeric dielectrics in MCMs it is essential to have basic knowledge in two very different areas:

1. Polymer science including polymer chemistry, polymer physics, polymer processing and an understanding how these areas relate to the observed properties, and

2. MCM engineering including electrical design, electronics packaging technology and an understanding of both thin film and thick film processes.

Unfortunately, the two areas rarely are interrelated. Available literature either is directed exclusively to the polymer science community or exclusively to the engineering community. Too often this results in costly misunderstandings and a lack of needed cross-fertilization. This chapter tries to fill the gap in the literature, attempting to allow the reader schooled in electrical engineering to understand polymer material aspects and to assist the chemist or physicist unfamiliar with polymers and/or electronics packaging in their development efforts. This chapter describes the basic properties required of polymeric dielectrics in high performance computer MCMs, the chemistry and properties of the most common polymeric dielectrics under consideration today (polyimides, fluorocarbon polymers, polyphenylquinoxalines (PPQ) and benzocyclobutenes (BCB)) and relates processing requirements to material properties. The reader should become better equipped to use vendor material data sheets to make more informed material selections for a particular application. Before describing the process variables in MCM manufacturing, the behavior of dielectrics in an electric field and their function in a package will be considered.

8.2 BEHAVIOR AND FUNCTION OF DIELECTRICS

There are three reasons why good dielectric properties are important for electronics packaging: signal speed, power consumption and wiring density. In the following section, the response of a dielectric in an electric field is described, explaining the dependence of signal speed, power consumption and wiring density on dielectric properties.

Dielectrics, materials that do not conduct electricity, are also called insulators. Exposure to a static electric field causes the electric charges present in any dielectric material as permanent electric dipoles and/or induced electric dipoles to be moved, polarizing the material. The equilibrium polarization remains a material constant for a given electrical field. However, it is the dielectric constant, ε, also symbolized by κ, that is used to characterize the dielectric properties of a dielectric, not the polarization. ε most commonly is defined as the ratio of capacitance in a capacitor filled with the dielectric material to capacitance of the identical capacitor filled with a vacuum. The magnitude of ε depends on the amount of mobile (polarizable) electrical charges and the degree of mobility of these charges in the material. Because the charge mobility depends on temperature, ε is temperature dependent. Time is required to establish the equilibrium polarization. Thus in a dynamic electric field the dielectric constant depends also on the frequency of the field change. In an AC field, the dielectric constant is complex (complex permittivity, $\varepsilon*$) and has two components: the real component (ε') called the dielectric constant of the material[1], or the relative permittivity, and the imaginary component (ε'') called the loss component or the dielectric loss factor. Instead of ε'' the dielectric loss tangent or dissipation factor (tan $\delta_\varepsilon = \varepsilon''/\varepsilon'$) is usually reported.[2] Loss or energy absorption occurs because energy and time are required to establish the polarization. Consequently, in a dynamic field the polarization of the material lags the applied field. A dielectric material with both the lowest possible dielectric constant and dielectric loss is desirable for applications as electrical insulators.

An electric signal moving through a dielectric can be described as an electromagnetic sine wave or as a superposition of different sine waves. As the signal advances it polarizes the dielectric. This interaction with the dielectric limits the speed of the signal. The speed at which an electromagnetic sine wave propagates through a medium is inversely proportional to the square root of ε, a factor relevant to the signal speed in an electronic package.

If the signal is a superposition of many different sine waves, as in a square wave signal pulse, the dielectric behavior at each of the component frequencies

---

[1] Commonly the ' is omitted in symbolizing the real component of the complex dielectric constant as the dielectric constant of interest is the one for dynamic (AC) fields.

[2] Incorrectly, tan δ is sometimes called the dielectric loss factor, the term reserved for ε''.

has to be considered. If the dielectric constants at the various frequencies differ, signal shape dispersion can result leading ultimately to weak signals unable to trigger the intended on/off event. To counteract signal dispersion higher energy signal pulses need to be used that directly influence the power consumption of a computer.

The final property to be discussed here concerns signal density. Electromagnetic signal waves traveling through a signal line can induce a signal in a parallel signal line. This phenomena is often called crosstalk. The magnitude of the induced signal depends on the magnitude of the original signal, the distance between the two conductor lines and the dielectric constant of the material between them. The lower the dielectric constant the closer the signal lines can be without inducing signals of significant magnitude.

8.3 MULTILEVEL THIN FILM STRUCTURES

Current state of the art MCMs in high performance computers consist of a multilayer ceramic (MLC) carrier which is combined with a polymer/metal thin film module (TFM)[3]. The MCMs are populated either with chips connected by wire bonding, tape automated bonding (TAB) or by flip chip as described in Chapter 9. TFMs consist of at least two signal planes but can have many more layers. All TFMs currently in production have one feature in common: they all use a polyimide as an interlayer dielectric [2] as described in later sections of this chapter.

8.3.1 Major Technical Challenges

It is desirable for any electronic component, particularly for the package, to avoid field failure. Package field failures, such as electrical shorts or opens, often occur because the package first fails mechanically, which then leads eventually to failure in conductor lines by corrosion, etc. Figure 8-1 shows a schematic drawing of a TFM. Highlighted in the drawing are areas responsible for the mechanical failure of the TFM:

1. Interfacial failure leading to local adhesion loss or to catastrophic delamination with complete detachment of a layer from the substrate

---

[3] This area is discussed from the perspective of IBM mainframe computers with emphasis on C4 chip connection technology.

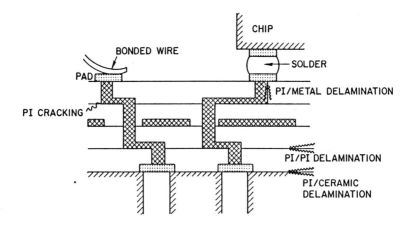

Figure 8-1 Schematic drawing of a thin film module (TFM).

2. Material failure such as cracking or crazing[4].

Interfacial and material failures are related to the presence of stress. In fully cured polymeric coatings, such stress is due nearly exclusively to the presence of materials with differing coefficients of thermal expansion (CTE). Therefore, one of the major technical challenges in the manufacture of TFMs is stress management, focusing on low stress materials.

Mechanical failure sometimes occurs because the properties of an initially good material degrade during processing. The severest demand on organic polymers in TFMs is the necessity to survive the high temperatures (up to 370°C) of the chip joining process. In this flip chip solder bump (FCSB) process, chips are connected to the substrate solder bumps that are used as electrical contacts between chips and signal and power sources. The solder bumps are made of a lead/tin alloy with a melting point on the order of 360°C. The requirement to withstand exposure to the many solvents and aggressive processes such as plasma cleaning which occur in the fabrication sequence also is very challenging.

Requirements on a polymeric dielectric for building reliable TFM structures lead to a wish list for the ideal TFM polymer. Table 8-1 shows such a list of

---

4   Crazing is a material failure which usually precedes crack formation in polymeric materials.

Table 8-1 Properties of an ideal packaging polymer.

Dielectric constant	2.0 to 3.0 frequency independent
CTE (ppm/C)	1 to 5
Thermal stability	at 400°C <0.1 wt. %/hr
Young's modulus	0.7 < x < 3.0 GPa
Elongation at break	>10%
Critical crack propagation	above 2.5% strain
Adhesion	good to metals, substrate and self
Application	from solution (spin, spray)
Planarization	> 90%
Shrinkage	< 10 vol. %
Water absorption	0.1 to 0.5 %
Solvent resistance	no swelling, no crazing
Etchability	RIE, laser, wet: no residue
Color	light, transparent

ideal properties. The requirements for a given application need to be ranked in order of their importance as determined by both the manufacturing process and the desired performance. While performance mostly is based on speed and density, both of which are related to the dielectric constant of the polymer, there are several options for the fabrication process. The range of these options may be limited if some processes are fixed, such as C4 chip connection process, so the material must be selected with that in mind. Some processes can be changed to suit the dielectric. These decisions depend on the extent that a company is committed to a given chip connection process or fabrication facility.

8.3.2 MCM Fabrication Processes

First, one has to decide on a chip connection technology as discussed in Chapter 9. Wire bonding and TAB require only moderate temperatures to make the connections and easily accommodate the CTE mismatch between chip and module due to the relative flexibility of the joints. Unfortunately, the number of I/Os and the cooling of TAB (flip TAB has no cooling problem) or wire bonded chips are serious limitations. Flip chips or C4s usually require a high chip joining process temperature. However, they allow much higher I/O densities.

Next, it is necessary to decide what substrate to use. Alumina multilayer ceramic (MLC) substrates have the advantages of a mature technology and good mechanical properties and the disadvantages of a high CTE and a high dielectric constant. Silicon, used as the substrate, matches the CTE of the chip perfectly, but silicon is relatively brittle and via formation and multilevel structures in silicon substrates are difficult to obtain. Glass ceramics, which have CTEs matched to silicon, have relatively low dielectric constants and can be used for manufacturing complex MLCs. However, their mechanical and thermal properties are limitations. These ceramics are discussed further in Chapter 6.

The relationship between the dielectric and conductor must be considered when selecting the conductor. The CTE of the conductor is usually higher than that of the substrate. The dielectric should be chosen to act as a cushion between the metal and substrate. Copper is often desired for its low resistivity $(1.7 \times 10^{-6} \ \Omega\text{-cm})$ and because it is possible to plate-up. Its disadvantage is that it has very poor adhesion to polyimides and requires capping with an adhesion layer such as chrome. Aluminum is sometimes used for its good adhesion to polyimides simplifying the processing, but it has a higher resistivity $(\approx 2.7 \times 10^{-6} \ \Omega\text{-cm})$ and can exhibit migration when heated repeatedly to typical polyimide curing temperatures. Such migration can lead to the formation of filaments that grow on the aluminum lines and can eventually cause shorts. Aluminum lines thinned by migration can rupture under the stress caused by the expanding polyimide.

The next decision concerns the design of the multilayer thin film (MTF) process. For instance, it is planar or non-planar as described in Chapter 7. The choice of design often determines the processes necessary to realize the product. A planar process might involve via formation, via laser ablation, blanket metal evaporation, chemical mechanical polishing, liftoff processing and metal etching. A non-planar process might involve reactive ion etching (RIE) of the polymer, metal sputtering and metal wet etching. Electro or electroless plating might be applied in either design. All processing steps possibly may have negative effects on the organic films or only with some polymers. In any case, the choice of design and processing steps significantly influences the choice of the dielectric.

In recent years photosensitive polyimides (PSPIs) have become available commercially. Figure 8-2 shows a comparison between a process utilizing PSPI and one with conventional polyimide. Obviously the PSPI process involves fewer process steps and, therefore, promises higher cost effectiveness.

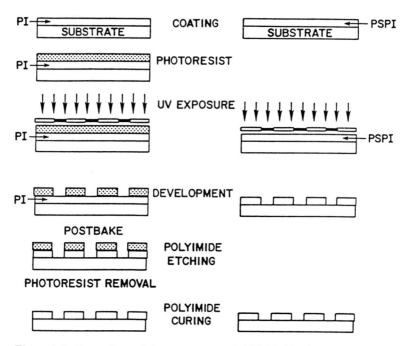

Figure 8-2 Comparison of the process steps in TFM built with conventional and photosensitive polyimide.

Unfortunately, PSPIs still have low photosensitivity, high stress, propensity to mechanical failure, high dielectric constants compared to their nonphotosensitive analogs, inferior thermal stability and often poorer shelf life. Because of the very substantial advantages PSPIs potentially have over conventional polyimides, research in this area is very intense.

8.3.3 Dielectric Processing

The most widely used methods to apply a dielectric are spin coating, spray coating, lamination and extrusion coating, all of which are described in Section 7.4.2. Most polymers require that a primer (adhesion promoter) be applied before polymer deposition. After drying, the polymer is cured to remove solvents. Chemical processes also can occur, producing volatiles species which also must be removed. The final properties of the dielectric are determined by the cure schedule. However, the properties also can be affected by further processing.

Patterning of the dielectric can be done by dry processes (RIE or plasma etching) or by wet etching as presented in Section 7.4.3. RIE or plasma etching, using oxygen plasmas, works for most polymeric dielectrics. Wet etching, usually with aqueous alkaline solutions, works only with certain polymer dielectrics such as polyimides and generally only if the polymer is undercured.

Liftoff (described in Section 7.4.3) is another process to which polymeric dielectrics may be exposed. The interaction of the hot solvent in the liftoff process with the underlaying polymeric dielectric can cause problems like swelling or solvent induced crazing.

After design and processing decisions are made, the polymeric dielectric is chosen. The combination of polymer properties must be appropriate for the user's particular application. Important properties for this consideration are dielectric constant, thermal stability, glass transition temperature (T_g), CTE, moduli, elongation at break, failure mechanisms, solvent resistance, adhesion to various materials, water uptake, planarization, curing conditions, etchability (plasma, RIE, wet etc.), shelf life and others (ionics content, batch to batch stability, barrier properties, processing conditions etc.). Some of these properties are related and knowledge of one property can give an indication of the quality of another. How to interpret data provided by a supplier or obtained from the literature will be discussed in the next paragraphs.

8.4 POLYMER PROPERTIES

8.4.1 Some Polymer Specific Terms

Polymers are either amorphous or semicrystalline. Amorphous materials do not exhibit any crystalline nor liquid crystalline order. They can be thought of as frozen liquids (glasses). Their softening temperature is the glass transition temperature above which no further transitions are observed. Semicrystalline polymers consist of amorphous and crystalline phases and also are called two phase or multiphase polymers. The degree of crystallinity can vary with processing conditions and film thickness. The phase structure of a polymeric material is called its morphology. Particles added to a polymer result in a filled polymer.

From a chemical standpoint, one distinguishes between linear, branched and crosslinked polymers. In a linear polymer, all repeat units, the building blocks of the polymer, are connected only to one or two other units. In branched or crosslinked polymers, some repeat units (branch or crosslink points) are connected to more than two other units. As the ratio between crosslink points and linear units increases, so does the crosslink density, the glass transition and, often, the brittleness of the material. Linear and branched polymers dissolve in appropriate solvents; crosslinked polymers only swell. Linear and branched polymers can be deformed plastically by heating them above their glass transition temperatures or their melting points as long as this is below their thermal decomposition temperature. Therefore, these materials are called thermoplastics. They are usually tough materials. Crosslinked polymers can be deformed elastically even above their melting or softening temperature. These materials usually are processed starting from small, soluble units that produce the insoluble, crosslinked polymer upon chemical reactions during cure. They are called thermosets. They tend to be brittle but are easily processed.

The chain of repeat units is called the backbone of the polymer. This backbone ranges from flexible to rigid. In the latter case one speaks of rigid rod backbones. Some polyimides have rather rigid backbones which are called rigid rod-like. Rigid rod-like backbones can align themselves with respect to a substrate. This phenomenon is called chain orientation and causes many properties to be anisotropic.

8.4.2 Dielectric Constant

The dielectric constant of a dielectrical material should be low (2 - 3.5) and constant over a very wide frequency range from DC to well into the GHz range. The dielectric loss tangent, $\tan \delta_\varepsilon$, should be frequency independent and below 0.01.

Routinely, dielectric measurements are made by parallel plate capacitance methods over a range of only 0.1 kHz - 1 MHz and over a wide temperature range. The dielectric constant does not vary significantly with frequency (on the order of 0.1) for many of the polymeric dielectrics used in electronic packaging. However, some fluoro-containing polyimides exhibit significant changes with frequency on the order of 0.3 - 0.5 [4]. The values obtained by parallel plate capacitance measurements on films depend very critically on the uniformity of the film. Consequently literature values are often inconsistent. Sources of error such as absorbed water, contact resistance, ion content, edge effects and/or pinholes can complicate further the measurement of dielectric constants. The combination of these error sources leads to the multitude of values found in the literature for identical materials; reported variations of as much as 0.7 for the same material are not uncommon. Tables 8-2 through 8-6 indicate ranges of values for properties of packaging polymers taken from product literature. Table 8-2 shows values for some classes of materials obtained by parallel plate capacitance measurements between 0.1 kHz and 1 MHz.

Many high temperature stable polymers exhibit rigid rod-like backbones. In the presence of a substrate such materials orient themselves with respect to the

Table 8-2 Ranges of Dielectric Constants.

Ceramics	PIs	Fluoro-polymers	BCBs	PPQs	PIQs
5 - 9	2.5 - 3.8	1.9 - 2.6	2.6 - 2.8	2.8 - 3.0	3.2 - 3.4

PIs = polyimides   BCBs = benzocyclobutenes   PPQs = polyphenylquinoxalines
PIQs = polyimide iso-indoloquinazolinediones

substrate and consequently many properties are anisotropic [5]. Particularly in rigid rod-like polyimides, the dielectric constant in the plane may be higher than that perpendicular to the plane [5], [7], [24]-[26]. Unfortunately measured by the usual parallel plate capacitance method only the lower value, that perpendicular to the plane, can be determined.

8.4.3 Thermal Stability

Sufficient thermal stability is essential for organic dielectrics because the polymer properties, particularly mechanical properties, must remain unchanged during high temperature processes, such as curing of subsequent layers, chip joining or rework.   Additionally, outgassing, which usually accompanies thermal degradation, can destroy the multilayer structure, by causing delamination.

Thermal stability is usually measured by thermal gravimetric analysis (TGA) which measures the weight of a sample versus temperature. Two modes of measurement are available. In the dynamic mode, the sample is heated with a given heating rate (typically 10°C/minute) and the temperature at which a percentage of weight (1%, 5% or 10%) has been lost is recorded. This method measures kinetic effects and the results are heating rate dependent: the higher the heating rate the higher the temperature at which the particular loss is observed.

In the isothermal mode, the sample is heated quickly, then held at a given temperature and the weight loss per hour is measured. The results are given as weight loss in unit of percent per hour (wt.-%/h). From an electronic packaging standpoint, the values obtained through the first method are of little value. Because many processing steps involve annealing at relatively high temperatures, weight loss data at a given temperature are much more useful in understanding the thermal stability of a material.

Contrary to common sense, a low weight loss per hour value does not necessarily describe high thermal stability. A material is only thermally stable when its properties at the temperature in question do not change.

Table 8-3 Isothermal Weight Loss Data (wt.-%/h in N_2).

Temperature	PIs	Fluoro-polymers	BCBs	PPQs	PIQs
350°C	negligible	0.006-too high	1	negligible	negligible
400°C	0.05-1.5	2-too high	10	0.02	0.1-0.5

PIs = polyimides   BCBs = benzocyclobutenes   PPQs = polyphenylquinoxalines
PIQs = polyimide iso-indoloquinazolinediones

Curing reactions are accompanied by weight loss caused by initial solvent loss or loss of condensation products. Additional weight loss can occur in polymers at high temperatures caused either by outgassing of residual solvent or by degradation reactions [8]. The latter can be of two types: chain scission accompanied by the formation of radicals and elimination of water, CO_2, HF, etc., leading to formation of unsaturated compounds (carbon enrichment). Whereas a small number of elimination reactions might alter the polymer properties only slightly, radical formation is usually only the first step in a series of reactions. The most common of them is crosslinking of the polymer. This can lead to embrittlement, changes in the glass transition temperature, lowered solvent uptake, lowered elongation at break, adhesion loss, etc. It is important to keep in mind that if one property changes usually others do as well. As desirable as a lowered solvent uptake might be, the parallel increase in brittleness might make such an improvement undesirable. In Table 8-3 weight loss ranges obtained from isothermal measurements are given for some classes of electronics packaging polymers.

8.4.4 The Glass Transition Temperature

The glass transition temperature, T_g, is the temperature at which glassy polymers soften. Above this temperature polymers that are not crosslinked can flow. If the polymer is a two phase material (consisting of crystalline or liquid crystalline and glassy domains), the T_g is the temperature at which the glassy phase softens. These materials behave like a crosslinked material and thus do not flow above T_g. The T_g is important because a number of properties change above this temperature. Above the T_g, stresses are released (due to a drop in modulus), the CTE increases (without increasing the stress as long as the material is not three dimensionally constrained due to the mentioned modulus drops above T_g), defects (such as crazes) might be healed, adhesion might increase and dimensions

Table 8-4   Ranges of Glass Transition Temperatures, T_g (°C).

PIs	Fluoro-polymers	BCBs	PPQs	PIQs
300-> 400	160-320	310-> 350	361	300-> 400

PIs = polyimides   BCBs = benzocyclobutenes   PPQs = polyphenylquinoxalines
PIQs = polyimide iso-indoloquinazolinediones

(line spacing) might change due to polymer flow. The magnitude of these changes depends on the amount of the glassy domains in the system. The most pronounced changes occur in purely glassy materials with a sharp glass transition and are less pronounced in multiphase systems.

For our discussion it is important to mention two important characteristics of the glass transition [27]. First, T_g is not a clearly defined temperature but rather a temperature range. Secondly, T_g is not a transition in an equilibrium thermodynamic sense, although the underlying phenomena might be such a (second order) transition. This has two consequences. First, measured T_g values depend on kinetic factors such as the frequency of the measurement and/or the heating rate of the sample. Secondly, the temperature given as T_g is chosen more or less arbitrarily to be an easily observable feature of a measured experimental curve. Typically T_g is measured by differential scanning calorimetry (DSC) or by dynamic mechanical methods such as dynamic mechanical thermal analysis (DMTA), dynamic mechanical analysis (DMA), or torsion pendulum. The T_g in DSC measurements is chosen to be either the onset or the midpoint of the observed transition. T_g values obtained by DSC are lower than the dynamic mechanical values, where the frequency is usually fixed at 1 Hz but higher frequencies are used as well. In dynamic mechanical methods, the T_g is usually taken to be at the maximum of the tan δ peak. Sometimes the temperature at the onset of the modulus drop is given.

The T_g can be affected by the degree of cure [10] and by the stress exerted on the polymer chains [11]. Crosslinking has a particularly strong effect on the glass transition especially at high degrees of conversion where small changes in the degree of crosslinking can lead to large T_g changes. A list of ranges of T_g for various classes of polymers is given in Table 8-4.

8.4.5   Coefficient of Thermal Expansion

Stress build up in electronics packages is due mostly to the different coefficients of thermal expansion (CTEs) of the materials encountered in a package [24]. An

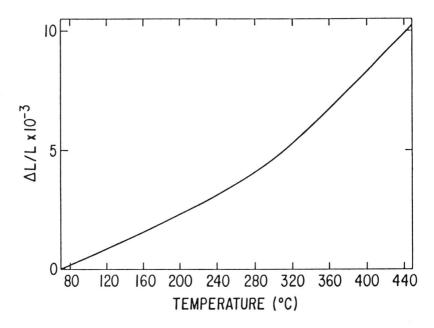

Figure 8-3 Relative dimensional change ($\Delta l/l$) versus temperature for a typical polymer dielectric such as BPDA-DA cured to 400°C at 5°C/min.

estimate of the magnitude of stresses in a package can be obtained knowing the CTE of the polymeric dielectric. As with other properties, the CTE of a polymer is a complex property depending on many factors. For example, it changes with temperature (Figure 8-3) and, therefore, the values of CTE should specify a temperature range. Furthermore, it can depend on the film thickness, the degree of cure, the heating (curing) schedule and the direction of the measurement with respect to the substrate plane [13]. If the polymeric material is isotropic, only one value is needed (for a given temperature range and cure state). However, many thin films, particularly of polyimides, are anisotropic. Therefore, the linear in-plane CTE can differ markedly from the linear out-of-plane CTE [13]-[14] (Table 8-5). Volume or bulk CTEs are rarely given.

Typically, the linear in plane CTE is measured on free standing films with a thermal mechanical analyzer (TMA) over a wide temperature range. Measurements on coatings also have been made by measuring wafer bending through x-ray analysis or optically with a laser (wafer bending instruments) [13]. In all the measurements on coatings, the CTE has to be extracted from the

Table 8-5 Linear Coefficients of Thermal Expansion (ppm/°C = 10^{-6} m/m°C).

Direction	PIs	Fluoro-polymers	BCBs	PPQs	PIQs
In-plane	2 - 60	90 - 300	65	40	3 - 58
Out-of-plane	60 - 100	90 - 300	65	40	n/a

PIs = polyimides BCBs = benzocyclobutenes PPQs = polyphenylquinoxalines
PIQs = polyimide iso-indoloquinazolinediones

substrate stress. The out-of-plane (often called z-directional) linear CTE can be measured by laser interferometry.

The thermal stress in a package is determined by the in-plane linear CTE, the film and substrate thickness, the relaxation modulus and the substrate CTE and modulus. The stress originating from the differences in film and substrate CTE is often called the thermal mismatch stress. If the one directional (tensile) stress, T, is known, one can calculate the stress in single layer coatings or thin multilayer films that is given by $T/(1 - \upsilon)$, where υ is Poisson's ratio, which in polymers is usually between 0.4 and 0.5. (Poisson's ratio is a measure of the compressibility of a material and is 0.5 for incompressible materials.) In thick multilayers, particularly in the presence of vias and signal lines, expansion perpendicular to the substrate is restricted. In such three dimensionally confined structures the stresses can become very high because the stress is now given by $T/(1 - 2\upsilon)$. The closer Poisson's ratio is to 0.5 the higher the stress and the higher the likelihood that mechanical failure (fracture) will occur.

The CTE is related to the elastic modulus of a material. The elastic modulus reflects the energy needed to change the bond length in the polymer chain by mechanical energy, the CTE reflects the same change effected by thermal energy. A steep energy profile for even the weakest bond(s) in the chain characterizes a high modulus, low CTE material; a shallow profile describes a low modulus, high CTE material. Because the thermal mismatch stress is determined by the product of the elastic modulus of the coating and the CTE difference between substrate and coating, a lower film CTE does not always lead to lower stress. Consequently, some low stress materials have been designed to exhibit lowered elastic moduli but higher CTEs. Another factor influencing the CTE, particularly in rigid rod polymers (like some low stress polyimides), is chain orientation. In these materials the linear in-plane CTE is lower the more the chains are oriented in-plane. Because the orientation is affected by the cure schedule as well as by the film thickness, the CTE depends on both. In general, thinner coatings have

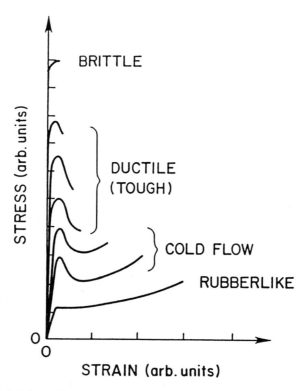

Figure 8-4 Schematic stress-strain behavior of a polymer at various temperatures.

the highest orientation and lowest CTE. Consequently, if thicker coatings are needed it is preferable to work with a low modulus, low stress material.

8.4.6 Mechanical Properties

The mechanical properties of the polymeric dielectric are of great importance for the package reliability in the short term, during handling and processing, as well as in the long term, during the lifetime of the package. The mechanical properties of polymers are a fascinating and intricate area in polymer science [9], [15]. This is due to the viscoelastic nature of polymers. The concept of viscoelasticity is developed from the fact that polymers respond to a mechanical stress by a combination of elastic (solid-like) and viscous (liquid-like) behavior. The ratio of the two behaviors depends on temperature and time (time referring

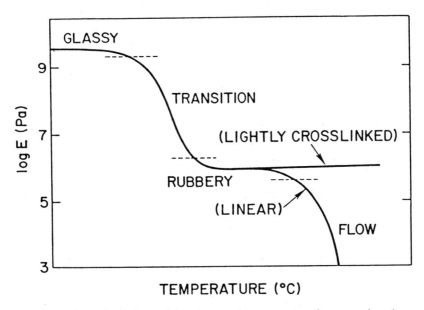

Figure 8-5 Schematic of the modulus change with temperature for a generic polymer.

to the time of application of a disturbance, such as the frequency of a dynamic mechanical measurement or the strain rate in stress strain experiments). The behavior of many polymers can change with temperature (or rate) from brittle to tough to cold flow to rubber like Figure 8-4. These responses also depend on structural features such as the degree of crosslinking, degree of order and chain orientation, presence of plasticizers such as solvent etc. The most important temperature with respect to the mechanical properties is T_g, which separates the region in amorphous polymers dominated by solid-like behavior from predominantly liquid-like behavior (Figure 8-5).

The data usually found in data sheets are Young's modulus, tensile modulus, linear elastic modulus, tensile strength, yield point, yield strength, elongation at break, and ultimate strength. These are usually extracted from stress strain curves at a given strain rate (Figure 8-6). The modulus is by definition the ratio of stress over strain (deformation) and is a function of rate and temperature. Therefore values of moduli are only useful if temperature and strain rate are given simultaneously. Fortunately, a number of the most important electronics packaging materials do not exhibit a strong strain rate dependence at room

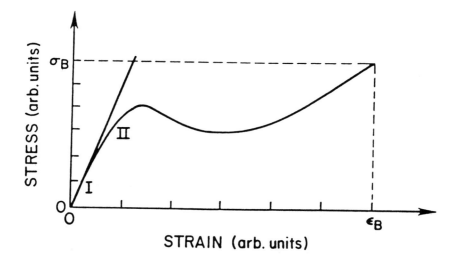

Figure 8-6 Schematic representation of a stress-strain curve for a tough polymer. The elastic limit is reached at *I*, the yield point is reached at *II*, and necking begins at the maximum of the curve. σ_B is the tensile strength and ε_B is the elongation at break. The Young's modulus is the slope of the tangent at small elongation.

temperature [4]. The Young's modulus is identical with the tensile modulus and the linear elastic modulus. They all represent the slope of the stress-strain curve at small strains (the linear region) in a tensile experiment (Table 8-6). In electronics packages, the strain usually encountered (from CTE mismatch) is small. The tensile strength is the maximum stress a material can carry. The yield point is the strain at which the polymer deforms irreversibly by flowing, developing a neck (a decrease of the cross sectional area) in the tensile specimen. This point depends on the accuracy of the measurement which is why ASTM Standard E 8-69 has been introduced. The yield strength represents the stress at which the yield point is reached. The elongation at break (Table 8-6) is the elongation at which a material breaks (maximum achievable strain). This property is watched closely because it gives an indication of the brittleness of the material which, in turn, is important for assessing the propensity of the material for mechanical failure such as cracking. A material is considered brittle when fracture occurs at the first maximum of the stress strain curve and the elongation at break is below 10%. The ultimate strength or ultimate tensile strength is the maximum load (at the elongation at break) divided by the initial cross sectional

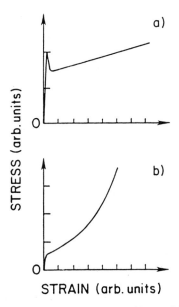

Figure 8-7 Stress-strain curves a) without and b) with consideration of the change in cross sectional area. Note the disappearance of the maximum in the experimental curve upon correction to the true test stress curve [16].

area. Necking in a material causes changes in the specimen. These changes are not reflected in the typical stress strain curves. Stress strain curves with and without correction are shown in Figure 8-7. A list of ranges of two mechanical properties for various packaging materials is given in Table 8-6.

8.4.7 Failure Mechanisms

Mechanical failure is a result of stress and leads to stress release. If the stress in a composite structure becomes too high, stress release occurs either by delamination (adhesive strength is exceeded) or by cracking, crazing or deformation zone formation (cohesive strength of the material is exceeded). A crack is a failure in which two new surfaces are created which are not connected across the crack, for example, a crack is not filled by any material. A craze, created by local yielding, is filled by oriented fibrils separated by voids [15]. Deformation zones are pre-crack zones occurring in materials which do not exhibit crazing, often high T_g materials. In these zones, shearing of the material has occurred; however, the density is not changed significantly.

Many material failures originate at stress concentrators that can be dust particles, vias, corners or edges of signal lines. Material failure can also be

Table 8-6 Young's Moduli and Elongation at Break (%).

	PIs	Fluoro-polymers	BCBs	PPQs	PIQs
Modulus (kpsi)	280 - 1300	200 - 235	480	500	550
Modulus (GPa)	1.9 - 9.0	1.4 - 1.6	3.3	3.45	3.8
Elongation	2 - 120	2 - 12	7	20 - 30	10 - 45

PIs = polyimides   BCBs = benzocyclobutenes   PPQs = polyphenylquinoxalines
PIQs = polyimide iso-indoloquinazolinediones

Table 8-7 Propensity to Failure Propagation.

DIELECTRIC	FAILURE PROPAGATION
Two phase polyimides	low
Amorphous polyimides	medium - high
Fluoropolymers	medium - high
BCBs	medium - high
PIQs	low - medium
PPQs	medium

PIs = polyimides   BCBs = benzocyclobutenes   PPQs = polyphenylquinoxalines
PIQs = polyimide iso-indoloquinazolinediones

initiated in areas where material property fluctuations lead to localized reduction of the yield stress. This can be particularly significant in the presence of plasticizers such as solvents or humidity (solvent induced crazing, environmental crazing). It follows that the propensity for forming failure zones will be greatest where stress concentrators are exposed to plasticizers. Failure can also occur due to fatigue; an example would be the cyclic application of stresses which, if applied only once, would not cause yield or fracture. Finally, processing steps like oxygen plasma treatments can cause weak areas. Again, stress concentrators are the most likely areas where accelerated damage occurs.

Crack propagation through a film can lead to catastrophic failure. The force propagating a crack in a coating has been shown to increase with film thickness and with the level of strain in the coating [17]. Crack healing can be attained by annealing the coating above T_g. However, if the cause for the mechanical failure has been thermal mismatch stress, annealing is only a temporary solution.

The propensity for mechanical failure depends strongly on the morphology of a polymer. Glassy, amorphous polymers, in general, will be more prone to failure than two phase systems such as filled polymers, block copolymers or semicrystalline polymers. This is due to the possibility of dissipating energy at the interphase between the phases.

Little data is available about avoiding mechanical failures of a coating. This is due in part to the fact that only a few methods are available to study crack growth in coatings or even in free standing thin films. Because of the lack of data in the literature, only a rough estimate of the propensity for failure propagation and crack formation for various materials is given in Table 8-7.

8.4.8 Chemical Resistance

Resistance to chemicals (solvents and solutions) depends strongly on the nature of the chemical and the polymer. Solvents used in the electronics industry typically include N-methylpyrolidone (NMP), diglyme, isopropyl alcohol (IPA), ethanol, fluorocarbons, 1,1,1-trichloroethane, cellosolve, and water. These solvents are used mainly for liftoff processes, cleaning steps, photoresist processes and flux removal. Solutions are typically aqueous such as plating solutions, caustic etchants, metal etching solutions, adhesion promoters. The problems encountered can be dissolution, swelling, crazing/cracking and degradation (etching). A few general rules exist. Partially fluorinated materials tend to dissolve easier in organic solvents than their hydrogenated counterparts. Perfluorocarbons, however, are the most solvent resistant. Crosslinked materials do not dissolve but they may swell. Swelling increases usually with solvent temperature and decreases with increasing crosslink density. Unfortunately, brittleness also increases with crosslink density. Linear polymers may also exhibit swelling, particularly if they are two phase materials (containing crystalline or liquid crystalline domains). Coatings, particularly of rigid rod-like materials, swell only in the thickness direction because the substrate hinders their expansion in the plane [5]. As with resistance to crack propagation, two phase systems exhibit higher solvent resistance. Solvent induced cracking or crazing is also more likely in amorphous polymers than in multiphase systems. To give a feeling of the variation in solvent uptake between the various dielectric polymers Table 8-8 lists the maximum uptake of NMP at about 80°C. The ranges indicate the dependence of the solvent uptake on the curing schedule.

Table 8-8 Maximum NMP Uptake (wt-%/hr) at 80°C.

PMDA-ODA	BPDA-PDA	Thermid 6015	Teflon® AF	PPQ
40	< 5	> 40 - 10	negligible	< 1%

Note: Uptake in thermid is highly cure dependent.

Resistance to caustic solutions like plating solutions depends strongly on the presence of groups susceptible to nucleophilic attack such as imides and esters. Again, multiphase systems are more resistant to attack, for example by KOH etching solutions. Furthermore, fluorine-containing polymers are usually more difficult to etch by aqueous solutions or are not attacked at all as in the case of Teflon® AF. This is due to the decreased wettability of fluoro-containing polymers compared to polyimides like Kapton®.

8.4.9 Adhesion

Destructive stress relief in a multilayer system occurs either when the cohesive strength of one of the materials is exceeded or when the adhesive strength is exceeded. The latter causes delamination. Often the interfacial strength is the weak link in a package. Fortunately, the adhesive strength between two materials can often be strengthened by using adhesion promoters. The choice of adhesion promoter depends on the substrate [18]. For surfaces with free hydroxyl groups (quartz, glass, alumina, metal oxides) silane coupling agents are used. For inert metal surfaces reactive metal layers are used.

Although partial delamination might not lead to structural disintegration or immediate electric failure of an electronics package, it is nevertheless grounds for concern because usually polymers pick up some amount of water. If a continuous water film can form over a two metal junction, a galvanic element is formed and corrosion occurs. A continuously adhering film acts as corrosion protection.

Adhesion depends on both materials connected in the adhesive link and on the way the adhesive bond is fabricated [18]. For instance the adhesion of copper deposited onto Teflon® AF is good but the adhesion of Teflon® AF deposited onto copper is poor. Particularly interesting is the behavior of polyimides. The polyamic acid precursors usually used can be quite strong acids and as such can dissolve metal oxides [19]. These oxides can act not only as catalysts in the thermal degradation of the polyimide [20], but also can increase the dielectric constant of the polymer. Once the precursor is imidized no acidity remains. However metal can now migrate into the film because of the lack of metal-polymer interaction. Such Cu diffusion and consequently dielectric constant increase is observed, for example, when Cu is slowly evaporated onto fully imidized polyimides [1]. The most important adhesion issues in MCM packages are the adhesion of the polymeric dielectric to itself, to metals (Cu, Cr, Al, Ti), to ceramics and to SiO_2 and the adhesion of metals to the dielectric.

The measurement of interfacial strength by adhesion tests, such as the very popular peel test, is ambiguous. Most methods depend on many factors such as the speed of peeling, width of peeled strip and the mechanical properties of the peeled material. Peeling of polymer films is particularly problematic because a substantial amount of the peel energy goes into the deformation of the peeled polymer and is not related to the actual adhesion strength. Another problem with most adhesion measurements is that they are destructive by nature. Only after the sample is destroyed, can it be known if its adhesion was good or not.

In listing peel strength data, test parameters usually are not indicated, making comparisons between peel test data impossible. Therefore, adhesion is often described in data sheets simply as "good" or "bad." It cannot be clear from such a generic description if "good" means "good enough for the intended application."

The complexity of adhesion issues becomes apparent when degradation of adhesion occurs over a period of time or after certain processing steps. This area is still under intense scrutiny and understanding of the process is at best limited. However, a few generalizations can be made.

1. Good wetting (good contact) between the two materials at an interface is a condition of good adhesion. Consequently, contamination of surfaces can severely decrease adhesive strength.

2. The influence of both temperature and humidity can severely degrade the adhesion, particularly between metal oxide or ceramic surfaces and many polymers.

3. If the metal oxide-ceramic surface is primed with an adhesion promoter, adhesive strength is retained over a significantly longer period of time.

4. If corrosion of the metal occurs due to humidity, severe adhesion loss is observed.

5. Polymers that exhibit a large modulus drop at T_g usually adhere well, particularly when heated above T_g during processing. This is caused by increased wetting.

6. Crosslinked polymers which start from low molecular weight precursors (thermosets) usually adhere well.

7. Thermal degradation often leads to decreased adhesive strength.

In the following, the generalized adhesion characteristics are given for a number of pertinent polymers:

1. Amorphous polyimides adhere well to themselves and metals such as Cu, Cr and Al.

2. Multiphase polyimides do not adhere well to Cu, but adhere well to reactive metals such as Al and Cr [1]. The adhesion to ceramics is good if aminosilane adhesion promoters are used. The self adhesion of multiphase polyimides is marginal. However, it can be increased by using appropriate curing schemes and/or surface treatments.

3. Fluoropolymers exhibit good self adhesion and good adhesion to ceramics and SiO_2. Also, the adhesion of metals evaporated onto these materials is good. However, fluoropolymers do not stick well to metals when deposited from solution. The adhesion to metals is improved markedly by exceeding T_g.

4. Information on adhesion of BCBs (benzocyclobutene) to various surfaces is not yet available but is expected to be good.

5. PPQs (polyphenylquinoxaline), which are in general amorphous, behave similarly to amorphous polyimides. They have excellent self adhesion and good adhesion to copper. No problems with metal corrosion are observed because PPQs are fully reacted.

6. PIQs (polyimide iso-indoloquinazolinediones) tend to exhibit two phase morphologies and show the same problems as two phase polyimides.

8.4.10 Water Uptake

Water uptake is of concern because it increases the dielectric constant and can lead to adhesion loss, corrosion, decreased stress in general [24] but increased stress in three dimensionally confined areas [4], molecular weight breakdown by hydrolysis, and blistering during heating. The quantity which is usually measured is the maximum amount of water uptake. The kinetics of drying is

Table 8-9 Maximum Water Uptake Values at 95% Relative Humidity and Room Temperature.

DIELECTRIC	WATER UPTAKE (%)
Polyimides	0.25 - 4.0
Fluoropolymers	0.01 - 0.1
BCBs	0.25
PPQs	0.1
PIQs	0.8 - 1.0

particularly important where fast heating rates and areas with continuous metal coverage are involved. Insufficient pre-drying in such cases might lead to blistering for instance, during chip joining [21].

Usually maximum amounts and kinetics of water uptake are measured by thermal gravimetric analysis (TGA), Kahn balance or by piezoelectric balances. The amount of water uptake depends on the degree of cure, on temperature and on relative humidity. These conditions must be given to compare values. It is also important to note that values obtained in steam are not necessarily identical to values obtained by immersion in water. Often the latter are lower. Table 8-9 gives ranges of maximum water uptake for various polymeric dielectrics.

8.4.11 Planarization

Planarization, the ability to produce a flat surface over features, is an important requirement for all TFM dielectrics mainly because good yields in the necessary photolithographic processes are obtained only when the resist is planar and uniformly thick. Sufficient uniformity and planarity are obtained only when the surface to be patterned is planar. Values for planarization are given as the percent of difference between the feature height before and after coating divided by the original feature height. Usually planarization is given for isolated features but sometimes planarization over double features is reported. In the latter case, planarization depends on the line spacing and is better for smaller line spaces.

Three properties determine the ability of a material to planarize. First, application of the dielectric should proceed from solutions with high solid content but low viscosity. Some solventless crosslinking systems can exhibit planarization values of over 95%. To obtain the necessary film thickness, the

Table 8-10 % Planarization, Solids Content and Shrinkage.

DIELECTRIC	PLANARIZATION (%)	SOLIDS CONTENT (%)	SHRINKAGE (%)
Polyimides			
PAAs	0.3 - 0.4	10 - 16	25 - 47
PAA-esters	0.3 - .05	20 - 30	25 - 35
PSPI 1	0.2 - 0.4	20 - 35	50 - 60
PSPI 2	0.3 - 0.5	10 - 30	10 - 30
Fluoropolymers	low	2 - 10	none
BCBs	up to 0.95	35 - 62	little
PIQs	0.2 - 0.7	12 - 50	20 - 35
PPQs	0.4 - 0.5	20 - 25	20 - 35

Note: PAA-polyamic acid; PSPI 1 - not intrinsically photosensitive polyimide; PSPI 2 - intrinsically photosensitive polyimide; BCBs can be powder coated.

viscosity of the solution used to apply the coating has to stay within given limits. That means for linear polymers a balance has to be found between a desired low viscosity for planarization and the desired mechanical properties that are best at high molecular weights. Modification of the precursors of the polymer is sometimes necessary. Thus, polyamic acid esters and polyisoimides both have higher solubility and lower viscosity than the corresponding polyamic acids.

Most high temperature stable, solution cast polymers contain appreciable amounts of solvent even after drying at temperatures around 100°C [4]. As a consequence these films shrink upon further heating. Additionally, shrinkage is observed when the curing reaction produces volatile by-products. This is particularly severe in photosensitive polyimides in which photopackages have to be burned out (not intrinsically photosensitive polyimides). Any amount of shrinkage is detrimental to planarization because the absolute shrinkage of the thinner film above a feature is less than that of the thicker film over the substrate.

Finally, polymer flow during cure can lead to increased planarization. This effect is only observed in thermoset materials because melt viscosities of high

Table 8-11 Propensity to Exhibit Cure Dependence of Properties.

DIELECTRIC	CURE DEPENDENCE
Polyimides	considerable
Fluoropolymers	little
BCBs	considerable
PPQs	little
PIQs	n/a

PIs = polyimides   BCBs = benzocyclobutenes   PPQs = polyphenylquinoxalines
PIQs = polyimide iso-indoloquinazolinediones

molecular weight polymers are very high. Table 8-10 gives ranges for percent of planarization, solids content and film shrinkage in the direction perpendicular to the film plane for the group of electronic packaging polymers under consideration.

8.4.12 Influence of Curing Conditions

Curing usually signifies any heat treatment. It always involves drying, and often involves chemical reactions and structural development (orientation, ordering, high temperature processes). In a number of important dielectrics the properties depend on the degree of cure. It is, therefore, very important to know what the best curing conditions are, how processing influences the degree of cure and what is the relationship between the cure schedule and the resultant properties. Some polymers, on the other hand, show only a very slight dependence on cure schedules. In general it can be said that all crosslinking materials (thermosets) and all materials in which a multiphase structure is formed exhibit properties that are cure dependent. Table 8-11 gives an indication of which dielectrics have cure-dependent properties.

8.5 POLYMER MATERIALS

After contemplating the most important properties of dielectrics for MCMs, the various materials are discussed in this section.

Table 8-12 Suppliers and Trade Names for Several Polyimides.

SUPPLIER	TRADE NAME
DuPont Corporation	Pyralin® PI-2545, PI-2525, PI-2611
Hitachi Chemical Company	PIQ L100, PIX L100, PAL, PIQ-13
Olin/Ciba-Geigy	Probimide® 200 series
National Starch	Thermid® EL-series
Amoco	Ultradel™ 4000 System
Ethyl Corporation	Eymyd® HP
Rogers Corporation	Durimide™ 100 and 120
Hoechst-Celanese	SIXEF® -44 and -33

Note: Trade names usually apply for a series of materials with varying structures.

8.5.1 Polyimides

Polyimides have been by far the most important materials for MCM packaging and are available from many sources (Table 8-12). They are a class of condensation polymers which have at least one imide group in their repeat unit [22]. The polyimides that are most widely used in the electronics industry all have two aromatic cycloimides in their repeat unit (Figure 8-8); the remainder of the repeat unit consists of aromatic hydrocarbons and connecting groups such as ether (-O-), carbonyl (C=O), 6F (CF_3 - C - CF_3 or hexafluoroisopropylidene). Polyimides are synthesized from dianhydrides and diamines. The convention for naming polyimides does not follow IUPAC rules. They are named after the dianhydride[5] and diamine from which they are synthesized. For example, the polyimide based on biphenyl dianhydride or (BPDA), and p-phenylenediamine (PDA) is referred to as BPDA-PDA polyimide. Some of the most common dianhydrides and diamines together with their common abbreviations are shown in Figure 8-9.

The chemical structure of the repeat unit determines the basic chemical and

---

[5] The names used for the dianhydrides and diamines rarely are IUPAC names which are very lengthy and confusing to the nonchemist.

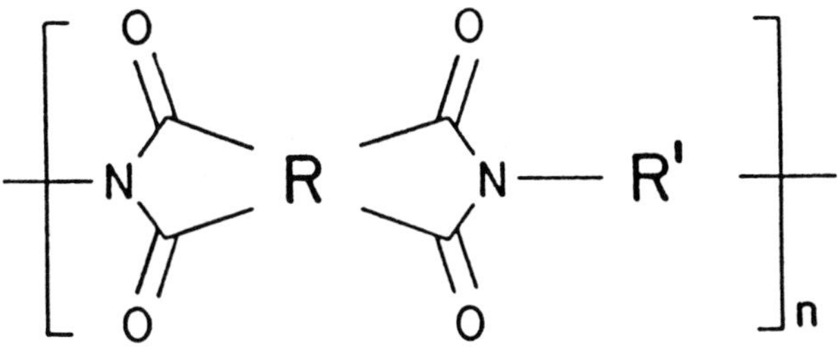

Figure 8-8 Schematic representation of polyimide with two aromatic cycloimides per repeat unit.

physical properties of the polyimide. The characteristics of the various polyimides can vary widely, although the polyimides used in packaging applications generally have a number of properties in common. These are summarized in Table 8-13. In general, the main advantages of polyimides are high thermal stability, good mechanical properties, and low CTEs. The main concerns are high water uptake and low adhesion.

In comparing polyimides it is of utmost importance to know the chemical structures or at least to know the exact commercial names and descriptions. Unfortunately, many polyimide trade names are confusing. Products with similar trade names are often altogether different materials. On the other hand, sometimes only the solvent or the concentration is changed. It is not surprising that people using these materials, but unfamiliar with the chemistry, tend to lump these materials together under the generic term "polyimides." However, this is comparable to describing electrical conductors as metals. Just as there are sharp differences between aluminum and copper, for example, there are sharp differences between polyimides based on PMDA-ODA (pyromellitic dianhydride and oxydianiline) and on BPDA-PDA.

Polyimides come in two basic variations:

- Polyimide precursors that are applied from solution and converted to insoluble polyimides by curing. These curing processes can be rather complex involving solvent and water removal, imidization, and often crystallization [10]. Only properly cured polyimides exhibit the desired properties.

- Polyimides that remain soluble in their fully imidized form. These include polyimides which contain siloxane segments. Curing of these materials involves only solvent removal.

PMDA	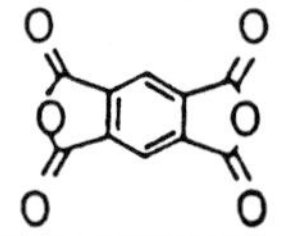
BTDA	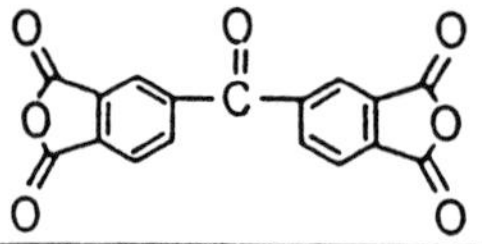
BPDA	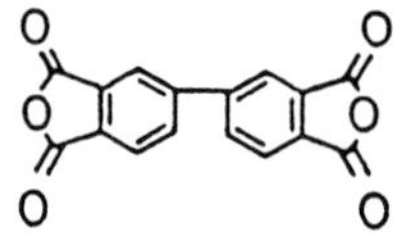
6FDA	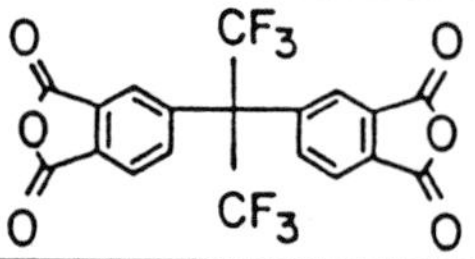

ODA	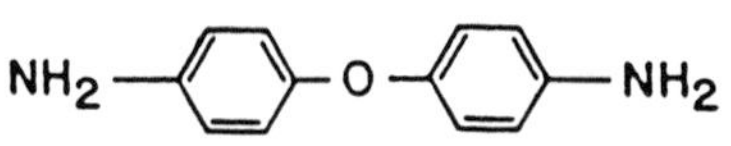
PDA	
6FDAM	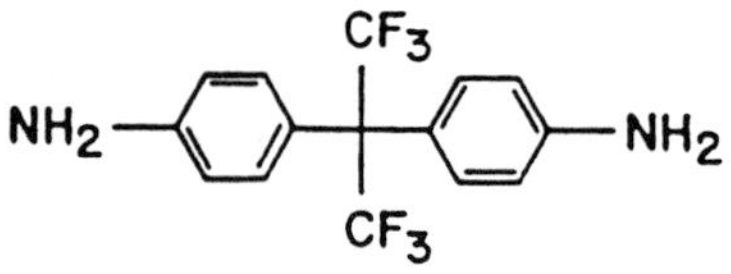

Figure 8-9 a) Common dianhydrides and b) diamines with their usual abbreviations.

Table 8-13 Typical Properties of Polyimides.

Dielectric constant	2.9 - 3.5 frequency independent (10^2 - 10^7 Hz); some PIs are anisotropic
CTE (ppm/°C)	in-plane: 5 - 45; in z-direction: 60 - 100
Thermal stability	at 400°C: < 0.1 wt.-%/h
Young's modulus (GPa)	2.5 - 10.0
Elongation at break (%)	10 - 120
Critical crack propagation	above 2.0% strain
Adhesion	marginal to metals, substrate and self; good with adhesion promoter
Application	spin or spray from precursor solutions
Planarization (%)	20 - 75
Shrinkage (vol. %)	30 - 47
Water absorption (%)	0.5 to 4.0
Solvent resistance	swelling in hot NMP, no crazing
Etchability	usually good, residue can cause problems

The most widely used precursor is polyamic acid (Figure 8-10a). More recently, esters (Figure 8-10b) and isoimides (Figure 8-10c) have found applications in the manufacture of MCMs. For example, the MCMs in the IBM System 390/ES9000 are fabricated using a PMDA-ODA ethyl ester precursor but have also been produced using an acetylene terminated BTDA-based isoimide precursor.

Polyamic acids are chosen over other polyimide precursors because they are the best known polyimides, widely used and cost effective, when compared to polyamic acid esters. Disadvantages include their poor adhesion, particularly their poor self-adhesion, and their propensity for attacking metals through their acid functionality. Polyamic acid esters generally have higher solubility and planarize better than the polyamic acids while retaining most of their good

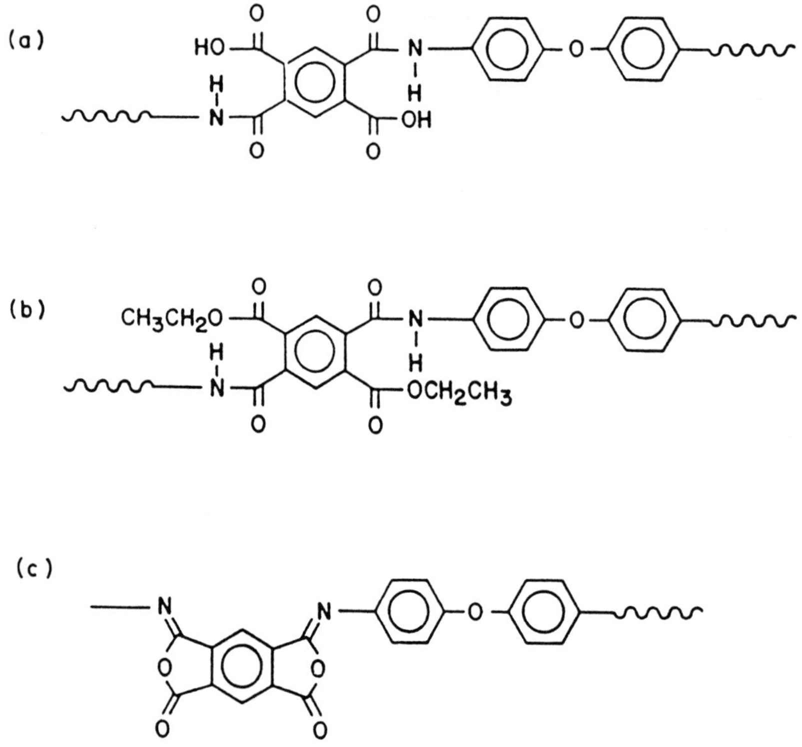

Figure 8-10 Precursors to PMDA-ODA polyimide: a) polyamic acid, b) polyamic acid ethyl ester and c) polyisoimide.

mechanical properties. Esters are difficult to synthesize and are therefore significantly more expensive. Adhesion is slightly improved.

In addition to classical polyimide precursors, thermosetting polyimides are offered, combining a fully imidized polyimide or isoimide oligomer with reactive end groups such as acetylene. Thermosetting polyimides have a number of advantages such as good planarization and good adhesion to themselves and to metal. Disadvantages include low elongation at break, low thermal stability at 400°C, swelling in process solvents and a tendency for solvent induced crazing.

Photosensitive polyimides (PSPIs) are photoresists based on polyimide chemistry. Usually, the photosensitive moiety is eliminated during the curing process and ultimately thermally degraded and volatilized. A smaller group of photosensitive polyimides are inherently photosensitive [23]. Inherently photosensitive polyimides are usually fully imidized and crosslinking occurs between the main chains. In both cases, sensitizers and other additives are part of the system. The main problems encountered with PSPIs are the short shelf life (some new materials, however, can have a shelf life of one year), the relatively low photosensitivity and the enormous amount of shrinkage (50% and more) incurred during curing, leading to high levels of stress. Furthermore, the resultant polymer tends to contain more voids than the corresponding polyimide without the photopackage. The higher void concentration causes increased water uptake. Inherently photosensitive polyimides often exhibit low thermal stability at 400°C. Exposure to such temperatures can increase the crosslink density leading to low elongation at break and a tendency to craze particularly in the presence of solvents.

PMDA-ODA

Until around 1985 PMDA-ODA based polyimides (Figure 8-11) were the most widely used polymeric interlayer dielectrics and today many companies still produce and use them. They were first commercialized by DuPont under the name Kapton® (in film form) and as Pyralin® 2545 or 2540 (as polyamic acid in NMP/hydrocarbon solution). The advantages of PMDA-ODA are its high thermal stability at 400°C in nitrogen, its excellent mechanical properties and excellent solvent resistance (no crazing). The disadvantages are its marginal adhesion to itself and to Cu, its limited planarization and its large water uptake.

BPDA-PDA

To overcome some of the shortcomings of PMDA-ODA, particularly with respect to its stress build up, BPDA-PDA (Figure 8-12) was first introduced by Hitachi in 1985. This polyimide is characterized by a rigid rod-like backbone (little flexibility in the chain) that imparts a high modulus and a low CTE to the material. The exact values depend on the degree of orientation with respect to the substrate surface, the film thickness and the cure conditions. Besides a lower CTE (5 ppm in 5 µm thick coating) and resulting lower stress, BPDA-PDA exhibits lower water uptake and lower swelling in hot NMP8. The dielectric constant of the dry material is 2.9 to 3.0, just below that of PMDA-ODA. Unfortunately, the adhesion is marginal and requires adhesion promotion. It is difficult to wet etch fully cured BPDA-PDA. Water degrades the mechanical and electrical properties of BPDA-PDA only minimally compared to PMDA-ODA. BPDA-PDA is available in film form (Upilex® S from Hitachi) and as polyamic acid solution in NMP (Hitachi's PIQ L100 and PIX L100, DuPont's Pyralin® 2610 and 2611).

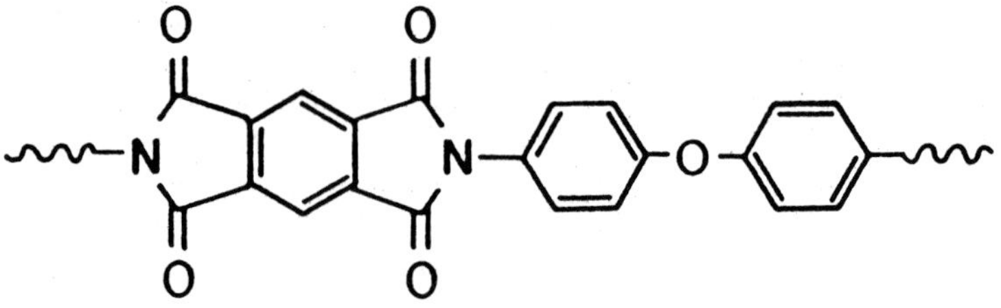

Figure 8-11 PMDA-ODA polyimide.

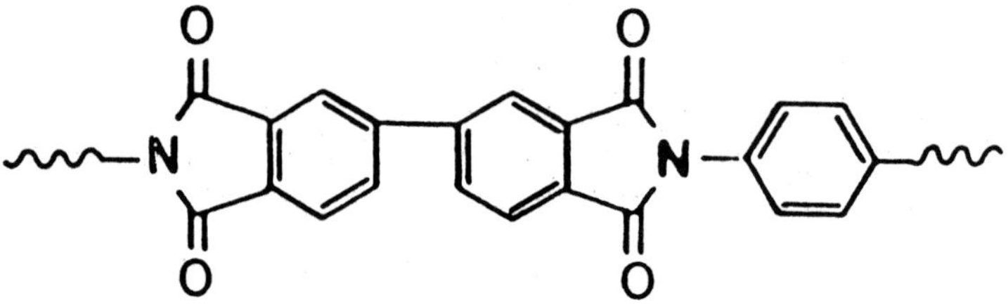

Figure 8-12 BPDA-PDA polyimide.

PIQ-13

PIQ-13 is Hitachi's trade name for polyimide-iso-indolo-quinazolinedione. It is also sold by Cemota. Its precursor is a polyamic acid copolymer based on a mixture of two anhydrides (BTDA and PMDA) and two diamines (4,4'- ODA and 3,4'- ODA with an amid group on the 4 position). It cures in two steps: cycloimidization at about 200°C and the oxime formation at 350°C. This material has been used by Japanese computer manufacturers. Its CTE is 58 ppm/°C for 5 μm films in the range of 30°C - 350°C.

BTDA-Based Polyimides

The most important examples in this group are Pyralin® 2525 and 2555, based on BTDA (benzophenone tetracarboxylic dianhydride) and a mixture of ODA (oxydianiline) and MPDA (m-phenylene diamine). This material is amorphous and therefore has limited solvent resistance compared to the two phase systems. It exhibits a sharp modulus drop at its T_g (about 310°C) which may lead to movement of thin lines above the T_g. The stress relaxation at that temperature leads to a lowering of the thermal mismatch stress at room temperature compared to PMDA-ODA. The advantage of BTDA-ODA/MPDA is its good adhesion to many substrates and to itself. It therefore has been used in one of AT&Ts MCMs.

Photosensitive Polyimides
There is a wide offering of photosensitive materials on the market (Table 8-14). However, their use today is limited because of drawbacks such as high volume shrinkage upon cure, high stress levels, high water uptake, brittleness, short shelf-life and low photosensitivity. Some of these problems have been overcome by the current generation of materials but they still exhibit one or two of these drawbacks. The development activity in this area is considerable.

Fluorinated Polyimides
These materials are the focus of much research driven by the search for lower dielectric constant materials. SIXEF®, based on diamines and dianhydrides containing the hexafluoroisopropyliden (6F) group, exhibit a dielectric constant of 2.65 and were semi-commercially available from Hoechst for a short time. The drawbacks of these materials are lowered solvent resistance, reduced mechanical properties and cost. The weight loss at 400°C is very small (0.1 wt.-%/h), but decomposition occurs above 350°C. Most other materials based on 6F chemistry exhibit very good adhesion, but none of them have been used in MCM packages because of their relatively high CTEs and their low solvent resistance.

Table 8-14 Suppliers and Trade Names for Several Photosensitive Polyimides.

DuPont Corporation	Pyralin® PI-2700 series
Olin/Ciba-Geigy	Probimides® 300 and 400 series
Amoco	Ultradel™ 7000 system
Asahi	Pimel® TL series

Table 8-15 Suppliers and Trade Names for Several Polymeric Dielectrics Other than Polyimides.

Cemota	Syntorg® IP 200
DuPont Corporation	Teflon® AF 1600 and 2400
Rogers Corporation	RO 2000 series
Dow Chemical	XU-13005-02L

8.5.2 Polyphenylquinoxaline

Polyphenylquinoxaline (PPQ) is sold under the name Syntorg® by Cemota (Table 8-15). Its use is limited by the fact that it dissolves only in m-cresol or similarly undesirable solvents. Aside from this drawback PPQs offer many good properties such as good adhesion, high thermal stability, low water uptake, good mechanical properties, and a relatively low dielectric constant of 2.9.

8.5.3 Fluorocarbons

From the standpoint of electrical designers, dielectrics with dielectric constants around 2.0 are highly desirable. Unfortunately, only a few materials combine thermal stability with low dielectric constants. Among them are fluorocarbons such as polytetrafluoroethane (PTFE, Teflon®) and the recently introduced amorphous Teflon® AF. PTFE in its bulk form has many drawbacks such as insolubility, very high CTE (>300 ppm/°C) and marginal mechanical properties. Composites of PTFE with inorganic fillers, usually quartz (available from Rogers Corporation), have interesting properties. However, standard processes such as spin application cannot be used. The amorphous fluorocarbons, Teflon® AF, are soluble in perfluoro solvents. Their dielectric constants, about 1.9, are very desirable. However, their T_gs (160°C and 240°C respectively) are too low for most practical MCM applications. Moreover, adhesion of these materials to metal is poor and their adhesion to SiO_2 degrades rapidly in the presence of water. However, adhesion of metal deposited onto these dielectrics is good.

8.5.4 Benzocyclobutenes

Benzocyclobutenes (BCBs) are the most recent addition to the growing number of dielectrics intended for the electronics industry. Currently, only one material is commercially available from Dow Chemical (Table 8-15). It contains divinyltetramethyl disiloxane (DVS) and is called DVS-bis-BCB. BCBs have very intriguing properties such as excellent planarization and low dielectric constant (2.65). However, they are plagued by low oxidation stability in air (at 150°C). The commercial product contains an antioxidant to alleviate this problem. The thermal stability of BCBs in nitrogen is good at 300°C but at 350°C weight loss is 1 wt.-%/h. Because these materials are new, further developments can be expected.

8.6 SUMMARY

Polymeric dielectrics offer very low dielectric constants as well as ease of processing and, therefore, are ideally suited for interlayer dielectrics in MCM structures. However, the choice of polymeric material is limited by the demanding thermal stability requirements. The most widely used materials, polyimides, usually require complex curing schedules which determine the ultimate properties of the polymer. The final film properties (CTE, elongation at break, modulus) can depend also on film thickness and, in some cases, on the nature of the substrate. These dependencies are still not well understood. PSPIs are still in the early stages of development although some use has been reported. Processing advantages are a strong driver for their further development. Very low dielectric constant materials (< 2.5) are currently just below the necessary thermal properties for high performance computer MCMs. The application of BCBs is limited to processing in nitrogen below 350°C and below 150°C in air. Finally, it must be stressed that many aspects of polymer dielectrics for packaging applications are not well understood and are the subject of intense research.

Acknowledgments

We would like to thank Gareth Hougham and Jeff Hedrick as well as many other of our colleagues for sharing and discussing their data and knowledge. We also thank our management at IBM for allowing us to undertake this project.

References

1   D. P. Seraphim, R. Lasky, C. Y. Li, ed., *Principles of Electronic Packaging*, New York: McGraw Hill, 1989.

2   R. J. Jensen, "Polyimide as Interlayer Dielectrics for High Performance Interconnections of Integrated Circuits," *Polymers for High Technology*, ACS Symp. Ser., vol. 346, pp. 466-483, 1987.

3   C. W. Ho, *et al.*, "The Thin Film Module: A High Performance Semiconductor Package," *IBM J. of Res. and Dev.*, vol. 26, no. 3, pp. 286-296, 1982.

4   C. Feger, M. M. Khojasteh, J. E. McGrath, ed., *Polyimides: Materials, Chemistry and Characterization*, Amsterdam: Elsevier Sci. Publ., 1989.

5   E. Gattiglia, T. P. Russell, "Swelling Behavior of an Aromatic Polyimide," *J. of Polymer Sci.: Part B: Polym. Phys.*, vol. 27, pp. 2131-2144, 1989.

6   S. Herminghaus, *et al.*, "Large Anisotropy in Optical Properties of Thin Polyimide Films of Poly(p-phenylene biphenyltetracarboximide)," *Phys. Rev. Lett.*, vol. 59, pp. 1043-1045, 1991.

7   C. Cha, S. Moghazy, R. J. Samuels, "Characterization of three Dimensional Surface and Bulk Anisotropy in High Refractive Index Polymers," *Soc. Plast. Eng. Tech. Pap.*, vol. 37, pp. 1578-1580, 1991.

8   H. H. G. Jellinek, S. R. Dunkle, *Degradation and Stabilization of Polymers*, New York: Elsevier, 1983.

9   J. J. Aklonis, W. J. MacKnight, *Introduction to Polymer Viscoelasticity*, New York: Wiley, 1983.

10 C. Feger, "Curing of Polyimides," *Polym. Eng. Sci.*, vol. 29, pp. 347-351, 1989.

11 F. W. Harris, *et al.*, "Organo-soluble, Segmented Rigid Rod Polyimides: Synthesis and Properties," *Proc. 4th Internat. Symp. on Polyimides*, (Ellenville, NY), pp. I-1. Oct. 1991.

12 J. Melcher, *et al.*, "Dielectric Effects of Moisture in Polyimide," *IEEE Trans. on Electrical Insul.*, vol. 24, no. 1, pp. 31-34, 1989.

13 H. M. Tong, L. T. Nguyen, eds., *New Characterization Techniques for Thin Polymer Films*, New York: Wiley & Sons, 1990.

14 G. Elsner, *et al.*, "Anisotropy of Thermal Expansion of Thin Polyimide Films," *Thin Solid Films*, vol. 185, pp. 189-197, 1990.

15 I. M. Ward, *Mechanical Properties of Solid Polymers*, New York: Wiley, 1983.

16 H. G. Elias, *Macromolecules: Synthesis, Materials, and Technology: Vol. 1*, New York: Plenum, 1984, pp. 451-453.

17 H. R. Brown, A. C. M. Yang, "The Propagation of Cracks and Crack-like Defects in Thin Adhered Polymer Films," *J. Matl. Sci.*, vol. 25, pp. 2866-2868, 1990.

18 S. Wu, *Polymer Interface and Adhesion*, New York: Marcel Dekker Inc., 1982.

19 Y. H. Kim, *et al.*, "Adhesion and Interface Investigation of Polyimide on Metals," *Adhesion Sci. Technol.*, vol. 2, pp. 95-105, 1988.

20 K. Kelly, Y. Ishino, H. Ishida, "Fourier Transform IR Reflection Techniques for Characterization of Polyimide Films on Copper Substrates," *Thin Sol. Films*, vol. 154, pp. 271-279, 1987.

21 D. D. Denton, H. Pranjoto, "Gravimetric Measurements of Moisture Uptake in Polyimide Films used in Integrated Circuit Packaging," *Mat. Res. Soc. Symp. Proc.*, (Boston MA), vol. 154, pp. 97-105, 1989.

22 D. Wilson, H. D. Stenzenberger, P. M. Hergenrother, eds., *Polyimides*, New York: Chapman & Hall, 1990.

23 K. K. Chakravorty, *et al.*, "Photosensitive Polyimide as a Dielectric in High Density Thin Film Copper-polyimide Interconnect Structures," *J. Electrochem. Soc.*, vol. 137, pp. 961-966, 1990.

24 J. T. Pan, S. Poom, "Film Stress in High Density Thin Film Interconnect," *Mat. Res. Soc. Symp.*, (Boston MA), vol. 154, pp. 27-37, 1989.

25 D. S. Soane, "Stress in Packaged Semiconductor Devices," *Solid State Techn.*, vol. 32, No. 5, pp. 161-171, May 1989.

26 H. Satou, *et al.*, "New Polyimide for Multichip Module Applications," *Proc. Electr, Comp. Conf.*, (Las Vegas NV), pp. 751-753, 1990.

9

CHIP-TO-SUBSTRATE (FIRST LEVEL) CONNECTION TECHNOLOGY OPTIONS

9.1 INTRODUCTION [1]

The selection of a first level connection technology for the connection of die to the multichip module (MCM) package or substrate involves several factors, each of which should be evaluated for specific MCM applications. This chapter discusses the basics of the three most widely used connection technologies, and together with die attachment, assesses their applicability to specific MCM constructions.

When choosing a chip connection technology, tradeoffs must be made between performance, density and cost. Performance here is defined as electrical (speed, design, testability) and mechanical (thermal dissipation, reliability, rework, repair). Density is related to chip size versus I/O count and chip to chip spacing on the module, both of which contribute to the overall size of the MCM.

---

[1] Editors' Note: The sections in this Chapter form the focus for individual author contributions. The Chapter as a whole, and the Introduction and the Summary were assembled by the Editors with written technical contributions by S. Bezuk, E. Larson and D. Haagenson. These authors and the editors appreciate also the helpful discussions pertaining to the chapter as a whole from C. C. Wong and T. Dixon Dudderar, AT&T Bell Laboratories.

Cost is the MCM unit cost which includes: chip and substrate costs, device preparation costs, manufacturing equipment costs and process costs related to cycle time and yield. Each process contains specific costs covered in the following sections of this chapter. The goal is to provide a marketable MCM product that is competitive in all these areas. To accomplish this goal, a thorough evaluation must be made during the MCM design phase to determine the correct connection technology selection(s) for particular MCM products.

9.1.1 Chip Connection Technologies

The three most prominent first level connection technologies discussed in this chapter are:

1. Wire bonding
2. Tape automated bonding (TAB)
3. Flip chip solder bump (FCSB)

These technologies also have been defined in Chapters 1 and 2. MCM designs may require one or more of these technologies to exist on the same module.

Wire bond is the oldest and most mature of the chip connection schemes. A die is attached rigidly to the surface of the package or substrate and each bond pad of the chip is individually connected to a corresponding pad on the package with a small (usually 1 - 2 mil diameter) wire. Wire bonding is discussed in detail in Section 9.3. TAB is a process that connects a chip to metal traces patterned in a metal frame or adhered to a polymer. Cutting or excising the chip from the frame with the metal leads attached then allows for its placement and attachment to the substrate. TAB comes in several varieties and is discussed in Section 9.4. Conventional TAB is similar to wire bond in that the chip is face up on the substrate. Flip TAB places the chip face down, presenting the back of the die for chip cooling and short TAB leads for better electrical performance. Placing the die in a cavity also can provide shorter TAB leads and permit the heat to be removed through the back of the die and substrate. Wire bond and TAB both use peripheral pad layouts (except for array TAB), limiting the I/O capabilities of these technologies. FCSB, on the other hand, uses an array of solder balls on the I/O pads of the chip to provide a soldered chip connection to the substrate. It has the shortest chip-to-substrate connection and also presents the back of the die for cooling. Controlled Collapse Chip Connection (C4), used by IBM since the 1960s, is one example of FCSB. Another example of flip chip includes the use of conductive polymers as the connecting medium. FCSB technology is discussed generically in Section 9.5 and a case study related to C4 is presented in Section 9.6.

Each of these chip connection technologies has advantages and disadvantages discussed in the chapter. Higher performance in any area generally means higher cost for the final package. Design issues must be considered carefully in choosing one technology over the other for a particular application. In some cases, a combination of connection technologies may be beneficial to maximize both the performance and cost of the final product.

9.1.2 Electrical Design

Design for electrical performance involves two areas of concern affected by the interconnection and packaging strategies. The first area of concern is I/O count versus chip and packaging density. The second area is pure electrical performance resulting from low resistance and inductance values.

When looking at the I/O capabilities of each of the available technology alternatives, the area array capabilities of FCSB offer the maximum number of connections for a given chip size. FCSB also offers the closest chip-to-chip spacing available, since no area is consumed around the periphery of the die. The maximum lead count available with TAB is approximately equal to those of wire bond but less than that for FCSB. TAB and wire bond also consume area outside the die for the outer connection points. This area varies with the selection of MCM materials and the resulting design constraints. The area required for these outer connection points is minimized if the pad pitch on the die is equal to the pad pitch on the substrate. Flip TAB designs can take advantage of these equal die and package pad pitches, resulting in shorter leads and in denser packaging of the die in MCMs.

When considering electrical performance in terms of RL properties, FCSB clearly has performance advantages relative to TAB and wire bond due to the very short connection length as discussed in Chapter 1. Since TAB leads generally have a larger cross section than wire bond wires, and TAB potentially can be designed with shorter lead lengths, TAB offers the next best performance electrically. One of the main disadvantages of wire bonding is the potentially high RL properties of the wires when compared to FCSB or TAB connections.

Electrical testability and burn-in are other areas of electrical design that should be considered. TAB offers the potential for testing and burn-in prior to committing the die to a substrate or module. Wire bond and FCSB chips must be committed to the module before the primary chip connection is tested, and before the chip is burned in.

9.1.3 Mechanical Design

Chip connection technologies serve to connect the chip electrically to the MCM substrate but they also can provide all or part of the mechanical connection to the substrate. The mechanical connection gives rise to reliability concerns. Stresses are generated when materials with different coefficients of thermal expansions (CTE) are joined together. Differences in CTE, due to the use of different materials, limit the chip size and I/O count for FCSB when different substrate and die materials are used, for example, silicon die on ceramic substrates. Substrate materials that have a CTE value closer to silicon generally are more expensive.

Cooling MCMs is another tradeoff in mechanical design. The denser the circuits are packed and the higher the chip operating power, the more heat is generated and must be dissipated per unit area. Each of the connection technologies allows different configurations for removing heat from the die. Mounting the chip face up, as with wire bond and conventional TAB, allows heat to be removed through the substrate. When designing with this in mind, the die bonding adhesive must be considered, since it contributes to the overall value of thermal resistance. For some conventional TAB applications, the TAB leads provide adequate structural strength, but an adhesive must be used between the die and substrate for thermal reasons only. Mounting the chip face down as with flip TAB and FCSB provides the back of the chip for heat removal. This can be very efficient but much more expensive. These package or MCM enclosures can be very dependent on particular designs.

9.1.4 Technology Comparisons

The maturity of the technology and the infrastructure existing in the industry are important issues. Wire bond is the oldest technology and is in widespread use. The process is well understood and documented. Equipment is available from many companies to support the technology. Chips and substrates can be purchased and used with none of the extra processing (bumping) required for TAB and FCSB designs. TAB is the second most widely used technology. This process, developed to speed the connection production process because of the ability to bond all the leads at once (gang bonding), is actually more involved than that for wire bonding. However, the intermediate step of attaching the chip to the TAB tape (Inner lead bonding - ILB) can allow for the testing and burn-in of the chips at this stage. Equipment and process information is available for TAB, but the infrastructure is less developed than that for wire bonding. Equipment, TAB tape and special chip bumping costs generally make this technology more expensive for low volume MCM applications. High volume production is required to make the bonding speed and testability of TAB pay off. FCSB, although it has been around since the early 1960s, has been championed by relatively few companies, most notably IBM. For all its longevity, FCSB is the least mature chip connection technology available on the open market. When chips are available in solder bumped form, from commercial chip vendors, this technology will attain its full potential.

Process yield and repairability are two major factors determining MCM cost. Repair is considered an absolute must for most MCMs. A finished MCM represents a significant monetary investment when compared to a single chip package. All of the chip connection technologies described in this chapter are repairable to some extent. A wire bond repair requires the manual removal of the die and all connected wires before the chip is replaced, and therefore becomes much less desirable as I/O counts on the chip increase. TAB and FCSB processes, on the other hand, lend themselves to less labor intensive and potentially more automated repair processes. This is especially true for solder metallurgies, where all the leads or bumps on the die can be reflowed simultaneously, allowing for one step removal of the failed device.

9.2 DIE BONDING AND PHYSICAL ATTACHMENT

Robert E. Rackerby

Die attach, the process by which the die is anchored to the substrate, is usually the first process performed in the assembly of microelectronic packages. Historically, the mission of single chip die bonding was to secure the die and provide a controlled electrical and thermal path to the substrate. This was accomplished by introducing an adhesive, usually a silver-filled epoxy or solder, between the die and substrate. The metal content or composition of the adhesive was manipulated until the desired thermal or electrical parameters were obtained. Die bonding is currently done [1] via two technologies: through the introduction of an adhesive between the die backside and substrate or by an electrical connection procedure, such as tape automated bonding (TAB) or solder bumping (see Figure 9-1), and the descriptions in Sections 9.4 and 9.5. In this section, we limit the discussion to those processes involving an adhesive between the backside of the die and the substrate.

The demands of MCMs have changed the conventional view of die attach. The complex interaction of materials, processes and cost often forces MCM designers into difficult processing tradeoffs. Frequently the die bonding material for an MCM is required to perform many different tasks which never were required for single chip applications. Rework is a primary example of a new addition to the list of bond material requirements. In the past, failures after die bond were not catastrophic because the individual part cost was low. Most high performance MCMs are quite expensive and require the ability to rework individual chip sites to remain cost effective.

9.2.1 Die Attach Material Choices

There are two classes of adhesive available for die bonding: organic and inorganic materials. These materials are divided further into their components as listed in Table 9-1. The inorganic materials, Au-Si eutectic, silver-glass pastes and soft solders have the benefits of low contamination byproducts (moisture or corrosive compounds), but suffer from relatively high processing temperatures. Some of the soft solders also suffer from fatigue problems [2] - [3]. The organic compounds, on the other hand, have nearly the opposite characteristics. Organic compounds are effective materials for use on large die because they easily withstand large strains, but may be inappropriate for hermetic environments due to outgassing.

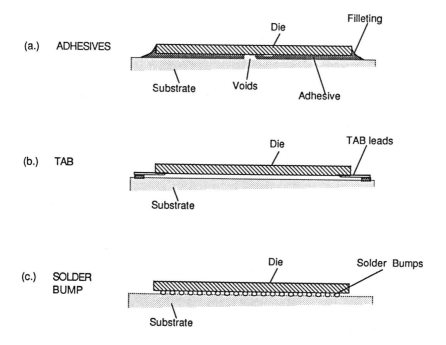

Figure 9-1 Die attach techniques showing two classes of die attachment: adhesive and electrical. Figure 9-1a Showing die attachment to substrate with adhesive. Figure 9-1b and 9-1c show an electrical connection method for mechanical attachment.

There are two classes of organic adhesives used for die attach, thermosets and thermoplastics. The thermoset compounds are characterized by epoxies and a few polyimide compounds. After they are cured, they remain solid with temperature increases up to their decomposition temperature. In contrast, thermoplastic compounds can be melted repeatedly to a liquid and resolidified with few or no side effects [4]. Thermoplastics decompose when heated past their boiling points.

Au-Si Eutectics/Silver-Glass Pastes/Soft Solders
The primary reason for choosing these materials is good thermal performance. High power applications have historically employed these materials. High processing temperatures [2]-[3], [5] usually rule out these materials for all but MCM-C. Rework is possible on an engineering basis with the eutectic and

Table 9-1 Inorganic and Organic Adhesives for Die Bonding.

DIE BONDING MATERIALS		
Inorganic Adhesives	Organic Adhesives	
	Thermoset	Thermoplastic
Gold-silicon Eutectic	Epoxy	Polyimide
Silver-glass Pastes	Polyimide	
Soft Solder Alloys	Urethane	
	Thermoset/Thermoplastic Mixtures	

solder materials, but is impossible with silver-glass. These constraints severely limit the usefulness of these materials.

The bonding mechanisms of the inorganic compounds are somewhat similar. The Au-Si eutectic and solder materials establish metallurgical bonds with the die and substrate [2]-[3], [5]. In the case of the Au-Si eutectic materials, a high degree of silicon diffusion from the die establishes a hard soldered joint [2]-[3]. Soft solders, containing some mixture of Pb, Sn, In, Ge or Ag, create a bond based on intermetallic formation [3]. However, the degree of silicon diffusion is somewhat less than with the Au-Si eutectic.

Silver-glass pastes bond by three mechanisms: metallurgical, chemical and mechanical. The metallurgical mechanism is accomplished by forming Au-Ag intermetallics between the gold plated substrate and the sintered silver matrix formed in the silver-glass paste [6]-[8]. The chemical mechanism is achieved by oxygen bonding between the metal oxides in the glass to the silicon on the backside of the die [2], [9]. Mechanical bonding takes place at both the backside of the die and the substrate when the glass re-vitrifies upon cooling [8]-[9].

Epoxies
Epoxies, like all the organic adhesives, are relatively easy to work with, are compatible with most surfaces and are amenable to rework. Epoxies (filled and unfilled) have a much broader use in the production of MCMs. In the last several years, epoxies have become clean enough for most MCM applications with the exception of some high reliability applications [2] found in the space or military industries.

Bleedout is a complication in the use of epoxies [10]-[11]. Bleedout is a condition characterized by resin flowing over portions of the substrate and coating unintended areas during the cure cycle. Experimental data indicate bleedout may be related to the surface energies of both substrate and resins [10]. Lack of surface cleanliness contributes to bleedout because it changes the surface energy of the substrate.

Chemically, epoxies are the dirtiest of the die bonding adhesives. The crosslinking agents, hardener and resin, react producing several byproducts in addition to the long epoxide chains. Ammonium hydroxide was a byproduct in the early development of epoxies. The ammonium hydroxide subsequently caused corrosion of the metal traces within the package. Advancements in epoxy technology have essentially eliminated this problem, leaving halogen and metallic cations such as Na^+ and K^+ present in minute concentrations (typically less than 10 ppm) [12] as dominant contaminants. Some epoxies may be suitable for environments requiring hermetic seals, however, they must be thoroughly tested and characterized for unwanted byproducts as a function of operating temperature.

The bonding mechanism for the epoxies is mechanical and chemical. When the epoxy cures, it undergoes a phase change and hardens. This phase change also establishes mechanical bonds where the epoxy sticks to microscopic fissures and crevices in the die and substrate. Chemical bonding, however, is responsible for the bulk of the joint strength. During cure, portions of the epoxy chains adhere to water adsorbed to the surface of both the die and substrate. The mechanism of chain attachment is through Lewis acid-base charge transfer [2]. The epoxy chains become electron acceptors while the surface film of water acts as electron donors. The bond sites for the epoxy chains are present uniformly across both the substrate and die.

Polyimides

Polyimides have processing characteristics similar to that of the epoxies [2]. However, they tend to be more thermally stable than the epoxies [4]. Filling polyimides with metals, such as silver or copper, tends to lower their thermal stability [13]. Polyimides can be formulated as either thermoplastics or thermosets. The condensation cure mechanism produces a thermoplastic compound whereas addition curing yields a thermoset [14]. Polyimides for die attach are chemically similar to thermoplastics, both have the ability to be remelted and cured (solidified) repeatedly [4]. Polyimides also are less susceptible to bleedout than the epoxies.

The polyimides are the cleanest of the organic adhesives for die bonding and are comparable to Au-Si eutectic attach [15]. Their major drawback is a tendency to absorb water (as much as 6% by weight). Measures to ensure against moisture induced failures include extra drying steps or inclusion of a desiccant. Polyimides adhere to surfaces through chemical and mechanical bonding similar to the mechanism for epoxies [2].

Thermoplastics

Thermoplastics have been used recently for die attach because of their ease of rework and their low glass transition temperature, T_g [4]. The adhesive is applied either as a hot melt or in a preform which is heated to its melting point and allowed to cool after die placement. Thermoplastics do not have high bond strengths and remain quite flexible or ductile relative to thermosets [4]. These properties make them less desirable as die bonding adhesives because they are less likely to pass the adhesive strength specifications required in the Mil-specs.

The bonding mechanisms for the thermoplastics are mechanical in nature, although chemical bonding can take place as well. The hot liquid penetrates the rough microstructure and, upon solidification by cooling, locks the die in place.

Mixtures and Organic Alloys

Mixtures of thermoset and thermoplastic materials represent a relatively new category of adhesives. The purpose of mixing these two polymers is to attempt to blend the best material properties of each into a single compound. The blend is accomplished by either mixing existing polymers together by blending (sometimes called alloying [16]) or chemically grafting the desired functional group to the backbone polymer. The chemical grafting technique holds the most promise for producing a material with desired properties.

9.2.2   Die Attach Processes And Process Control

The die attach process can be broken down into three steps:

1. Dispensing of the adhesive
2. Alignment and placement of the die
3. Curing of the adhesive

In the first step, the adhesive is dispensed in an amount which ensures the proper bondline (thickness of the cured adhesive under the die). Au-Si eutectics, solders and several of the organic compounds are dispensed using a rigid preform. The alternative to using preforms is using a low viscosity paste. All the organic materials are available in a paste or liquid form.

The second step of the process includes die alignment and placement. The die is acquired from its carrier and properly oriented. The die is placed onto the adhesive and set in place producing a fillet of material around the edge of the die (refer to Table 9-2). During the filleting process, alignment is performed for three axes of rotation and three translations simultaneously - resulting in six degrees of freedom. Many die bond problems such as low bondline, edge voids, undercutting, material on die, over filleting (refer to Table 9-3 for a complete list), are traceable to lack of control in die alignment.

Table 9-2 Parameters Defining Die Attach Process and Process Control.

DIE ATTACH PROCESS AND CONTROL PARAMETERS		
STEP 1: Dispense Adhesive	Paste	Preform
Dispense quantity and repeatability	2	1
Pattern uniformity and repeatability	2	1
Rheology	2	0
Pot life	2	1
Bondline thickness—before and after cure	2	1
Bleedout	2	0
0 = NO ISSUE 1 = MINIMAL ISSUE 2 = PRIMARY ISSUE		
STEP 2: Alignment and Placement of Die		
Six degrees of freedom placement accuracy Repeatability		
STEP 3: Cure of Adhesive		
Time-temperature sensitivity Humidity Shrinkage Post bonding thermal and mechanical testing (Sage, die pull and shear, SLAM) Fillet inspection		

The actual die bond operation can be performed manually, semiautomatically or fully automatic. Manual bonding is performed usually by hand under a microscope. The six degrees of freedom, as well as the dispense parameters, are controlled poorly by manual processes. Semiautomatic processes are somewhat better than manual processes. Typically manufacturers of die bonding equipment pick and choose which parts of the process to automate and which to leave manual. Die bonding equipment is specific to the bonding material, equipment

Table 9-3 Parameters Associated with Die Bonding Quality Control.

DIE BONDING PROCESS PROBLEMS					
	Dispense	Place	Package Handling	Cure	Rework
Low Bondline	X	X			
Edge Voids	X	X			
Undercutting	X	X			
Material on Die	X	X		X	
Overfilling	X	X			
Flaking	X				
Bridging	X				
Cracking			X	X	
Disbanding			X	X	
Bleedout				X	X
Misorientation Misalignment		X			
Die Chips/Scratch		X	X		

for bonding organic adhesives is different than that used for bonding silver-glass pastes. Machines used to dispense silver-glass pastes have mixers and sophisticated pressure dispensers which are unnecessary on epoxy machines.

Curing of adhesives is usually the easiest to perform and control. It is necessary to control time, temperature and environmental parameters such as humidity or air flow rate.

Quality Control Parameters

Table 9-2 depicts the die bonding process and the list of parameters needing process control. Tables 9-4 through 9-6 compare each control property for each material choice (organic and inorganic). The tables show relative difficulty for controlling each property.

Consider Table 9-4, which shows the relative ease of control for die adhesive dispense parameters. Notice that the paste-based materials (silver-glass

Table 9-4 Relative Ease of Control of Adhesive Dispense Parameters.

ADHESIVE DISPENSE CONTROLLABILITY							
Property	Au-Si Eutectics	Ag-glass Paste	Solder	Epoxy	PI	TP	M
Quantity	3	1	3	2	2	2	2
Uniformity	3	1	3	2	2	2	2
Consistency	3	1	3	2	2	2	2
Pot life	1	1	2	1	1	2	1
Bondline thickness	3	1	3	3	2	2	2
1 = DIFFICULT 2 = MODERATE 3 = EASY							

PI = polyimide   TP = thermoplastic   M = mixture

and organics) are more difficult to control than the metallic preforms (Au-Si and solder). Silver-glass pastes are the most difficult to dispense and control. Because of the degree of fluidity of the Au-Si or solder metals, excess material is not likely to impact the overall result as is the case with the organic adhesive. Pot life is a concern with all the materials. As the adhesives sit around prior to use, controlling their environment is important, especially with the thermoset compounds. Oxidation of the solder compounds has an impact on the overall strength of the resulting bond. In the case of Au-Si eutectics, oxidation on the backside of the die creates difficulty in achieving a proper bond. Formation of the eutectic melt is hindered.

The complexity of the MCM determines the overall importance of the die alignment and placement step. Table 9-5 compares the level of difficulty encountered during die placement. Because the eutectic and solder processes usually are done manually, controlling these parameters is difficult. The process for silver-glass pastes is difficult to control because of its stringent bond thickness requirement. Silver-glass pastes shrink upon firing which impacts die alignment accuracy and repeatability. The organic compounds are all easier to control at this step than the inorganics. In MCMs, where the placement requirement is stringent, organic materials may be the only materials.

In Table 9-6 the process parameters for adhesive cure are addressed. Adhesives which require a separate cure usually can be cured in one step. The exception, silver-glass pastes, is due to the fact that its firing profile tends to

Table 9-5 Relative Ease of Control of Die Alignment and Placement.

DIE ALIGNMENT AND PLACEMENT CONTROLLABILITY							
Property	Au-Si Eutectics	Ag-glass Paste	Solder	Epoxy	PI	TP	M
Placement Accuracy	1	1	1	2	2	2	2
Repeatability	1	1	1	3	3	3	3
1 = DIFFICULT 2 = MODERATE 3 = EASY							

PI = polyimide   TP = thermoplastic   M = mixture

Table 9-6 Relative Ease of Control of Adhesive Cure.

ADHESIVE CURE CONTROLLABILITY							
Property	Au-Si Eutectics	Ag-glass Paste	Solder	Epoxy	PI	TP	M
Time-temperature	2	1	2	3	3	3	3
Environmental	2	1	2	3	3	3	3
Shrinkage	3	2	3	2	2	3	3
Electrical Test	3	3	3	3	3	3	3
Mechanical Test	3	3	3	3	3	3	3
Inspection	3	3	3	3	3	3	3
1 = DIFFICULT 2 = MODERATE 3 = EASY							

PI = polyimide   TP = thermoplastic   M = mixture

have several temperature plateaus. Environmental control (atmosphere H_2, N_2 or humidity) is important for the success of the inorganic materials. Again, the eutectic and solder materials are moderately easy to bond and control because of their manual nature.

9.2.3 Die Attach Equipment

Selection of die attach equipment depends on the degree of control needed in the process. MCMs are designed usually with very tight tolerances requiring tight die attach tolerances. The small volume production lot sizes typical of MCMs requires even greater machine flexibility.

MCM-C and MCM-D are likely to be found in low volume, custom part strategies, such as high performance computer systems. MCM-L is likely to be found in high volume applications such as consumer electronics (pocket calculators, cameras, etc.). Trends in die bonding systems are towards higher computerization, automation, robotics and integration [17]. Recent advances have pushed machine-attributable defects below current visual inspection screening ability [17].

9.2.4 Issues for MCMs

Manufacturability and Rework
Reworkability is an important consideration in the manufacture of MCMs because of their cost. Reworking MCM-C and MCM-L is comparatively easy. By choosing organic materials for die bonding, it is possible to manage both the temperature processing hierarchy and material interaction issues. In some cases, it is possible to mix die attach materials to optimize performance and cost. For MCM-C, silver-glass paste materials are used on the high power components of the MCM while organic adhesives are used for the lower power dice (see Figure 9-7). The resulting module may be hermetically sealed provided the epoxy die attachment can withstand the process temperatures required for sealing.

Organic bond adhesives are the only choice available for the MCM-L structure. The inorganic adhesives require process temperatures, usually too high for the resin-based substrates. Solders may be sued in MCM-L for die attach, but care should be taken to ensure it is not accidentally reflowed in later phases of assembly. A bonding material must be chosen whose processing temperatures do not irreversibly distort a resin-based MCM-L substrate. Adhesives with high T_g temperatures or high melting temperatures are unacceptable. The thermoplastics and alloyed organic adhesives are better choices especially if rework is desired because of their low process temperatures and low T_g.

Fabrication of MCM-D structures is mostly compatible with organic adhesives as well. Polyimides are used frequently as dielectrics in the fabrication of MCM-D substrates. It is necessary to consider the materials brought in contact with the polyimide because of its propensity to absorb moisture and epoxies susceptibility to outgas it. These two materials may be incompatible for hermetic applications. In situations where encapsulants are used to protect the die, epoxies and polyimides may prove compatible.

The easiest MCM to rework is MCM-C followed by MCM-L, with MCM-D being the most difficult. All organic adhesives are reworked readily with MCM-C, even some of the inorganic materials can provide a limited degree of rework coverage. MCM-L and MCM-D are best served by the thermoplastics and polyimide compounds as shown in Table 9-7.

The major cause of rework is defective die. Electrical defects in the die are detected during functionality tests during manufacture. There are several types of defect which render a module unrepairable or rejected. The most common defect in this category is bleedout onto adjacent bond pads. The cured resin is not removable without damaging either the substrate or the other good die in the module.

Failure Mechanisms and Reliability

Many tests are available to analyze the susceptibility to failure for an MCM. Table 9-8 indicates seven of the most frequently used analytical techniques for characterizing adhesives in electronic packaging. In addition to these basic analytic techniques, there are environmental and mechanical stress tests to perform. These tests have their origins in military specifications. Current stress tests are applicable mostly to single chip packaging implementations. MCMs tend to be significantly larger than single chip packages making existing stress tests very harsh on MCMs. Because the Mil-specs are stringent, a few organizations have attempted to write commercialized versions with relaxed endpoints. These efforts have generally been passed over in favor of using the more universal and familiar Mil-specs. The issue of relaxed endpoints for stress tests on MCMs or MCM specific stress tests must be addressed by industry. Some environmental stress tests with practical application in MCMs are described in Table 9-9.

The usual procedure for testing an MCM is to characterize materials in their cured and uncured states using some of the analytic testing techniques (Table 9-8). Some environmental and mechanical stress tests (Table 9-9) can be performed and the results evaluated by using the analytic tests again (Table 9-8). Comparisons can be made among the different materials as a function of different cure procedures. The information gained in stress testing can be used to predict failure modes and estimates of failure rate. A list of failure modes and the tests likely to predict them is given in Table 9-9.

The overall reliability of an MCM is a function of failure mode coverage. Designs that account for the worst failure modes do better than designs that do not. Mechanical failure modes such as delamination or disbanding are handled through material selection which minimizes high temperature exposures during

Table 9-7 Relative Ease of Reworkability for MCM Technologies.

REWORKABILITY							
Property	Au-Si Eutectics	Ag-glass Paste	Solder	Epoxy	PI	TP	M
MCM-C	0/1	0	0/1	3	3	3	3
MCM-L	0	0	0	2	2	3	3
MCM-D	0	0	0/1	1	2	3	2
0 = IMPOSSIBLE 1 = DIFFICULT 2 = MODERATE 3 = EASY							

PI = polyimide   TP = thermoplastic   M = mixture

Note: Inorganic adhesives generally are not amenable to rework.

Table 9-8 Analytical Characterization of Adhesives Used in Die Attach.

ANALYTICAL TESTING SERVICES	
DSC Differential Scanning Calorimetry	Determination of % cure and T_g. Thermal stability, Reactivity. Phase change versus temperature.
TGA Thermal Gravimetric Analysis	Determination of % cure. Thermal stability. Reactivity. Weight loss versus temperature—Used in conjunction with DSC.
TMA Thermal Mechanical Analysis	Precise determination of T_g. Mechanical performance versus temperature.
RGA Residual Gas Analysis	Identification of impurity and gas composition in hermetic modules. Identification of breakdown products of organic adhesives.
SEM Scanning Electron Microscopy	Determination of failure mode, microstructure of die attach adhesives.
Die Pull	Adhesive strength.
Die Shear	Adhesive strength.

Table 9-9 Tests for Evaluating Adhesive Performance.

ENVIRONMENTAL STRESS TESTS	
Environmental Test	**Identifiable Problems**
Thermal Shock	Poor adhesion properties. Brittle materials. Accelerates fatigue and crack generation and propagation.
Temperature Cycle	Poor adhesion properties. Brittle materials. Accelerates fatigue and crack generation and propagation.
HTSS High Temperature Steady State	Temperature related electrical sensitivity. Corrosion sensitivity due to residual contaminates within module.
HAST Highly Accelerated Stress Test	Temperature related electrical sensitivity. Corrosion sensitivity due to moisture entering the module.
Burn-in	Temperature related electrical sensitivity. Corrosion sensitivity due to residual contaminates within module.
Mechanical Shock	Mechanical stability and strength. Poor adhesion.
Mechanical Vibration	Dynamic mechanical stability. Poor adhesion.
Solvent Resistance	Process vulnerability to other process steps and solvents. Chemical nature of adhesives.

cure. Excess thermally induced stress should be avoided [18]. Environmental failure modes (moisture induced contamination, corrosion due to outgassed byproducts and moisture entering the module, due to loss of hermeticity) are serious problems. Die reliability has a large impact in overall MCM reliability. Because it is difficult to test thoroughly all electrical functions of a die, faulty ones slip through. Some companies have designed internal electrical test functionality into the logic function of the MCM [19] to assist in detecting faulty die. Bad die are replaced using a die rework process. Ideally the rework process should leave the substrate in a like new condition, making it difficult to distinguish between it and a non-reworked site. A major element of MCM reliability is how the rework process changes the substrate. MCM-C is least affected by rework processes. These substrates can be reworked several times with little damage. MCM-L and MCM-D on the other hand contain intrinsic organic components. When removing the organic adhesive used to bond the die, it is possible to alter permanently or to contaminate the intrinsic organic components of the substrate.

9.3 WIRE BONDING

Erik N. Larson

Wire bonding has a long and successful history as a microelectronic connection method because of its ability to adapt to increasingly complex packaging challenges, its well established reliability, its infrastructure and its potential for low cost. Multichip Module (MCM) technologies are able to leverage the experience gained in both the semiconductor and the hybrid industries using wire bonding techniques. Several forms of the wire bonding process are being used for various single chip and multichip applications. The first objective of this chapter is to supply information to potential users which enables them to decide if a specific wire bonding connection process is appropriate for their MCM application. The second objective is to supply the user with the tools necessary to develop a successful MCM wire bond process. Additional information which benefits the user, including package design, wire bonding equipment, electrical performance, reliability and rework also is covered.

9.3.1 Wire Bonding Methods and Procedures

There are three fundamental wire bond methods which have been developed over the years in the semiconductor industry. These methods are identified as thermocompression (T/C), ultrasonic (U/S) and thermosonic (T/S) bonding. Several wire materials are available, but those used most commonly are gold and an aluminum alloy. Available bonding techniques are identified as "ball" and "wedge" bonding. A bonding process is defined as a combination of one of these methods with a technique. The characteristics of each method along with available processes are shown in Tables 9-10 and 9-11. The technique selected is normally a function of one or more of the following: package design and environment, desired wire density, production volume, surrounding package or module clearance and the metallurgical properties and characteristics of the wire. Common die and package or substrate metallizations are aluminum and gold. Successful wire bonding requires the die to be attached rigidly to the MCM substrate surface using suitable die attach materials.

Each wire bonding method involves three steps to bond the wire to a surface. Initially, there is an applied force holding the wire firmly against the bonding surface. The second step involves the application of bonding energy. To form the thermocompression bond, thermal energy is used. For the ultrasonic bond ultrasonic energy is used and for the thermosonic bond, combined thermal and ultrasonic energies are used. Wire flow during the weld to the bonding pad

Table 9-10 Comparison of Wire Bond Methods.

METHOD	COMMON METALLURGY	TEMPERATURE	PRESSURE	TIME
Thermo-compression	Au wire Al or Au pad	300°C - 400°C	High	High (40 ms+)
Ultrasonic	Al alloy wire Al or Au pad	Ambient	Low	Low (20 ms)
Thermosonic	Al wire Al or Au pad	150°C - 225°C	Low	Low (20 ms)

Table 9-11 Comparison of Wire Bond Processes.

PROCESS	COMMON WIRE/PAD METALLURGY	AVAILABLE SINGLE ROW PAD PITCH	APPROX. MACHINE SPEED
Thermocompression Ball	Au wire Al or Au pad	4.0 mil	2 wires/sec
Ultrasonic Wedge	Al alloy wire Al or Au pad	3.0 mil	4 wires/sec
Thermosonic Ball	Au wire Al or Au pad	4.0 mil	10 wires/sec
Thermosonic Wedge	Au wire Al or Au pad	3.0 mil	3 wires/sec

is achieved, in all cases, by a combination of applied force and applied energy. Material flow during the application of bond energy is responsible for generating a clean interface at the bond zone, which is required for the weld to occur. Many high reliability wire bonding processes include a plasma bond pad cleaning prior to wire bond to enhance bondability and reliability at the interface. The third parameter is the bond time that energy and force are applied. Table 9-10 shows the relative parameters and materials for each wire bonding method.

Wire bonding uses a range of bonding wire diameters from small (< 25 μm, sometimes termed fine wire) to large (> 500 μm) wires. The larger wire (> 75 μm) is used primarily for power devices with relatively low wire counts and coarse pad pitches (> 10 mils).

Ultrasonic and Thermosonic Wedge Bonding

The wedge bonding technique is used for both ultrasonic aluminum wire bonding and thermosonic gold bonding processes. Ultrasonic wedge bonding is a low temperature process (typically at ambient temperature) in which the welding is accomplished by force and ultrasonic energy. The stages of the ultrasonic wedge bonding process are shown in Figures 9-2a and 9-2b. The ultrasonic energy at a frequency of about 60 kHz is applied to a wire, held in contact with a bonding pad by the wedge, or bonding tool. The combination of pressure and ultrasonic energy forms a metallurgical weld without the addition of significant thermal energy. The thermosonic gold wedge bonding process basically is identical to ultrasonic aluminum, with the addition of component heating. The most common wedge bonding process is ultrasonic aluminum. And the major reason is the lower process cycle time, and the reduced equipment complexity achieved with the absence of component heating. In some cases, there can be favorable metallurgical combinations gained by using aluminum wire.

Normally, the first bond is made to the die and the second bond is made to the MCM substrate. This is referred to as forward bonding. It is preferred because it is far less susceptible to edge shorts between the wire and die. This is discussed in more detail later. The procedure for ending the wire after the second bond takes several forms. For the small wire (25 - 50 μm), clamps can be used to break the wire after the second bond while machine bonding force is maintained on the second bond (clamp tear, see Figure 9-2a), or the clamps can remain stationary and the bonding tool can be raised off the second bond to tear the wire (table tear - see Figure 9-2b). The clamp tear process typically offers a slightly higher yield and reliability potential than the table tear process due to the force maintained on the second bond during the clamp tear motion. The clamp tear process also offers a slight speed advantage over the table tear process due to fewer required table motions. However, the table tear process, with a higher wire feed angle capability and a stationary clamp, has the potential to provide slightly more clearance from package obstructions such as a bond shelf or pin grid. The metallurgical characteristics of the ultrasonic wedge bond are shown in Figure 9-3. For large wedge bonding wire (> 75 μm), other methods have been used such as a cutting blade or the placement of the wire into a channel in the wedge for wire termination.

The wedge bond configuration shown in Figure 9-3, is characterized by a directional first-to-second bond alignment. The toe of the bond is the end nearest the severed wire while the heel is at the end of the bond where the wire bends from the bond surface. A negative feature of the wedge bonding

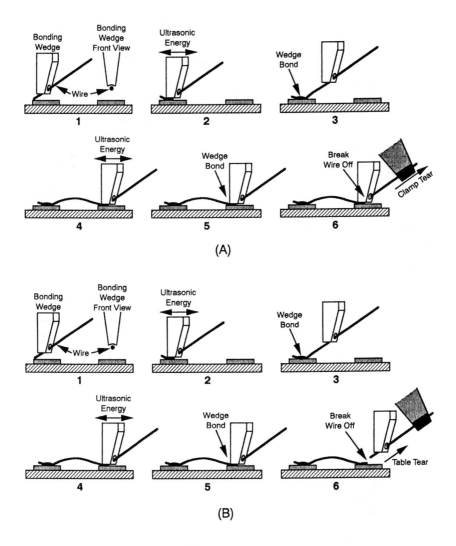

Figure 9-2 (a) Wedge bonding - clamp tear process. (b) Wedge bonding - table tear process.

technique is the design and machine demands required to maintain this directional first-to-second bond alignment. The requirement to rotate and align the die and substrate with the direction of the wire can restrict the use of wedge

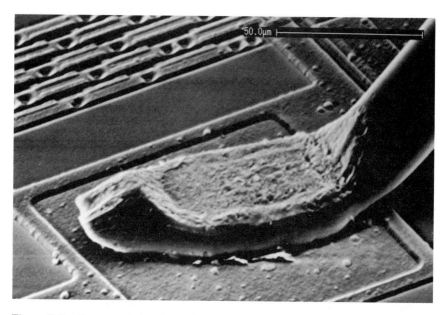

Figure 9-3a Features of the ultrasonic aluminum wire bond to the device. There is a tradeoff between bond development and wire deformation. The quality of the bond integrity cannot be determined from a visual examination of this type. Pull tests must be used to confirm bond strength.

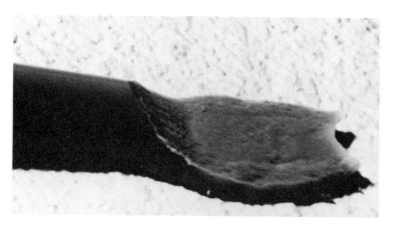

Figure 9-3b The major difference between the second bond and the first bond to the chip is the wire tear feature which is typical of a second ultrasonic bond. (Photos courtesy of Motorola.)

bonding on large MCMs (> 3 inches on a side) due to equipment limitations. An advantage of wedge bonding is the narrow bond configuration allowing a tight spacing between bonds. Aluminum wedge bonding has been demonstrated down to 75 µm by the author. In summary, the ultrasonic and thermosonic wedge bonding processes have the advantage of allowing a small bond pad pitch. However, factors based on machine rotational movements make the overall speed of the process less than that of thermosonic ball bonding.

Thermosonic Ball Bonding

Currently, thermosonic ball bonding operations almost always use gold wire. Relatively small wires (< 75 µm) are used, so that the capillary tool can sufficiently deform the second bond for easy wire separation. The thermal bond (thermocompression or thermosonic) is normally a ball bond process for the first bond position at the die. Originally the ball was formed by using a hydrogen torch, but modern bonders use an electric flame off (EFO) to melt a small portion of the wire extending beneath the capillary. Figure 9-4 shows the stages of the thermosonic ball bonding process. In Figure 9-5, ball bond features for both the first and second bond positions are shown. The vertical nature of the ball bond at the die reduces the possibility of edge shorts to the die. At the outer lead bond, the capillary leaves a characteristic circular pattern termed a crescent

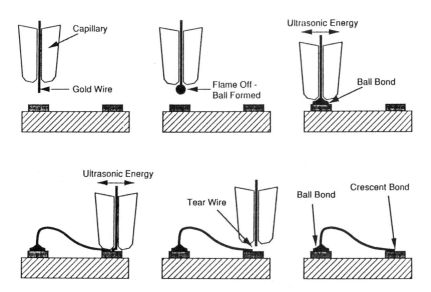

Figure 9-4 Thermosonic ball bonding process.

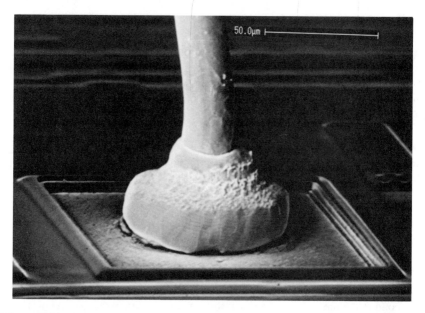

Figure 9-5a Features of the thermosonic ball bond to the device. Shear testing of the ball structure to the pad is becoming a standard method of bond quality evaluation.

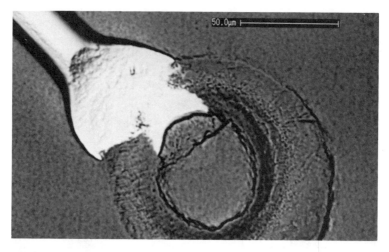

Figure 9-5b Unlike the ultrasonic bond method, the second bond of the "ball bond" method is distinctly different in appearance from that of the first. This is called a "crescent" bond. (Photos courtesy of Motorola.)

bond. The highly flowed gold in this second bond area generates a clean surface area for good interfacial atomic bonding. Once again, precleaning with a plasma process often is used to enhance the reliability of the interface. The round configuration of the first bond allows bonding machine motion to be made very rapidly in any direction. Modern automated wire bonders provide highly controlled targeting accuracy to the center of the bond pad for both ball and wedge bonding. The major pad pitch limiting feature of the ball bond method is the capillary as shown in Figure 9-6. Ball bonding is generally used in applications where the pad pitch is greater than 100 μm.

Thermocompression Ball Bonding
Thermocompression bonding is conceptually the simplest of the wire bonding techniques. Gold wire is used most commonly and deformed plastically by pressure and by bonding surface heat (300°C - 400°C) to form a metallurgical weld with the aluminum or gold bond surface. The high temperatures and long bond times required for thermocompression bonding have reduced its popularity since the development and integration of ultrasonic energy in assisting bond formation.

In summary, of the wire bonding alternatives, thermosonic ball bonding is expected to constitute the greatest wire bond MCM product volume. The vertical feed and looping features of ball bonding enable it to provide the highest chip to chip placement density. The reduced die pad spacing capability and potential metallurgical advantages of aluminum wire promote the ultrasonic wedge bonding process. It also eliminates another concern with ball bonding - the complexity of applying thermal energy to advanced MCM substrate materials and configurations. In low cost MCM applications where glob top encapsulation is used, gold wire has established excellent reliability. This is due to the high strength and ductility of the gold wire.

9.3.2 Wire Bonding Processes and Configurations: Geometry, Density and Package Design Considerations

Wire bonding has been adapted for a wide range of applications and, as a result, several single chip and MCM packaging geometries have been developed. The objectives of these various geometries are to increase chip and package density, electrical performance, and manufacturability while reducing cost. Wire bond density is limited mainly by its requirement for periphery I/O pads on the chip. The minimum I/O pad spacing on the chip results from adjacent pad metal reliability concerns, as well as from the limitations of test probing.

Differences in silicon and packaging manufacturing capabilities, in reliability concerns, as well as in the capabilities of test and assembly equipment, have

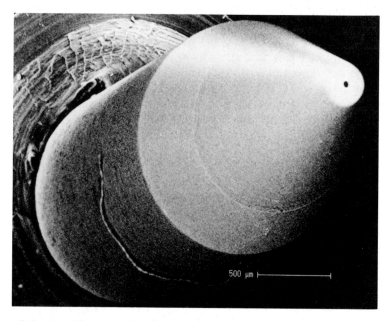

Figure 9-6a A capillary used for thermocompression bonding is shown mounted in the SEM fixtures. The size of hole for the wire is smaller than the capillary itself, yet in bonding, the capillary is not to contact previously bonded wires. The capillary diameter and not the wire diameter controls ball bond spacings.

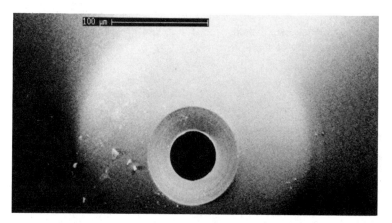

Figure 9-6b The capillary configuration at the wire exit region influences the shape of the ball bond. (Photos courtesy of Motorola.)

placed individual density limitations on each. As an example, cofired ceramic packaging technologies using screen printing have limited adjacent pad pitch to approximately 200 μm. On the other hand, chip wafer level test equipment capabilities have limited die pad pitch to approximately 85 μm. Therefore, to take advantage of the maximum densities of each, a fanout or pitch expansion method must be used. Some expansion techniques used are: multiple shelf ceramic PGAs with orthogonal and radial designs, and staggered multiple row package pads on a single shelf. Of the geometries developed, the most popular applications which use wire bonding are listed below:

- Single and multitiered cofired ceramic pin grid arrays (PGAs), single chip and multichip
- Ceramic and plastic quad flat packages (CerQuads and PQFPs)
- Chip on board (COB)

For these applications, the pad pitch on both the chip and package are being pushed downward as the wire count and I/O per device are being increased. QFP technology has been limited with this packaging technique to approximately 300 I/O, where multitiered package technology enables over 500 I/O. An example of where the high density available currently for both T/S-gold and U/S-aluminum wire has been useful is the Unisys MRAM multichip module with a 252-pin PGA package. Both the multichip module and the single chip package utilize a double shelf and radial fanout technique. Ultrasonic aluminum wedge and thermosonic or gold ball bonding can be used for the processor, which has 252 wires at 5.2 mil die pitch. The memory chips are bonded using a thermosonic gold ball bonding process.

Another example is a single chip, fine pitch 323-pin two-tiered cofired ceramic PGA package which incorporates a double shelf orthogonal design. 388 wires at 4.25 mil pitch were wire bonded on a 0.517" square CMOS die using ultrasonic aluminum wedge bonding with a 30 degree clamp tear process. The selection of a die pad pitch of 4.25 mils on the 323 PGA was based on studies using an experimental test vehicle designed to establish aluminum wedge bond process limitations. The test vehicle employed 3, 4, 5 and 6 mil die pad pitch, double shelf cofired ceramic package construction with an orthogonal wire layout.

As both MCM and single chip packaging density increase, the area surrounding the wire bond positions on the chip and substrate is reduced. This requires that the package designer and wire bond process engineer analyze both the constraints of the wire bond equipment and the process related issues with high density wire bonding. The dimensions of the wire bond machine hardware and the wire bond process motions are necessary to ensure a successful design when developing a high density package or MCM. This information can be obtained through equipment vendors, or can be attained by direct equipment measurements. Once the machine dimensions and parameters have been established, the package designer can integrate these with the potential package variations to verify adequate clearance for a given wire bond process.

When designing an MCM, die attach area and placement tolerances should be analyzed to ensure adequate area for adhesive fillets, as well as adequate area surrounding the die. The reason for this is to provide adequate machine clearance as well as to protect the substrate or package wire bond pads from die attach foreign material contamination. Die attach material splattering, wicking or bleedout can contaminate bond pads and reduce the bondability or reliability of the surface.

When considering wedge bonding, there are several combinations of both the table tear and clamp tear processes which differ in both performance and clearance requirements. Table 9-12 shows clearance considerations for several available wire bond process combinations. Figure 9-7 shows a cross section of a typical multitiered, single chip ceramic package with the wedge-clamp-wire profile for ultrasonic wedge bonding. Wedge bonding can be used in either the forward mode (first bond on die, second bond to package or substrate), or reverse mode (first bond on package or substrate, second bond on die). The most common and preferred wedge bonding direction is forward bonding. This

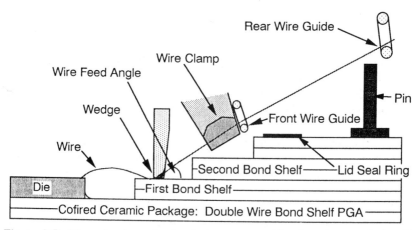

Figure 9-7 View showing wire bond clearance considerations for a typical cofired ceramic PGA.

is because forward bonding delivers wire loop shapes that provide more clearance from the edge of the die, preventing wires from edge shorting to active material or to the base silicon on the die periphery. Thermosonic ball bonding offers the most machine clearance, since there is no clamp mechanism below the transducer and the wire is fed vertically through the tool. For COB MCM applications, the spacing between devices, as well as the height of the devices, must be analyzed. If heat can be supplied through the MCM substrate, ball bonding can provide two advantages for these applications: reduction of die to die spacing because of the vertical wire feed and security from edge shorting. An example of this type of design is shown in Figure 9-8, where die are mounted on a thin film multilayer ceramic substrate. In this MCM example, thermosonic gold ball bonding is used to connect the chip pads to the to the thin film substrate pads. Also shown in Figure 9-8, are the thermosonic gold ball bonding looping profile features. A disadvantage of ball bonding with COB is the requirement and complexity of heating the board or substrate. The repairability and rework potentials of each process differ, and are discussed in a later section.

In a multitiered or multiwire length fine pitch application, wire-to-wire clearance should be analyzed to guard against shorting. Shorting may occur with one or more of the following: die shift, bond placement drift, loop variation (all due to machine accuracy and repeatability limitations) and package dimensional fluctuations. By combining empirical studies with theoretical modelling, wire-to-wire clearance can be analyzed. Wire loop shapes are determined by bonding to a test vehicle with equivalent geometry as the planned application. These wire loops can be cross sectioned, plotted and curve fitted using a representative polynomial equation. The curves then can be integrated into a three-dimensional algorithm which will calculate wire-to-wire spacing at any point along the wire. The vertical and horizontal components of the space between wires are calculated independently. The total vector is calculated, indicating the gap between wires. An example of this is shown in Figures 9-9a and 9-9b.

9.3.3 Wire Bonding Equipment

Wire bond successes owe a great deal to the dedicated efforts of equipment suppliers. The massive equipment infrastructure of wire bonding cannot be overemphasized. For example, as other connection technologies have gained popularity, wire bond equipment vendors have maintained their competitive position by continuing to offer increased bonding rates as well as the ability to decrease bond pad pitch.

Requirements are now in place for wire bonding to be evaluated statistically (high strength with low variability). Low variability is achieved and maintained

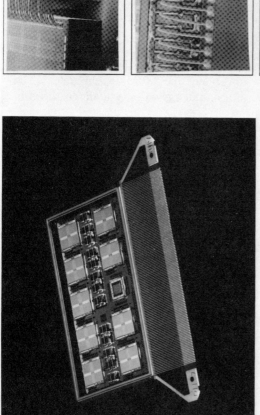
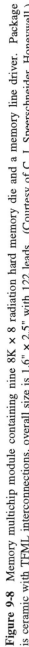

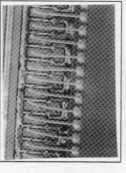

- Memory Multichip Module - TFML Interconnect
- Module is 1.6" x 2.5" with 122 Leads
- Module Contains Nine 8K x8 Rad Hard Memory Die and One Memory Line Driver
- Gold Wire Bonds (1.0 mil dia. wire)

Figure 9-8 Memory multichip module containing nine 8K × 8 radiation hard memory die and a memory line driver. Package is ceramic with TFML interconnections, overall size is 1.6" × 2.5" with 122 leads. (Courtesy of C. J. Speerschneider, Honeywell.)

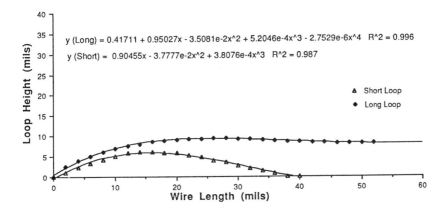

Figure 9-9a Curve fitted wire loop profiles for a typical two-tiered, wedge bonded package.

by monitoring every phase of the bonding process. Many types of analytical equipment are used to monitor wire bond machine performance and to identify undesirable conditions (often not apparent). For instance, equipment is available to measure the impedance of the bonding tool in free air and as the bond is formed [20]. These impedance measurements can be used to maximize the performance of the ultrasonic system and to troubleshoot the system.

Wire bonding equipment, in competition with the emergence of tape automated bonding (TAB) technology, has been pushed with respect to bonding speed and density. For example, it is possible in tenths of a second for a ball bonder to automatically form the ball, to position the capillary, to apply power and force for the requisite dwell time, to loop the wire to the ending pad forming that bond and then to excise the wire. Speeds of 4 wires/sec. for wedge bonding and 10 wires/second for ball bonding are common. Fully automated equipment is manufactured by several U.S., European, and Japanese companies. U.S. companies include: Hughes Aircraft, Kulicke & Soffa, and Orthodyne [21]. This automation has been made possible by the addition of microprocessor control and optical vision systems to the equipment. The vision system uses fiducials or standard features on the die and package to accurately determine their location in space. Vision and positioning systems having location accuracy within a few tenths of a mil have allowed wedge bonding of parts in the 3 mil pitch range. Vision systems also accommodate package variations inherent in cofired package processing. Another wire bond equipment feature beneficial to

VARIABLE DESCRIPTIONS	VARIABLES
Short Bond Wire Length (nominal).................	40.00 mils
Long Bond Wire Length (nominal).................	52.00 mils
Bond Wire Diameter (nominal)............................	1.25 mils
Lower Shelf Misalignment (due to artwork shift).....	+/-2.00 mils
Upper Shelf Misalignment (due to shelf shift).........	+/-1.00 mils
Die Placement Tolerance (X-Y)...............................	+/-1.00 mils
Die Placement Rotation Tolerance.........................	+/-0.50 Deg's
Die Pad Pitch (nominal)...	5.00 mils
Die Size (longest side measurement)....................	460.00 mils
Wire Bond Placement Tolerance............................	+/-0.40 mils
Distance From Die Pads (for detailed error info).....	12.00 mils

VERTICAL SPACE COEFFICIENTS	VALUES
Short Wire Formula (3rd order equation)	
Coefficient of x	9.05E-01
Coefficient of x^2.......	-3.78E-02
Coefficient of x^3.......	3.81E-04
Long Wire Formula (4th order equation)	
Coefficient of x	9.50E-01
Coefficient of x^2.......	-3.51E-02
Coefficient of x^3.......	5.20E-04
Coefficient of x^4.......	-2.75E-06

ERROR SUMMARY	ERRORS
Spacing Error Due To Bond Placement Tolerance	0.80 mils
Spacing Error Due To Upper Shelf Shift Tolerance	0.23 mils
Spacing Error Due To Lower Shelf Shift Tolerance	0.60 mils
Spacing Error Due To Die Shift Tolerance.............	0.07 mils
Spacing Error Due To Die Rotation........................	0.13 mils

RESULTS SUMMARY	RESULTS
Vertical Wire Spacing At The Critical Distance.......	1.12 mils
Horizontal Spacing At The Critical Distance...........	1.92 mils
Horizontal Spacing After RSS Analysis..................	2.71 mils
Total Worst Case Wire Spacing........................	**2.22 mils**
Total Wire Spacing After RSS Analysis..............	**2.94 mils**

Distance Measured From Die Pad	Wire Spacing
1.00 mils	2.86 mils
2.00 mils	2.78 mils
3.00 mils	2.70 mils
4.00 mils	2.62 mils
5.00 mils	2.54 mils
6.00 mils	2.47 mils
7.00 mils	2.40 mils
8.00 mils	2.34 mils
9.00 mils	2.29 mils
10.00 mils	2.25 mils
11.00 mils	2.23 mils
12.00 mils	2.22 mils
13.00 mils	2.23 mils
14.00 mils	2.27 mils
15.00 mils	2.32 mils
16.00 mils	2.40 mils
17.00 mils	2.50 mils
18.00 mils	2.63 mils
19.00 mils	2.77 mils
20.00 mils	2.94 mils
21.00 mils	3.12 mils
22.00 mils	3.33 mils
23.00 mils	3.54 mils
24.00 mils	3.77 mils
25.00 mils	4.02 mils
26.00 mils	4.27 mils
27.00 mils	4.53 mils
28.00 mils	4.80 mils
29.00 mils	5.07 mils
30.00 mils	5.35 mils
31.00 mils	5.63 mils
32.00 mils	5.92 mils
33.00 mils	6.20 mils
34.00 mils	6.48 mils
35.00 mils	6.76 mils
36.00 mils	7.04 mils
37.00 mils	7.30 mils
38.00 mils	7.56 mils
39.00 mils	7.81 mils
40.00 mils	8.05 mils

Figure 9-9b Three-dimensional wire spacing analysis using curve fitted loop profiles and significant packaging tolerances.

MCM manufacturing is the capability to change quickly from one product to another with minimal tooling and software changes. Software programs for a wide variety of products can be written by the user and stored on disk. A product change can be made in as little as 20 minutes. An example of a commercial ball bonder is shown in Figure 9-10.

Figure 9-10 Fine pitch 132 lead device bonded in a K & S 1484 automatic gold ball bonder. (Courtesy of Kulicke & Soffa Industries.)

Table 9-12 Wire Bond Process Clearance Requirements.

WIRE BOND PROCESS	WIRE TEAR METHOD	WIRE FEED ANGLE	WIRE TEAR/FEED MOTION REQUIRED	CLAMP CLEAR-ANCE REQUIRED
Thermocompression Ball Bonding	Vertical tear	90°	Yes, vertical	None
Ultrasonic Wedge Bonding	Table tear	45°-90°	Yes	Yes
Ultrasonic Wedge Bonding	Clamp tear	30°-45°	Yes	Yes
Ultrasonic Wedge Bonding	Clamp tear with swipe motion	45°-60°	Yes, + swipe	Yes

A consideration when selecting wire bond equipment for MCM assembly is the required table travel to bond the desired module. If the working area of the wire bonder is sufficient to bond the entire MCM, production cycle time is reduced. For wedge bonding, the dimensions of the MCM and die placement on the module should be considered, since the machine must rotate the part or the machine bond head to maintain first and second bond alignment.

9.3.4 Electrical Performance

Wire bonding ranks behind TAB and flip chip bonding when it comes to high speed performance where low inductance is critical. The electrical performance of a wire bonded package or module can be optimized if the package designer understands the capabilities and limitations of the wire bond process. Wires can be shortened to the minimum length without sacrificing manufacturability. Wire geometry can be optimized to reduce inductance. If the MCM is designed creatively for short wires, the inductance of the wire bonds can be close to, if not less than, conventional TAB leads. However, flip TAB can be designed with shorter lead lengths, providing significantly less inductance than most wire bond designs [22]. A typical inductance value for wire bonds approximately 0.100" long (1.00 - 1.25 mil diameter wire, Au and Al) is on the order of 2.5 nH. A typical DC resistance value for 1.00 mil diameter wire of this length would be approximately 0.14 Ω for Al and 0.11 Ω for Au wire. In comparison, a flip

TAB lead 50 mils long designed for 4 mil pitch (0.002 sq. in. × 0.0013 sq. in. cross section Cu) would have approximately 1 nH inductance and 0.013 Ω resistance. Area array flip chip designs supply resistance, inductance and capacitance values much lower than either wire bond or TAB. This is due to the short connection lengths and high I/O capabilities without periphery connection restrictions.

9.3.5 Reliability as Applied to MCMs

A major advantage of wire bonding for MCM connections is its solid base of reliability from bond strength studies to time and temperature design factors. MCM technologies, however, bring new metallization schemes and deposition processes which may affect reliability. A diligent effort must be made to assure that any new, or previously unproven conditions (residual chemicals, different metals, or metal thicknesses) are evaluated thoroughly to ensure a reliable MCM environment.

The major reliability concerns with wire bonded structures are the intermetallic reactions occurring with time and temperature [23]. The condition of intermetallic formation between dissimilar metals, such as in gold wire with an aluminum alloy bond pad, or with aluminum wire on a gold bond pad, should be studied and understood. Intermetallic diffusion can occur in which the failure mode is an increased joint electrical resistance or low pull strengths due to eventual voiding at the interface. For the gold ball bond on an aluminum alloy pad, the thickness of the aluminum is the main concern. With insufficient aluminum, the pad can be consumed by an intermetallic formation, eventually causing voiding. With the aluminum wedge bond on a gold pad, the main concern is the ratio between the bond heel thickness and the gold thickness on the pad.

Contamination from wafer processing, plastic package outgassing or epoxy die attach outgassing and bleedout affects bondability and reliability. While these surface impurities may affect bond yields, it is the reliability factor after bonding which is of concern [24]. Contaminates accelerate failures by reducing the onset temperature or the time to failure due to intermetallic growth. Plasma cleaning using argon or argon-oxygen mixtures can be used in most cases to clean the interface prior to bonding [25]. The effects of pre- and post-processing temperature excursions, away from those involved directly in the bonding process, should also be analyzed, as they may reduce initial bondability or accelerate intermetallic failure.

Some actual reliability concerns related to the various metal systems and processes incorporated in different packaging methods are listed in Table 9-13. Cofired ceramic single and MCM packages with hermetic high temperature

Table 9-13 Intermetallic and Reliability Concerns for Various Package Types.

Package Type	Package Pad Metal	Die Pad Metal	Wire Type	Lid Sealing Method	Intermetallic and Reliability Concerns
CerDip	Al	Al	Al	Hermetic; high temperature	None
Cofired Ceramic	Plated Au/Ni	Al	Al	Hermetic; high temperature	Management of gold thickness on package.
Cofired Ceramic	Plated Au/Ni	Al	Au	Hermetic; high temperature	Management of assembly, process times, temperature, bond pad cleanliness.
Plastic	Plated Ag/Cu	Al	Au	Nonhermetic; low temperature molding	Chemical impurities, mechanical stress on wires.

sealing have been designed and assembled using gold and aluminum wire. It is necessary to accurately define the post processing and burn-in time at temperature exposures during the early stages of the package development.

Wire pull tests for wedge bonds and shear tests for ball bonds should be used to determine that complete interfacial bonding is achieved. Environmental tests, such as accelerated time temperature and temperature cycling, must be conducted to assure that long term materials compatibility is maintained in the product, especially if new materials or processes are introduced.

9.3.6 Yield and Repairability as Applied to MCMs

Wire bond yield potential and repairability factors may influence the selection of wire bonding for an MCM chip connection solution. Industry yield values exceeding 99% for single chip devices with over 300 wires have been achieved for both the ultrasonic and thermosonic bonding processes. This yield may be obtained without repair or rework, and correlates to a 34 ppm defect rate on an individual wire basis. Table 9-14 shows process ppm defect rates versus connected chip yield for various wire counts. Rework can be performed in most cases for individual wires by manually removing the defective wire and bonding a new wire in place. In some cases, if a reworkable die attach material is used, individual die can be removed and reworked. If individual die removal and replacement is necessary, the substrate and die attach materials (FR-4, ceramic or thin film) should be carefully considered, as die rework may, or may not, be feasible.

When considering rework for wire bonds on a die, the pad dimensions and bonding process should be considered. Multiple wedge bonds can be placed on a pad if the new bonds do not overlap the old bonds, or if the new bonds only partially overlap the old bond impression. Thermosonic gold ball bonds can be stacked once or twice over previous ball bonds to rework individual wires. Military Specification 883C provides criteria for rework.

9.3.7 Wire Bond Process Development

When developing and improving a production wire bond process for single or multichip packaging, several key stages must be addressed. These stages include initial process design and development, process characterization, process control and process optimization. To refine a process, these stages form a continuous loop between characterization and control, with periodic optimization and development, as shown in Figure 9-11.

During the initial stages of process design and development, process capabilities should be understood to set achievable goals. For wire bonding, information is available through the published literature, as well as from industry and equipment vendor experience. Vendors for wire bond machines, wedge and capillary tooling, and bonding wire are excellent sources of information needed to develop a successful process.

The process characterization stage should include the installation of a data collection system, identifying specific defects on a per wire level. By breaking this down to the lowest level, and by identifying defects on a per wire basis, the major contributions to yield loss can be identified. Measures can be taken to resolve the specific problems. An example of the break down of defects (in ppm), as well as the common defect categories for a typical wedge bonding process, are shown in Table 9-15. The defects can be identified with individual machines, programs, products etc., to enable a full characterization of the process.

Process control also is important if a process is to be developed successfully. To define the performance level of a process accurately, a stable operation must first exist. Operating variables must be minimized and strict regulation of the existing variables must be established. Variables such as bond program

Table 9-14 Production Yield for Specific Wire Counts and Process Defects (ppm).

PRODUCT YIELD (%)	WIRE COUNT (CHIP OR MCM)							
	20	50	100	200	400	600	800	1000
99.9	50.0	20.0	10.0	5.0	2.5	1.7	1.2	1.0
99.5	250.0	100.0	50.0	25.0	12.5	8.3	6.3	5.0
99.0	500.0	200.0	100.0	50.0	25.0	16.7	12.5	10.0
98.0	1000.0	400.0	200.0	100.0	50.0	33.3	25.0	20.0
97.0	1500.0	600.0	300.0	150.0	75.0	50.0	37.5	30.0
96.0	2000.0	800.0	400.0	200.0	100.0	66.7	50.0	40.0
95.0	2500.0	1000.0	500.0	250.0	125.0	83.3	62.5	50.0
94.0	3000.0	1200.0	600.0	300.0	150.0	100.0	75.0	60.0
92.0	4000.0	1600.0	800.0	400.0	200.0	133.3	100.0	80.0
90.0	5000.0	2000.0	1000.0	500.0	250.0	166.7	125.0	100.0

Note: Allowable process defects (ppm) = $(1 - \text{desired yield}) \times 10^6/\text{wire count}$ (chip or MCM).
Entries represent 1 bad wire per defective chip (or MCM).

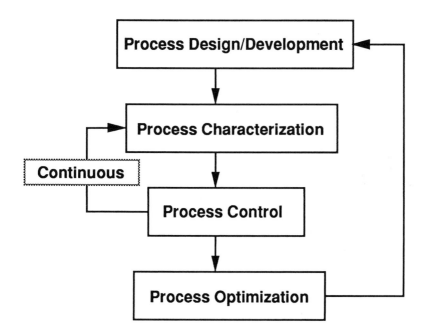

Figure 9-11 Typical wire bond process development flowchart.

parameters, machine setup and operation procedures, bonding tool installation, wire pull procedures and product change overs are examples of the areas needing focus and control. Consistency across the process, such as in the training of operators also must be established.

If the previous stages of process development are in place, process optimization can be performed using standard Taguchi or design of experiment techniques and focusing on the resolution of specific problems. Once a process is operating in a production environment, statistical process control (SPC) can be applied to such items as wire pull to monitor the process and to minimize process drift [26].

9.3.8 Wire Bond Process Costs

The cost, reliability and yield potentials of wire bonding make it a logical option when developing a MCM facility. Since wire bond processes involve no chip

Table 9-15 Defect Tracking System Summary.

Bond Program	Total Wires Bonded	Total Defects	Total Defects (ppm)	SPECIFIC DEFECTS (PPM)				
				Bond Off Center	Bond Not Sticking on Die	Bond Not Sticking on Package	Broken Wire	Foreign Material
MCM 1	200,000	4	20	5		10		5
MCM 2	300,000	7	23		10	3	10	
MCM 3	800,000	20	25	10		5		10

modifications and the equipment has an established base of competitive development, wire bonding is lowest in cost as an MCM connection method.

Current equipment costs are roughly $140,000 for wedge bonders and in the $120,000 range for ball bonders. Die attach equipment costs roughly $80,000 to $100,000. Support equipment, such as wire pull stations, plasma etchers and storage facilities, contribute to the remaining significant costs for wire bond processing. If die attachment requires very accurate and repeatable placement (±1 mil), costs rise significantly. With wire bonding, substantial non-recurring engineering (NRE) charges and tooling charges are minimized while equipment availability is maximized. In some cases, wire bond is unaffected by midstream module design changes that require redesigning TAB tape.

Cost differences become evident when competitive studies are made. In one study, a three chip module, based on ceramic substrate technology, was constructed using wire and flip chip methods. Wire bonding was done using conventional high volume, single chip equipment. An additional cost factor in using a flip chip technology, was the increased substrate costs. Tolerances required for flip chip bonding were more stringent than those required for wire bonding, therefore increasing costs. From this study, it became obvious that each application may require a separate cost evaluation.

Cost analyses include volume and individual process cycle time predictions, equipment costs, tooling costs and non-recurring tooling and engineering costs. In some cases, multiple wire bonders are set up in parallel, providing volume comparable to a single gang bonding TAB operation at an approximately equivalent equipment cost. The cost associated with the flexibility required for engineering and product design changes, which occur and should be anticipated as a part of the planning cycle, also should be evaluated.

9.3.9 Comparison to Other Connection Techniques

TAB pitch limitations are similar to those in wire bonding. However, smaller pitches in the range of 50 μm have been cited for TAB [27]. Area array flip chip solder bump and area TAB are current alternative technologies. They have the potential to exceed peripheral pitch limitations to I/O density occurring with wire bonding. Electrical performance is a key issue necessitating a switch from wire bonding to TAB or to flip chip solder bump technologies. TAB, in general, can exhibit less inductance and capacitance than an equivalent wire bonded package. Also, TAB can provide a pretest capability for die prior to installation in the package or on the module. An advantage of wire bond relative to flip TAB or to flip chip is the ability to perform a complete visual inspection of the connection after the module is assembled. The wire bond processes also offer limited abilities in repair.

Of three current alternatives for MCM chip connection applications (wire bond, TAB and FCSB), wire bonding offers the most mature, flexible and inexpensive assembly method. A mature infrastructure is in place. Extensions of the process have permitted wire bonding to meet new challenges. There is a history of reliability data and general industry experience. This base of technology and experience establishes wire bonding as a viable alternative for current as well as for future MCM development. TAB and FCSB provide benefits in electrical performance and density relative to wire bond, but at higher cost and with more extensive development requirements.

9.3.10 Summary

When selecting a chip connection method for an MCM application, several considerations are important, including cost, time to market, manufacturability, electrical performance and reliability. The applicability of wire bonding to MCM packages is dependent on the substrate material, requirements in geometry, secondary process thermal requirements, material compatibility, chip removal and repair requirements, and pre test requirements.

The optimum wire bond process needs to be determined relative to the application. For example, when the substrate is heated, thermosonic gold ball bonding offers the best method for COB-type MCM applications, where components are placed as close together as possible. Aluminum wedge bonding offers the finest pad pitch and the highest wire density, together with potentially favorable metallurgical conditions and processing at ambient temperature.

When designing an MCM, careful attention must be given to the space and clearance requirements for the wire bond process chosen, as well as to metallurgical and materials compatibility. Once a process is selected, and the MCM is designed, process development tools should be applied to enhance the process by increasing yields and product reliability, as well as reducing the overall product cost. The well established infrastructure for wire bond is utilized to reduce the product time-to-market and to maximize the process efficiency. Although the wire bond approach does not include a pretest possibility like TAB or a low impedance like flip chip solder bump, its specific advantages should warrant its consideration when developing a MCM facility or product.

Acknowledgments

The author would like to acknowledge with sincere appreciation the following individuals for their assistance in developing a successful wire bond process at Unisys, and for their help and guidance in preparing this section: S. Bezuk, D. Haagenson, J. Nelson and T. Severn from Unisys, D. Robinson and B. Winchell from Motorola, G. G. Harman from NIST and D. Leonhardt from K&S Industries. The wire bond SEM samples were supplied by D. Silvas and R. P Padmanabhan of Motorola. The SEM of the memory MCM was provided by C. J. Speerschneider of Honeywell. These contributions also are acknowledged with sincere appreciation.

9.4 TAPE AUTOMATED BONDING

Dean R. Haagenson, Steve J. Bezuk, Rajendra D. Pendse

9.4.1 Introduction

Tape automated bonding (TAB) is a connection technique for attaching semiconductor die to a variety of packaging media, including single chip and multichip packages. The connection is made by bonding a patterned conductor to the corresponding I/O pads on the die and package (See Figure 9-12). TAB parts can be assembled (see Figure 9-13) in three basic configurations: conventional TAB, flip TAB and cavity TAB.

Conventional TAB consists of mounting the die with the backside attached to the substrate or package. Leads are formed to facilitate the connection from the plane of die to the plane of the substrate or package. Flip TAB consists of mounting the die with the active surface facing the substrate or package. The bonding pitches on the die can be equal to those on the substrate or package with little or no leadforming required. This results in leads that are short relative to

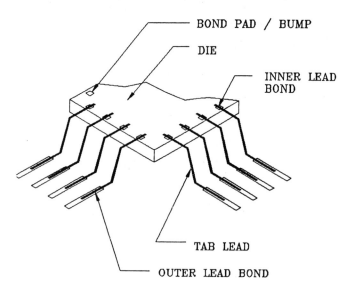

BOND PAD / BUMP

DIE

INNER LEAD BOND

TAB LEAD

OUTER LEAD BOND

Figure 9-12 Schematic illustration of the TAB concept.

CONVENTIONAL TAB

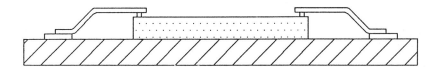

FLIP TAB

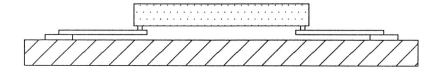

CAVITY TAB

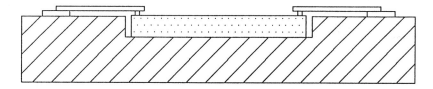

Figure 9-13 TAB assembly configurations.

conventional TAB. Cavity TAB is similar to flip TAB in that the die and package pitches can be equal with short leads. Flip TAB and cavity TAB provide for the denser TAB assembly of MCMs compared to conventional TAB.

Cavity TAB, however, is difficult to implement due to package material and design limitations. Another, more recent, TAB configuration provides for routing the leads to an array of solder bumps on the tape frame, which can then be mounted on the substrate or package. This is called array TAB.

TAB originally was envisioned as an alternative and eventual replacement for wire bonding in the 1960s. However, steady improvements in wire bond technology and the high costs associated with TAB implementation have limited the use of TAB to high volume consumer electronics products that require low profile and high density assembly.

The interest in TAB as an alternative to the wire bond and surface mount assembly of VLSI chips arises because its application eliminates a level of packaging. The TABed die can be bonded directly to a PWB or to an MCM substrate (See Figure 9-12). Additionally, TAB offers the following advantages over traditional wire bonded packages:

1.   Superior electrical performance
2.   Assembly of high lead count, fine pitch devices
3.   Device testability prior to commitment to package
4.   Repairability and
5.   The dense assembly of low profile components.

Despite the inherent advantages of TAB, its implementation in the high performance and MCM areas has been slow. This is the result of an inadequate infrastructure for advanced equipment and materials needed, as well as cost. Lower volumes and custom tooling required for the die and package designs have contributed to these cost issues.

9.4.2 Basic Process Flow for TAB Packaging

Figure 9-14 shows a process flow for a typical TAB packaging application. In this section, the specific steps in the TAB packaging flow and assembly process are discussed.

Wafer Bumping

Typically, metal standoffs, or "bumps," are added to the I/O pads on an IC. These bumps are normally gold plated on the wafer, although solder and copper bumps also are used. The purpose of these bumps is to provide the proper metallurgy for the inner lead bond (ILB), to provide a standoff preventing the TAB lead from shorting to the edge of the die, and to protect the underlying aluminum from corrosion or contamination. A typical bump configuration is shown in Figure 9-15. An under bump metallization (UBM) is required to

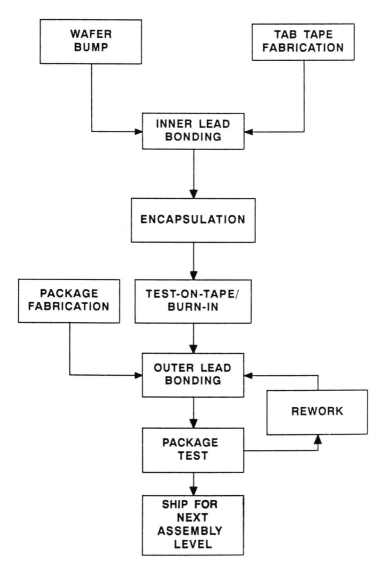

Figure 9-14 Process and materials flow for TAB packaging.

provide a diffusion barrier between the gold of the bump and the aluminum pad. The bottom metal must have good adhesion and the top metal should be inert and plateable. A typical choice of UBM is 0.2 μm Ti-W over the aluminum,

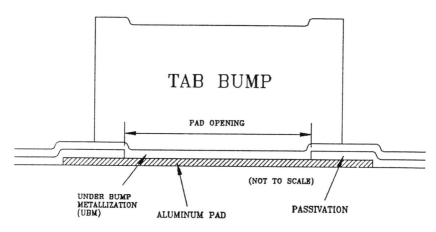

Figure 9-15 A typical bump configuration showing the pad opening and the under bump metallization (UBM).

covered by 0.1 μm Au. The bump is plated over this thin gold layer. The extension of the bump and UBM beyond the pad opening and over the passivation layer, facilitates the sealing of the aluminum pad from moisture and external contamination (see Figure 9-15). Bumping technology, and alternatives to wafer bumping, including the criteria for UBM selection are reviewed elsewhere [28]-[29]. Alternatives include bumped tape, bumps deposited by gold ball wire bonders and bumpless configurations [30]-[32].

TAB Tape Processing

The important features of a typical TAB tape design are illustrated in Figure 9-16. MCC has developed CAD tools specifically for the expeditious design of tape. Attempts have been made to standardize outer lead bond (OLB) footprints and tape formats by the JEDEC and EIAJ committees. Standardization of pad patterns for PCBs has been pursued by the Institute for Interconnection and Packaging of Electronic Circuits (IPC). Standardized pad patterns help reduce the costly tooling associated with TAB, but also can limit the advantages in density and electrical performance.

Tape is manufactured in continuous reels or in panel arrays. The tape is supplied for use on inner lead bonding (ILB) equipment in reels or singulated and mounted into individual slide carriers. Slide carriers are becoming popular, particularly for higher lead count die. The carrier format eliminates mechanical

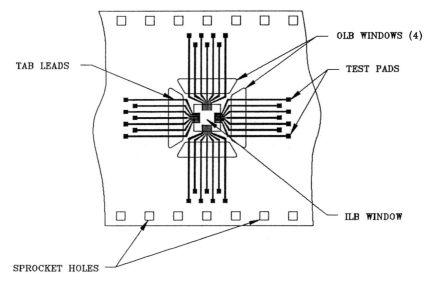

Figure 9-16 Important features of a TAB tape design.

damage resulting from the bending stresses encountered during the rolling process, and also reduces the likelihood of handling damage.

Inner Lead Bonding
The patterned TAB tape frame is connected to corresponding I/O pads on the die during the inner lead bonding (ILB) process. This process involves the metallurgical bonding of the tape leads to the bumps or bond pads. A variety of process options exist for ILB that depend on chip size, tape and bump metallurgies and other process constraints.

Encapsulation
Following ILB, the IC is typically encapsulated using one of many possible methods [28] (see Figure 9-17). The purpose of encapsulation is to protect the die chemically and mechanically. Epoxies and silicones are popular materials used for TAB encapsulation. Material properties that are important considerations for encapsulants include:

- Low alpha particle emission
- High temperature stability (governed by the glass transition temperature)

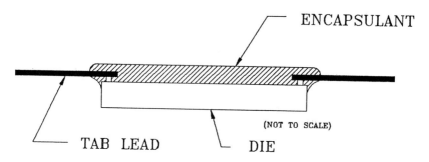

ENCAPSULANT

(NOT TO SCALE)

TAB LEAD   DIE

Figure 9-17 An encapsulated TAB chip.

- Minimal hardening or embrittlement at low temperatures (-55°C)
- Low moisture absorption
- Low ionic content
- Low modulus of elasticity and
- A coefficient of thermal expansion (CTE) minimizing the stress on the die or cracking of the encapsulation.

Test chips can be designed to evaluate and qualify encapsulants. A variety of machines are available for the dispensing and curing of these materials. Other encapsulation schemes include overmolding the inner lead bonded device in an injection mold, much like a wire bonded device on a leadframe [33].

Test-on Tape and Burn-in
One of the major advantages of TAB is the ability to test an inner lead bonded and encapsulated die prior to committing it to a package, especially a multichip module (MCM). A number of vendors sell TAB test and burn-in sockets designed to work with JEDEC and EIAJ standard tape and carrier designs. Use of these sockets may require a larger tape format for larger lead counts in order to fit the test pads into the standard test pad layout. This can increase the cost per frame of the tape. Alternatively, designing and fabricating custom test sockets, at finer test pad pitches than those of the standards, can increase costs.

Outer Lead Bonding
The outer lead bonding (OLB) process connects the die to the package and subsequently, to the outside world. This process is actually a set of sub-processes that includes:

- Excise and form
- Flux and die attach
- Align and place
- Actual bond process
- Clean

As with ILB, a variety of OLB bonding methods exist.

Final Test and Assembly
Following OLB, the packages are inspected and tested. Assembly errors or electrically failed die can be reworked at the OLB level, another advantage of TAB. The packages are then shipped for system assembly.

9.4.3 TAB Tape Considerations

Choosing the proper TAB tape, from among various configurations, material options and formats, is a crucial step in its application. This section explains the issues involved with tape materials and formats and their impact on the assembly options.

TAB Tape Materials
A variety of material options are available to tailor a particular TAB tape design to a specific application. Chemical, metallurgical and mechanical properties must be balanced to optimize the design with respect to the manufacture of the TAB tape, the subsequent TAB assembly processes and the predicted environmental exposures of the package.

Conductor. Copper is the predominant choice for the conductor material of TAB tape, because it best meets the primary requirements of a TAB conductor: satisfactory electrical, mechanical and chemical properties. Two types of copper are used in TAB applications: rolled and annealed (R&A) and electrodeposited (ED). The two kinds of copper have different microstructure and mechanical properties that influence the bond formation mechanism at the inner and outer leads. The use of these coppers depends on the desired application and the type of tape (see *Tape Types* later in this section) being used. Some important material properties of the copper are shown in Table 9-16 [34].

Dielectric. The dielectric should be resistant to high temperatures, mechanically stable, have low ionic contamination and show high moisture resistance. Polyimides have been the material of choice for the dielectric. Two variations of the basic polyimide chemistry, sold under their respective brand names, are in common use: Upilex, manufactured by Ube Industries in Japan, and Kapton, manufactured by DuPont. Upilex, a stiffer material, is not suitable

Table 9-16 Typical Properties of Coppers Used in TAB Tape.

	COPPER TYPE	
PROPERTY	**ROLLED AND ANNEALED**	**ELECTRO DEPOSITED**
Resistivity $(10^{-6}\ \Omega\text{-cm})$	1.72	1.77
CTE $(10^{-6}/^{\circ}\text{K})$	16.6	16.6
Thermal Conductivity (W/m-K)	392	392
Tensile Strength (MPa)	34 - 40	45 - 55
Elongation (%)	20	12
Elastic Modulus $(10^6\ \text{psi})$	17	16

Microstructure
  Grain structure
  Grain size
	flat and elongated
	in rolling plane
	1 µm, typical
		equiaxed
		>5 µm, typical

Table 9-17 Typical Properties of Polyimides Used in TAB Tape.

	POLYIMIDE TYPE	
PROPERTY	Kapton H	Upilex S
Dielectric Constant, ε_r	3.5	3.5
CTE $(10^{-6}/^{\circ}\text{K})$	20 - 36	15 - 20
Moisture Absorption (%)	2.0 - 4.0	1.0 - 2.0
Tensile Strength (MPa)	25	57
Elongation (%)	70 - 75	30
Elastic Modulus $(10^6\ \text{psi})$	0.4 - 0.5	1.14
Zero-strength Temperature (°C)	> 600	> 600

for wet etching, limiting its use to three layer tapes. The generic properties of the two are listed in Table 9-17 [35]. A range is indicated for certain properties because of the variability in data received from different sources.

Adhesive. Adhesives are used in three layer tapes to attach the copper conductor to the dielectric. Mechanical stability and compatibility with the other materials and processes making up the laminate is important, as described in Section 9.2. Electrical resistance at various relative humidities also is important. Acrylic or epoxy based materials, usually with proprietary formulations, are most commonly used.

Plating. The finish plating on the copper leads provides the desired bond metallurgy and environmental protection to the copper. Common plating metals include tin, gold and solder with tin and gold being the most popular. A nickel barrier between the copper and the cover plating also can be included as a metallurgical barrier. Selective plating of different cover materials also is a possibility, depending on tape design, and enables the user to tailor the ILB and OLB metallurgies. Gold is electroplated on the copper leads, necessitating a bus structure to connect all of the leads. Typically tin is plated electrolessly and does not require bussing of the leads.

TAB Tape Types

TAB tape is a patterned conductor, free standing or laminated to a dielectric layer. The pattern matches the I/O pattern on the semiconductor die and provides for the connection of that die to the package. These tapes are generally categorized by their number of distinct, material layers. The three most common types, with one conductor layer, are called one, two and three layer tapes (see Figure 9-18). Multiconductor layer tapes (see Figure 9-19) also exist; those with two conductor layers are becoming more common, and can furnish the package designer with enhanced electrical properties. Attempts have been made to standardize the various tape formats based on overall dimensions, socket hole pitch and OLB patterns. Overall dimensional outlines of 35 mm, 48 mm and 70 mm are JEDEC and EIAJ registered tape formats. Adoption of these standards allows the material, equipment and tooling suppliers to minimize costs.

One Layer Tape. This type of tape consists of only the patterned metal conductor (see Figure 9-18). The material is typically R&A copper, chemically etched to define the pattern. One layer tape has advantages in low cost, high volume manufacturing applications, typically molded plastic packages. The principal disadvantages, which preclude it from most MCM applications, are lack of testability on tape and the limited tape format size (usually 35 mm).

Two Layer Tape. This type of tape refers to a construction in which the patterned metal conductor adheres to a supporting layer of dielectric (see Figure 9-18). The conductor material is additively plated to a seed layer, usually

ONE–LAYER TAB TAPE

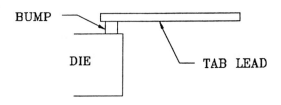

TWO–LAYER TAB TAPE

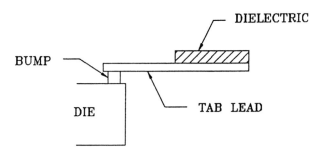

THREE–LAYER TAB TAPE

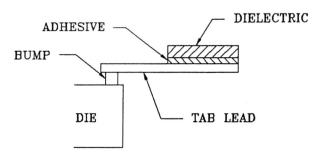

Figure 9-18 TAB tape constructions.

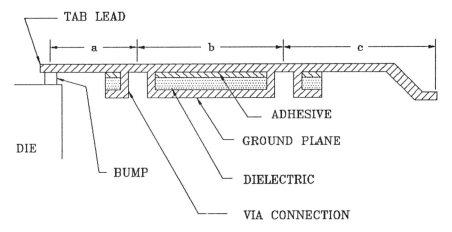

Figure 9-19 Two metal layer tape construction.

chrome, on the dielectric. The conductor can be patterned during additive plating or by using subtractive etching following a full plating of the conductor layer. Patterning while plating provides for finer pitch capability since it is free from the undercutting effects of the subtractive etch process. Conductor material in two layer tape is limited to ED copper, reducing flexibility in tailoring the tape materials to the demands of the application. Patterns in the dielectric, usually a polyimide, are chemically etched and limit the thickness of this layer to 0.002". Flexibility in choosing desirable chemical and mechanical characteristics of the dielectric are limited. The leads in two layer tape are isolated individually by this dielectric, which is not the case for one layer tape. This isolation allows for device testability. This characteristic, along with its fine pitch capabilities, make it an attractive candidate for MCM applications.

Three Layer Tape. Three layer tape is similar to two layer tape, with the exception that the adhesion of the conductor to the dielectric is accomplished by means of an obvious adhesive layer (See Figure 9-18). The dielectric, usually a polyimide, is 0.005" thick and is mechanically processed, or punched, prior to lamination of the conductor. The windows, sprocket holes and other features are formed during this process. An adhesive is applied and the conductor foil, R&A or ED copper, is laminated to the dielectric carrier. The conductor is patterned by a subtractive etch process. Three layer tape provides the user with greater flexibility in material choices, superior adhesion of the conductor to the dielectric and better flatness.

Table 9-18 Inner Lead Bonding Processes and Associated Metallurgies.

BONDING METHOD	METALLURGY		BONDING TECHNIQUES	
	Tape	Die	Gang	Single Point
Thermocompression	Bare Cu	Cu bumps	Yes	Yes
	Bare Cu	Au bumps	Yes	Yes
	Bare Cu	Capped	Yes	Yes
	Bare Cu bumps	Au bumps	Yes	Yes
	Au plated	Au bumps	Yes	Yes
	Au bumps	Al pad	Yes	Yes
Thermosonic	Au plated	Al pad	No	Yes
	Au plated	Au bumps	No	Yes
Eutectic and Reflow	Sn plated	Au bumps	Yes	Yes
	Au plated	Solder bumps	Yes	Yes
	Au plated	Sn-plated Au bumps	Yes	Yes
Laser	Au plated	Au bumps	No	Yes
	Au plated	Sn-plated Au bumps	No	Yes
	Sn plated	Au bumps	No	Yes

9.4.4 Inner Lead Bonding

The ILB process consists of the metallurgical attachment of the inner tips of the TAB leads to the semiconductor I/O pads. Bumps are supplied on the IC pads or by the TAB leads. An exception is the bumpless ILB process, in which the lead is deformed mechanically into the passivation depression of the bond pad and bonded. The plated TAB leads combine with the bump or pad metallurgy to form the ILB. Table 9-18 lists the different types of bonding methods and metallurgies that have been used to date [36]-[43]. For example, thermocompression bonding can be used with gold tape and bumps, and can be bonded by both single point and gang techniques. The different bonding methods and techniques will be explained in more detail below.

ILB Process Flow

Normally, the die is placed on a heated stage, the tape leads aligned over the bumps and the bond formed. Die can be picked up either from waffle packs or from sawed wafers on expanded tape while the tape can be fed reel to reel or in slide carrier format. As is indicated in Table 9-18, a variety of processes exist for forming the ILB. These processes are the product of mating a bonding technique with a bonding method and are discussed below.

ILB Techniques

ILB can be accomplished by two techniques. The first is gang bonding, in which all of the leads are bonded simultaneously to the die. The second is single point bonding, where each lead is bonded serially to its corresponding die pad or bump. Both techniques have advantages and disadvantages, depending on the application.

Gang ILB. ILB of all the leads simultaneously necessitates a tool that is very flat and evenly heated around the desired bonding area. Gang bond thermodes consist of two types: constant heat or pulse heated hot bar. The former is set at a specific process temperature, while the latter can be programmed to achieve a specific temperature profile during the bonding process. Constant heat thermodes are amenable to thermocompression bonding, while the hot bar thermodes are better suited to eutectic reflow. To achieve uniform bonding for both types of modes, tight temperature control, thermode flatness and planarity of the thermode to the bonding plane are critical.

A typical constant heat thermode consists of a polished diamond face, slightly larger than the die, captured in a metal shank heated by means of a cartridge heater. These thermodes hold their flatness well and generally, depending on size, have a uniform temperature distribution.

Hot bar thermodes, by design, change temperature during the bond process. These thermodes, applicable to lower temperature soldering applications, show greater deformation for the higher temperature ILB processes. This reduces the flatness of the tool and increases the temperature variability due to nonuniform contact with the part.

Single Point Bonding. Bonding one lead at a time to the die bumps eliminates the temperature uniformity and planarity problems of gang bonding, especially with larger die. Although the thermocompression, eutectic, reflow or laser methods all can be thought of as single point bond methods, the most common is thermosonic and involves the use of modified wire bonders to accomplish the task. Single point bonding requires less force per bond than thermocompression, reduces the need for custom tooling, opens up the possibility for repair of unbonded leads and may even enable the user to bond directly to the aluminum.

ILB Methods

ILB, using one of the techniques described above, can be accomplished using one or combinations of the following four methods:

- Thermocompression
- Eutectic/reflow
- Thermosonic
- Laser

Selection of the proper method also depends upon the tape and bump and pad metallurgies (see Table 9-18).

Thermocompression Bonding. Thermocompression (TC) bonding involves the application of heat and pressure to form the metallurgical bond between the TAB lead and bump or pad metallurgy. The most common technique for accomplishing TC bonding is gang bonding. The gang bond is accomplished using a constant heat thermode, as described above. The bonding mechanism requires plastic flow of the bulk copper. The mechanical properties of the copper, particularly its tensile strength and hardness, are of critical importance. Thermocompression bonding of Au plated leads to Au bumps, probably the most popular version of the TC bonding, has the advantage of being an established process with a wide process window, although other metallurgies have been demonstrated. Typical parameters for TC bonding are 450°C - 550°C bonding temperature, 75 - 150 g/lead force, and 100 - 300 ms bond duration. The range in parameters is due to the different metallurgies.

Eutectic and Reflow. Eutectic and reflow ILB pertains to the local melting at the bond interface, induced under heat by suitable choice of tape and bump metallurgies. Tin plated tape on gold bumps, gold plated tape on tin capped bumps, as well as gold plated tape on solder bumps, are some typical combinations. Gang and single point techniques can be used, although gang is the most common. An estimate of the required bonding temperatures can be obtained by reviewing the phase diagrams for the metals involved. Bonding forces are generally less than those needed for TC bonding as force is needed only to maintain thermal contact. Thermode materials, not amenable to wetting of the liquid phase formed at the bond, are required. Temperature profilable thermodes that maintain contact pressure on the bond until solidification of the metallurgy is complete are required.

Thermosonic. Thermosonic bonding couples the heat from a die on a heated pedestal (150°C - 250°C) and the energy supplied by ultrasonics to form the bond. This method is limited to single point techniques. The single point tools are designed to scrub the lead and are made with geometries, finishes and materials enabling the tool and lead to couple during the bonding process. Less force is required than for TC bonding, although ultrasonic energy can be potentially damaging to the silicon.

Laser. Laser bonding is a relatively recent development. It uses a laser beam to locally heat the bond interface. The wavelength of light used must be chosen with regard to the specific materials to be heated. A Nd:YAG laser, operating at its fundamental wavelength of 1.064 μm, is a good choice for tin tape and gold bump or solder bonding applications because tin and tin/lead solders absorb 40 - 50% of the laser energy. The Nd:YAG laser, operating at the same wavelength, is a poor choice for gold to gold bonding since only 2 - 5% of the energy is absorbed. A frequency doubled Nd:YAG laser, operating at a wavelength of 0.532 μm, is a better choice for gold to gold since 40 - 50% of the energy is absorbed. Laser bonding, like reflow bonding, requires lower forces making it a good choice in applications for more fragile chips, such as GaAs.

ILB Equipment Issues and Availability

The predominant sources of ILB equipment supply thermosonic single point and constant heat gang bonders. The gang bonders are used predominantly for lower lead count, smaller die applications while single point bonders are popular in higher lead count, finer pitch work. Recently, laser bonders have become viable alternatives, but currently are generally limited to tin tape applications.

Gang Bonding. The major attribute of gang bonding is that it is a fast, one step process, taking only a few seconds per die irrespective of lead count. This allows for high throughputs. The major disadvantage of this process is the high tooling costs for each part type. Typically, the thermode, die holder and tape clamp are unique for each die. Process concerns include the planarity of the die, tape and thermode, as well as the temperature uniformity across the thermode. As die become larger than 10 mm, these issues become more pronounced and single point bonding becomes an attractive alternative [44].

Single Point Bonding. Thermosonic bonding is probably the most popular method associated with single point bonding. Thermosonic bonders are gaining some degree of automation comparable to their gang bond counterparts. An advantage of single point bonding is that the equipment senses the vertical locations of the lead and bump individually and, therefore, can tolerate much larger non-uniformities in tape thickness and bump height than gang bonders. Other advantages of single point bonding are minimal set up time and consistent, controlled parameters for each bond. Changing from one part type to another involves a software change and easily realigned tooling. Single point thermosonic bonders have been, in general, derived from wire bonding equipment. The wire bond companies have marketed TAB versions of some of their wire bond platforms. Automatic alignment stages have been added to orient the tape over the bumps with minor software changes. A variety of tip configurations are available for the bonding of different lead or pitch configurations. A typical single point thermosonic system is shown in Figure 9-20 and an example of fine pitch (0.004") bonding in Figure 9-21. Metallurgies are limited to gold to gold for thermosonic single point bonders. Laser assisted (lasersonic) single point bonders are new additions to the equipment available for single point TAB. These bonders are well suited for reflow bonding (solder or tin) at their standard wavelength and frequency doubled versions are being developed. The cost of semiautomatic single point thermosonic TAB bonders is approximately $180,000 (1992 dollars) with die and tape automation adding another $100,000 (1992 dollars).

9.4.5 Outer Lead Bonding

The final TAB connection of the semiconductor die involves a process known as OLB. OLB can be performed successfully on a variety of substrate materials, including FR-4, ceramic and silicon using both metallurgical and mechanical bonding mechanisms.

OLB Process Flow
OLB can be considered as a group of sub-processes, including excise and form, fluxing and die attach, alignment and placement, the actual bonding process itself and cleaning.

   Excise and Leadform. The inner lead bonded IC typically is removed from the TAB tape prior to placement and outer lead bonding in a process known as excising. Forming of the TAB leads also may take place during this step. The purpose of the leadforming is to bring the leads from the plane of the ILB to the plane of the OLB (as with conventional TAB) and to provide for thermal and mechanical stress relief. Excise and form is a critical part of the OLB process and becomes increasingly difficult as lead dimensions and pitches decrease.

   Fluxing and Die Attach. Depending on the TAB application and the bonding mechanism, flux and/or die attach may be needed prior to the placement of the device. Flux can be applied to the OLB site or to the outer leads. Die attach substances can be dispensed on the package surface prior to placement to permit transfer of the package from the placement equipment to separate reflow or bonding equipment.

Figure 9-20 Single point thermosonic TAB bonder. (Courtesy of Hughes Aircraft Company.)

Figure 9-21 Single point thermosonically bonded TAB ILB 0.002" leads on 0.004" pitch. (Courtesy of Hughes Aircraft Company.)

Alignment, Placement and Bonding. After excise and form, the part is staged for pick up by the placement head. Alignment is critical to reliable bonding, necessitating accurate mechanical and vision systems. OLB pitches down to 0.004" need to be placed without misalignments that can cause shorting.

The bond process follows the placement of the device and can be accomplished by a variety of methods.

Cleaning. Solder OLB processes that use flux may need cleaning to remove the flux residues. This cleaning can be done using traditional surface mount technology (SMT) processes although attention should be given to the vulnerability of the assemblies to the cleaning solution.

Bonding Processes

Bonding mechanisms can generally be classified as being metallurgical or mechanical in nature. Technologies for achieving these bonds can be classified as mass, single component gang and single point bonding. Table 9-19 lists the metallurgical possibilities for the OLB and can be read for a specific process to arrive at the possible OLB and tape metallurgies. For example, using a single component hot bar gang process, only a solder OLB pad should be used and the tape metallurgy can be gold, tin or solder plate. Explanations of the process options are explained below.

Mass Bonding. Mass OLB methods refer to the IR reflow and vapor phase processes commonly associated with surface mount [45]-[46]. These processes are applicable to larger pitch devices (> 0.015") and do not require significant deviation from the existing assembly culture for fine pitch SMT. The common metallurgy for these processes is solder, although conductive epoxy dispensed at individual pad locations is another possibility. Issues, such as lead coplanarity and the ability to screen or dispense controlled amounts of solder paste, force the user of fine pitch TAB devices (< 0.015") to consider alternate bonding methods.

Single Component Gang. Gang bonding techniques for OLB include hot bar thermocompression, hot bar reflow, hot gas and focused IR. Thermocompression gang bonding, like that of ILB, uses heat and pressure to form the metallurgical bonds, generally between gold leads and gold pads, or copper leads and copper pads. Limitations of this process include planarity of the leads and substrate, tool flatness and thermode temperature control.

Hot bar reflow, currently the most popular method of performing OLB, is a gang bonding technique typically applied to solder metallurgies. Flux is used to enhance solderability. A thermode with four heated blades is brought into contact with the leads and OLB pads. The temperature of this thermode is profiled to control reflow and resolidification. Contact pressure eliminates the coplanarity problem and provides for the even transfer of heat.

Hot gas gang bonding uses a heated gas, focused on the bonding area to the device. Like hot bar reflow, the temperature is profiled. An added bonus may be the gas acting as a substitute for fluxing. Disadvantages may include controlling the effects of the hot gas on neighboring components.

Focused IR uses infrared energy focused on the bond area to achieve

Table 9-19 Outer Lead Bonding Processes and Associated Metallurgies.

BONDING PROCESS OPTIONS			OLB PAD MATERIAL	
			Solder	Gold
Mass Reflow		Infrared Reflow	Au, Sn, Solder	No
		Vapor Phase Reflow	Au, Sn, Solder	No
Single Component	**Gang**	Hot Bar Reflow	Au, Sn, Solder	No
		Hot Gas Reflow	Au, Sn, Solder	No
		Focused IR Reflow	Au, Sn, Solder	no
		Thermocompression	No	Au
	Single Point OLB	Thermosonic	No	Au
		Thermocompression	No	Au
		Eutectic and Reflow	Au, Sn, Solder	No
		Laser	Au, Sn, Solder	Au, Sn

TAB Tape Metallurgy:

bonding. Close control of temperature profiles and the reproducibility of bond strength have presented a problem. Controlling the effects on neighboring components is also a concern of this technology.

Single Point Bonding. Single point bonding methods include thermosonic, thermocompression and laser. Thermosonic single point bonding, as explained in the ILB section, uses the combination of the heat and ultrasonic energy to form the bonds. Substrates and die typically are heated as high as 250°C, causing potential temperature exposure problems to the MCM user, where exposure times are a function of the number of leads needing bonding. Advantages of this method include less stringent planarity requirements and the possibility of reworking bad bonds.

Thermocompression bonding, like that of it's gang bonding counterpart, uses heat and pressure to achieve the bond. This method may be applicable where thermosonic methods fail, due to ultrasonic damping or material differences.

Laser methods require consideration of laser wavelength (see Section 9.4.4) and lead hold down techniques, but promise fast, clean methods of performing OLB. One of the challenges in implementing laser OLB technology is finding a method that consistently holds the leads in contact with the pad. This is needed to facilitate the energy transfer between the two and form the bond.

Mechanical Bonds. Demountable TAB (DTAB) provides an easily repairable OLB connection and is a recent development [32]. The connection is made by pressure contact directly between the leads on the TAB tape and the vias on the PC board. DTAB offers a fluxless, solderless system with the added advantage of unlimited rework. Also, no excise and form are required.

Another quasi-mechanical bond involves the use of anisotropic epoxies (z-axis conductive epoxies). These materials eliminate the need for accurate screening since lateral bridging is not possible. A bonding method that applies pressure to activate the z-axis conductivity is needed however. Currently, these methods are used only in low current applications such as LCDs, since the joints are typically highly resistive compared to solder joints.

OLB Equipment Issues and Availability

Gang hot bar reflow and single point thermosonic bonders are the dominant equipment available. Generally the gang bonders have systems integrated onto one platform, enabling the entire OLB process to be performed on one piece of equipment, while the single point bonders perform only the bonding portion. The ability to place fine pitch components is critically dependent on both the motion and vision systems [47]-[48]. OLB pitches down to 0.004" pitch require linear accuracies of ±-0.0002" and rotational accuracies of ±-0.003°. The vision systems that drive such robotic placement systems have submicron resolution and should have the ability to make alignments based on pad locations, as opposed to fiducials. When selecting TAB OLB equipment, level of integration, placement accuracies and bonding capabilities must be considered.

Single Point Thermosonic. Thermosonic outer lead bonding is performed using modified wire bond equipment to bond the Au-plated TAB leads to the Au-plated OLB pads. These bonders, designed primarily for ILB applications (see Section 9.4.4), perform only the bonding step and do not have the capabilities integrated on one platform to perform the previous steps in the OLB process. This dictates that the excise and form through alignment and placement steps take place on a separate station. The device can be die attached or tack bonded on pick and place or gang bond equipment and then transported to the single point bonder and bonded. The die attach, or tack, prevents movement of the device during the initial bonds and transport. Table travel on these bonders, originally designed for objects the size of semiconductor die, can cause problems for multichip users with larger package sizes. Tool life and slower production cycle times are also issues. The equipment infrastructure for single point ILB exists, but is somewhat immature for OLB applications. The cost of manual or semiautomatic thermosonic TAB bonders is around $200,000. The cost for placement, excise and other tooling needed for a complete OLB process can easily triple that amount.

Gang Hot Bar Reflow. The gang hot bar reflow method of performing OLBs is the most common technique of accomplishing OLB. Solder is the metallurgy of choice. This has allowed an infrastructure of bonding equipment vendors to supply fully integrated OLB machines based on this method. The most capable of these machines, equipped with component feeders, excise and form stations, fluxing, die attach dispensing and hot bar reflow thermodes, are integrated on precise, accurate robotic systems, capable of placing devices with pitches down to 0.004" (see Figure 9-22). Thermode improvements have been made, allowing the user better flatness and temperature control. Mechanisms are integrated into the bonding head, allowing for real time planarity adjustments, resulting in better solder joint consistency. The most advanced of these machines costs between $400,000 and $500,000 (1992 dollars).

9.4.6 Single Chip and MCM Implementations of TAB

Implementations of TAB in electronic packaging are classified into three broad categories: TAB as a buried interconnection, TAB on board and TAB in MCMs.

TAB as a Buried Interconnection
In a buried interconnection scheme, the ILBs and OLBs are buried inside a conventional package such as a plastic dual in-line package (PDIP), plastic leaded chip carrier (PLCC), plastic quad flat pack (PQFP) or pin grid array (PGA) similar to that shown in Figure 9-13 for cavity TAB. The TAB tape is inner lead bonded to the chip and outer lead bonded to a leadframe (PDIP, PLCC and PQFP) or to a metal land (PGA). The TAB part of the assembly is invisible to the final user of the package. National Semiconductor Corp. successfully used this approach in the 1970s as an alternative to wire bonding for the high volume production of low lead count PDIPs. Presently, this approach is envisioned as a way to overcome the pitch limitations of wire bonding without changing the external outline of packages such as PQFPs. This application has been called the TAB interposer.

TAB-on-Board Single Chip Packages
The TAB-on-board is basically a chip-on-board (COB) application. The TAB tape essentially comprises the package itself as the outer leads are directly attached to matching pads on the PWB. The most extensive application of this packaging scheme has been in the field of consumer electronics, such as LCDs, watches and calculators. Several proprietary packages have been developed by

Figure 9-22 Hot bar OLB gang bonder. (Courtesy of Universal Instruments.)

companies for packaging high pin count VLSIs. The following are examples of
TAB-on-board applications:

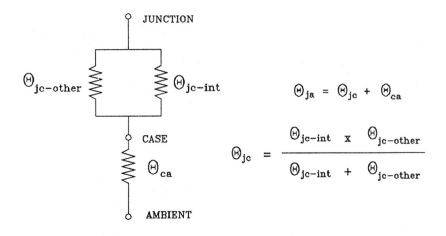

Figure 9-23 The role of interconnect in the thermal performance of a package.

1. Tape Pak™, National Semiconductor
2. Micropack™, Siemens
3. Tape Quad Flat Pack (TQFP), LSI Logic
4. TAB Pak™, Texas Instruments [33]
5. "Clamshell" ceramic package, Digital Equipment Corp. [39], [49]
6. TAB package with thermal spreader, Hewlett-Packard [50]
7. Demountable TAB (DTAB), Hewlett-Packard [32]

TAB MCM Implementations
Bull Corp., in France, first used TAB to package low lead count devices to thick film substrates. Honeywell reported the use of TAB with copper thick film substrates in the early 1980s, using a flip TAB format. Later work by Honeywell utilized a copper/polyimide thin film technology (TFML) and conventionally mounted TAB devices, discussed in Chapter 7.

A recent TAB application for MCMs is the DEC VAX-9000, discussed in Chapter 17. This application utilizes a copper/polyimide interconnect on a base substrate and tin-plated tape with solder OLB joints.

9.4.7 Thermal and Electrical Performance

The techniques used for thermal analysis and for thermal management of conventional single chip packages apply to TAB packages as well. Thermal performance can be analyzed and enhanced using the same general techniques described in Chapter 12. One feature unique to TAB is the presence of copper beam leads that are massive when compared to wire bonded leads. These copper leads influence the heat flow out of a chip. As an example, the effective thermal resistance of a copper lead 0.050" long, 0.002" wide and 0.0014" thick is approximately 29 times lower than a looped gold wire 0.001" in diameter and 0.100" long. A simplified thermal resistance model of a TAB bonded chip is shown in Figure 9-23 [51]. The connection is shown as a separate heat flow path in parallel with other existing paths. The beneficial effects of TAB are realized in cases where Θ_{jc} dominates the overall thermal performance and $\Theta_{jc\text{-}int}$ is much smaller than $\Theta_{jc\text{-}other}$. The performance of two metal tape is slightly better than single metal tape since as the ground plane acts as an additional path for heat flow.

The general techniques for improving thermal performance, such as the use of heat spreaders, heatsinks, forced air flow and liquid cooling apply to TAB packages in much the same way they would any other package. Extremely high performance TAB packages, both in single chip packaging and in MCM implementations, have utilized one or more of the above techniques [32].

Thermal characteristics are specific to the particular TAB configuration (conventional, flip, etc.) selected for a particular MCM application. Conventional and cavity TAB conduct heat out of the back side of the die and through the substrate material, similar to that of wire bonding. Flip TAB, on the other hand, requires more elaborate heat dissipation structures since the back side of the die is not directly attached to any thermal conduction path. These structures include finely toleranced and expensive heat spreaders as well as liquid cooling methods.

Electrical

The electrical performance of any IC package is analyzed in terms of the environment it presents to signal transmission and to the distribution of power, as discussed in Chapter 11. The electrical characteristics of a TAB package also are best presented in this form. A schematic illustration of the signal environment presented by a TAB lead is shown in Figure 9-24 [52]. It is assumed that the TAB lead terminates in a controlled impedance transmission line representing the environment outside the package. The lead represents an uncontrolled impedance between the chip and the outside environment. The values of relevant electrical parameters applicable to the case of a typical TAB lead (assuming a 48 mm TAB tape format, copper leads 0.002" wide and 0.0014" thick at 0.002"/0.002" line and space in the ILB area) are shown in Table 9-20. The values have a range based on the longest and shortest leads typical for this tape format. Table 9-20 also shows a comparison between one and two metal tapes. In two metal tape (see Figure 9-19), a substantial portion

Table 9-20 Electrical Parameters for a Typical 48 mm TAB Circuit.

	TAB TAPE TYPE	
PARAMETER	**Single Conductor Layer**	**Two-Conductor Layer**
Inductance, L (nH)	2.0 - 5.0	0.3 - 0.5
Capacitance, C_{self} (pF)	0.7 - 0.9	0.2 - 0.3
Resistance, R (mΩ)	50 - 200	50 - 200

of the signal line is shielded by the ground plane to form a controlled impedance line segment matching the one shown outside the package in Figure 9-24. Small portions of the lead, such as from an inner via to the chip pad and an outer via to the substrate pad, remain unshielded electrically by the ground plane and give rise to non-zero parasitic inductance and capacitance.

An equivalent circuit for the power environment is illustrated in Figure 9-24. The inductance of the power trace and its capacitance to ground have been included in the equivalent circuit. It is common practice to design the power leads wider than the signal leads to minimize their inductance. The values shown in Table 9-20 have assumed this to be the case. The actual TAB configuration selected has a significant impact on the electrical performance of the MCM. Flip TAB, with its inherently shorter leads, minimizes these electrical penalties.

9.4.8 Reliability

General reliability principles apply to TAB assemblies, with the possible exception that the die and leads may be exposed to more mechanical and environmental exposure. The sources of infant and service life failures are assembly and material related, while the failure modes are physical or chemical in nature. The type of packaging selected, hermetic or nonhermetic, encapsulated or not, and the choice of bonding metallurgies, also play a major role in TAB reliability. This section focuses on a few, critical risk areas including the TAB tape, the ILB joints and the OLB joints.

TAB Tape
TAB tape is a material source with respect to reliability and its contributions may be physical or chemical in nature. An important, and perhaps most well studied

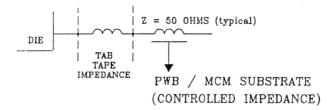

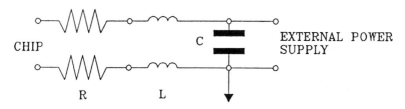

Figure 9-24 Electrical characteristics of TAB: (a) signal environment, (b) power environment.

failure mode is that of the TAB lead itself and the OLB joint. These failures are mechanically induced by temperature cycles [53]. Transient stresses result from rapid temperature excursions and steady state stresses from the inherent differences in the CTEs among the principal materials. Failures are investigated using liquid to liquid thermal shock (LTS) or thermal cycling tests. Finite element analysis has identified regions of high stress in given outer lead configurations. These studies reveal the TAB leads as the apparent weak link in the mechanical assembly. This is due primarily to the relatively smaller cross sectional area of the TAB lead compared to that of surface mount leads. Mechanical stresses from shock and vibration result in similar failure modes. When performing shock and vibration testing, it is important to perform such tests in the final box configuration, since the failures are particularly sensitive to resonant frequencies that may be specific to the design of the box. Other mechanical areas of concern, with regard to reliability of the TAB tape are metal to dielectric delamination and dielectric expansion. Chemically caused failures may relate to the moisture absorption of the dielectric or plating inconsistencies. These are closely related to ILB and OLB failures and discussed below.

ILB Joints
Failures occur due to the gradual degradation of the ILB. A variety of modes are possible:

- Separation between the bump and die pad due to insufficient cleaning of the Al prior to the deposition of the bump metallization
- Failure of the under bump metallization to prevent interdiffusion between the Au and Al
- Separation between the bump and TAB lead due to contamination at the bump and lead interface and
- Separation between the TAB lead and bump interface due to intermetallic formation, particularly in reflow bonded joints.

Physical causes may be induced by assembly process issues (such as very restrictive process windows) and material issues (such as poor lead frame design and material selection). Chemical corrosion can be induced by moisture penetration through the encapsulant or passivation and is largely due to the presence of halides, remnants of the encapsulation or OLB (flux) processes. Moisture is absorbed by the dielectric material during tape manufacturing and handling. Plating issues, such as contaminants, variable thicknesses and shelf life also contribute to the ILB failures listed above. The majority of the above failure modes is characterized using HAST (highly accelerated stress test) and LTS testing. The HAST test uncovers chemically induced failures while the LTS uncovers mechanical failures.

OLB Joints
OLB metallurgies are predominantly solder or Au to Au. As mentioned previously, studies have determined that the TAB lead is more likely to mechanically fail than the solder joint [28]. The point of issue is the quality of the solder joint itself. Gang reflowed solder joints show non-uniformity based on thermode temperature variations or planarity and flatness problems. This can result in good reflowed joints, as well as incomplete reflowed solder joints on the same device which, in turn, can result in mechanical failures in the poor quality solder joints. Chemically, the choice of tape plating effects the quality of the solder joints. Tin-plated tape is subject to shelf life and oxidation concerns and results in poor solder joints if used beyond it's expected life or is not stored properly. Gold plated tapes used with tin-lead solder systems can form brittle intermetallics in the solder joint. This necessitates control of the gold content in the solder joint and may even preclude the use of gold solder OLB systems in finer pitch TAB devices. The maturity of gold to gold thermosonic bonding for OLB is low and reliability information comparable to that available for solder OLB is scarce, if not nonexistent.

Another interesting failure, chemical in nature, involves the migration of metal between adjacent leads under an applied potential, leading to low resistance leakage paths or even complete electrical shorts by bridging. This phenomenon has also been identified in the field, caused presumably by the existence of constant voltage across adjacent leads in certain logic devices. The failure has been found to occur profusely between leads with exposed copper over a polyimide surface. It can also occur in tin plated leads and is termed tin migration. Metal migration can be largely prevented by suitably plating leads to seal any exposed copper and by close control of the polyimide quality during tape manufacture. Generally, the finer the pitch and the higher the voltage gradient between adjacent leads, the higher the probability of metal migration failure. Potential failures due to metal migration can be typically uncovered using the HAST test with the application of a bias across nearest neighbor leads while monitoring any leakage currents. Tin plating also causes a potential reliability problem due to a phenomenon known as tin whiskering. Needle-like whiskers grow from the plating as a result of stresses introduced during the plating process. Especially on fine pitch devices, this can potentially cause lead to lead shorting.

9.4.9 Reworkability

The failure of even one device on a MCM can render the package useless; the high cost of an assembled MCM prohibits simply throwing it away. This necessitates processes that can rework TAB assembly errors and actually remove and replace failed TAB devices. This rework effort is usually focused on the OLB level of the assembly. ILBs are, generally, not considered candidates for repair unless the high cost of the die necessitates it. Even then, rework is limited to repairing open leads using single point methods. Considerations prior to deciding whether and what is to repair include the effort and costs involved, the equipment and methods available and the effects on neighboring die on the MCM.

Reworking Assembly Errors
The rework of TAB OLB assembly errors usually involves the bonding of unbonded leads, removing solder that is shorting between two pads, removing solder balls and repositioning misaligned leads [28]. Although methods and equipment are available for making these repairs, with regard to solder OLB metallurgies, the processes for these corrective actions may be labor intensive. Solder joint inspection systems, in some cases, can be integrated with single point reflow equipment to perform some of these repairs. Correcting assembly problems related to gold to gold thermosonic bonds can be more difficult. While repairing an open lead may be a relatively easy task for the single point thermosonic bonder, repositioning misaligned leads may prove challenging due to the nature of the gold to gold bond.

Removal and Replacement

The process of removal and replacement of failed TAB devices is justified, especially in MCM applications, when the cost of the remaining good die and package outweigh the cost of discarding the assembly. Solder OLB systems and mechanically bonded MCMs present the greatest potential for repairs of this type. The general process flow for this type of process involves: removing the component, cleaning and dressing the OLB pads with new solder and positioning a new component.

Removing the Component. With regard to solder OLB systems, removing the component involves mass reflow of the OLBs followed by the lifting of the device when the solder is at a liquid state. The die attach bond (epoxy), if present, also has to be broken during this process. The TAB leads separate from the OLB pads and, together with the die, are lifted from the package surface, preferably intact. The removal process can be viewed as a reversal of the original gang bond process and is done with hot bar, hot gas and infrared heating. Hot bar is currently the best option, due to it's superior thermal transfer and reduced effects on neighboring devices.

Redressing OLB Pads. Following the device removal, the OLB area must be cleaned and re-dressed. This may include reflowing the solder left on the OLB pads or actually replacing solder removed with the leads of the original TAB device. Various methods, some proprietary, have been proposed for redressing the solder pads and the choice depends on the package types and OLB pitch [54].

Component Replacement. The new component can be bonded to the repair site following the clean and redress step using the original OLB process.

Rework Issues

Critical decisions made early in the design phase of MCMs should include the TAB configuration (conventional, flip etc.) and tape and substrate metallurgies. Flip TAB designs lend themselves to an easier repair strategy since the use of a die attach material is usually absent. Conventional TAB designs, on the other hand, require the added step of breaking the bond of the die attach material. This can be a difficult and messy task. Metallurgical problems can also make the removal process difficult. Tin plated tapes contribute copper-tin intermetallics to the solder joint, raising the reflow temperature of the solder and making the removal difficult and sometimes impossible after high temperature/time exposures. Gold plated tapes form a brittle gold-tin intermetallic when soldered, not only affecting the reliability of the joint but raising the reflow temperature of the solder. The removal and repair of gold to gold thermosonic bonds is a more difficult process. Reasons relate to the original bonding methodology (single point) and the related bond metallurgy.

This makes simple, cost effective, and one step (gang) removal virtually impossible. Due to some of the inherent problems associated with tin tape, however, and the need for a fine pitch TAB assembly technology using gold tape, some companies are currently developing removal and repair processes for gold to gold bonds. Mechanical OLB bonds, a recent development, offer a great potential for rework and repair.

9.4.10 Manufacturing Issues and Costs

The manufacturing of TAB devices, regardless of the packaging scheme, depends on an industry infrastructure that provides the proper materials (TAB tape) and processing equipment (bonders, tooling etc.) to form a reliable and robust process. Historically, the capabilities of TAB equipment vendors have focused on low lead count devices for consumer electronic products. This focus has limited the application of the equipment in areas of high lead count, fine pitch TAB assembly, needed for multichip applications. As a result, the infrastructure for advanced TAB materials and assembly equipment needs further development, particularly in the United States. Equipment suppliers, tape manufacturers and contract assembly houses must cooperate to fill this technology gap. Consideration of the manufacturing processes, material choices and temperature exposures are issues needing attention in the successful implementation of TAB. This section will highlight a few issues and costs that should be given some attention.

TAB Tape
The major issues associated with TAB tape include consistent tape quality, the cost for advanced circuits and shelf life for tin tape.

The most crucial aspect of a robust TAB process rests on the material consistency of the TAB tape. Dimensional requirements must be met repeatably and depend upon the design of the tape as well as on the material selection and tape processing practices. Fine pitch tapes, for example, with long, unsupported cantilevers can cause dimensional problems during tape processing and handling. Consistent plating is required to enable the ILB and OLB processes to maintain controlled process windows.

The cost of the TAB leadframe influences the decision to pursue a TAB application and greatly depends on the design of the leadframe and the challenges it presents to the tape manufacturer. Tape costs range from $5.00 per frame to over $50.00 per frame (1992 dollars), depending on the complexity of the design (one metal layer tape versus multi-metal layer tape) and the volumes ordered.

The shelf life of the tape is a determination factor for the low volume user. These users must balance the low volume versus cost versus shelf life to justify the use of TAB. One possible option is the use of a JIT approach to receiving the tape material.

Bumping. The cost of wafer bumping can have a negative impact on the implementation of a TAB process, as this cost is concentrated only on the yielding die. As a result, large, low yielding die can bear the burden of wafer bumping costs totaling as much as $100 per wafer. Wafer bumping also requires expensive semiconductor processing equipment and cleanroom facilities.

Many users, buying die from merchant semiconductor companies, do not have the option of having them bumped. This, along with cost issues, has led to bumping alternatives requiring less capital intensive methods, such as transfer bump TAB (TBTAB), gold ball bumping etc. These methods, however, may not have the ultra-fine pitch and reliability capabilities of wafer bumping and are, therefore, limited in application.

Inner Lead Bonding. Inner lead bonding equipment costs can total anywhere from $200,000 to $400,000 (1992 dollars), depending on the level of automation and the type of bonder. As mentioned earlier, single point bond technology has yet to realize full automation and use of this equipment can result in slower production times. Automated gang bonders have limited application to larger, fine pitch die and require die specific tooling.

Outer Lead Bonding. The cost of setting up an OLB process can be as much as $500,000 (1992 dollars), depending on the desired bonding technology and level of TAB complexity. Fine pitch applications require very accurate placement systems which contribute the bulk of the cost. Single point bonding systems may require the purchase of separate placement and bonding systems, whereas gang bonders usually have the bonder and placement functions integrated. Custom tooling is also a cost concern, as the TAB standards do not take full advantage of the density attributes of TAB.

Package quality is an important consideration and is reflected in the choice of bonding technology. Critical among these considerations is metallurgy consistency and flatness (for gang bonding). Flux contamination may be a reliability issue. Cleaning the flux is also, currently, a major environmental issue.

Temperature Hierarchy. Encapsulation, test on tape/burn-in and repair are other steps in a TAB assembly process that should be looked at closely, especially from a temperature exposure perspective. Each of the TAB processes affect both previous and future processes. Burn-in can be detrimental to the solderability of tin plated tapes by causing accelerated growth in copper-tin intermetallics. The OLB process temperatures can have an effect on the encapsulation materials. These issues force the broad consideration of the entire process flow and the effects it may have on materials and other processes. A coordinated approach with design, development and manufacturing is necessary.

9.4.11 Comparison with Other Connection Technologies

The choice of which connection technology (wire bond, TAB or flip chip) to use on MCM applications is based on a balance between cost and performance. TAB is most cost effective in high volume, low product mix environments. At lower volumes and higher part mixes, TAB becomes an expensive alternative. This expense can be justified where performance (electrical, thermal, etc.) of TAB justifies the use of a more expensive alternative.

TAB inherently adds cost to the assembly process. The additional step of wafer bumping, the cost of each TAB frame and custom tooling all contribute to a higher assembly cost for TAB. These costs can make TAB unattractive unless the user has a high volume product.

Cost considerations have lead to the development of new materials and methods that may enable TAB to compete with the other chip connection technologies. Developments in single point and laser bonding techniques help eliminate ILB and OLB pattern specific gang bond tooling. TAB standards help minimize the number of excise and form tools needed, as well as the hard tooling for tape fabrication.

Electrical performance is improved through the use of TAB assembly. As the number of I/Os on the die increase, the corresponding pad pitches decrease. This leads to the use of finer wire in wire bond applications that eventually limit electrical performance. TAB leads, at these pitches, using fine pitch, peripheral leaded die can achieve the desired electrical performance. These designs, however, must be carefully tailored to the electrical environment and may necessitate flip TAB configurations with short TAB leads or two metal layer tapes to attain the desired electrical goals. The result can be a balancing of high performance needs versus cost requirements since the use of short TAB leads goes against the current TAB standards and would require expensive custom tape designs and tooling. Likewise, two metal layer TAB tape is more costly than single metal layer tape.

Thermal management issues also affect the connection decision. While conventional TAB configurations conduct heat through the substrate, as with wire bond, flip TAB designs require novel heat removal designs. Flip TAB designs also incorporate shorter lead lengths that help electrical performance and increase density. Again, the extra cost must be balanced against performance issues.

Reliability studies of TAB assemblies indicate failures unique to TAB, while other concerns are common to all connection methods. These unique failures include TAB lead and solder joint failure which is controlled or eliminated with proper TAB design and material choices.

Testability, and burn-in and repair are important attributes of TAB, especially for the MCM user, and constitute more advantages of TAB over the other connection methods. An inner lead bonded die can be tested or burned-in prior to committing that die to an expensive module, enabling the user to screen bad die out of the assembly flow. Die that do fail after attachment to the MCM can be removed and replaced at the OLB level, depending on the metallurgy. This process is difficult, if not impossible with wire bonded devices and is not a trivial process with flip chip.

Flip chip applications are natural and, probably, inevitable extensions of packaging connection assembly. While the flip chip eliminates the tape frame and custom tooling required by TAB, bumping still is required. Presently, the infrastructure and assembly know-how for this technology is immature to the world at large. This makes TAB, especially flip TAB, a natural alternative. When the flip chip infrastructure has matured sufficiently to handle the advanced assembly needs of the MCM user, knowledge gained from flip TAB may be applied to future flip chip applications.

9.4.12 Summary

TAB assembly for MCM applications offers the multichip designer an opportunity to utilize high lead count, fine pitch semiconductors and create densely packed modules with superior electrical performance. The inner lead bonded die also can be tested or burned-in prior to bonding to the package, enabling the user to screen out bad die prior to their commitment to an expensive MCM. Die also can be removed at the OLB level and replaced with new die, preventing the loss of an expensive assembly.

Materials and equipment for implementing a TAB process are available, allowing the use of TAB in design and manufacturing.

Tape material and assembly method options have presented the manufacturing engineer with a variety of process options, enabling a tailoring of designs for different uses and environments. Reliability data for high performance TAB applications, a relatively young practice, has begun to filter through the electronic packaging industry.

The choice of assembly methods for multichip applications, TAB, wire bond or flip chip, depends on the balance between cost and the desired performance. New methods and materials allow TAB to reduce its cost structure and compete, favorably, with the other assembly technologies.

9.5 FLIP CHIP CONNECTION TECHNOLOGY

Chee C. Wong

9.5.1 Introduction

Flip chip connection technology as a first level chip to package connection option traditionally is regarded as being synonymous with the Controlled Collapse Chip Connection (C4) process pioneered by IBM more than 20 years ago. The C4 process has set the highest record in I/O density, chip packing density and electrical performance, and establishes the industrial benchmark in field reliability. Details of the C4 process are presented in Section 9.6.

This section presents the flip chip connection technology as a generic technology with the C4 process as a subset example of one particular application. Emphasis is placed on the concepts behind the design of the flip chip connection configuration, the material options for the connection medium, the processing options in implementing flip chip and the cost and manufacturability issues in the context of inherent process limitations and existing infrastructure. The objective here is to introduce an overall perspective on flip chip connection technology beyond the C4 process to enable a judicious comparison to be made between the several flip chip variants and between flip chip connection and other chip connection options.

This is a section on concepts rather than details. It offers an organized format of questions to be asked and provides a framework for answering those questions on an individual basis. Hopefully, it will stimulate the reader to evaluate the applicability of flip chip technology for his or her product goals.

9.5.2 The Basics

Definitions
Flip chip is defined by the schematic in Figure 9-25 which shows a bare IC device flipped upside down with its active area or I/O side attached to a substrate via a connecting medium. In this generic description, the device may be a silicon microelectronic IC or any other monolithically integrated active functional block. The substrate in Figure 9-25 may be any of the MCM substrates providing an interconnection network between the flipped active device and other active, or even passive devices. The connecting medium may be any suitable interface serving the various needs of the matchmaking between the flipped device and the underlying structure. Each member of this flip chip ensemble is examined in later sections.

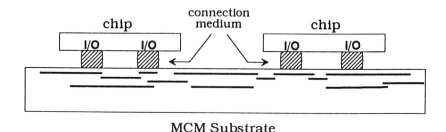

Figure 9-25 Flip chip configuration consisting of chip, connection medium, and substrate.

Why Flip a Chip?

What is the greatest distinction of the flip chip configuration? *Why a flip chip?* Flip chip is the only connection configuration that allows assembled active chips to approach the form in which they were originally created, namely, the form of a wafer. This is an advantage because it provides superior electrical and thermal performance.

To understand this point one must realize that the goal of any packaging scheme is to allow each chip to perform at its peak and to allow the system as a whole to take full advantage of the peak performance of each individual component. Circuit speed on the bare chip level is the highest speed achievable. As soon as the chip leaves its original wafer form and enters the first level of packaging, its performance begins to suffer. Why then don't we build entire systems or subsystems on a single wafer? This approach, called "wafer scale integration," has met with little success. The problem is poor yield. While most parts of the system may function as designed, functional failure of a single part can "doom" the entire system. Hence, the current approach in hybrid microelectronics follows the modular concept, namely, to break a big system into smaller systems, build many small systems on a wafer, isolate and package the functioning small systems (IC chips) and reassemble them back into a big system. Figure 9-26 shows a schematic comparison between monolithic wafer scale integration and two modular alternatives, an unpackaged flip chip version and a packaged surface mount version.

Since wafer scale integration has proven to be impractical because of yield issues, the next best thing in terms of performance is to build a separate interconnection structure of commensurate interconnection density (such as an MCM) and to assemble the unpackaged IC chips back onto the MCM substrate so as to resemble their original wafer form as closely as possible. This translates

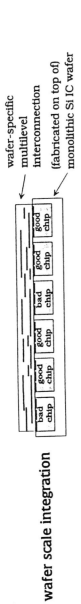

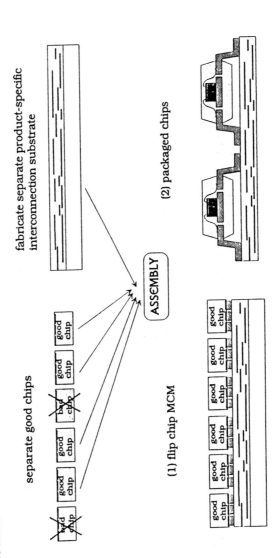

Figure 9-26 Comparison of wafer scale integration and modular approaches to multichip packaging, highlighting the packing density achievable by flip chip MCMs.

into a requirement for a connection technique which permits the closest possible chip proximity and a connection medium whose dimensions are contained within the area of the chip. Also, the connection medium should be amenable to short connection lengths to minimize electrical parasitics. By flipping a chip and directly attaching its I/Os via a connecting bump of controllable height onto the substrate as shown in Figure 9-25, the maximum footprint that the chip requires is that of its own. No fanout is required. This constitutes the distinct advantage of flipping a chip. As shown in Figure 9-26, the packing density of a flip chip MCM could, in principle, approach that achieved in wafer scale integration.

Members of the Ensemble
The characteristics of the individual members in a flip chip ensemble are presented as follows:

1. **IC Chips**. Most silicon IC chips presently are designed for perimeter wire bonding. The I/O pads, on the order of 4 mils, are finished with an Al metallization and surrounded by a passivating layer of dielectric. The degree of perfection of this passivation layer is inadequate in providing mechanical and environmental protection for the chip. The number of I/O pads on IC chips could range from several tens to several hundreds. There is no industrial standard in the spatial arrangement of I/O pads. Silicon chips generally are not readily available in bare wafer form. Instead, they are readily available only in die form.

2. **Substrate**. The substrates for MCMs could be in the form of cofired or thin film ceramics, thin film silicon, printed boards or flex circuits. These various structures are reviewed in Chapters 1, 5, 6 and 7. MCM substrate design, being considerably less mature compared to chip design, is more tolerant of the needs of chip connection. To make full use of any chosen MCM platform, design of the substrate and design of chip connection need to be carried out in parallel.

3. **Connection Medium**. The connecting medium couples the chips to the substrate to form a functional and reliable MCM. TAB and wire bonding techniques achieve connection using leads which fanout from the chip I/O to the corresponding pad on the substrate. Flip chip bonding achieves electrical and thermal connectivity using bumps (before joining), called joints after joining. These joints provide mechanical support for the flipped chip on the substrate. The C4 process uses bumps made of solder or solder-coated copper balls. Organic conductors are new candidates for the connection medium.

Note that out of the several functions of the IC chip package discussed in Chapter 1, flip chip connection has fulfilled only the functions of electrical and thermal connection. Mechanical protection generally is delayed until the module packaging level. If the module level package is not considered adequate for environmental protection, then chip encapsulation techniques are used to protect the chip from operating ambients. The issue of chip testability in its bare chip or wafer form prior to module level assembly is considered a major inadequacy of flip chip technology.

Mechanical support of the chip itself, which was not an issue in the case of unflipped IC chips die bonded onto the package, is now provided by the joints and introduces an important new variable in fatigue-related reliability. Hence, to complete the picture of using flip chip as a connection technique for MCMs, at least three new members have to be added to the ensemble depicted in Figure 9-25: a chip testing capability for the bare or bumped die prior to assembly, an encapsulation technology after assembly [55], and proper designs for minimizing susceptibility to thermal fatigue [56]-[57].

Why flip a chip revisited

As mentioned, flipping a chip onto an MCM could achieve wafer level packing density by eliminating fanout. There is another feature unique to having the active side of the chip face the top of the interconnecting substrate. Since the I/O pads on the chip also are fabricated on the active side, the layout of these pads easily can be expanded into an array covering the entire inner area of the chip, rather than being confined onto the perimeter. Area arrays offer a way of increasing I/O density without taxing other technologies for a finer I/O pitch. For example, for a chip size of 5 mm and a constant I/O pad spacing of 100 μm, a perimeter array could accommodate about 200 I/Os while an area array could accommodate about 2000 I/Os, a tenfold increase. Only the flip chip configuration provides the ability to achieve higher I/O density without decreasing I/O pitch.

By having the active side down, flip chip bonding also offers the shortest possible leads with the lowest inductance, maximizing the operating frequency. This consideration alone could justify the usage of flip chips in high performance systems. The actual length of the lead, or, in this case, the standoff height of the joint, could be controlled by any of the flip chip techniques in use, and is usually chosen to optimize other criteria rather than being predicated by the geometry of the unflipped chip. Joint height is usually designed for better fatigue endurance and chip underside cleaning.

Flip chip assembly is inherently a batch bonding process. This contrasts with wire bonding which proceeds serially. The throughput advantage of batch bonding is obvious at high I/O densities. Batch bonding also is offered in beam lead bonding and gang bonding versions of TAB. While TAB could sidestep the high tooling cost of TAB gang bonding techniques by temporarily employing single point bonding techniques, the flip chip technique has to confront the assembly issue of high speed batch bonding directly. Once the initial barriers of developing versatile and cost effective flip chip batch bonding machines are overcome, batch bonding could become the bonding method of choice for most current chip sizes. As maximum chip size increases, tooling for batch bonding becomes more difficult because of issues in planarity, heat distribution and wider disparity in chip sizes. At that stage, the benefits of batch bonding can be realized only by investing in more costly tooling.

The robustness of flip chip connections has set the reliability benchmark in the connection industry. The absence of leads makes the IC chip rugged and easier to handle. Issues of thermal fatigue have so far been adequately addressed by proper joint design (see Section 9.6). As chip sizes increase, issues of fatigue life again will dominate the question of joint mechanical integrity. This point argues in favor of using silicon MCM substrates for flipped silicon chips to circumvent the detrimental effects of coefficient of thermal expansion (CTE) mismatches and the resultant fatigue phenomenon.

9.5.3 Connection Medium (I): Solder Bumps

The connection medium between the flipped silicon chips and the underlying MCM substrate is the key in realizing any flip chip technology. This is where the issues of manufacturability and cost are defined. The material and fabrication method chosen for the connection medium influence chip and substrate reliability, yield and throughput in assembly.

The two connection media currently in use for flip chip MCMs are solder and organic conductors. This section presents an overview of solder bumping (bumping using organic conductors is deferred to another section), with emphasis on the comparative strengths of different techniques and the basic roles of various materials used in the formation of the bump. One of the techniques mentioned is the C4 process, treated in greater detail in Section 9.6.

Bump Location
The three possibilities of locating the solder bump on the substrate, on the chip, or both, are shown in Figure 9-27. The decision is based on the following:

- Which is the easiest to do
- Which side is likely to experience less impact going through a solder bumping process
- Which is better for handling and storage

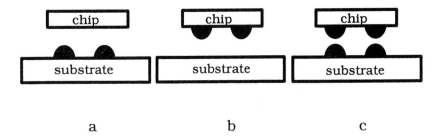

Figure 9-27   Three possibilities of locating bumps connecting chip to substrate.

- Which is the easier to assemble
- Which is the easier to repair and
- Which is the most cost effective in terms of overall yield.

To answer these questions, consider the members chosen for the flip chip ensemble. Silicon chips are produced in wafer form, regardless of type and function. MCM substrates could range from silicon wafers (making them equivalent to silicon chips in terms of processing) to flexible copper polyimide circuits. The equipment for handling and processing these various substrates are equally varied, each specializing in optimized processing for that particular substrate. If different substrates for different applications are envisioned, it would be practical to concentrate on placing bumps on silicon wafers only, since the wafer form of silicon chips remains invariant. The placement of solder bumps onto chip wafers turns out to be advantageous in areas such as testability and repair also. Details of the rework process for solder bumped chips are discussed in Section 9.6.

There is also an overall yield advantage for placing bumps on chip wafers. Any silicon chip on any MCM substrate is always smaller than the substrate. For a given unit area being processed, more chips are produced than substrates. Let's take an example of a five chip set MCM of 1 inch square. Here, in the same inch square area being solder bumped for one MCM substrate, five chips could be processed. Assume also that the defect level is such that one bad bump (short or missing) is produced per inch square on average. This one bad bump on the substrate would render the entire MCM useless, whereas it would only disable one of the five chips. A healthy chip from a neighboring group could be substituted in place of the bad chip to still produce a good MCM. The moral here is that, for a given defect level, smaller objects are more tolerant of defects than larger objects in terms of overall yield. Since chips are always smaller than their mating substrates, it is justified to place bumps on the chip wafer.

Placing the solder bump onto the substrate (Figure 9-27a) is justified in the case where chip wafers cannot survive the temperature or the physical and chemical environment of a solder bumping process. For example, an aggressive backsputtering step (for via cleaning) in a solder bumping process may damage CMOS chips sensitive to strong radiation. Bumping substrates also is justified if chips are not available in wafer form. Placing bumps on both the chips and the substrates adds more cost; however, this scheme becomes mandatory if the solder height limitation inherent in any single bump architecture is insufficient to meet joint height specifications.

Bump Shape
Schematics of a solder bump before and after reflow are shown in Figures 9-28a and 9-28b, respectively. Reflow is a heating process which takes the solder bump through a solid to liquid to solid transition, allowing the solder to consolidate its bonding with the connecting interface. While the geometry of the bump in Figure 9-28a could differ depending on the processing technique, the reflowed bump shape in Figure 9-28b is universal, governed solely by the forces of surface tension, gravity and the tendency of the liquid solder (during reflow) to assume a shape of minimum surface energy. For small bumps where the effect of gravity can be neglected, the equilibrium shape is a spherical segment.

The surface onto which the solder bump is fabricated consists of two distinct areas in the vicinity of the bump: a wettable and a non-wettable area. The wettable area is the bonding interface to the solder. Usually the chip's I/O pad with an Al finish is located directly below the wettable area for electrical

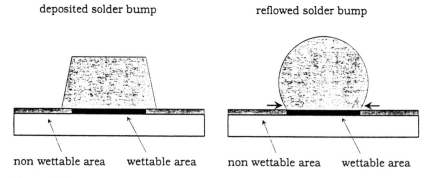

Figure 9-28   Schematic of a solder bump (a) before and (b) after reflow.

connection, although dummy bumps whose wettable areas are not connected to I/O pads also could be fabricated for reasons of mechanical robustness or improved thermal performance. The non-wettable area is necessary to confine the solder within its allowable area, thus controlling the final height of the bump for a given volume of solder.

The shape transformation from Figure 9-28a to Figure 9-28b is determined entirely by the volume of the deposited material and the area of the wettable region. The original area of the solder deposit may or may not correspond to the area of the wettable base, depending on the technique and the design. The final footprint of the reflowed bump, however, corresponds exactly to the wettable area, assuming complete wetting during reflow. In other words, if the solder deposit area is smaller than the wettable area, then solder spreads outward during reflow; conversely, the solder footprint shrinks back. This fact, coupled with the knowledge that the equilibrium shape of a reflowed solder bump is that of a spherical segment, allows us to predict the final shape and height of the reflowed bump.

The plot in Figure 9-29 shows the reflowed bump height as a function of the height of the solder deposit. The plot is delineated into three regions, corresponding to different shapes of the spherical segment. Figure 9-29b shows a perfect hemisphere, while Figures 9-29a and 9-29c show spherical segments which are smaller (sub-hemisphere) and larger (super-hemisphere) than the hemisphere, respectively. The line of unity slope in the plot is given as a yardstick to distinguish the region where the reflowed height is larger than the deposited height (above this line) and the region where the opposite is true (below this line). The reflowed height of the solder bump has an approximately cube root dependence on the height of the solder deposit. This curve crosses the straight line in the region where the shape of the bump is a super-hemisphere. This means that as long as the bump shape is that of a hemisphere or smaller, the deposited volume of solder is used efficiently in building the height of the bump. When the shape crosses into the regime of super-hemisphere, a further increase in the deposited volume makes little contribution to the height of the reflowed bump, most of the material going toward enlarging the waist of the bump instead. This can be seen in the shape of the super-hemisphere where the largest cross sectional area of the bump is no longer at the base; rather, it has migrated toward the middle. This has an effect of creating a reentrant corner at the point where the solder bump meets the wettable area. Similar considerations regarding solder joint geometry have been presented by Goldman [57], for which the same conclusions apply. Unfortunately, such reentrant corners would concentrate strain at the base of the joint, rather than distributing it throughout the solder volume. In other words, the solder joint would not be used efficiently as a mechanical support. The point of this discussion is that geometrical design

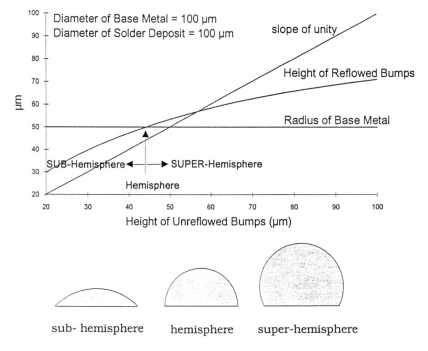

Figure 9-29   Plot of reflowed solder bump height as a function of unreflowed bump height, using the spherical segment approximation. Regions corresponding to a bump shape of sub-hemisphere, hemisphere and super-hemisphere are delineated.

is an important factor in optimizing processing efficiency as well as in mechanical integrity. The influence of joint shape in fatigue life is further discussed in Section 9.6.

Solder Bump Material Choices

Three classes of materials are of interest in making a flip chip solder bump: the solder itself, the wettable area (base metals) and the non-wettable area (solder dam). Most solders currently used are of the lead-tin (Pb-Sn) family, although the momentum toward limiting usage of Pb may necessitate consideration of other families of solders. Pb-Sn solder materials and their characteristics are reviewed by Wassink [58]. The exact composition chosen and any additives thereof depend on the desired reflow temperature, fatigue performance, corrosion susceptibility and ease of fabrication. The use of solders with 95% Pb and 5% Sn in the C4 process for many years has generated the most extensive research and development and field reliability database for this family of solders.

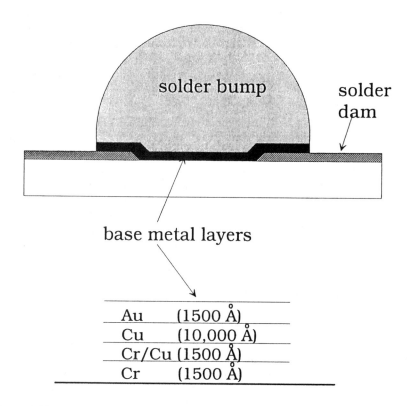

Figure 9-30 Cross section of a solder bump showing the multilevel base metal layers.

The base metallurgy for a 95Pb/5Sn solder bump fabricated using the C4 process is shown in Figure 9-30. Other metallurgies are reviewed elsewhere [59]. Invariably, the base metallurgy has to provide three functions: adhesion to the underlying surface, wettability by the solder and a barrier between the solder and whatever lies underneath. The first function is typically filled by one of the classical glue layers of titanium (Ti) or chromium (Cr). These metals stick well to other metals and most dielectrics. This point is important in chip wafer bumping because, as shown in Figure 9-30, the base metal footprint typically extends beyond the area of the I/O pad to provide a sealing function to the last remaining unpassivated area of the chip. To do this well, glue layers must have excellent adhesion to metals (I/O pads) as well as to dielectrics (solder dams). The Ti or Cr adhesion layers also can provide the barrier function; however, they are not solderable.

Solderable metals such as copper (Cu), nickel (Ni) and silver (Ag) have been studied and used in other applications involving 95Pb/5Sn solders. They become natural choices for wettable base metal layers in flip chips with one important distinction - the Cu in a base metal multilayer structure is much thinner than the Cu on a printed wiring board. Thin films of these materials may be completely consumed or exhausted during multiple reflows, resulting in dewetting of the solder from the base. Although a thin film of Cu is still being used as a solderable layer, an additional layer of phased, or codeposited, Cr/Cu is now interposed between the Cu and the underlying Cr. This mixed layer performs a dual function: part of it is a barrier metal (Cr) and part of it is a solderable metal (Cu). The result is that, at the proper compositional ratios, the mixed layer exhibits sufficient wettability to prevent non-wetting but limits chemical interactions to prevent material exhaustion and dewetting during multiple reflows.

The final top layer of the base metals shown in Figure 9-30 is a thin layer of gold (Au). The nonoxidative properties of Au and its excellent solderability preserve the wetting function of the base metals throughout material handling and storage.

The reflow process initiates aggressive chemical interactions between the liquid solder and the base metals. Au is dissolved almost instantaneously. The dissolution rate of the common solderable materials has been reported by Bader [60]. The chemical interactions add a new member to the multilayer structure - Sn-based intermetallics. The brittleness of these intermetallic compounds has been shown in some cases to introduce a weak link into the structure. Reviews on the formation and effects of intermetallics can be found elsewhere [61]-[63].

Lastly, the role of the solder dam can be filled by most dielectrics which do not interact with molten solder and which do not degrade at the reflow temperature (about 330°C for 95Pb/5Sn). The finishing passivation layer on the surface of a chip generally is adequate as a solder dam, although an additional protection layer such as polyimide can be used.

Although the above discussion makes use of material examples taken from chip wafer solder bumping, the roles of the different layers remain unchanged in the case of substrate solder bumping. If the MCM substrate were a silicon wafer, then solder bumping materials and procedure could be identical to that of the chip wafer. Otherwise, in the case of other substrates such as ceramics, wettable base layers and solder dams can be chosen from materials and techniques mature for the processing of that particular class of substrates. The reader is referred to Tummala's work [G1] for a review of various substrate pad structures.

Bump Formation

This section discusses several common techniques currently used in solder bumping. Understanding the functions of individual steps is critical in the design of a process sequence suitable to the needs and expectations of each application. In essence, process sequences are not unique; many different sequences may achieve the same goal.

Steps in making solder bumps consist of:

- Deposition of base metal layers,
- Deposition of solder
- Patterning of the base metals and solder layers.

The goal here is to create a structure similar to the one shown in Figure 9-29.

Deposition of Base Metal Layers. Such deposition is usually accomplished by sputtering, evaporation, electroplating or a combination of these methods. Deposition can be done under two different conditions: blanket deposition, which covers the entire surface uniformly with the deposit or patterned deposition which deposits the material through a masking layer containing the pattern to be transferred. The former option requires only that the object (chip wafer or substrate) be compatible with the deposition environment, while the latter option extends the requirement to the masking material. For sputtering and evaporation techniques, the environment is one of high vacuum and heat; for electroplating, the environment is reactive chemical solutions. In the case of electroplating, a conductive plating base is required to initiate and sustain the plating reaction. This plating base often is deposited using either sputtering or evaporation. Thus, one of the two vacuum techniques is a prerequisite to base metal deposition unless electroless plating methods could be developed for the materials discussed in the previous section.

Both sputtering and evaporation (electron beam or thermal) have been used extensively in the thin film industry. They are well characterized, well controlled and well supported by mature and sophisticated equipment. These processes also can be automated. The materials suitable as base metal layers for solder bumping fall within the range of capability of this equipment. Their major distinction lies in the end requirement for uniformity. The surface of uniformity for evaporation is that of a sphere. The size of the sphere determines the deposition rate, the number of wafers that can be processed together and the required size of the vacuum chamber. For sputtering, the surface of uniformity is roughly that of a plane of area to the target. Wafers receive deposition serially, supported on a rotating carousel. As wafer size increases, evaporation techniques require larger vacuum chambers whereas sputtering requires larger targets and larger sputtering guns. In both cases, substantial increases in cost would be incurred. Beyond this distinction, both techniques are well suited to the deposition of base metal layers, including the codeposited layer discussed in the previous section.

Electroplating of base metals has been used in the deposition of the solderable layer (Cu, Ni, for example) and the Au finish [64]. The ability to electroplate a mixed layer (Cr and Cu, for example) of a microstructure similar to that attained by evaporation or sputtering has yet to be demonstrated. The wide use of electroplating in industries, such as circuit board manufacturing, has resulted in a mature technology for coating large surfaces. In applications requiring large area processing with large feature sizes not requiring precise thickness control, electroplating can be the most cost effective.

Deposition of Solder for Flip Chips. Deposition could be carried out using thermal evaporation or electroplating. Sputtering is not suitable for low melting point materials such as solder. Screen printing of solder paste, although widely used in the surface mount industry, does not to meet the fine pitch requirements of flip chips.

Evaporation is the more common technique for solder deposition. It is the technique used in the C4 process. The primary difference between evaporation of base metals and evaporation of solder is the thickness of deposit. The evaporation of base metals could follow standard IC processing of thin films (1 μm), but the evaporation of much thicker solder deposits (tens of μm) requires special tooling. Pounds of material are needed for the charge, requiring special crucibles and special power supplies. The long periods of deposition needed to achieve large thicknesses tend to generate much higher temperatures on the wafer surface which, in an uncontrolled situation, could lead to melting of the solder coating. If a patterned deposition technique is used, care must be exercised to ensure that the masking material does not deform under these temperatures. These complexities notwithstanding, the evaporation process produces solder coatings of high purity, high uniformity and consistent composition at high throughput.

Solder can be deposited using an alloy or two elemental charges (for Pb-Sn). In an alloy charge, the different vapor pressures of the two elements cause the Pb layer to be deposited first, followed by that of Sn. Elemental charges could reverse the position of these two layers if desired. An alloy charge is preferable because it occupies the central position in the chamber, the optimized location for uniformity. Two elemental charges cannot occupy the same ideal location.

Electroplating of solder has been reported in [59] and [65]. There are two primary concerns in the plating of solders: bath chemistry and electrode design. Additional factors are: the need to ensure efficient mass transport inside the bath, the maintenance of bath composition and the control of electrical current density. These factors combined determine the purity and uniformity of the solder deposits. Since equipment for electroplating are not as automated as those used in vacuum deposition, the human error factor becomes more pronounced. Electroplating of solder often is used in a patterned deposition, rather than a blanket deposition scheme. In this case, the masking material has to be evaluated with regard to each different plating chemistry being considered.

Patterning. Patterning is the key to forming solder bumps of the desired geometry. In general, patterning techniques fall into two categories: additive (patterned deposition followed by liftoff) or subtractive (blanket deposition followed by patterned etching), as discussed in Chapter 2 and 7. Liftoff techniques have to contend with residues while etching techniques have to contend with unwanted chemical attack. In either case, a suitable masking layer has to be inserted at some point within the process sequence, and removed at some other point. The different ways to insert and remove one or more masking layer(s) among the two prerequisite deposition steps (base metal and solder) constitute the different process sequences unique to each flip chip solder bumping technology. Only two cases are presented as examples.

The C4 process uses a double liftoff technique for the base metal and the solder. The masking layer in this case is a metal mask (physically distinct from a glass mask used for patterning photoresists) with openings corresponding to the I/O patterns of the chip wafer. The mask is aligned with the wafer and both are mechanically held together in a fixture. After base metal evaporation, the mask is removed (first liftoff), a second mask with larger openings put in place, and solder is deposited through the second mask (second liftoff). This is a purely subtractive process; no etching is involved. The second liftoff step is not necessary if the design does not call for a solder footprint larger than the base metal footprint. A polymeric mask, such as dry film resist used in the circuit board industry can be substituted for the metal mask and the same sequence applied. Polymeric masks have the virtue of conformal coating to the wafer surface, avoiding the misregistration issues in mechanical fixturing of metal masks. On the other hand, metal masks do not introduce residues since they were never adhered onto the surface. In terms of resolution for fine pitch I/O, photosensitive polymeric masks are more broadly applicable since they conform to standard photolithographic practices in IC manufacturing.

The baseline electroplating process discussed in [65] uses liftoff for solder and etch patterning for base metals. The base metals are blanket deposited so that they would form the plating base for solder. A dry film resist is patterned onto the base metals before electroplating solder through the via openings. Following removal of the resist (liftoff for solder) the base metals are chemically etched using the solder bumps as a mask (etch patterning). In this case, the etching chemistry has to be chosen such that etching is preferentially done on the base metals, leaving the solder intact.

One other issue critical to patterning using liftoff is that of via cleanliness. This usually is accomplished by chemical etching or a physical means such as backsputtering or ion cleaning. Failure to clean vias could lead to nonplating of solder in the case of electroplating or non-adhesion of base layers in the case of the C4 process.

Attachment

The final step in creating solder connections for flip chip MCMs is reflow and assembly. As mentioned before, reflow allows the molten solder to form chemical bonds with the base metal layers of both the chip and the substrate. The heating is done either in a reducing atmosphere such as H_2 or in the presence of fluxes to remove oxides. A common practice is to reflow solder bumps prior to assembly to consolidate the solder into its equilibrium shape and to confirm its bonding with the underlying interface, whether it be a chip or an MCM substrate. After assembling the flip chip onto the substrate, a second reflow is carried out to turn solder bumps into solder joints. Following cleaning (if flux has been used), the flip chip solder MCM is ready for the next level of testing or packaging.

Common problems detected at the reflow stage are solder dewetting from the base and the formation of solder voids. Dewetting is a result of improper design or processing of base metals. A solder bump with this problem is useless. Voids, on the other hand, arise from a variety of sources, and come in a variety of sizes. Voids due to trapped fluxes or impurities can create huge hollow spaces within the bump, making it mechanically unsound. Shrinkage voids due to natural material freezing phenomena can be small compared to the dimension of the bump and have not been shown to cause deleterious effects, provided they do not coalesce to form a big void.

An important benefit derived during the second reflow stage due to the nature of molten solder is self alignment - the ability of the chip to center itself onto the mating footprint on the substrate regardless of placement misregistration. This is illustrated in Figure 9-31. The solder bump seeks its equilibrium shape during the first reflow, and the solder joint during the second reflow. The tendency of the joints to assume their equilibrium shapes provides the driving force to center the bump pads footprint form the chip onto the corresponding substrate. This eases the requirement for high assembly accuracy. An ideal assembly of flip chip MCMs approaches the standard set by surface mount assembly, since both processes involve batch bonding using a solder connection medium. To approach this ideal for flip chip MCMs, some considerations are:

1. Heat tacking (application of heat during placement in addition to pressure) is required to mounting flip chip solder bumps. Tacky flux (flux which glues the chip onto the substrate in preparation for reflow, discussed in Section 9.6) also may be used. Both procedures assist in the self alignment process. Pick-up tools currently used in pick and

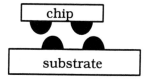

misregistered placement

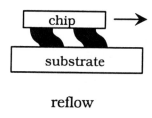

reflow

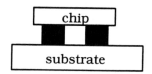

self aligned assembly

Figure 9-31 Self alignment of solder bumped components during reflow.

place machines incorporate heating and heat distribution elements to ensure uniform heating of the chip. Alternatively, heating may need to be applied to the entire substrate. Depending on the choice of base metal and solder materials, heat may have to be applied to both.

2. Solder pads on MCMs are at least a factor of two smaller than those on surface mount boards. Even though self alignment eases the tolerance on registration, placement accuracy must be sufficient to ensure that all

of the solder pads on one surface at least touch the appropriate features on the mating surface.

3. Unlike chip carriers which have a number of standard sizes, there is no emergent standard on IC chip sizes, MCM substrate types or MCM substrate sizes. A "standard" MCM assembly tool for flip chips needs to handle disparate IC sizes and substrate sizes and types. Although custom tooling always is available, standardized tooling is the key to lowering costs.

The infrastructure for assembling flip chip solder bumped MCMs currently lags that available for fabricating solder bumps. Solder bumping techniques, despite various methods, follow well established procedures in processing, while the assembly of MCMs is particular to each technology and its attendant chip connection method. Given sufficient demand, flip chip MCM assembly could approach the cost and performance standards of surface mount assembly.

9.5.4 Connection Medium (II): Conductive Polymers

Electrically conductive epoxies are used extensively as an alternative to eutectic die bonding, as discussed in Section 9.2. The conductivity in these polymers is isotropic. Such polymers are suited only for making a single connection such as ON the backside of a die to a package leadframe. An extension of this technology has led to the development of conductive polymers whose conductivity is anisotropic. These polymers conduct current preferentially in the z-direction (normal to the plane of the polymer film) while maintaining electrical isolation in the xy-plane of the film. This characteristic qualifies films such as a multi-I/O connection medium. They are referred to as anisotropic conductive adhesive films (ACAF) [66]-[69].

Most conductive polymers are formed by dispersing metallic particles into the polymer film so that current is conducted through the polymer via the bridging of the particles. ACAFs are prepared by controlling the dispersion of conductive particles, placing a sufficient concentration to enable conduction in the z-direction only. This is accomplished using the single particle bridging concept illustrated in Figure 9-32 which shows a schematic of a bare die flip bonded onto the substrate using ACAF. The metallic particles commonly used are made of Ag, Ni or Au. By controlling particle size, ACAFs successfully connect chips with 4 mil I/O pads [66].

An ACAF has one important difference from solder bumps - it does not need to be patterned and, thus, becomes more versatile. Only a blanket

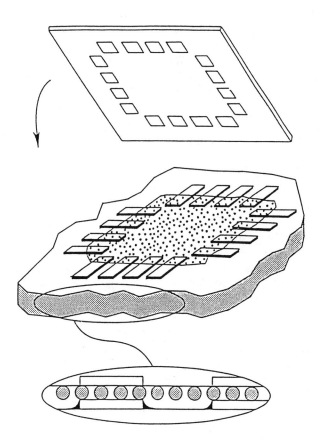

Figure 9-32 Schematic of chip connection using an ACAF material. Inset shows particle bridging between conductors. (Reprinted from [15], courtesy of A. M. Lyons)

application of this material onto the appropriate I/O surface (preferably Au) is needed. ACAFs can, in principle, be applied onto any chip wafer and any MCM substrate. Assembly consists of aligning and placing the chip onto the substrate with concurrent application of heat and pressure to ensure good adhesion and curing of the polymer. Unlike solder bumps assembly, a reflow step is not present. Because of the absence of reflow, the benefits of self alignment are not available; meaning that ACAF-type flip chip assembly has to rely on placement machinery with higher accuracy than those needed for flip chip solder bumps.

Electrically, ACAFs show negligible inductance since the connection length is on the order of the diameter of a conductive particle, which ranges from 0.5 - 30 μm [66]. The performance of this material at high frequencies is under investigation.

The usage of ACAFs in flip chip technology is relatively immature compared to that of solder bumps. Formulations of ACAF materials need to be standardized and individually tested. Performance, reliability, yield and throughput data have yet to be collected. It is likely that high end systems based on rigid substrates will continue to use solder bumping, while flex circuits and other compliant substrates, will utilize ACAFs. Also, the versatility of ACAFs could very well extend into other hierarchies of packaging beyond the MCM level.

9.5.5 The Whole Picture

Flip chips have a unique geometry, allowing them to be packed at a density higher than that of any other chip connection technology. This feature is best exploited when matched with MCM substrates with fine pitch features. Thin film MCM substrates are therefore the best candidates for utilizing this particular strength of flip chip technology. Thin film substrates, because of their high electrical performance, also are well matched to the low inductance, high performance connections of flip chips. In terms of reliability, the susceptibility of solder joints to fatigue due to CTE mismatch argues for the use of silicon MCM substrates. Other substrates must be CTE-matched to silicon before large IC chips can be used in high reliability flip chip MCM products.

In terms of cost, the expense of solder bumping has to be viewed on a cost per connection basis since solder bumping cost is based on wafers (for chip wafer bumping) with different size chips and variant numbers of I/O. The total number of I/Os per wafer is a product of the number of I/Os per chip and the number of chips per wafer. For 4" wafers, at small chip sizes (< 5 mm) and high chip I/O counts (> 100), the cost per connection for solder bumping becomes extremely competitive. The impending increase in I/O density certainly favors the use of solder bumped flip chips. This advantage is maximized for chips designed with area array pads since flip chips can accommodate higher I/O density without incurring the risks of a finer pitch technology.

A similar viewpoint on cost also can be adopted for the issue of assembly because of batch bonding which allows all connections to be made at once, regardless of number. The forgiving nature of self alignment enables us to project the future status of flip chip assembly as close to the cost and throughput standard of surface mount assembly. Thus, low cost flip chip assembly can be envisioned, though not currently realized.

Given the unquestioned performance and foreseeable low cost of flip chip attachment, the only issue that remains is that of availability. Besides IBMs C4, other companies are beginning to install manufacturing capabilities based on flip chip solder bumping, notably, Intel, Motorola and AT&T. As higher levels of IC integration demand higher performance levels in chip packaging, expect to see increasing activity in flip chip bumping manufacturing.

A final view of the whole picture extends beyond the role of flip chip in hybrid circuit technology to the role of hybrid circuits in electronic systems in general. Hybrid circuit functions can be replaced by advanced ICs. However, these advanced ICs will require advanced level hybrid packaging to connect them to other advanced ICs to build yet more powerful systems. The unique contributions of flip chip connection to high performance hybrid packaging schemes will ensure its role in the continuing evolution of electronic systems.

Acknowledgments

I wish to thank King Tai for introducing this subject to me, Alan Wachowicz for partnering with me to mass produce solder bumps, T. Dixon Dudderar for his patience and insights, Alan Lyons and Rich Bentson for providing me with useful references and Flora Tsai for lending me her Macintosh in a time of need.

9.6 FLIP CHIP SOLDER BUMP (FCSB) TECHNOLOGY: AN EXAMPLE

Karl J. Puttlitz, Sr.

9.6.1 Introduction

IBM first introduced flip chip solder bump (FCSB) interconnections in 1964 as an integral part of solid logic technology (SLT) hybrid modules utilized in the System/360. It still remains as the primary chip to substrate connection technique practiced by IBM. The connection technique consists of chips with solder bump terminals and a matching set of solder wettable pads on the substrate. The chip is placed upside down (flip chip), each solder bump aligned with its matching substrate pad, as discussed in Section 9.5. All the solder joints are processed simultaneously by reflowing the solder in a furnace. The surface tension force and the confinement of solder volume between solder wettable terminals pads, prevent chip collapse during reflow. Therefore, this type of chip connection often is referred to as a controlled collapse chip connection (C4). Early joints contained a copper ball standoff. Some physical changes have evolved over the several generations of module programs since its inception. FCSB technology is usable with several types of chip carriers: ceramic substrates with thick or thin film metallization, directly attached to polymer cards or flex circuits and on silicon.

FCSB technology is extendible to meet the requirements of future high performance chips. Since FCSB connections are arranged over an area, much larger numbers of I/O terminals can be accommodated in comparison to wire bonding or TAB, whose pads are arranged peripherally on a chip. The feasibility of fabricating dense 128×128 area arrays (25 μm bumps, 60 μm centers) has been demonstrated and further advances are anticipated.

Initially, the driving force behind FCSB was to develop a competitive practice to manual wire bonding in the area of cost reduction, increased reliability and productivity. This technology is characterized by self alignment during the chip joining process, high joint strength, ruggedness and the ability to make large numbers of bonds simultaneously. The technology also provides high yield, high chip connection density and high reliability. In recent years, a significantly increased level of activity in FCSB technology has been reported in the literature by major component manufacturers. The demand for increased I/O capacity and higher connection density imposed by VLSI and ULSI fuels this interest.

Various aspects of FCSB technology and its use in multichip module (MCM) applications are discussed in this section. Fabrication processes of the technology are described first, followed by topics including manufacturing viability, extendability to increased I/O density, reliability and chip replacement capability.

9.6.2 Fabrication, Process Flow, Tools and Hardware

This section describes how the chip connection features unique to FCSB technology are accomplished (namely the solder bump terminals on chips), matching solder wettable areas on the substrate and method of joint formation. An overview of the general process flow is shown in Figure 9-33.

Chip Terminations
Chip terminals are defined during the final stage of chip fabrication. Thus, the majority of a chip's fabrication is the same as chips with terminals other than solder bumps. In the case of FCSB and chips bumped for TAB bonding, the similarity goes even further. Most chips are passivated with a suitable material, such as rf-sputtered quartz, polyimide etc., through which holes or vias are etched to provide an external communication path. Opened vias must be sealed hermetically by evaporating consecutive layers of chromium, copper and gold (Figure 9-34) through a molybdenum mask. Chromium and other materials promote adhesion to the passivation layer and serve as a reaction barrier between the aluminum-based chip metallization and the solder. Copper or some other solder wettable metal is required for subsequent solder reflow. The chromium and copper are phased (a few hundred angstroms of each metal is coevaporated to avoid separation at their interface since theses metals do not exhibit mutual solubility). A thin gold layer (< 0.1 μm) protects the solder wettable copper from oxidation when exposed to room ambient. The solder evaporation which follows is performed in a different evaporator to avoid contamination. To ensure a proper seal, the evaporated pad diameter is made larger than and concentric with the etched vias in the passivation layer. The thin film pads are the only solder wettable regions on the chip. These regions are referred to as the ball limiting metallurgy (BLM).

To complete the chip terminal, solder is evaporated through the same mask used for the BLM films or another mask. A solder alloy is selected whose melting point is sufficiently high to assure compatibility with subsequent processes and assembly.

High lead content Pb/Sn alloys (95 Pb/5 Sn) often are used [71]-[72]. Choice of alloy, however, may be driven by application or assembly-specific requirements. For example, the process temperatures of lead-indium alloys are

Chip Terminal	Substrate Terminal
• Open passivation layer to expose chip contact	• Form solder wettable pads (TSM) on a suitable substrate (ceramic, organic, silicon, or composite)
• Form solder-wettable terminal pad (BLM)	- Several methods: post-fire thick films, metal thin films, metal/insulator thin film combination
• Form solder bump	
• Reflow / Test / Dice wafer	- Common to all methods: substrate pads must match chip solder bump pattern

Chip Attach

- Flux and place chip
- Form joint by solder reflow
- Test
- Continue Assembly / Rework

Figure 9-33 General FCSB (C4) technology process flow to create chip and substrate terminals and to achieve chip reflow.

about 100°C less than standard lead-tin FCSB alloys. This, coupled with their superior fatigue resistance, makes them candidates for high thermal mismatch applications, such as direct chip attach (DCAs) to organic chip carriers.

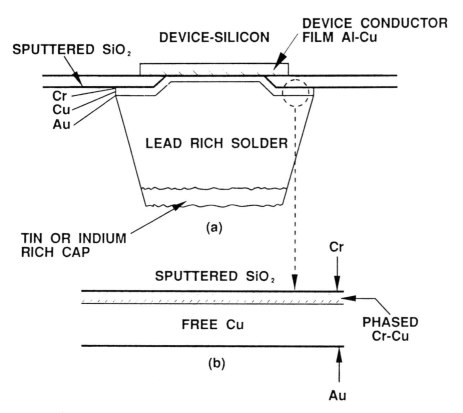

Figure 9-34 As evaporated FCSB solder pad and BLM structure. (a) Cross sectional view of the device terminal metallurgy, (b) BLM layered structure (taken from [70]).

Due to a considerable vapor pressure difference between the elemental constituents, as-evaporated solder deposits consist of two discrete layers (see Figure 9-34a). Evaporated lead-tin and lead-indium solder pads therefore are reflowed to achieve homogeneity. Each chip is electrically tested prior to dicing.

FCSB: Compatible Substrates
Materials. The choice of substrate material to attach FCSBs is limited by the need to withstand solder reflow temperatures exceeding at least 200°C and more typically in the range of about 340°C. Additionally, the material must possess sufficient strength and stiffness (high Young's modulus) to satisfy handling and fabrication requirements, have good thermal conductivity to avoid thermal

gradient induced cracks and be chemically benign to process fluids. Ceramics ideally meet these requirements. Alumina (Al_2O_3) has been the material of choice. Other ceramic materials are used for applications requiring a better thermal expansion match with the chip to reduce joint stress, and improved thermal conductivity. However FCSB also can be joined to polymer-based chip carriers (polyimide) for DCA applications such as flex circuits and CTE-matched cards and boards (such as laminates with copper and Invar layers).

Terminals (TSM). The only substrate surface feature required for flip chip solder bump connections is an array of solder wettable pads which match the chip solder bump I/O pattern. These terminals to which FCSBs are attached (referred to as the top surface metallurgy—TSM), each must be isolated from other solder wettable areas to prevent chip collapse when joined by solder reflow. Depending on substrate type, these pads are formed in various manners.

Postfired Thick Films. Thick film technology most often is used to form circuits on staked pin, pressed ceramic substrates with materials such as Ag, Pd, Au Pt and Ag Pd Au. Ceramic substrates are required to withstand the thick film firing temperatures. TSM pads are formed by screening of thick film glass dams near the ends of the circuit lines (lands) as shown in Figure 9-35a. Being confined, molten solder is restrained from flowing along the lands during reflow, preventing chip collapse.

Metal Thin Films. A metal thin film structure can be used to form TSM pads. This method is applied in the IBM metallized ceramic (MC) technology, which consists of a three layer structure (Cr-Cu-Cr). The bottom Cr layer provides adhesion between the ceramic substrate and thick copper (50K Å) on conductor layer. As shown in Figure 9-35b TSM pads are defined by selectively etching the top Cr film to expose the underlying solder wettable copper. The chromium acts as a solder stop, preventing chip collapse during reflow. Since metal thin films (deposited by evaporation or sputtering) are applied at much lower temperatures compared to thick films, the method has application to substrate materials other than just ceramics. Thin films also have the advantage of permitting narrower lines and spaces, increasing the allowable number of lines per channel for inboard pad escapes.

Multilayer Ceramic Microsockets. The fabrication of multilayer ceramic (MLC) substrates is described in Chapter 6. Therefore, only the via structure, essential to FCSB technology is discussed here. Some of these vertical columns, formed by stacking metal-filled punched holes in an individual green sheet, intercept the substrate top surface and are referred to as microsockets. Substrate microsockets are patterned to mate with chip solder bumps (I/Os). A refractory metal, such as molybdenum or tungsten, is required to withstand the high sintering temperatures of cofired ceramic substrates. The metals also must exhibit good electrical conductivity. Refractory metals, are not solder wettable

• Post-fire Thick Films

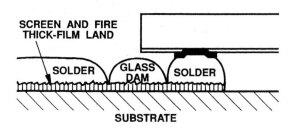

• Metal Thin Films

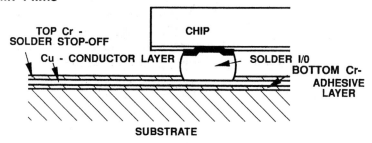

• Multilayer Thin Films (Metal / Insulator)

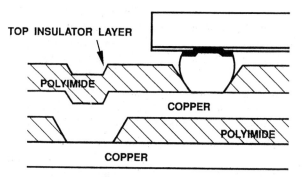

Figure 9-35 Side view construction illustrating how FCSB connections are made to various substrate schemes. (a) Postfired thick films, (b) metal thin films, and (c) multilayer thin film structure (not to scale).

and thus, are plated with electroless nickel (about 2 μm), followed by immersion gold for corrosion protection [73].

Multilayer Thin Film Structure. Various combinations of metal and insulator thin films (usually polyimide) are used to define substrate TSM pads to which FCSB can be solder reflowed. As shown in Figure 9-35c, TSM pads are defined by selectively opening holes in the top insulator film by laser ablation or other suitable means. This exposes the underlying solder wettable metal to which the chip solder pads are reflowed.

Chip Connection
The FCSB connection technology is similar for any of the substrate types described in the previous section. That is, a nonactivated flux (containing no halide species) is applied to the chip site prior to chip place. Water-white rosin (mostly abietic acid) [74], dispensed in xylene and water soluble fluxes, have been found suitable for standard lead-rich alloys, and lower melting FCSB solders respectively. Flux promotes solder wetting to the TSM pads by cleansing the molten solder bumps of oxides or other surface contaminants. Using pattern recognition or split optics tooling, chips are placed face down so that their solder I/O pads align with mating substrate pads. Chip reflow is accomplished on an individual basis, using a local heat source or simultaneously using an oven or belt furnace to reflow all the chips placed on a substrate [74].

Tools and Hardware
The tools required to form solder bumps on chips are commercially available (wet and reactive ion etch tools, evaporation and sputter deposition tools etc). However, bare die are not available for bumping. In fact, completely processed chips with FCSB terminations generally are not available to the merchant market. Sources for flip chip solder bump will come about in response to demand. But, for the near term, flip chips solder bumps will generally be available only on a contract basis from chip manufacturers with this capability (IBM, Intel, Motorola). IBM, a leader in this technology, has recently made its device and electronic packaging technology available to the merchant market. This should aid the industry in moving along the learning curve more quickly.

Obtaining FCSB compatible substrates is much less difficult, since some infrastructure already is in place. Much of the technology is very mature and the tools readily available. Therefore, MCM manufacturers to place circuits on vendor substrates with metal thick or thin films. These technologies should be sufficient for most applications. Tool costs and process complexity increase dramatically with multilayer thin film technology and is justified only for special applications.

Manual chip placement tools are available for flip chip but automatic production tools are not. Belt furnaces for chip attach reflow also are available, but post join flux clean tools, designed to remove flux residue from beneath the chip, are not.

Alternative Approaches

Electroplated copper bumps capped with sufficient solder for reflow bonding is an alternative to solder bumps [75]. The copper bumps act as standoffs to prevent chip collapse. Several versions of solderless flip chips are described in the literature. The bump to substrate connection typically is made with a conductive epoxy or conductively filled polymer. The chip bumps consist of copper or conductive polymer formed by screen printing. The comparatively low process temperature for these materials, in comparison to solder reflow, allows for a diversity of substrate materials, including many plastic and low temperature composites. Conductive polymer joints are more compliant than their solder counterparts owing to their low elastic modules [76]. However, electric current and moisture-induced silver migration is a concern with this type of material.

A solderless flip chip connection system recently has been demonstrated where chip and substrate are not physically attached [77] , as illustrated in Figure 9-36. LSI chips with gold bumps (3 μm high, 10 μm pitch, 2330 I/O) are held in contact with a substrate by the contractile forces generated within an ultraviolet light setting resin. Chips are replaceable since the resin is soluble, and is therefore extendable to ULSI (10,000 I/O).

9.6.3 Manufacturability

The ease with which chip connections are achieved, with acceptable yields in a manufacturing environment, is a major consideration. This section discusses several features, unique to FCSBs technology, that enhance manufacturability.

Self Alignment

The self alignment feature of FCSBs allows coarsely or misplaced chips to be pulled into position (centered) by surface tension forces when the solder is molten. A misalignment of up to approximately one-quarter pad can be tolerated and still assure self alignment of the final joints. Self alignment can be used to address a variety of issues, such as aligning small pad sizes (24 μm) and low cost flip chip soldering in which large alignment pads are used to align smaller functional joints without requiring precision placement as described in Section 9.5.

Ruggedness/Strength

Unattached flip chips are sufficiently rugged to permit handling, including automated testing and chip placement. Unlike TAB and wire bonds, which are

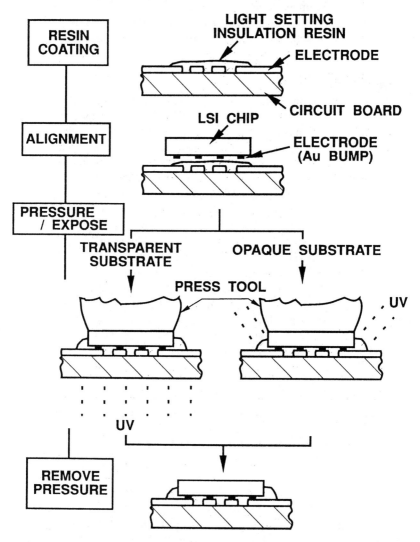

Figure 9-36 Process flow for achieving flip chip connections by resin-generated compressive forces (taken from [77]).

both rather delicate and exposed, FCSB joints are located beneath the chip and therefore are protected. Also, the joint tensile strength is high, compared with other chip connections, ranging from about one pound for the original three terminal SLT chip to over 80 pounds for some current VLSI chips. Electrical shorts sometimes occur in wire bonded devices when the encapsulating material is poured around them. This is not a problem with FCSB joints as they are quasi-rigid.

Yield/Throughput

It has been pointed out that the high cost of chips ($2 to $100 per IC), combined with the large number of total I/O from several attached chips (which may total several thousand), necessitates a very high yield chip to substrate connection technology. For example, an MCM package with 4000 chip I/O, with only a 90% overall yield, requires a 99.997 per I/O attachment yield [78].

Flip chip technology provides the highest bond yield and joint reliability. Also, being a batch process, throughput does not depend upon the total chip I/O on a module. Identifying defective chips prior to chip attach is becoming increasingly more important in meeting MCM yield objectives. Burn-in can be achieved with solder bumped flip chips, but not with the versatility of TAB. A rework capability with a high degree of control is fundamental to attaining yield levels necessary for economic viability. This is another key attribute which makes FCSB technology particularly well suited for MCM applications.

9.6.4 Rework

In the case of single chip modules (SCM), rework costs usually are not warranted; unsatisfactory parts are simply discarded. However, if one or more chips require replacement on multichip carriers, cost considerations prohibit discarding these carriers and their functional chips.

Among the more common reasons for replacing chips prior to shipping include: incorrect orientation (chip is rotated), wrong chip at a particular site, defective chip (electrical test escapes, lower solder volume causing an electrical open or partial wet), mechanical damage due to handling and changes necessary to satisfy customer requests. Also, performance upgrades are achieved by replacing chips on modules from the field with new state of the art chip sets.

In general, the replacement of components is considered a significant technical hurdle [78], particularly ultrasonically bonded TAB and wire bonded chip formats which do not easily lend themselves to replacement. Wire bonded chips usually are removed by manually severing individual wires at their bonds, a very time consuming operation, especially for high I/O count chips. A TAB assembly is removed by heating and remelting solder reflowed outer lead bonds (OLB). Heating also sufficiently weakens the die attach adhesive to permit chip removal. The following sections discuss several techniques for replacing flip chip with solder bumps.

Chip Removal

Mechanical methods offer the most direct approach to chip removal. However, sufficient space must be available to grasp the chip in a suitable manner. For example, a chip with less than 300 I/O may be removed by a few back and forth rotations (torque approximately 2 degrees in each direction), as depicted in Figure 9-37. Chips with larger numbers of FCSB joints, however, can, be safely removed from multichip carriers using ultrasound methods. A transducer coupled to a target chip in the 20 - 40 watt range is sufficient for most applications. Removal is rapid, requiring only a few seconds or less, depending on I/O count.

Melting the solder joints allows chips to be lifted from the surface with a vacuum pencil. Heating the entire module above the solder joint liquidus temperature is an obvious approach. Localized heating is preferred to prevent the formation of intermetallics, which can weaken the solder joint interfaces.

Multichip carriers are normally pre-heated (referred to as bias heating) from the I/O side to a temperature approximately midway between room ambient and

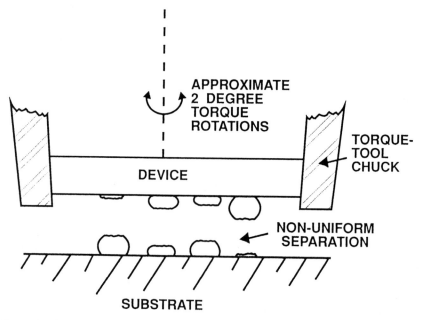

Figure 9-37 Cross sectional schematic of FCSB mounted device removed from a chip carrier by mechanical means (torque) (not to scale) (taken from [104]).

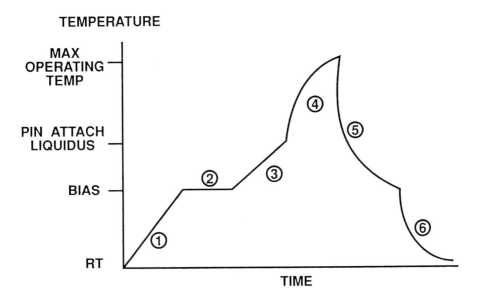

Figure 9-38 Generalized profile for thermally-enhanced, chip removal techniques at various stages of heating and cooling. Stages represented are: (1) heating from ambient to bias temperature, (2) temperature equilibration, (3) grading the temperature between a replacement site and surroundings to prevent sharp thermal gradients, (4) locally spiking the temperature at the replacement site, (5) slow convection cooling and (6) forced cooling. (Time and temperature not to scale.) (Taken from [103].)

solder solidification temperature. Additional energy from another source is directed to locally heat only the chip targeted for removal.

A general thermal profile for localized, thermally enhanced, chip replacement techniques is shown in Figure 9-38. Figure 9-38 depicts typical stages of heating and cooling.

When direct conduction techniques are utilized, they usually consist of a heated element in contact with the back side of a chip. Maintaining the required planarity and parallelism is difficult, due in large part to heat element warpage. Convective techniques, such as directing a flow of heated gas (nitrogen gas) to a target chip, are not sensitive to planarity. Typically, these methods require ample chip to chip spacing to avoid melting the joints of neighboring chips. Radiation methods, such as focused infrared or laser sources, also are suitable for removing closely spaced chips.

Solder Dress

Residual solder is left on substrate pads after the chips are removed. Site dress is the process of removing residual solder prior to replacing a chip. Site dress prevents electrical shorts due to solder volume buildup at sites which experience multiple chip replacement cycles. Since dressed sites are planarized, replacement chips are placed with the same accuracy as initially joined chips. Also, there is a strong relationship between fatigue life and solder volume of FCSB joints. Chip solder pads are evaporated to possess both the required uniformity and optimum solder volume when fabricated. Attaching these chips to undressed sites can have a significant adverse effect on reliability. Adjusting replacement chip solder volume would result in a substantial proliferation in chip part numbers, complicating manufacturing logistics. Solder dressing eliminates these concerns, rendering substrate pads to a near original condition.

Residual solder can be shaved mechanically, however, this procedure should only be used to remove debris such as chip fragments occasionally left after mechanical removal (tensile pull). Mechanical shaving potentially can damage the substrate.

The industry has long practiced ways of removing molten solder (solder suckers, solder wicks etc.). Similar techniques have been developed to remove molten residual solder from FCSB sites. As with other localized heating processes described earlier, the entire module is preheated to reduce the potential for introducing thermal gradient-induced stress cracks. Also, as before, external heat sources are utilized to raise the temperature at replacement chip sites. One such source is a semiautomatic hot gas site dress (HGD) tool. A tip with pedestals surrounds the target site and contacts the substrate as indicated in the sketch of Figure 9-39. A heated inert gas (nitrogen) directed at the substrate, heats and melts the residual solder. The velocity is sufficient to propel the molten residue into the gas stream removing it from the site [74].

Although HGD is not considered extendible to large footprints (2.5 mm square array is about maximum), there are techniques which are relatively footprint insensitive. One technique utilizes a solder wick chip [80]. The residual solder spreads over a grooved silicon surface prepared by an anisotropic etch process, and in followed by a solder wettable film deposit. Another technique utilizes a porus metal slug fabricated from any solder wettable metal by standard powder metallurgical practices. The molten residual solder is absorbed into slug by capillarity.

9.6.5 Reliability

Thermal Fatigue

Flip chip solder bump joints experience a displacement with each machine on and off cycle due to strains resulting from a (CTE) mismatch between silicon

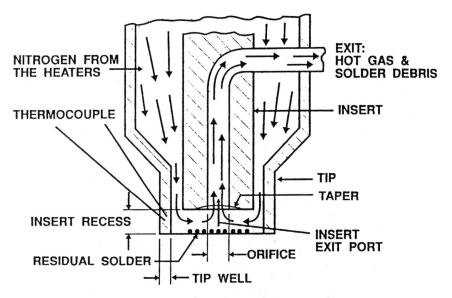

Figure 9-39 Side view construction indicating the direction of gas flow within the probe tip of the hot gas solder dress tool (taken from [74]).

chips (2.8×10^{-6}/°C) and the alumina substrate (6×10^{-6}/°C). The degree of strain experienced by a particular solder joint depends directly on its distance from the zero displacement point or neutral point (DNP). Normally the neutral point is located at or close to the geometric center of a chip. Thus, corner joints in square arrays experience the greatest strain, as illustrated in Figure 9-40. The concern for premature failure due to CTE mismatch often is cited as a key reason FCSB technology has not gained wide spread acceptance.

The joint is susceptible to thermal fatigue failure since it is the pliable member separating chip and substrate, two relatively rigid elements. Accordingly, a major function of FCSB joints is strain accommodation to avoid premature failure and loss of functionality. The accumulated plastic (permanent) deformation over a lifetime can be severe, exceeding a 1000% in some cases.

Design Considerations

The shape of FCSB joints, best approximated as a truncated sphere, depends on three geometric factors: the terminal radius between the joint and chip, r_c (BLM) terminal radius between the joint and substrate, r_c (TSM) and joint height, h

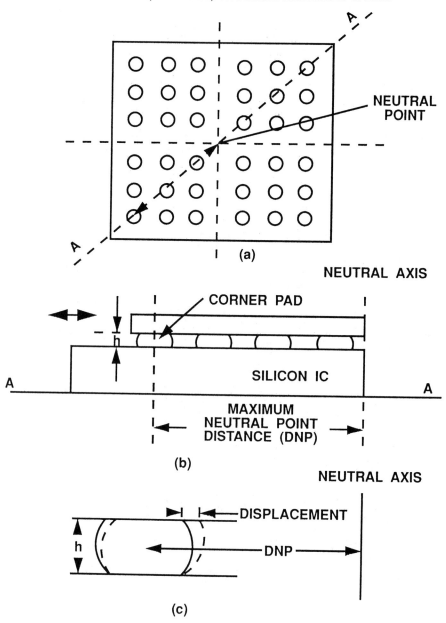

Figure 9-40 Schematic illustrating the displacements of FCSB joints from the neutral point due to the mismatch in thermal expansivity between chip and chip carrier.

(Figure 9-41). Chip weight has little effect on joint shape. The I/O count of current chips is more than sufficient to exert the surface tension forces necessary to resist collapse when the joints are molten.

The procedure normally followed is to optimize the substrate terminal dimensions and joint solder volume with respect to the chip terminal dimensions, usually fixed at the time of substrate design. The effect that several radius ratios, $R = r_s$ (TSM) $/r_c$ (BLM), have on the maximum local strain experienced by FCSB joints as a function of solder volume is shown in Figure 9-41. Actual strain values depend upon temperature, so the graph has been normalized to indicate relative differences. The solder volume corresponding to the lowest strain, has been assigned a value of unity. That condition occurs, as Figure 9-41 illustrates, when the terminals have the same dimensions $(R = 1)$. Under this optimized condition, the strain is distributed symmetrically, and the maximum strain is relatively insensitive to solder volume variations which may occur during processing. However, for any set of terminal dimensions, there is an optimum solder volume for which the joint strain is a minimum. That minimum represents the most favorable balance between two competing conditions. Increasing joint solder volume increases joint height, which in turn reduces the shear strain. however, the joint horizontal cross section also increases, effecting strain distribution. Selecting matched terminals (R=1) only ensures optimum joint fatigue life if failure occurs within the bulk solder or that the solder-bond strength at both terminals is about equal. Otherwise the radius ratio must be shifted to appropriately compensate for a weak interface. Joint fatigue life is optimized by balancing these design parameters such that the predominant mode of failure is within the bulk solder, not at the terminal interfaces.

Models, Tests, Field Experience
The Coffin-Mansion relationship is widely utilized in predicting low cycle fatigue life (10K cycles) of metals. Elastic strain contributions, small compared to plastic strain, are neglected. But, since operating conditions are above the homologous temperature (typically one-half the absolute melting point) for solder materials, Norris and Landzberg [82] introduced frequency and temperature parameters to account for two plastic flow failure mechanisms in FCSB joints. One mechanism is characterized by intragranular deformation, while the other is characterized by relative movement of grains along their boundaries with attendant creation of intergranular voids. Both processes cause defects within the solder from which fatigue cracks can initiate and propagate. The cycle condition determines which mechanism predominates. Accelerated laboratory tests are used to predict the fatigue life of FCSB chip connections under field conditions

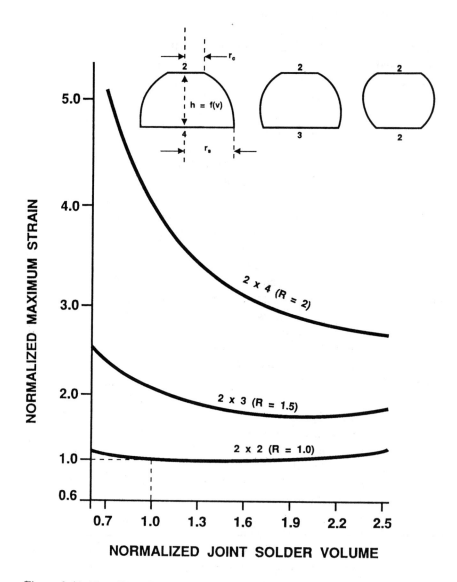

Figure 9-41 The effect of volume on the maximum strain on FCSB joints as a function of solder volume for various terminal ratio combinations [81].

on a per joint basis. Accordingly, fatigue life in the field (N_F) is expressed as:

$$N_{F(x)} = N_{T(X)} \left(\frac{\gamma_T}{\gamma_F} \right)^n \left(\frac{f_F}{f_T} \right)^{\frac{1}{3}} \exp \left\{ 1414 \left(\frac{1}{T_F} - \frac{1}{T_T} \right) \right\}$$

(9-1)

where: γ = maximum shear strain, in./in.

n = empirical constant, approximately 2.0

f = frequency, cycles per day

T = maximum temperature of the excursion

F = field, T = test indicates percentage of joints failed in $F_{(x) \text{ and } T(x)}$.

The following example illustrates the application of Equation (9-1). Assume the FCSB joints of a chip mounted on an MCM are expected to experience an average of six thermal excursions between room ambient (25°C) and 60°C per day resulting in a 0.005 in./in. maximum plastic shear strain. We also assume the average fatigue life, $N_{T(50)}$ of FCSB joints mounted on similar parts subjected to a laboratory thermal cycle test to be 3,000 cycles. The temperature variations on this laboratory test are taken to be between 0°C and 100°C, 72 times per day, resulting in a maximum plastic strain of 0.02 in./in. Accordingly, the projected FCSB connection fatigue life for the field conditions is stated as:

$$N_{F(50)} = 3000 \left(\frac{0.02}{0.005} \right)^{1.9} \left(\frac{8}{72} \right)^{\frac{1}{3}} \exp \left\{ 1414 \left(\frac{1}{333} - \frac{1}{373} \right) \right\}$$
$$\approx 31,500 \text{cycles}$$

(9-2)

Stress relaxation, which takes place in plastically deformed FCSB joints eventually establishes a state of equilibrium. A minimum value of six is used for the frequency terms to account for this condition. Below this value there are no additional physical changes which take place. Also comparisons between field and test only can be made at the same failure percentage, (use $N_{T(50)}$ to determine $N_{F(50)}$) Utilizing their model, Norris and Landzberg predicted that the end of life (EOL) failure rate for logic chips utilized in the IBM System/370 would not exceed 10^{-7}% per 1000 power on hours (POH) on a per joint basis, a projection verified by field data [83]. Thus, FCSB connections also set the

Even so, there have been two lingering concerns. Manufacturers and users have been reluctant to depend upon process control to assure the reliability of FCSB array interior joints which cannot be inspected visually. Additionally chip dimensions have steadily increased with an accompanying increase in neutral point distances. That is, in the nearly two decades since the initial FCSB reliability projections were made, chip area has increased about 90 fold, accompanied by a more than 200 fold increase in I/O (based on a 6.5 mm logic chip, 27 × 27 solder pad array). However, over that same period, there has never been a single FCSB joint that failed in the field which represents many billions of power on hours (POH) - 30 billion POH for MCM-C packages alone [84] for which shipping began in 1978. These results should not be interpreted as suggesting that thermal fatigue of FCSB joints is not a concern. They do, however, indicate that within the design and use condition parameters exercised to date, FCSB joints continue to be the most reliable chip to next level of assembly connection in the industry. Additionally, it underscores the fact that a rather simple model, given the metallurgical complexities and dynamic nature of the joint, has been remarkably accurate in predicting field behavior and, therefore, is a very useful design guide. Model modifications may become necessary in the future as chip sizes and attendant DNP's continue to grow.

Fatigue Enhancements
The strain experienced by FCSB joints is directly proportional to the difference between chip and substrate CTE, ambient and maximum operating temperature and distance from the neutral point. But joint strain varies inversely with joint height, h as shown in Equation 9-3.

$$\gamma \propto \frac{(\Delta CTE)(\Delta T)(DNP)}{h} \tag{9-3}$$

The following steps can be taken to reduce the strain experienced by FCSB connections.

- **Eliminate High DNP Terminals.** Truncating the corners of large square arrays eliminates high DNP pads which are most prone to failure this results in a significant improvement with only a slight reduction in I/O density.

- **Reduce CTE Difference.** Various alternatives to alumina-based substrates have been reported in the literature. For example, Greer [85] using a polyimide Kevlar substrate achieved dramatic improvements in using a polyimide Kevlar substrate achieved dramatic improvements in FCSB thermal fatigue life by matching the substrate (CTE) to that of silicon, as shown in Figure 9-42. Matched card and board materials, such as copper-Invar-copper, are now commercially available, providing the opportunity for FCSB. CTE matched ceramic materials have also been pursued. Most notable among these are silicon carbide, aluminum nitride and glass-ceramics [86]. Also, silicon on silicon applications, which provide a direct CTE match, are becoming more prevalent [87].

• **Increase Joint Height.** Several means of increasing joint height to decrease joint stress (Equation 9-3) have been demonstrated. Among these methods are stacking solder pads to form a column or solidify elongated joints which are stretched while molten. Stretched joints have been fabricated by centrifugal methods and also using two solders whose melting points vary only slightly. Surface tension forces, and thus volume, of the higher melting point solder must be sufficient to stretch the functional pads. This method pertains only to applications wherein chip real estate availability is not an issue. Regardless of how achieved, stretching provides FCSB joints with an additional benefit - an enhanced shape factor. The as-reflowed barrel shape of FCSB joints. (Figure 9-43a) concentrates most of the strain deformation and accompanying fatigue deformation in planes close to the joint interfaces. However, stretched joints with slightly barrel or slightly hour glass shapes (Figures 9-43b and 9-43c distribute the deformation more uniformly throughout the joint volume. Improvements of 10 times and more are not unusual compared with standard barrel-shaped joints. Fatigue life is, however, adversely affected if the degree of stretch is excessive (Figure 9-43d), with early failure occurring within the middle portion of the joint.

• **Conformal Coating and Encapsulant Materials.** A wide range of materials are utilized to protect electronic devices from moisture, ionic contaminants, radiation and hostile environments. They also are installed in the gap between chip and substrate to enhance fatigue resistance. In one approach, the gap is only partially filled, conformally coating the substrate chip site, FCSB joint and chip bottom surfaces as illustrated in Figure 9-44. Low stiffness materials have been used in this application. Among these are dispensed liquids, such as amide-imide polyimide (AIP), which are subsequently cured [91] and chemical vapor-deposited Parylene [92]. Only peripheral (high DNP) pads need be coated to counter the ill effects due to chip bending, a factor whose importance increases with chip size [91]. Fatigue life enhancements of 1.5 - 3.0 times are typical.

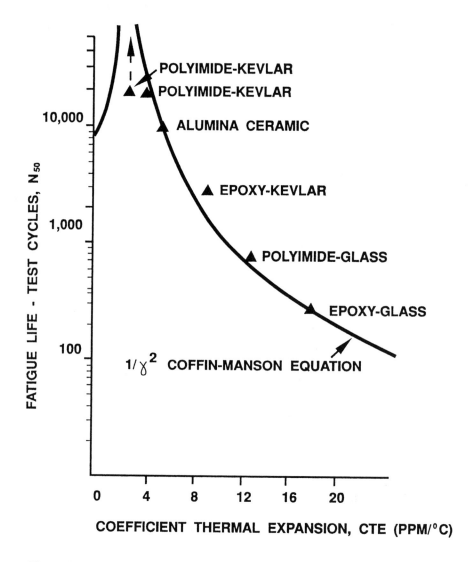

Figure 9-42 Effect of the coefficient of thermal expansion of a chip carrier on the fatigue life of FCSB mounted silicon chips (taken from [85]).

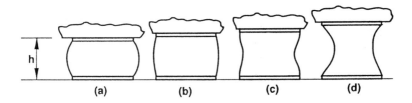

Figure 9-43 Cross sectional schematic of FCSB joint profiles with increasing degrees of stretch, (a) as-reflowed, barrel, (b) near column (c) hour glass, desirable for enhanced fatigue resistance, (d) highly concave, which, if excessive, has an adverse effect on fatigue life (not to scale).

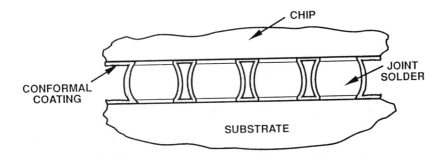

Figure 9-44 Cross sectional schematic of conformally coated FCSB joints (not to scale).

Yet another approach is to completely fill the chip-to-substrate gap with a filled epoxy resin (Figure 9-45). Dramatic fatigue life improvements (greater than 10 times are not uncommon, see Figure 9-46) can be achieved by this method. Improvements are optimum when encapsulant and joint solder CTEs are matched. This development greatly extends the allowable DNP and operating temperature ranges for FCSB mounted devices, freeing designers from the imposition of those constraints.

Currently, none of the available matched CTE materials is reworkable, thus limiting their use in MCM applications. However, the expectation is that reworkable materials will be available shortly.

• **Joint Solder Materials**. Lead indium alloys offer an opportunity for significant fatigue life enhancement over 95 Pb-Sn, a commonly used

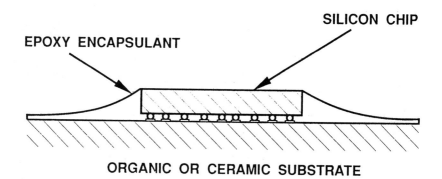

Figure 9-45 Cross sectional schematic representation of a FCSB mounted chip whose joints are encapsulated (taken from [93]).

FCSB solder. A parabolic-like dependence exists between indium concentration and fatigue life, with a minimum at 10 - 15% In. Improvement factors of 2, 3 and 20 times relative to 95 Pb-Sn correspond to indium concentrations of 5%, 50% and 100% respectively. Corrosion susceptibility increases with In content [94] and was recognized as a key concern for nonhermetic package applications [70]. Under these conditions the potential for corrosion fatigue exists as well. Additionally, 50 Pb-In FCSB joints and Pb-In solders, in general exhibit a much greater thermomigration rate in comparison to Pb-Sn alloys. Howard [94] has recommended Pb-low In (3% - 5%) alloys as candidates since their susceptibility to corrosion is virtually nonexistent.

Ambient Control. Recrystallization, creep and microstructural coarsening processes, which occur at high homologous temperatures, all favor fatigue damage. Even under normal conditions FCSB solder joints operate at over two-thirds of their absolute melting points [95]. Environmental effects are generally greatest under low cycle fatigue conditions [96]. An order of magnitude enhancement was observed in early fatigue studies of lead tested in a vacuum compared to air. Similar enhancements were reported for solder joints protected from oxidation by vacuum grease. In a comparative study, 95 Pb-Sn FCSB joints hermetically sealed in packages under a dry nitrogen atmosphere exhibited no failures, even when cycled to twice the fatigue life of similar joints not hermetically packaged [95]. Closely controlling the packaged environment is yet another factor that can be exploited to enhance fatigue life.

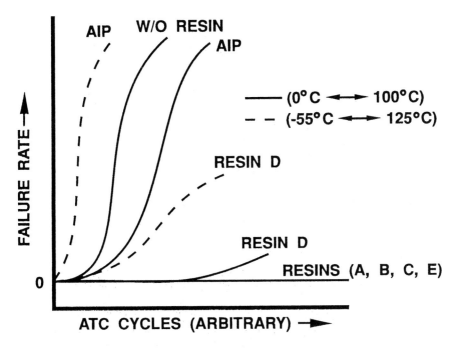

Figure 9-46 Effect of temperature cycle range and encapsulation on thermal fatigue life of FCSB mounted silicon chips (taken from [93]).

9.6.6 Performance

Cooling Capability
Initially FCSB technology's major heat dissipation path was through the chip solder joints. These early applications typically required less than 25 I/O (peripheral), offering limited capacity to dissipate heat in comparison to back bonded chips. At the time, however, total module dissipation requirements were also low, rarely exceeding 0.5 watt. Cooling capability has been significantly improved due to VLSI, which has driven the technology to large and dense (27 × 27, and 9 mil pitch respectively) area arrays, greatly increasing the area of contact with the substrate. This has made it possible to maintain chip junction temperatures of FCSB utilized in variety of IBM cost/performance products at or below 85°C using standard convection cooling. Typical of these is a 28 mm module with four chips (4.6 mm, 121 I/O) and a module heat dissipation rating of 5.4 watts utilized in the IBM System/38 computer.

Various techniques are employed to enhanced cooling efficiency which address both the back and front (I/O side) sides of flip chips. For example, a direct thermal path is established by introducing a thermal compound (TC) [97] [98] in the gap between the chip and cap (Figure 9-47) or die bonding the chip to the cap [99]. Thermal compounds, sometimes referred to as thermal greases, are commercially available with conductivities ranging from about 0.5 - 1.1 W/m-°C. They consist of thermally conductive but chemically benign material such as ZnO or BN dispersed in a suitable carrier (such as silicone oil). This method of cooling is possible with FCSB technology because the solder joints are located and protected beneath the chip, compared with delicate wire and TAB bonds which are exposed on the surface. The cooling advantage provided by FCSB/TC package with two standard back bond configurations (wire bond connections) is shown in Figure 9-47. The conditions are the same for all cases: chip and substrate size, chip thermal dissipation and external convection cooling. The external package resistance (heatsink resistance) also is the same for all cases compared. Figure 9-47 show the relative differences in internal thermal resistance for the conditions noted. This difference is responsible for the ability of the FCSB/TC configuration to maintain the chip junction temperature (T_j) at 69°C when compared to 81°C and 99°C for the back bond cavity down and cavity up configurations respectively. Although the values may change with other examples, the trends are the same because FCSB technology provides the lowest internal thermal resistance, taking into account the total path length and bulk thermal conductivities of materials in the heat path between the chip and heatsink. FCSB/TC have a short chip to heatsink path, typically about 5 mils. As noted earlier, heat also is dissipated through the solder joints into the substrate. The ability to dissipate heat from both sides of the chip provides FCSB technology with a distinct advantage.

The cooling requirements of some demanding applications have been achieved by directly contacting the chip backside with a spring loaded piston [86], [99]-[100]. Pistons have been used on modules containing up to 133 chips [81]. Chip front side (I/O side) thermal enhancement also is achieved with high thermal conductivity ceramics such as silicon carbide and aluminum nitride [87], [101] whose conductivity is 6 to 10 times better than alumina (Al_2O_3).

Electrical

Chip connections have a significant impact on the electrical performance of MCM packages. It is desirable to minimize the chip to substrate lead length and distances between chips. FCSB technology allows chips to be placed in close proximity on MCMs, with a greater than 90% packaging density possible [102].

(a) SBFC (C-4) / Thermal Compound

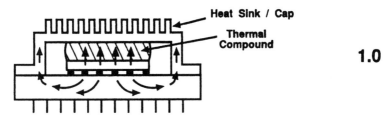

Heat Sink / Cap

Thermal
Compound

1.0

(b) Back-bonded, cavity down

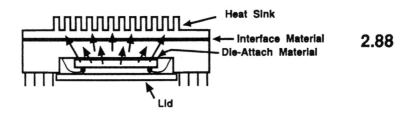

Heat Sink

Interface Material
Die-Attach Material

2.88

Lid

(c) Back-bonded, cavity up

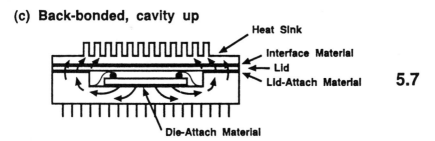

Heat Sink

Interface Material
Lid
Lid-Attach Material

5.7

Die-Attach Material

Conditions

- Alumina (Al$_2$O$_3$) substrate, 32mm
- Chip, 10mm dissipating 5 watts
- Convection cooling, 150 W/m^2 - C°

Figure 9-47 Cooling comparison between FCSB mounted devices with thermal compound and back bonded chips. (Courtesy of R. Sherif [104].)

Trends of increased device performance coupled with smaller chip and substrate feature dimensions is forcing a movement away from wire bonded, single chip modules to MCM utilizing TAB connections and ultimately to flip chip solder bump systems to minimize inductance. A key factor is the optimal geometry of FCSB joints (3 mil high, 5 mil diameter typically) in comparison to the smaller cross sections and much longer length characteristic of TAB and wired bond connections. The inductance of TAB and wire bond connections is typically in the range of 0.5 - 2.0 nH compared with FCSB joints, which are several orders of magnitude less. Lead inductance affects the number of simultaneous switch operations a chip driver may execute. The lead inductance also influences the maximum operating frequency. Under the best conditions this frequency is judged to be several hundred MHz for TAB. Although better than wire bonds, it does not compare favorably with FCSB joints, which operate in the GHz frequency range. Also, since FCSB are area array connections, power is brought close to the point of application when compared with either TAB or wire bonded chips where power is brought to the chip periphery. This also results in flip chips having shorter chip lines. These factors all serve to enhance the electrical performance of FCSB mounted devices.

9.6.7 Summary

This section notes that FCSB technology is capable of I/O extendibility, greatest connection density, pre-attach burn-in, test and rework. Thus, progress achieved in moving from a VLSI to an ULSI era makes it apparent FCSB technology is well suited to meet both I/O and performance (electrical, thermal etc.) requirements for current and future MCM applications. However, wider practice will require significant effort on the part of the electronics packaging industry and device manufacturers since the infrastructure currently is not in place.

9.7 SUMMARY

Wire bond, TAB and flip chip (FCSB) will be the dominant chip connection technologies in the foreseeable future. Each of these chip connection methods has advantages and disadvantages. The technologies are maturing partly to meet industry requirements and also because of their competition with one another. Tables 9-21, 9-22 and 9-23 provide a current comparison of the relative strengths and weaknesses for each of these technologies. The categories and ratings in these tables are summarized below and are based on previously discussed material.

9.7.1 Cost

Many factors contribute to the cost of an MCM. The choice of a chip connection technology can have a significant influence on the cost of an MCM. Some of the considerations involved that impact on the total cost are discussed in this section and in the following section where achievable performance also is related to cost.

Direct Costs
Each chip connection technology has costs directly related to its implementation and use. Equipment and tooling costs are examples of direct costs and are common to all three methods. Wire bond equipment is relatively mature in comparison to TAB and FCSB, resulting in an infrastructure that provides equipment at more competitive costs (refer to Table 9-21). TAB equipment can be very expensive due to the fine geometries, the degree of automation and the number of qualified vendors. FCSB assembly equipment is available, but is automated to a lesser degree than that of wire bond or TAB. Die specific tooling costs are a major issue with TAB, especially for custom, non-standard designs. Material and device preparation costs also are important. TAB and FCSB may require expensive wafer bumping processes. TAB also requires tape frames. These costs are reflected in the chip connection and device preparation cost comparisons in Table 9-23.

Indirect costs
The choice of chip connection technology also can affect other design and manufacturing considerations. Methods of approaching the cooling of MCMs are affected by the choice of flip TAB or flip chip over wire bond. Densities of chips on a module, or the ratio of silicon area to package area also has an effect on cost. FCSB has the potential for the densest assembly while conventional TAB and wire bond designs consume the most area. Repair and rework

Table 9-21 Comparison of Chip to Package Connection Technologies.

CONNECTION TECHNOLOGY	INDUSTRY USE	MATURITY	INFRA-STRUCTURE	DESIGN COMPLEXITY
Wire Bond	H	H	H	L
TAB	M	M	M	M
FCSB	L	L	L	M

L = Low   M = Medium   H = High

requirements, especially in MCMs, are critical factors when making a choice between these technologies. While TAB and FCSB are amenable to repair in certain configurations, wire bond does not lend itself easily or inexpensively to repair processes, as compared in Table 9-22.

9.7.2 Performance

The choice of a chip connection technology can have a significant influence on the performance of an MCM. Thermal performance, electrical performance and reliability are components of the general performance of the module and are related closely to the cost the MCM manufacturer is willing to pay to achieve specific performance objectives.

Thermal performance is influenced by the type of chip connection technology selected. Wire bond and conventional TAB configurations allow for the die to be cooled by conducting heat through the package. Flip TAB and FCSB, on the other hand, may require novel and potentially more expensive cooling methods. The desired thermal performance probably can be achieved if the MCM manufacturer is willing incur the expense.

Electrical performance also is a function of the chip connection choice. High performance applications may rule out wire bond because of the limitations of the fine wire. Conventional TAB also may be limited, unless a two metal layer tape is used. Only flip TAB with its short leads and FCSB with its leadless connections offer the greatest potential for optimizing electrical performance.

Wire bond is the most understood of the chip connection technologies. The documented reliability of wire bonded packages is plentiful. TAB, and especially FCSB, are not as mature or as well documented. Special considerations need to be made for the various metallurgical systems and materials allowed by TAB.

Table 9-22 Comparison of Chip to Package Connection Technology Design Issues.

CONNECTION TECHNOLOGY	DEVICE AVAIL- ABILITY	SILICON DENSITY	I/O DENSITY	ELECTRICAL PERFORMANCE	REWORK POTENTIAL
Wire Bond	H	L/M	L/M	L	L
TAB	L	L/M	L/M	M	M
FCSB	L	H	H	H	M

L = Low   M = Medium   H = High

Table 9-23 Comparison of Chip to Package Connection Technology Costs Assuming Low Volume MCM Production.

CONNECTION TECHNOLOGY	CAPITAL INVESTMENT	TOOLING COST	CONNECTION COST	DEVICE PREPARATION COST
Wire Bond	L	L	L	L
TAB	H	H	H	H
FCSB	M	L	M	H

L = Low   M = Medium   H = High

FCSB must be concerned with material thermal mismatches that contribute to thermally induced mechanical stress.

9.7.3 Tradeoffs: Making a Choice

Choosing a chip connection technology involves compromise. There is no one perfect solution for all products. The ultimate choice over time can be determined only by matching requirements against the available solutions. Careful evaluation of the cost tradeoffs, together with the performance requirements, should be made during the design phase of the MCM so that an optimum choice can be made. High performance applications may be able to absorb the higher cost associated with TAB and flip chip for the added performance. Low performance or cost driven products, on the other hand, may look solely at the cost issue.

References

1   T. L. Hodson, "Bonding Alternatives for Multichip Modules," *Electr. Packaging and Prod.*, vol. 32, no 4, pp. 38-42, April 1992.

2   R. K. Shukla, N. P. Mencinger, "A Critical Review of VLSI Die-Attachment in High Reliability Applications,' *Solid State Technology*, vol. 28, no. 7, pp. 67-74, July 1985.

3   T. A. Bishop, "A Review of Repairable Die Attach for CuPI MCM," *Proc. Internat. Electr. Packaging Conf.*, (Marlborough MA), pp. 77-89, 1990.

4   P. Burggraaf, "Polyimides in Miroelectronics," *Semi. Internat.*, vol. 11, no. 4, pp. 15-17, March 1988.

5   H. K. Charles, Jr., "Applications of Adhesives and Sealants in Electronic Packaging," *Proc. Internat. Soc. Hybrid Microelectr. Conf.*, (Orlando FL) pp. 139-146, 1991.

6   M. N. Nguyen, "Low Stress Silver-Glass Die Attach Material," *IEEE Trans. Components, Hybrids and Manufact. Techn.*, vol. 13, no. 3, pp. 478-483, Sept. 1990.

7   T. L. Herrington, G. G. Ferrier, H. L. Smith, "Low Temperature Silver Glass Die Attach Material," *Johnson Matthey Corp.*, San Diego CA.

8   S. M. Dershem, C. E. Hoge, "Effects of Materials and Processing on the Reliability of Silver-Glass Die Attach Pastes," *Quantum Materials, Inc.* 9988 Via Pasar, San Diego, CA 92126, 1992.

9   E. Razon, Y. Tal, "Silver Glass Die Attach," *Kulicke and Soffa Industries, Inc.*, Israel Div.

10 J. E. Ireland, "Epoxy Bleedout in Ceramic Chip Carriers," *Medtronic Corporation*, Tempe, AZ.

11 M. L. White, "The Removal of Die Bond Epoxy Bleed Material by Oxygen Plasma," *Bell Laboratories*, 555 Union Boulevard, Allentown, PA 18103.

12 R. C. Benson, T. E. Phillips, N. DeHaas, "Volatile Species from Conductive Die Attach Adhesives," *IEEE Trans. Components, Hybrids and Manufact. Techn.*, vol. 12, no. 4, pp. 571-577, Nov. 1989.

13 T. E. Phillips, *et al.*, "Silver-Induced Volatile Species Generation from Conductive Die Attach Adhesives," *Proc. 42nd ECTC*, (San Diego CA), pp. 225-233, May 1992.

14 R. H. Pater, *et al.*, "LaRC-RP41: A Tough, High Composite Matrix," *NASA Tech. Briefs*, vol. 15, no. 9, pp. 82, Sept. 1991.

15 R. K. Lowry, K. L. Hanley, "Volatility Behavior of Organic Die Adhesive Materials," *Harris Semiconductor*, P. O. Box 883, Melbourne, FL 32901.

16 C. J. Lee, J. Chang, "Reworkable Die Attachment Adhesives for Multichip Modules," *Proc. ISHM*, (Orlando FL), pp. 110-114, 1991.

17 MMT Staff, "Selection Guide: Die Bonding," *Microelectr. Manuf. and Testing*, pp. 16-20, July 1990.

18 K. Chung, *et al.*, "MCM Die Attachment Using Low-Stress, Thermally Conductive Epoxies, *Proc., IEPS*, (San Diego CA), pp. 167-175, Sept. 1991.

19 M. Akhavain, "Cost Effective Multi-Chip Modules Utilizing High Temperature Cofired Multilayer Ceramic Technology," *NEPCON West Proc.*, (Anaheim CA), pp. 1422-1430, Feb. 1992.

20   "Monitoring of Ultrasonic Wire Bonding Machines," *Hewlett-Packard Application Note 393*, (1990).

21   P. H. Singer, "Hybrid Wire Bonding Advances," *Semiconductor Internat.*, vol. 7, no. 7, pp. 22-25, 1984.

22   D. Herrell, H. Hashemi, R. Smith, W. Weigler, "Electrical Performance of Tape Automated Bonding," *Handbook of Tape Automated Bonding*, J. H. Lau ed., New York: Van Nostrand Reinhold, 1992, pp. 44-84.

23   E. Philofsky, "Design Limits when Using Gold-Aluminum Bonds," *9th Annual Proc. Reliability Physics*, (Las Vegas NV), pp. 11-16 (1971).

24   H. A. Schafft, "Testing and Fabrication of Wire-Bond Electrical Connections—A Comprehensive Survey," *NBS Tech. Note 726*, 1972.

25   G. G. Harman, "Reliability and Yield Problems of Wire Bonding in Microelectronics—The Application of Materials and Interface Science," Reston VA: ISHM, 1989.

26   H. B. Eisenberg, I. Jensen, "When Control Charting is Not Enough: A Wirebond Process Improvement Experience," *Proc. Internat. Soc. for Hybrid Microelectr. Conf.*, (Chicago IL), pp. 61-66, 1990.

27   K. Fukuta, T. Tsuda, T. Maeda, "Optimization of Tape Carrier Materials," *Proc. 4th Internat. TAB Symp.*, (San Jose CA), pp. 283-312, 1992.

28   A comprehensive handbook on TAB technology is, J. H. Lau, ed., *Handbook of Tape Automated Bonding*, New York: Van Nostrand Reinhold, 1992.

29   A. S. Rose, F. E. Scheline, T. V. Sikina, "Metallurgical Considerations for Beam Tape Assembly," *Proc. IEEE 27th Electr. Computer Conf.*, (Arlington VA), pp. 130-134, 1977.

30   K. Hatada, *et al.*, "New Film Carrier Assembly Technology - Transfer Bump TAB," *Proc. IEEE Internat. Electrical Manuf. Techn. Symp.*, pp. 122-127, 1986.

31   C. Montgomery, "A Low Cost High Performance Interconnection Technique for GaAs Devices," *1991 VLSI Packaging Workshop*, (Tucson AZ), pp. 9-12, Sept. 1991.

32   R. D. Pendse, *et al.*, "Demountable TAB - A New Path for TAB Technology," *Proc. 4th Internat. TAB Symp.*, (San Jose CA), pp. 9-24, 1992.

33   A. Reubin, B. Bohrn, R. Smith, "Transfer Molded TAB Package," *Proc. NEPCON West*, (Anaheim CA), pp. 894-905, 1990.

34   J. H. Lau, S. J. Erasmus, D. W. Rice, "Overview of Tape Automated Bonding," *Circuit World*, vol.1, no. 2, pp. 5-24, 1990.

35   J. M. Smith, S. M. Stuhlbarg, "Hybrid Microcircuit Tape Chip Carrier Materials/Processing Tradeoffs," *IEEE Trans. Parts Hybrid Packaging*, vol. PHP-13, no. 3, pp. 257-268, 1977.

36   V. Iyer, R. Pendse, "A Novel Inner Lead Bonding Technique for TAB," *Proc. 40th Electrical Computer and Techn. Conf.*, (Las Vegas NV), pp. 754-756, 1990.

37   D. Walshak, "The Effects of Bonder Parameters on Au-Au TAB Inner Lead Bonding," *Proc. NEPCON West '90*, (Anaheim CA), pp. 906-913, 1990.

38   C. J. Speerschneider, J. M. Lee, "Solder Bump Reflow Tape Automated Bonding," *Proc. ASM Internat. Electr. Material and Processing Cong.*, (Philadelphia PA), pp. 7-12, 1989.

39   D. A. Field, "A Tin Based TAB Assembly Process," *Proc. 1991 IEEE Internat. Manuf. Techn. Symp.*, (San Francisco CA), pp. 31-35, 1991.

40   P. J. Spletter, "Au/Au Inner Lead Bonding with a Laser," *Proc. 4th ITAB Symposium*, (San Jose CA), pp. 58-71, 1992.

41   E. Zakel, G. Azdasht, H. Reichl, "Investigations of Laser Soldered TAB Inner Lead Bonds," *Proc. 41st Electr. and Computer Techn. Conf.*, (Atlanta GA), pp. 497-506, 1991.

42   A. Emamjomeh, *et al.*, "TAB Inner Lead Process Characterization for Single Point Laser Bonding," *Proc. 1991 IEEE Internat. Manuf. Techn. Symp.* (San Francisco CA), pp. 21-26, 1991.

43   E. Zakel, R. Leutenbauer, H. Reichl, "Reliability of Thermally Aged Au and Sn Plated Copper Leads for TAB Inner Lead Bonding," *Proc. 41st Electr. and Computer Techn. Conf.*, (Atlanta GA), pp. 866-876, 1991.

44   W. T. Chen, *et al,*. "A Fundamental Study of the Tape Automated Bonding Process," *J. of Electr. Packaging*, vol. 113, no. 3, p. 216, 1991.

45   M. Wong, "TAB Outer Lead Bonding and SMT," *Proc. NEPCON West*, (Anaheim CA), pp. 785-789, 1988.

46   J. M. Altendorf, "SMT-Compatible, High Yield TAB Outer Lead Bonding Process," *Proc. NEPCON West*, (Anaheim CA). 233-249, 1989.

47   D. J. Arnone, J. J. Tong, "Determination of Placement Accuracy Requirements for Fine Pitch and Very Fine Pitch Component Assembly," *Proc. NEPCON West*, (Anaheim CA), p. 2154, 1991.

48   M. Y. F. Kou, "Assessing Fine Pitch Placement Machine From a System Standpoint," *Proc. NEPCON West*, (Anaheim CA), p. 2125, 1991.

49   L. Fox, "High Performance TAB Package," *1991 VLSI Packaging Workshop*, (Scottsdale AZ), pp. 21-22, Sept. 1991.

50   J. Deeny, D. Halbert, L. Nobi, "TAB as a High Lead Count PGA Replacement," *Proc. Internat. Electr. Packaging Conf.*, (Marlborough MA), pp. 660-669, 1990.

51   M. Mahalingam, J. A. Andrews, "TAB vs. Wire Bond - Relative Thermal Performance," *Trans. IEEE-CHMT*, vol. CHMT-8, no. 4, pp. 490-499, 1985.

52   D. Herrell, D. Carey, "High Frequency Performance of TAB," *Trans. IEEE CHMT*, vol 10, no.2, pp. 199-203, 1987.

53   J. H. Lau, D. W. Rice, G. Harkins, "Thermal Stress Analysis of TAB Packages and Interconnections," *Trans. IEEE CHMT*, vol. CHMT-13, no.1, pp. 152-187, 1990.

54   M. J. Bertram, "Repair Method for Solder Reflow of TAB OLB," *Proc. 2nd ITAB*, (San Jose CA), pp. 147-164, 1990.

55   D. S. Soane, Z. Martynenko, "Encapsulation and Packaging of Integrated Circuits," *Polymers in Microelectronics*, New York: Elsevier, 1989, p. 213.

56   K. C. Norris, A. H. Landzberg, "Reliability of Controlled Collapse Interconnections," *IBM J. Res. Develop.* vol. 13, no. 3, p. 266, 1969.

57   L.S. Goldman, "Geometric Optimization of Controlled Collapse Interconnections," *IBM J. Res. Develop.*, vol. 13, no. 3, p. 251, 1969.

58   R. J. Wassink, "Solder Alloys," *Soldering in Electronics, 2nd ed.*, Scotland: Electrochemical Publications Limited, 1989, p. 135.

59   T. Kawanobe, *et al.*, "Solder Bump Fabrication by Electrochemical Method for Flip Chip Interconnection," *Proc. 31st Electr. Computer Conf.*, (Atlanta GA), p. 149, May 1981.

60   W. G. Bader, "Dissolution of Au, Ag, Pd, Pt, Cu and Ni in a Molten Tin-Lead Solder," *Welding J.*, vol. 48, no. 12, p. 551, 1969.

61   P. J. Kay, C. A. MacKay, "Barrier Layers Against Diffusion," *Trans. Inst. Metal Finishing*, vol. 54, part 4, p. 169, 1979.

62   K. N. Tu, R. D. Thompson, "Kinetics of Interfacial Reaction in Bimetallic Cu-Sn Thin Films," *Acta Met.*, vol.30, p. 947, 1982.

63   D. Olsen, R. Wright, H. Berg, "Effects of Intermetallics on the Reliability of Tin Coated Cu, Ag and Ni Parts," *13th Annual Proc. of the Reliability Physics Symp.*, (Las Vegas NV), p. 80, April 1980.

64   A. vanderDrift, W. G. Gelling, A. Rademakers, "Integrated Circuits with Leads in Flexible Tape," *Phillips Technical Review*, vol. 34, no. 4, p. 85, 1974.

65   E. Yung, I. Turlik, "Electroplated Solder Joints for Flip Chip Applications," *IEEE Trans. on CHMT*, vol. 14, no. 3, p. 549, 1991.

66   D. Chang, *et al.*, "Design Considerations for the Implementation of Anisotropic Conductive Adhesive Interconnection," *Proc. NEPCON West*, (Anaheim CA), 1992.

67   H. Yoshigahara, *et al.*, "Conductive Epoxies for Attachment of Surface Mount Devices," *4th Internat. SAMPE Electr. Conf.*, (Albuquerque NM), p. 255, 1990.

68   B. Sun, *Connection Technology*, vol. 20, no. 8, p. 31, Aug. 1988.

69   I. Tsukagoshi, *et al.*, *Hitachi Technical Report*, No. 16. p. 23, 1991.

70   K. J. Puttlitz, "Corrosion of Pb-50 In Flip Chip Interconnections Exposed to Harsh Environment," *IEEE Trans. CHMT.*, vol. 13, no. 1, pp. 183-193, March 1990.

71   V. C. Marcotte, N. G. Koopman, P. A. Totta, "Review of Flip Chip Bonding," *Proc. 2nd ASM Internat. Elec. Mat. Proc. Congress*, pp. 73-81, April 1989.

72   C. Dostal, M. Woods eds., *Electronic Materials Handbook*, Materials Park,6 OH: ASM International, 1989, vol. 1, p. 231.

73   A. H. Kumar, "Corrosion of the Joining Metallurgy in Multilayer Ceramic Substrates During Processing," *Proc. 40th Elec. Comp. Techn. Conf.*, vol. 1, pp. 89-93, May 1990.

74   K. J. Puttlitz, "Flip Chip Replacement within the Constraints Imposed by Multilayer Ceramic (MLC) Modules," *J. of Elec. Mat.*, vol. 13, no. 1, pp. 29-46, Jan. 1984.

75   K. G. Heinen, W. H. Schroen, O. R. Edwards, *et al.*, "Multichip Assembly with Flipped Integrated Circuits," *IEEE Trans. CHMT.*, vol. 12, no. 4, pp. 650-657, Dec. 1989.

76   F. W. Kuleszsa, R.H. Estes, "Solderless Flip Chip Technology," *Hybrid Circuit Techn.*, pp. 24-27, Feb. 1992.

77   K. Hatada, H. Fujimoto, "A New LSI Bonding Technology, Micron Bump Bonding Technology," *Proc. IEEE 39th Elec. Comp. Conf.*, pp. 45-49, May 1989.

78   M. Bartschat, "An Automated Flip-Chip Assembly Technology for Advanced VLSI Packaging," *Proc. 38th Elec. Comp. Conf.*, pp. 335-341, May 1988.

79   B. Inpyn, "Practial Considerations for Tape Automated Bonding," *Circuits Manufacturing*, vol. 6, pp. 42-43, June 1989.

80   N. Basavanhally, S. Gahr, J. Liu, H. Nguyen, "Flip Chip Repair Process," *Proc. 41st Elec. Comp. Techn. Conf.*, pp. 779-782, May 1991.

81   P. Lin, J. Lee, S. Im, "Design Considerations for a Flip Chip Joining Technique," *Solid State Techn.*, pp. 48-54, July 1970.

82   K. C. Norris, A. H. Landzberg, "Reliability of Controlled Collapse Interconnections," *IBM J. Res. & Devel.*, vol. 13, no. 3, pp. 266-271, May 1969.

83   P. A. Tobias, N. A. Sinclair, A. S. Van, "The Reliability of Controlled-Collapse Solder LSI Interconnections," *Proc. 1976 Internat. Microelectr. Symp.*, ISHM, pp. 360-363, Oct. 1976.

84   S. Ahmed, R. Tummala, "Packaging Technology for IBM's Latest Mainframe Computers," *Adv. Techn. Workshop '91 on Multichip Modules*, (Oqunquist ME), June 1991.

85   S. E. Greer, "Low-Expansivity Organic Substrate for Flip-Chip Bonding," *Proc. 28th Elec. Comp. Conf.*, pp. 166-171, May 1978.

86   R. R. Tummala, H. Potts, S. Ahmed, "Packaging Technology for IBM System 390/ES9000, Models 820 and 900 Mainframe Computers," *Proc. 41st Elec. Comp. Techn. Conf.*, pp. 682-688, May 1991.

87   K. K. Hagge, "Ultra-Reliable Packaging for Silicon-on-Silicon WSI," *IEEE Trans. CHMT.*, vol. CHMT-12, no. 2, pp. 170-179, June 1989.

88   M. Matsui, S. Sasaki, T. Ohsaki, "VLSI Chip Interconnection Technology Using Stacked Solder Bumps," *Proc. IEEE 37th Elec. Comp. Conf.*, pp. 573-578, May 1987.

89   K. Puttlitz, T. Reiley, "Centrifugal Stretching of C-4 Joints," *IBM Tech. Disc. Bulletin*, vol. 27, no. 10B, pp. 6198-6200, March 1985.

90   R. Satoh, *et al.*, "Development of a New Micro-Solder Bonding Method for VLSI," *Proc. 3rd Internat. Elec. Pkg. Conf.*, pp. 455-461, Oct. 1983.

91   K. Beckham, *et al.*, "Solder Interconnection Structure for Joining Semiconductor Devices to Substrates that have Improved Fatigue Life, and Process for Making," U.S. Patent No. 4,604,644, Aug. 5, 1986.

92   H. M. Tong, L. Mok, K. Grebe, H. Yeh, "Parylene Encapsulation of Ceramic Packages for Liquid Nitrogen Application," *Proc. 40th Elec. Comp. Techn. Conf.*, vol. 1, pp. 345-350, May 1990.

93   D. Suryanarayana, R. Hsiao, T. P. Gall, J. M. McCreary, "Flip Chip Solder Bump Fatigue Life Enhanced by Polymer Encapsulation," *Proc. 40th Elec. Comp. Techn. Conf.*, vol. 1, pp. 338-344.

94   R. T. Howard, "Packaging Reliability and How to Define and Measure It," *Proc. 32nd Elec. Comp. Conf.*, pp. 376-384, May 1982.

95   R. T. Howard, "Optimization of Indium-Lead Alloys for Controlled Collapse Chip Connection Application," *IBM J. Res. and Develop.*, vol. 13, no. 1, pp. 29-46, Jan. 1984.

96   K. J. Lodge, D. J. Pedder, "The Impact of Packaging on the Reliability of Flip Chip Solder Bonded Devices," *Proc. 40th Elec. Comp. Techn. Conf.*, vol. 1, pp. 470-476, May 1990.

97   H. Ewalds, R. Wanhill, *Fracture Mechanics*, London: Edward Arnold Publishers Ltd., 1985, p. 180.

98   A. Bar-Cohen, "Thermal Management of Air and Liquid-Cooled Multichip Modules," *Proc. 23rd ASME Nat. Heat Transfer Conf.*, (Denver CO),, 1985.

99   S. Oktay, R. J. Hanneman, A. Bar-Cohen, "High Heat From a Small Package," *Mech. Engr.*, vol. 108, no. 3, pp. 36-42, March 1986.

100 T. Hatsuda, H. Doi, T. Hayasida, "Thermal Strains in Flip-Chip Joints of Die-Bonded Packages," *Proc. Internat. Electr. Pkg. Conf.*, IEPS, vol. 2, pp. 826-832, Sept. 1991.

101 T. Watari, H. Murano, "Packaging Technology for the NEC SX Supercomputer," *Proc. 35th Elec. Comp. Conf.*, pp. 192-198, May 1985.

102 T. Horton, "MCM Driving Forces, Applications and Future Directions," *Proc. NEPCON West*, (Anaheim CA), pp. 487-494, 1991

103 K. Puttlitz, "An Overview of Flip Chip Replacement Technology on MLC Multichip Modules," *Proc Internat. Electr. Packaging Conf.*, vol. 2, pp. 909-928, Sept. 1991.

104 R. Sherif, Private communication, 1992.

10

MCM-TO-PRINTED WIRING BOARD (SECOND LEVEL) CONNECTION TECHNOLOGY OPTIONS

Alan D. Knight

10.1 INTRODUCTION

Electronic packaging involves many levels of connections for components, subsystems and systems. The connections between the individual die and the multichip module (MCM) substrate forms a level 1 connection. A level 2 connection joins the MCM to the printed wiring board (PWB), either directly or indirectly. This should not be confused with level 1 (MCM) and level 2 (PWB) packages, as defined in Chapters 1 and 3.

There are two basic purposes of the level 2 connection. The first is to provide the electrical paths between the MCM and PWB to transfer electrical signals. The second is to provide the electrical power required for the chip to function [1]-[3].

A connector is often used to provide a level 2 connection from the MCM to the PWB. As well as providing the electrical path, a connector incorporates structures that help to provide a solid contact between the conductors on the MCM and PWB and the conductors in the connector. A connector is manufactured separately from the PWB and MCM and used when all three are assembled together.

In some cases, the second level connection is relied upon for physical attachment and for retention of the MCM to the PWB. This eliminates the cost of attachment hardware and assembly labor. However, this approach should be considered only for small, lightweight MCMs used in shock and vibration free environments and is not recommended for most applications.

Because level 2 connections are the interface for the MCM to the rest of the system, the method of connection requires careful consideration. The choice of connection and connector technology directly influences electrical performance, manufacturability, reliability and cost. As with the MCM itself, the technology chosen must address appropriately any mechanical and electrical performance issues that affect signal transmission quality.

The ideal connection is cost effective, easily processed and compatible with the thermal performance and test requirements of the MCM. Just as important, the connection should be as electrically transparent as possible - signals through the connection should not be distorted, attenuated, reflected or delayed. The ideal connection also would be capable of providing for a very large number of connections. The number of I/O pins of a large MCM can be very high, even greater than 10,000.

10.1.1 Second Level Connection Alternatives

The basic second level connection choices are summarized in Figure 10-1. The choices are differentiated by the geometry, the lead type and the connection method. There typically are two geometrical patterns used to distribute the connections: area array and edge array.

An area array geometry places connections on a square grid over the bottom, or most of the bottom, surface of the MCM. An identical pattern is placed on the PWB. Since the entire bottom surface may be used for the connections, this pattern provides either the greatest quantity of connections or the maximum spacing between connections. However, all connections are concealed beneath the MCM and, as such, may not be visually inspectable. An exception may be an MCM with pins soldered to through-holes in the PWB.

Edge arrays have all connections distributed on one, two, three or four edges of the MCM in one or more rows per edge. The same or expanded pattern is provided on the PWB. This configuration offers the largest variety of the connection means, most of which are inspectable. However, as the pin count increases, edge arrays generally require closer spacing of the connections than an area array.

For either area or edge geometries, the MCM package can include leads or be leadless. If leads are used, they are either formed flat leads or pins. Formed leads are used only with a single row edge geometry. (Forming means that the leads are bent.) Leaded MCMs are similar to surface mount packages used for

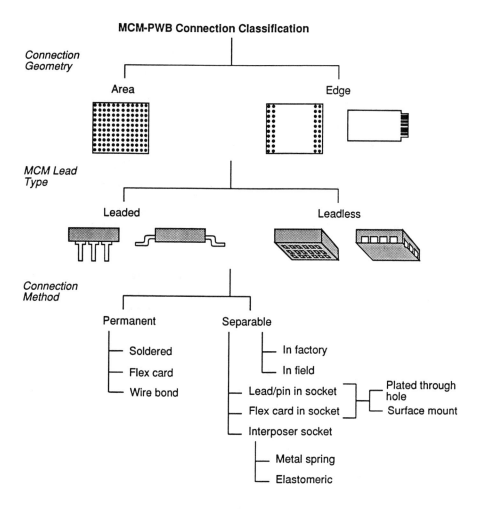

Figure 10-1 Classification of MCM-PWB (second level) connection geometries and methods.

single chip packages such as quad flat packs (QFP). Pinned packages are often found in the pin grid array (PGA) format. A common form of leadless array is the land grid array (LGA) in which pads are arrayed over the bottom of the package. Leadless packages also may be configured in the edge array format.

The methods by which MCMs are attached to PWBs fall into two basic categories: permanent and separable connections [4]-[5].

Permanent connections are typified by the use of a metallic bond, such as solder, to create the electrical connection between the MCM and PWB. As such, the MCM becomes an integral part of the PWB and requires special factory operations for removal and replacement. Thus replacing the MCM in the field usually requires replacement of the PWB as well. There are a number of additional options. In addition to direct solder attachment, the MCM also can be connected by a flex card (a flexible PWB). In this case, the wires at each end of the flex card are attached permanently to the MCM and PWB. Another alternative is to wire bond the MCM directly to the PWB. If the MCM has pins, these pins are inserted into matching holes on the PWB and soldered into place. A new method of attachment known as a pressure contact connection, uses gold bumps on the MCM contact pads and matching gold pads on the PWB. The MCM is glued to the PWB using a special adhesive which contracts as it cures and maintains a compressive force on the MCM-PWB interface. In many cases, leaded and leadless MCMs are soldered to the surface of the board (surface mount). This allows MCMs to be placed on both sides of the PWB. Out of the above methods, only the flex card is an example of a connector; the others consist of components (leads and pads) manufactured together with the MCM and/or the PWB.

Separable connections use a metallic element of some sort creating an electrical connection through mechanical means. Accordingly, the MCM can be installed and separated from the PWB at anytime during assembly and test or after shipment and installation. All separable connection methods require the use of a connector, which, together with any retention mechanism, is referred to as a socket.

Separable connectors are classified in terms of where the connection is made and by their type. With an in-factory separable connector, the connection is made, broken and reformed only in clean conditions. With a field-separable connector, the connection can be broken and reformed anywhere. The connector must have the ability to displace particulates and penetrate deposited films to ensure a good and reliable connection.

There are three types of separable connectors. In the first type, the MCM leads or pins are fitted into corresponding gaps or holes in the socket. In the second type, flex cards which have been connected permanently to the MCM are used to make a separable connection. Instead of making a flex card connection by soldering the leads to the PWB, the leads are force fit into a socket on the board. Both of these types of sockets are classified further by how they attach to the PWB. They either are pinned and soldered into holes in the board, much like a permanently mounted PGA is soldered, or they are soldered into place using surface mount techniques, just like a QFP or LGA is soldered into place. One large advantage of the surface mount alternative is that it allows MCMs to be placed on both sides of a PWB. Another alternative to mounting the socket to the PWB is to make the pins larger than the holes in the PWB and force fit the socket body onto the PWB without a soldering step.

The third type of separable contact or socket involves use of an interposer [6]. In an interposer, the connections are provided by a sheet of contacts placed (interposed) between matching pads on the MCM and PWB. This is used with leadless MCMs that have I/Os on the bottom such as in a LGA. The MCM and interposer are clamped to the PWB. Since there are no pins to fit into the socket, interposers allow MCMs to be used on both sides of the PWB. There are two types of interposers: those based on metal springs and those based on wires supported in a rubber material (elastomer).

Another connection scheme, not shown in Figure 10-1, is to bypass the MCM directly and combine the first level and second level connections by attaching the bare die directly to the PWB. This is referred to as chip on board (COB).

This chapter discusses the basic mechanical and electrical issues that affect connection choice. The different connection methods then are discussed in further detail.

10.2 BASIC ISSUES AFFECTING CONNECTION CHOICE

10.2.1 Mechanical and Materials Issues with Permanent Soldered Connections

There are a number of considerations that must be taken into account when soldering leaded and leadless MCMs onto PWBs.

The choice between peripheral or area array connection usually is based on the nature of the MCM package. Since peripheral pads are placed along the edges of the MCM, there is less space for connections than in array area configurations which make use of the entire bottom side of the package. In MCM applications where there are often a large number of connections, the peripheral approach usually requires tight center line spacing. Currently it is difficult to space solder connections closer than about 0.3 mm (12 mils). Area arrays allow more connections in the same area with wider pad-to-pad separation.

Area array contacts are not without their disadvantages, however. Traditionally, there has been some concern about the wisdom of soldering leadless MCMs where the pads are underneath the MCM [7]. However, the technology of reflow soldering LGA modules has improved dramatically in recent years. The techniques used to reflow solder LGAs are similar to the techniques used to reflow solder bump arrays described in Sections 9.5 and 9.6.

The soldered joint must be able to absorb the stresses that arise due to different levels of thermally induced expansions of the MCM and PWB. A coefficient of thermal expansion that closely matches that of the board and the MCM is highly desirable. The larger the package and substrate, the higher the levels of stress. Stress levels are reduced by using a leaded MCM with compliant (bendable) leads. This is why the leads in leaded packages have bends (gull wing or J-lead). Straight leads have little compliancy. In leadless packages the only compliancy provided is that of the solder itself. Again, the amount of compliancy there is small. This is discussed in Chapter 9 with respect to flip chip solder bump arrays.

Most of the thermal stress arises during the actual reflow step when the solder is heated and flows to wet (cover) the metal surfaces being connected. The temperature must be well below the glass transition temperature of the PWB. If the dielectric materials in the PWB start turning into glass during the reflow step, the thermal stresses will be much higher.

10.2.2 Separable Connector Interface Physics

The physics of electrical contacts in separable connections is a broad, complex subject [8]. The following discussion highlights some of the main issues dealing with MCM connections. For a more complete discussion, consult reference [1]. There are two areas of concern in the quality of an electrical connection: the DC performance and the AC performance. The DC performance relates to the ability of the connection to carry the required current with a minimum voltage drop. The AC performance relates to the distortion or delay experienced by high frequency signals as they cross the connection. DC issues are discussed first in terms of bulk properties of the metal and the surface properties of the interface.

Contact Conductivity
The contact must have low bulk resistance, meaning a substantial cross section of high conductivity metal. A high conductivity (low resistivity) material minimizes the voltage drop across the contact and minimizes the possibility of incorrect signal levels. Generally, materials with good spring properties do not have good conductivity. Beryllium-copper or beryllium-nickel, for example, often chosen for their spring properties and their ability to withstand high

Table 10-1 Resistivities of Typical Contact Metals.

CONTACT METAL	BULK RESISTIVITY ($\mu\Omega$-cm)
Silver	1.6
Copper	1.7
Gold	2.2
Aluminum	2.8
Tungsten	5.3
Nickel	6.8
Titanium	43
Beryllium-copper	5.6 to 8.5*
Lead	22
Tin	11
Tin/lead solder (63/37)	15
Tin/lead solder (5/95)	21.5

* Depending on composition.

temperatures over the long term, have low conductivity.

The need for low resistivity contacts is increasing due to shrinking semiconductor geometries, bringing about a reduction in IC working voltage - from 5 V to 3.3 V and even as low as 1.6 V - while leaving the current the same. As the working voltage decreases, the voltage drop across the contacts becomes a greater percentage of this voltage and thus is a greater concern. Table 10-1 lists the resistivities of metals typically used in both separable and permanent connections [8].

Contact Reliability: Force and Wipe

In addition to good bulk properties, a good contact must have a low contact resistance. Whenever two dissimilar surfaces are placed into contact, there will be some resistance to the flow of current from one material to the other. A reliable contact interface must have a very low end of life electrical resistance - in the range of 10 - 20 mΩ at 60°C after 15 years. To obtain a low and stable contact resistance, the contact must penetrate and fracture any oxide or nonconductive film which has been deposited, chemisorbed (chemically bound adsorbed layer) or adsorbed on contact surfaces or the substrate contact area. This surface layer is removed by a combination of wipe and normal force or normal force alone. Typically, the two contacts are pressed together and motion occurs in the plane of the contact surface. This combination of motion and normal force cleans the surface just as one cleans a dirty dish with a brush. If normal force cleans the surface just as one cleans a dirty dish with a brush. If there is no lateral motion, cleaning is completed through normal force alone (a force perpendicular to the plane of the surface). This force cleans the surface through a compressing action similar to squeezing dirty water from a household sponge. As a normal force requirement alone does not account properly for the effects of wipe, connector requirements often are expressed in terms of the stress in the contact surface. A contact surface stress in excess of 150,000 psi is required to clean the surface. Insertion force is the force required to insert the part into the socket and is created by a combination of normal force and wipe [9].

The required force or stress is highly dependent on the metallurgy of the contact surfaces. Metallurgies are classified as either reactive metallurgies, such as copper, brass, beryllium-copper, tin or tin-lead, or as noble metallurgies, such as gold, palladium or palladium-nickel.

Reactive metallurgies are low cost and easy to obtain. However, they react with atmospheric gases producing non-conducting compounds such as oxides. When these compounds are present, not only is it difficult to make a good metal to metal contact during insertion, but they add to the thickness of the base material. This increased volume increases further if airborne moisture is absorbed. If this volumetric expansion occurs after the material is mated, it may eventually overcome the normal force of contact, producing intermittent or open circuit contact. This corrosion of the contact is one example of a time dependent failure [9].

A second potential problem can arise if the connector is plugged and unplugged many times. This repeated action grinds the non-conductive material into the base metal, increasing the resistance. Also, channels, through which gas and moisture reach the reactive material under the final contact point, may be created. This also leads to intermittent contact or non-contact, excessive heating and total connection failure. This is another time dependent failure mechanism.

Noble metals, while more costly to obtain, have the advantage of providing a non-reactive, stable surface for the connection. However, when placed directly onto copper or alloys of copper, commonly used for the PWB printed wiring patterns and contact pads, the noble metal is absorbed by the copper, producing a reactive contact surface thus negating the rationale for using noble metals. Therefore, diffusion barriers, such as nickel, should be present between the base contact material and the final metal surface. A further caution must be exercised when using noble metals. The presence of the noble metal does not guarantee high performance reliability. The metal coating must be thick enough to provide a continuous surface in the wear track as well as at the final contact area. Wear-through due to multiple insertions or minute holes in the plating, called "pores," may result in corrosive products of the base or diffusion barrier metal to appear on the surface.

Contact surfaces often are formed by adding another complication. Microscopic hydrogen bubbles may cling to the metal being plated, with the result that the plating is deposited around the bubbles. As the plating thickness increases, the bubble migrates away from the substrate. A fistula or pin hole (Figure 10-2) results, frequently containing plating solution, wetting agent or organic complex or residues. Osmotic pressure often holds these deposits temporarily inside the bubbles, even at high temperatures.

When the pin holes are small, the contacts may pass accelerated environmental tests such as humidity and temperature cycling, porosity tests and fuming nitric acid tests. Two or more years later, when the moisture inside the bubbles has dissipated, gases enter the void, initiating corrosion. The accumulation of corrosion products separates the contacts, even those with high contact forces. The combined action of wipe and normal force displaces surface layers of the metal and burnishes it to seal the pores in the mating area and eliminate the problem. During subsequent mating cycles, wiping cleans and seals the mating areas [10].

The second reason for wipe is the need, particularly in dirty environments, for dislodging, removing and displacing dust particles. IBM, for example, has

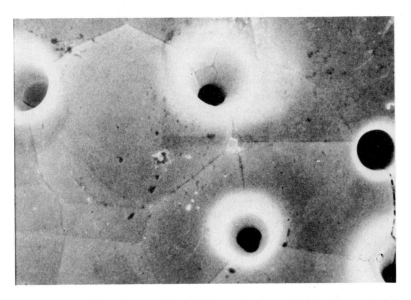

Figure 10-2 Pinholes in plating form a failure mechanism that can be controlled by contact wipe. (Courtesy of AMP.)

recognized the threat of such particles to system performance and has created artificial dust to help judge contact reliability. To pass the test, contacts must have a local high stress point and substantial wipe in a very small area.

Wipe is provided in different ways. For example, when a socket is used where pins are plugged into holes, the wipe and normal force often are provided by having the pins slightly larger than the holes. In this case, if all the pins are inserted at once, considerable insertion force might be required to press fit all of the pins simultaneously. Conversely, considerable withdrawal forces might be required to separate the parts. If these forces are large, manual insertion and withdrawal might be difficult and mechanical aids such as cams, wedges and screws must be provided. In most cases, wipe is provided by the displacement of either the MCM lead or a spring loaded contact inside the socket in such a way as to provide a scrubbing action between the two surfaces.

When little wipe occurs, the required normal force depends on the configuration of the contact forces. Two clean, flat contact surfaces (pads) require considerably more force for a reliable metal-to-metal contact than a half sphere shape on a flat pad. A configuration with half spheres provided on both surfaces requires the least force. In the last two configurations, geometry acts to concentrate the force onto a small area, creating a high surface stress. Typical force values for a well designed contact with geometric force concentrators are 100 grams per contact when noble metals are used and 200 grams per contact for non-noble metals, though success is achieved with smaller forces.

Failure to address the need for normal force and wipe leads to reliability problems. Consider, for example, a proposal consisting of a gold plated flexible circuit in direct contact through an elastomer with a gold plated PWB surface. This approach provides low forces and no wipe, based on the theory that gold to gold interfaces require only low contact force and no wipe. Such an approach, however, results in low reliability and high field failures, since contaminants and oxidation products migrate into micro pinholes formed during the plating process, particularly if products sit in inventory for any length of time. A classic, and painful, example occurred in a program which incorporated low force connection of gold plated flat pack leads to a gold plated multilayer PWB. The system passed all accelerated tests, but failed after less than two years in the field. Other approaches, such as anisotropic and conductive gel connectors, generally have poor reliability because of the absence of sufficient force and wipe.

Contact Compliance in Interposers
The contact must have a high degree of mechanical compliance, with an interposer separable connector, to compensate for any lack of coplanarity and flatness in the MCM and the PWB. The contact also must be compressed easily. A working range of 0.02" (0.5 mm) is desirable, although difficult to achieve.

In a metal spring interposer, this is achieved through bending the spring. With an elastomer interposer, this is achieved through compression of the rubber-like material that supports the metal wires. Interposers are discussed in detail later in the chapter.

10.2.3 Electrical Performance Issues

AC Performance
A driving force in electronics is higher packaging densities at all levels, from semiconductors to systems. This increased density is required to reduce the overall size of the end product and to decrease the electrical path lengths in the circuit. As the clock speed of digital electronics increases, there is less time for signal propagation between components. Clock speed in a digital circuit is akin to frequency in an analog circuit and, as the clock speed increases, a digital circuit takes on radio frequency circuit attributes requiring consideration of electrical impedance and coupling of circuit elements. What constitutes high speed depends on the application, with the electrical length of the circuit playing a part in the determination.

Impedance and crosstalk control are of special concern at high speeds; in many cases, both are treated through similar design approaches. Impedance mismatches and disturbances cause reflections, distorting signals. To ensure maximum transfer of a signal, all the components in the transmission path should have the same characteristic impedance. Crosstalk, which couples energy from one signal line to another through either capacitive or inductive coupling, must also be considered. The AC performance issues in a second level connection are similar to those on the MCM itself (see Chapter 11).

Transmission Line Performance
The performance of an connection is related to its resistance, capacitance and inductance. The capacitance and inductance of the lines affect the high speed performance. Interactions occur both between signals and from signals to power and ground lines. As the length of a line increases, both its capacitance and inductance also increase resulting in signal delays and distortions. At high speeds, circuit designers must treat a circuit as a transmission line, using the distributed rather than the lumped properties of all components in the design [11].

Transmission line rules generally become applicable when the length of a conductor approaches one-quarter of the signal wavelength, $\lambda/4$, or the length of a conductor approaches 1/100 of the rise length of a digital signal. (Rise length is the distance a signal travels during its rise time.)

The length of a circuit element, whether a wire, circuit trace or socket contact, is described in electrical terms. In high speed applications, electrical length is the major consideration - how the length of the circuit element compares to the signal wavelength. If the electrical length of the circuit element is long, impedance control becomes important. For a very short length, conventional lumped element circuit analysis is sufficient. In between those lengths is a gray area where the decision depends on other system-related factors, such as acceptable noise margins and cost and performance tradeoffs.

Rise time is the "turn on" time of a pulse, measured between the 10% and 90% (or 20% and 80%) points of amplitude. The rise time is used to calculate an equivalent frequency for the signal based on Fourier time-to-frequency analysis. This frequency then is used in the modeling of the performance of the electrical system.

In a high speed system, the designer has two options: either to use a controlled impedance connection or to keep the circuit element short enough to eliminate the need for impedance matching. If the circuit element is considerably smaller than one-quarter of a wavelength, impedance matching may not be necessary. This is the approach usually taken for second level connections as it is less expensive. However, the circuit element still presents a discontinuity in the transmission lines passing through the connection. This discontinuity disturbs the high speed signals propagating through it and reflects part of the signal. This is measured through time domain reflectometry (TDR). With this measurement technique, a high speed signal is sent down a transmission line with a second level connection placed somewhere near the middle and the resulting signals are observed at both ends of the line using an oscilloscope.

Crosstalk

Crosstalk results when signals from one line on a circuit couple onto another line. Crosstalk causes signal loss on one line, as well as contaminating the signal on the other line. Energy transferred from one line to the other produces a signal on the second line which might create an unexpected change of state. Crosstalk increases with the frequency of the signal and decreases as the separation between the two lines grows. Crosstalk is minimized by ensuring that lines operating at high frequency are placed sufficiently far apart or by using shielding to isolate the lines. Power and ground lines are used to shield the signal lines by placing each signal line close to a ground or power line. The coupling then occurs between the signal line and ground, rather than between adjacent signal lines.

Propagation Delay

Propagation delay is the time a signal takes to pass through a section of circuit. Differences in propagation delay between lines cause signal skewing in parallel transmissions typically found in bus configurations. A good connection delays all lines in a bus by the same amount to minimize skew. Since the path through a second level connection tends to be short, propagation delay is not usually a problem. As the frequency of systems increases, this becomes a greater issue and drives the size of connectors smaller.

Electromagnetic Modeling

The complex electromagnetic environments of a high speed, high density system means that verifying logic design is not enough. Component electromagnetic properties must be modeled early in the design process and tested to verify the modeling. Modeling must include the transmission line effects of the connection: reflections, crosstalk and any other effects that may disrupt a signal passing through the connection.

The connection does not exist electromagnetically alone in space. The influence of surroundings must be included, such as the capacitance of a plated through-hole solder connection, and the effect of the signals themselves. Signal rise times, for example, directly influence the number and placement of ground lines. As the signal rise times decrease, a larger number of power and ground connections usually are needed to minimize ground bounce during switching.

Properly modeling a connection is a difficult task, particularly with a socket, because of its complex structure. The model should evaluate the distributed values of capacitance, inductance and resistance, and should allow multicontact analysis through a matrix model [12]. (Contact capacitive and inductive coupling effects are not limited to adjacent contacts. A matrix model also studies the effects of simultaneous switching on several lines.)

A frequent and significant mistake is to ignore or to simplify the connection when using a circuit analysis program such as SPICE. Because the connection can have a dramatic impact on system performance, it must be included.

10.3 BASIC APPROACHES TO MCM
LEVEL TWO CONNECTIONS

As described earlier an MCM can be permanently and directly attached to the PWB or attached using a socket, so that they are easily separated again. Cost evaluations should consider component costs and total system costs. System costs include such factors as field replaceablity, upgradeability, troubleshooting and inventory costs in both the factory and repair depot. The cost of the lowest field replaceable unit (FRU) varies significantly depending on whether it is an entire printed circuit assembly or an easily swapped MCM. The reliability of a socketed MCM depends on a number of factors, including the quality of the mating connections and the protection provided from atmospheric elements.

Permanent, direct attachment generally simplifies reliability considerations since there are fewer contact interfaces and the interfaces are in more intimate contact. It is this same feature that makes field replacement of a directly attached MCM very difficult or impossible. Directly attached MCMs still face harsh environments during assembly on the soldering line and in the field. In some cases, the MCM may be encapsulated after mounting on the PWB to protect the unit and its connections.

10.3.1 Direct Attachment Through Soldering

Soldering is used with PGAs, and peripheral leaded and leadless MCM packages. PGA pins are soldered to plated through-holes, while peripheral devices are surface mounted. The pins on a PGA often are spaced at 0.05" or 0.1" intervals while the leads on a surface mount device often are spaced at intervals as small as 0.012".

Because this permanent, direct connection avoids the cost of a socket, it presents obvious component part cost savings. Any decrease in the number of connections tends to improve the reliability. Direct attachment through solder presents only one connection, while a socket presents two interfaces.

Another attractive feature of soldering is that it allows processing of MCM connections simultaneously with other components. The growing popularity of surface mount components and processing requires that the MCM withstand the high temperatures of reflow soldering. In a typical surface mount process, a solder paste is applied to copper lands on the board. Components are mounted to the board with the pads or leads on the paste. When the board is then passed through the soldering equipment, the solder melts (reflows) around the contacts and cools to form a solder joint. An alternative method of soldering is wave soldering which passes a wave of molten solder over the areas to be connected. Wave soldering does not subject the MCM or PWB to the full temperature of the molten solder and may be preferable in some cases. In all soldering, if a flux is used to promote surface wetting, it must be carefully cleaned off to prevent corrosion of the connection.

A device does not have to be a surface mount component to be processed by surface mount assembly techniques. Even a pinned area array MCM, which uses through-hole mounting, can be processed in a reflow soldering line. To be compatible with generic surface mount processing, a component must have the following characteristics:

- Ability to be mounted to the PWB by pick and place equipment
- Ability to be mounted on EIA-481 compliant tape
- Ability to be reflow soldered by infrared or forced air convection
- Ability to withstand aqueous solvent cleaning
- Allow visual inspection of solder joints

A component with characteristics different from these usually requires special equipment or techniques. Leadless area array MCMs obviously do not allow visual inspection of solder joints, although inspection can be performed with x-rays. Mass production soldering of such packages has been difficult. On the other hand, MCMs often are used in applications where standard high volume processing techniques are not required because of the specialized nature of the application. This lessens the need for compatibility with standard surface mount assembly equipment.

The decision to solder affects the choice of plating for the MCM leads. The most common form of solder is tin-lead which easily joins to gold and other noble metals. The most preferable arrangement is to have the leads of the MCM pre-tinned (coated) with tin-lead solder to simplify the joining process. In some configurations, it may be necessary to solder more than one connection between the MCM and PWB. For example, a lead frame may be soldered first to the MCM and then to the PWB. In this case, a soldering temperature hierarchy is required. The solder used to join the lead frame to the MCM is chosen to melt at a higher temperature than that used for the leads to PWB connection so that soldering the leads to the PWB do not melt the connections to the MCM.

Wire Bonded Modules

This form of permanent connection is an extension of chip wire bonding techniques (see Section 9.3). The MCM typically is bonded to the PWB with an adhesive. Wire bonding equipment is used to connect module edge mounted pads with corresponding pads on the PWB. The number of connections is limited by the pitch of the pads on the PWB. Visual inspection and rework of the resulting connection is performed readily and usually no post process cleaning is required. However, wire bonds from the MCM top surface to the PWB can be fairly long. This results in extra inductance in the connection and adversely affects the electrical performance as described earlier [14].

Flex Connected Modules

In flex connected modules, the discrete wires used in the wire bonded example are replaced with a series of flex ribbon cables. These ribbons act as jumper cables and usually consist of parallel copper lines placed on a thin dielectric film, with or without a fanout pattern at the PWB end. The circuit lines generally

Typical assembly consists of attaching the pads on one edge of the flex to the MCM. The MCM and flex assembly then is precisely aligned and bonded to the PWB with adhesive. The pads on the free edge of the flex circuit are attached to the PWB, usually accomplished by soldering or compression bonding techniques (see Section 9.3). When solder attachment is used, attention must be given to the cleaning process to assure complete removal of all flux residues between the flex and the MCM or PWB for long term joint reliability.

Inspection and rework are performed readily with flex connected modules. Reliability is generally high, since thermal expansion differences are absorbed by flexure of the flex circuit. Electrical performance is enhanced through the use of alternating signal and reference (power or ground) lines on the flex as well as through the addition of a ground plane.

10.3.2 Chip-on-Board Connections

The COB technology merges level 1 and level 2 connections by attaching chips directly and permanently to the PWB [15]. This approach has been used for years in applications requiring maximum space conservation. Intel Corp., for example, used COB technology to create an IBM XT-compatible personal computer on a board slightly larger than a credit card. The microprocessor was mounted directly to the board.

The COB approach is accomplished in one of three ways, each described in Chapter 9. These are: wire bonding, flip chip solder bump and tape automated bonding.

COB applications typically encapsulate the chip in a protective silicone or epoxy material to prevent moisture and gases from degrading bonding wires and joints. While encapsulants do not provide a true hermetic seal, they provide very good moisture sealing in high reliability applications. As thermal vias are rarely provided in COB PWBs, the thermal performance generally is poor. Both the materials above and below the chip are poor conductors of heat. Due to limitations on the minimum pad size available on a PWB, COB generally is not used for chips with a large number of I/Os since long wires would be needed to connect from the chip to the much larger pattern on the PWB.

10.3.3 Separable Connections (Sockets)

Sockets permit removal and reinstallation of the MCM to test the device, isolate parts of a circuit or to swap modules. They also permit PWB integrity to be checked prior to installing expensive components. Sockets allow fast, easy field replacement of faulty MCMs and eliminate the need for special desoldering equipment.

replacement of faulty MCMs and eliminate the need for special desoldering equipment.

If new MCMs are not available in full quantity, sockets allow the manufacturing process to proceed through testing by using a small quantity of on-hand modules to test the boards. When late deliveries arrive, they are tested and plugged into the board. Since MCM cost is relatively high, inventory costs are minimized by installing the MCM at the last moment rather than carrying it in inventory for months. Additionally, portions of a system are assembled at lower cost offshore and imported at a low tariff. Assembly of the MCM and its socket can then be completed domestically.

Sockets for MCMs include those designed for peripheral and area substrates, both leaded and leadless. The selection of a socket depends largely on the MCM packaging approach. Some sockets are essentially standard IC sockets, while others are designed especially for MCMs and similar high density, high pin count applications.

Peripheral Versus Area Array

The heart of the socket is the electrical contact, which serves as the electrical interface between the MCM and PWB. While several contact geometries have evolved to meet packaging needs, a given contact design can be adapted to many different socket configurations. Typical contact types include leaf springs and pogo pins. A leaf spring contact consists of a conductive member configured as a cantilever beam which applies force through the displacement of the beam. A pogo contact consists of a pin assembly containing an integral coiled spring which applies the required force.

The choice between peripheral and area array MCM and socket combinations presents several considerations. For a given number of MCM I/Os, the peripheral package presents the least area for contacts. Consequently, one either enlarges the package for large lead counts or places the contacts on close center lines.

Tight center lines are more rigorous in their requirements for alignment and maintenance of exacting tolerances. But peripheral packages have an attractive advantage: they permit easy access for visual inspection and for probes during testing. The area array package, with its greater area for I/O, relaxes the demand for tight center lines at contact points but at the expense of access for inspection or testing.

10.3.4 Sockets for Leadless MCM Substrates

Spring Contact Sockets for Peripheral and Area Array MCMs

The INTERPOSER™ contact (Figure 10-3) is an example of a basic contact structure that forms the basis for a variety of sockets [16]. It is a beryllium

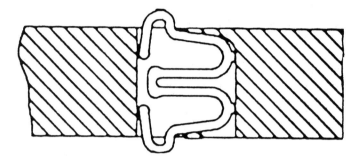

Figure 10-3 Basic INTERPOSER™ contact. (Courtesy of AMP.)

copper contact, featuring a double spring design with two internal and two external wiping surfaces that provide wiping action on the module, the board and the contact. When the contacts are interposed between the two mating surfaces and compressed, they exert an outward force to achieve and maintain a gas tight interconnection. The contacts are assembled into a rugged liquid crystal polymer housing which forms the basic module from which a variety of sockets are made. This solderless approach relies on a top and bottom plate on either side of the board held together by screws. A strong cover plate is required for a socket with a large number of contacts.

A smaller alternative of this socket for high speed, high density applications features a 0.090" contact suited to both peripheral and area applications [4]. Compression forces are reduced. The contacts are beryllium-copper with plated gold over nickel for mating with gold-plated interfaces, or made using palladium alloy (PAL-7). An example is an area array connector made by AMP using an 86 × 86 array with contacts on a true 1.2 mm grid. Each individual contact resides in its own plastic housing, and these housings are packed into a metal housing. At 120 grams per contact, the connector requires over 2,000 pounds for compression, presenting two problems. First, sufficiently rigid hardware is required with screws or clamps to compress the contacts. Second, the MCM substrate may be deformed or flexed by the clamping forces sufficiently to lower the reliability of the chip connections.

Contact Array Sockets for Area Array MCMs
The socket seen in Figure 10-4 shows an approach that further reduces contact height and required compression forces, complementing the drive toward higher

Figure 10-4 Contact area array socket. (Courtesy of AMP.)

density systems packaging by reducing the size of the socket. The compressed contacts are less than 0.010" high, and compression forces are low at 50 grams. The low compression force makes the contact suited only to noble metal applications.

At the heart of the socket is the contact array (Figure 10-5). A sheet of beryllium copper is chemically etched to produce the contact pattern. The contacts are then formed and plated with gold over nickel. Plastic insulating sheets are laminated to the array; this laminated structure is punched to isolate the contacts electrically. This provides an array of contacts on an 0.050" grid. The contact array is assembled into the socket [16].

A new type of interposer manufactured by Cinch consists of an array of conductive "fuzz" buttons which bridge the gap between pads on the MCM and matching pads on the PWB. The fuzz button (called a button board connector) accommodates variations in planarity and insures a good contact. The button board connector is made by twisting a small diameter wire of a known length and having it collapse on itself to form a small cylindrical shape of dense wire (Figure 10-6), which is then inserted into the pre-drilled holes in the polymer material to form the connector (Figure 10-7) [17]-[18].

Figure 10-5 AMPFLAT™ socket contact array. (Courtesy of AMP.)

SIMM Sockets for Peripheral MCMs

The socket used for single in-line memory modules (SIMM) also may be used in laminate and some cofired MCM applications. Figure 10-8 shows a representative SIMM socket. SIMM contacts must furnish enough force (over 200 grams) to provide a reliable, gas tight connection for tin-to-tin interfaces. Because of the popularity of SIMMs, SIMM sockets are manufactured in very high volumes. As a result, they offer the benefits of low cost and configuration and option flexibility such as vertical, angled or right angle SIMM insertion; one or two rows; 22 to 84 positions; gold or tin-lead plating; and 0.100" or 0.050" center lines.

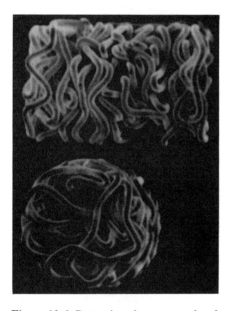

■ Multiple Springs
  – Cantilevered Beams
  – Columns
  – Torsion Springs

■ Random Wire Formation Provides
  – Low Inductance
  – Low Resistance
  – Unmatched Durability

■ Low Compression Force
  – Redundant Contact
  – High Contact Pressure
  – Mechanical Wipe

Figure 10-6 Button board connector showing connections to random wire configuration. (Courtesy of Cinch Connectors, a division of Labinal Components and Systems.)

usually feature a high temperature plastic that allows processing in surface mount soldering lines. SIMM sockets eliminate secondary operations such as wave soldering.

The main disadvantage of the SIMM socket in MCM applications is the low pin count of standard sockets (84 maximum) which may not suffice for many MCMs. Because only one side of the MCM can plug into the socket, routing on the module also is more complex. In addition, there are practical limits to extending the socket for larger sizes; the assembly becomes too long. These sockets also present a longer electrical path, making them less suited to high speed applications.

Elastomeric Connectors for Peripheral and Area Array MCMs
Elastomeric connectors consist of two components. First is the elastomeric element, a rubber material in which unidirectional conductors are formed. The elastomeric element is a form of interposer. Second is the rest of the connector

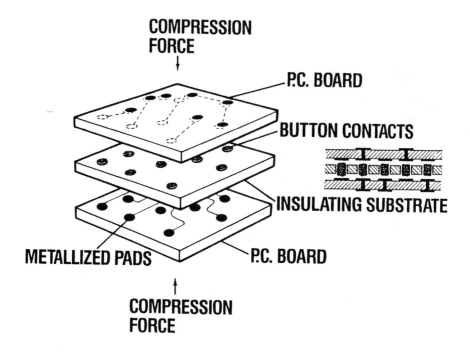

COMPRESSION
FORCE

P.C. BOARD

BUTTON CONTACTS

INSULATING SUBSTRATE

METALLIZED PADS   P.C. BOARD

COMPRESSION
FORCE

Figure 10-7 CIN:: Apse™ button board connections to printed wiring board. (Courtesy of Cinch Connectors, a division of Labinal Components and Systems.)

elastomeric element is a form of interposer. Second is the rest of the connector consisting of a holder to retain the elastomeric element. The hardware used to mount the element, attach it to the MCM and PWB and provide the necessary force is also part of the holder [19].

The basic advantages of elastomeric connectors include their high contact density, low profile, very low resistance, tolerance of wide temperature ranges and hostile environments, shock and vibration resistance, ease of installation and replacement and packaging versatility. Disadvantages arise from the fact that most elastomeric element conductors provide for minimal or no wiping action (through a compressing action) and that the elasticity of the rubber reduces with age. Normal force also is reduced as the elastomer ages. This is referred to as permanent set. As a result, the initial deflection (compression) and the initial force provided to do this must be increased in compensation for reduction of normal force [20]-[21].

Figure 10-8 SIMM (single in-line memory module) socket. (Courtesy of AMP).

Resistance and current carrying ability in these connectors vary with the geometry of the elastomeric elements and the substrate pads. The choice of element type depends on the maximum resistance that can be tolerated, the contact area, the possible clamping pressure and the cost [22].

The elastomer interposer connector group encompasses numerous configurations and compositions. Almost all use a form of silicone rubber, either as a solid or a foam of open or closed cell construction. This is used not only as the contact carrier but also to provide the energy storage that springs provide in more traditional contact systems. To create conductivity, a host of materials are used in either particle or continuous conductor form [23].

Conductive particles dispersed in the silicone rubber are one form of elastomeric interposer connectors. Typically the particles are copper or nickel and may be plated with silver or gold.

Compression of a thin silicone sheet (20 mils or less) with the correct density of particles displaces the silicone between the particles until adjacent particles touch. This creates conductive paths only in the compressed axis direction, making an electrical connection from the top to the bottom of the elastomeric sheet. Because of the thinness of the silicone sheet, this embodiment has an operating range of only a few mils, requiring very flat connecting surfaces.

Subjecting nickel-based particles to a magnetic field of varying intensity during solidification of the silicone produces a somewhat ordered, columnar structure of particles in the area of high magnetic flux. Generally, these columns can be placed on the desired grid creating conductive paths. These silicone sheets tend to be relatively thicker to accommodate the column of particles and, therefore, have a larger operating range.

Similar fabrication techniques have been used to produce elastomer connectors with solid wires embedded in the silicone rubber. These wires are made of copper, silver or gold and are straight or buckled. Upon compression, the film of silicone rubber, which may be over the ends of the wires, must be displaced to provide conductivity. These systems have the advantage of eliminating the multiple contact points between particles and, therefore, exhibit lower bulk resistance per contact path.

Layered elastomeric interposers are constructed by laminating alternating sheets of silicone and conductive layers together, creating rows of conductive paths. The conductive layer may be a solid conductor or a series of parallel line segments on a typical fine pitch of 5 mils. Thus, many parallel paths are used per contact providing redundancy. Constructions made by Elastomeric Technologies, illustrated in Figures 10-9a and 10-9b, typify this connector technique.

Continuous, parallel conductive strips, placed on a this sheet of silicone or polyimide and wrapped around a silicone rubber core, producing another version of elastomeric interposer connectors. These are normally produced on a fine line pitch to produce multiple conductive paths per contact. Typical values are 5 mil lines on 10 mil centers.

The metal on elastomer (MOE) is a robust elastomeric connector. The element consists of multiple rows of gold ribbon conductors across a silicone rubber medium along the z-axis. The components which make up the connector are silicone rubber and solid gold. These materials are resistant to harsh environments which make the second level connection ideal for space and military applications. The Matrix MOE connector which relies almost entirely on silicone rubber to apply the contact force useful when required to make a connection between nonplanar surfaces [17].

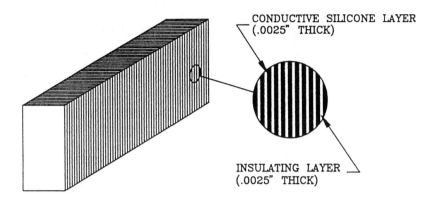

Figure 10-9a Layered elastomeric connector formed by alternating layers of conductive and nonconductive silicone rubber.

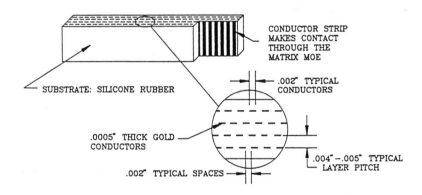

Figure 10-9b Details of a matrix metal on elastomer (MOE) construction. (Courtesy of Elastomeric Technologies.)

Another variety of elastomeric connector, produced by the Rogers Corp., uses rigid metal contacts formed in a S-type shape. These individual contacts are inserted into slits in a sheet of elastomer on the desired grid. During compression, the S-shape contact rotates, depressing the elastomer beneath the top and bottom portions of the S, producing both wipe and normal force on both top and bottom interfaces.

Most elastomer interposer connectors require high compressive loading to produce a limited operating range. They also have poor debris and film penetration ability. Normal force is lost with time as the result of stress relaxation (from 15 - 50% at end of life) and the particle filled elastomers have greater internal resistance than the solid conductor versions.

Connector holders generally are used to position mating substrates before clamping pressure is applied to the elastomer element. Such holders also provide a deflection stop; the slot in the holder must be wide enough to accept the deflection in width and length of the element. Parallel, coplanar, perpendicular and offset configurations are available. The height of the holder is equivalent to the desired substrate separation. Mechanical features built into the holder fit into matching features on substrates to ensure alignment to pads.

Most commonly, deflection force is applied with machine screws or eyelets. Plastic and metal edge clips or molded plastic snaps provide clamping along the entire side of the assembly. Pressure is the predominant method for establishing and maintaining contact, so, a stable rigid backing is needed to ensure reliable, long term operation. Thus a stable rigid back up plate is always essential. The PWB is not rigid enough to provide this function.

In many designs, connections are made as the package is assembled. The clamping force is maintained by latching with interference fits of the mating package sections. If the connector is unlikely to be removed, plastic posts can be molded onto the package or adhesives can be used to hold parts in place for surface mount applications.

The contact area is pliable and conforms to irregular surfaces. The resiliency of the connector core also helps to create a sealed area around the contacts providing environmental protection at the contact interface. A wide variety of cross sectional configurations and lengths are available. The circuit is assembled into a housing and offered as a completed connector assembly.

The basic elastomeric strip easily is adapted to a variety of connector styles for both area and peripheral devices. Figure 10-10 shows an example of a socket designed for area array applications, while Figure 10-11 shows a socket for peripheral applications.

Because the elastomeric member can have a free height diameter as small as 0.040", it presents a short electrical path causing minimum signal disturbance. The main concern with such connectors is contact resistance. A wide variety of elastomeric member or conductor combinations have been devised with resistances ranging from low tens of mΩ to several hundred mΩ. Low resistance designs are available, for example, offering a resistance of 20 mΩ, inductance of 1 nH and a capacitance of 0.2 pF. A typical propagation delay is 12 ps [23].

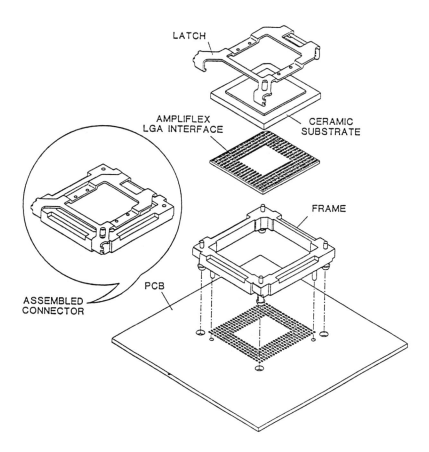

Figure 10-10 Elastomeric socket for leadless area array applications. (Courtesy of AMP)

10.3.5 Sockets for Leaded MCM Substrates

Leaded MCM Sockets for Peripheral MCMs
A leaded MCM socket, developed for leaded single and MCM packages, is shown in Figure 10-12. This socket doesn't need the clamping forces associated with other sockets discussed above. Instead, it uses flexible circuit leads that extend from the perimeter of the MCM over tuning fork-shaped contacts. The wedge-shaped protrusion of the socket cover forces the MCM leads into the

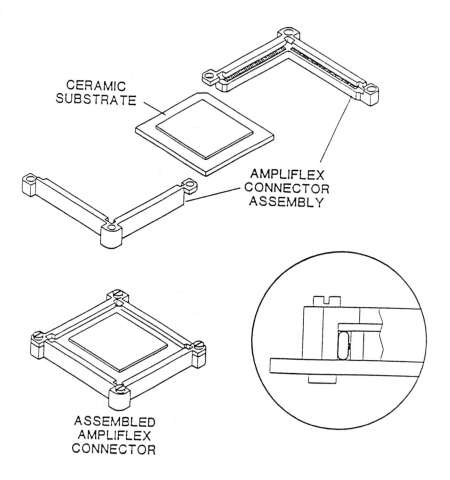

Figure 10-11 Elastomeric socket for leaded peripheral applications. (Courtesy of AMP.)

contacts. The configuration allows the MCM substrate to expand and contract without transferring micromotion to the contact interface. Motion is absorbed elastically at the base of the contact. While this approach seems simple, it requires particular attention to the alignment and registration of the flexible circuits on all four sides of the MCM. Tolerance analysis of the leads, connector components and PWB reveals the precise positioning and alignment required for a reliable connection.

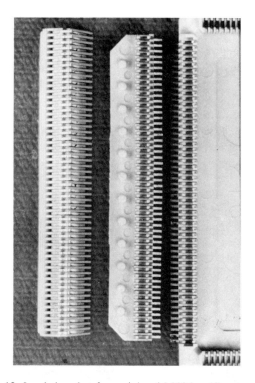

Figure 10-12 Leaded socket for peripheral MCMs. (Courtesy of AMP.)

Prototype implementations of this socket use several short 54-position strips with contacts on 0.5 mm center lines. An alignment plate positions the strips on the PWB, the strips are reflow soldered and the alignment plate is replaced with a nest. The nest serves the function of a bottom housing, a seat for securing the MCM. The MCM is placed in the nest and the cover is installed and pushed down.

Socketed Flex Connected MCMs
A socket similar to the SIMM socket is used with a flex card connector to make a separable connector. In this case, the flex leads are attached permanently to the MCM and also attached to the PWB through a variation of the SIMM socket with narrower openings. The flex leads are wedged into these contact openings.

Quad Flat Pack Sockets for Peripheral MCMs
QFP sockets for peripheral MCMs were originally designed for JEDEC QFPs with gull-wing shaped leads. The sockets (Figure 10-13) use a two piece arrangement. Normal force is created by the cover pressing the MCM leads into

the contacts. These sockets generally offer very high packaging density and a low profile. Their main attraction is that they are production tooled for compatibility with high volume commercial semiconductors. Inexpensive and readily available, they also are compatible with reflow soldering.

Current versions use through-hole mounting to the PWB with contact legs arranged on a 0.100" × 0.75" grid. Through-hole mounting makes trace routing on the board more difficult. On a PWB, plated through-holes used for component mounting also may serve as vias. (Vias interconnect different layers.) The difference is that normal interlayer vias are small (about 0.013") and plated through-holes are large (0.035"), making routing of conductors on all layers of the board more difficult. An additional disadvantage of through-hole mounting is the effect on performance driven applications. The contacts typically offer higher inductance and resistance as well as a longer electrical path.

Surface mount versions of QFP sockets are available, with a footprint identical to the device. Through-holes are eliminated to make routing easier. The QFP sockets make it easier to mount MCMs on both sides of the PWB.

Pin Grid Array Sockets for Pinned Area Array MCMs
PGA sockets for pinned area array MCM, like the QFP and SIMM sockets, are production sockets originally designed for commercial, high volume application.

Figure 10-13 Quad flat pack (QFP) sockets for peripheral MCMs. (Courtesy of AMP.)

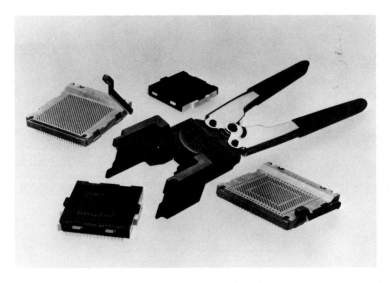

Figure 10-14 Zero insertion force (ZIF) sockets for pin grid array (PGA) packages (Courtesy of AMP.)

These sockets are available in both low insertion force (LIF) and zero insertion force (ZIF) versions. LIF sockets use contacts staggered on two heights to lower the engagement forces, while ZIF sockets (Figure 10-14) employ a camming or spring mechanism to hold the contacts open during MCM insertion and then close the contact around the pin. ZIF sockets are further divided by actuation method - handle or tool.

PGA sockets, with grids up to 20 × 20 (400 positions), are suited to high pin count MCMs. Versions of ZIF sockets with high temperature materials particularly are well suited to burn-in applications since the absence of insertion force accommodates high cycle life.

The main disadvantages of the PGA socket are the same as for the QFP socket: through-hole mounting on the PWB makes routing more difficult and can present undesirable high inductance, resistance and propagation delays in some applications.

A surface mount zero insertion force pin grid array socket is shown in Figure 10-15 and 10-16. This socket was used in the NEC ACOS 3900 [24] to connect an MCM containing 9440 contacts and, being surface mount, it allows two sided mounting. The contact shown in Figure 10-16 is 4.3 mm high and has a 0.4 mm solder terminal. The contacts are soldered to the PWB pads which are arranged on a 2.54 mm staggered grid (0.1"). The contacts are separated from

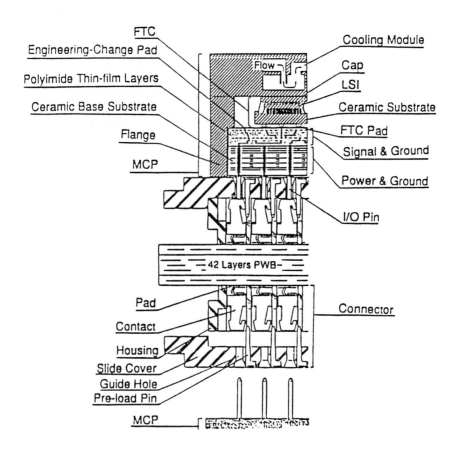

Figure 10-15 Packaging cross section of NEC ACOS 3900 showing the complete connector [24].

each other by a plastic molded housing. During assembly, the MCM pins are inserted into the slide cover (Figure 10-15) and the slide cover (with MCM pins) is inserted into the housing in one motion. As the MCM pin enters the contact, the pre-load pin compresses the contact around the MCM pin.

10.4 STANDARDS ACTIVITIES

Standards provide stability to industry, hasten acceptance and implementation of new technologies, reduce development costs and break down trade barriers [13].

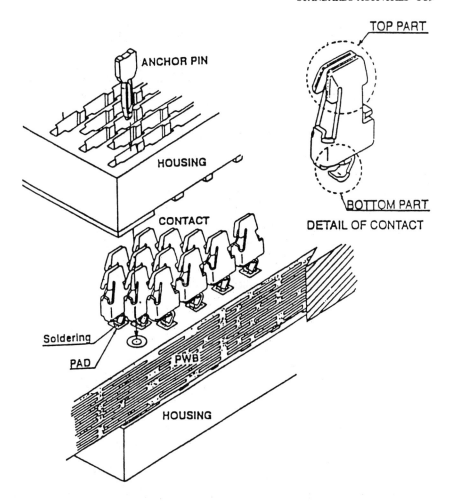

Figure 10-16 Details of the surface mount contact and housing for the NEC ACOS 3900 [24].

As with any standard, those affecting MCMs (and consequently their level 2 connections) are evolving at a rapid pace. With the worldwide emphasis on microminiaturization and surface mount technology, the job of standards groups has been made even more critical. The need for faster, denser and more cost effective designs brought to market in a short time complicates the challenge of developing and issuing standards. Several groups are involved in standards

Table 10-2 IEEE Task Force Recommendations for MCM Package Sizes.

SUBSTRATE SIZE (mm)	PACKAGE SIZE (mm)	
	Peripheral	Area Array
40	55	52
48	65	62
62	80	77
81	100	97
97	115	112

activities as discussed below. Some recommendations from these groups are given in Table 10-2.

10.4.1 Supporting Groups for Standards

The IEEE Computer Society Technical Committee on Packaging
This IEEE (Institute of Electronic and Electrical Engineers) committee appointed a special task force in the fall of 1990 to "seek early consensus on the need to standardize MCM sizes, and to propose some possible sizes." Its efforts promoted the development of standards limiting the number of MCM package sizes, around which an infrastructure could be built economically.

The task force recommended five sizes of peripheral packages with leads spaced on 0.5 mm center lines and five sizes of pin and pad area array packages using established PGA center lines. These are summarized in Table 10-2.

The EIA JEDEC Committee 11 (JC-11)
The Electronic Industries Association (EIA) Joint Electronic Device Engineering Council (JEDEC) committee establishes voluntary standards for the mechanical outlines of solid state and related products. Early in 1992, a task force developed MCM packaging standards, also in an attempt to stem a proliferation of many different package sizes. Its efforts include PGA and ceramic QFPs.

JEDEC Publication 95 contains published outlines prepared by the JC-11 Committee.

The EIA Committee on Sockets (CE-3.0) has established voluntary standards for sockets [13].

The Institute for Interconnecting and Packaging Electronic Circuits (IPC)
The IPC has established standards for printed circuit land patterns for electronic products.

10.5 SUMMARY

10.5.1 Recommendations

MCMs do not necessarily represent a new need in second level connector technology. The leadless area MCM has its counterpart in an LGA package used for packaging microprocessors. Likewise, tight center line spacing is as common with high volume ICs as with MCMs. Challenges unique to MCM-to-board connections include very large I/O counts, large component size and high speed performance. These require careful analysis for a satisfactory connection.

Speed, in particular, must be carefully analyzed. In any high speed system, the characteristics of the level 2 connection must be included in any performance model since these connections can significantly affect signal transmission quality. The connection is more than hardware or simple mechanical connections. It is an electrical connection and part of the transmission path. The often irregular geometry and relatively long electrical length of any connection can be a significant portion of the path. The adverse influence of the connection is best minimized either through some degree of impedance matching or by keeping the connection length short enough so that it is insignificant to signal transmission.

The choice of connections covers a wide range of options, including direct connection through soldering, standard high volume sockets and specialized sockets. Any choice involves a tradeoff between cost and performance. But failure to calculate this tradeoff rigorously runs the risk of degrading the rest of the system.

10.5.2 Future Trends

The growing preference for surface mounting is contributing to a trend away from PGA packages to LGAs and to various styles of QFPs. Socket makers are accommodating these changes. To address the continuing popularity of PGAs, designers of surface mount boards can avoid making through-holes for the PGAs by placing the packages in surface mount sockets. However, socketed PGAs present a very high profile. The vulnerability of the pins to damage is another disadvantage of socketed PGAs. In contrast, LGAs are a practical alternative. With no pins, they are less vulnerable to damage and provide very short electrical paths. The trend is toward smaller and smaller packaging to meet the needs of hand-held computers and communications devices.

As the speed of components continues to increase, new forms of packaging are required. Optical connects can be expected to play an important role as clock speeds approach the GHz level. This will create a whole new range of problems including the need for very precise alignments and a quick and cost effective separable connector containing many fibers. Future connectors may even include lasers to communicate with remote components in the system.

Another trend in socketing follows the transition in the United States toward metric dimensioning of flat packs. New flat packs generally have contacts on 1 - 0.3 mm pitches. New package outlines submitted for the JEDEC registration process must be hard metric, although extensions to existing families based on inch measurements may be acceptable to JEDEC.

In response to customer demands, package designers have created a proliferation of outlines for flat packs, more than 100 in Japan, and about 60 registered with JEDEC. This diversity represents a challenge to socket manufacturers to their product offerings, and both surface mount and through-hole versions of their products. JEDEC and EIA Japan are working together to develop standards.

Acknowledgments

The author gratefully acknowledges the significant technical and organizational help of the editors. He would also like to thank AMP for allowing him to use a number of their photographs and diagrams, and for giving him access to technical descriptions of many of their connectors. Sincere thanks are given to Dr. Leonard Buchoff of Elastomeric Technologies, Inc. for similar contribution. He also thanks Dr. Thomas C. Russell and Zeev Lipkes of Alcoa Electronic Packaging for their technical assistance in preparing the final version of this chapter.

References

1   *Connectors and Interconnections Handbook, Vol. 1* , G. Derman, ed. Deerfield MA: Int. Inst. Connect. & Connect. Tech., 1990.
    A. J. Bilotta, *Connections in Electronic Assemblies*, New York: Marcel Dekker, 1985.
2   W. E. Gilmour, "Connector and Interconnection Devices," *Electronic Packaging and Interconnections Handbook*, C. A. Harper ed., New York: McGraw Hill, 1991, Ch. 3.
3   J. B. Gillett, B. D. Washo, "Connector and Cable Packaging," *Microelectronics Packaging Handbook*, R. R. Tummala, E. J. Rymaszewski, eds., New York: Van Nostrand Reinhold, 1989, Ch. 14.
4   M. Freedman, "MCM Interconnection Options," *Proc. NEPCON West,* (Anaheim CA), p. 1527, 1991.

J. H. Whitley, "The Mechanics of Pressure Connections," AMP Inc., Harrisburg PA, Dec. 1964.

5   W. H. Knausenberger, "Multichip Module Connections," *Thin Film Multichip Modules*, G. Messner, I. Turlik, J. W. Balde, P. E. Garrou, eds, Reston VA: ISHM Press, 1992, Ch. 10.

6   D. Grabbe, "ModuPak: A New, Low Cost, High Speed MCM Socketing System," *Proc. NEPCON West*, (Anaheim CA), p. 1537, 1991

7   H. Kent, "Multi-chip Module Connectors and Sockets: High Density and High Speed," *Proc IEPS Conf.*, (Marlborough MA), p. 726, Sept. 1990.

8   R. Holm, *Electronic Contacts*, New York: Springer-Verlag, 1967.

9   S. S. Simpson, M. E. St. Lawrence, "A High Density Land Grid Array (LGA) Connector with Wipe and High Compliance," *Proc. IEPS Conf.*, (Marlborough MA), pp. 951-955, Sept. 1990.
     J. H. Whitley, Anatomy of a Contact: A Complex Metal System," *Insulation/Circuits*, vol. 8, p. 39, Fall 1981.

10   R. S. Mroczkowski, "Materials Considerations in Connector Design," Technical Paper, AMP, Inc., pp/ 31011-1388, 1989.

11   J. A. Defalco, "Reflection and Crosstalk in Logic Circuit Interconnections," *IEEE Spectrum*, vol. 11, pp. 44-50, July 1970.

12   D. Royle, "Rules Tell Whether Interconnections Act Like Transmission Lines," *EDN*, vol. 23, pp. 131-160, June 1988.

13   *Microelectronics Standards MS002-MS008*, JEDEC Publication no. 95, Washington DC: EIA Eng Dept., 1989.

14   G. G. Harman, *Wire Bonding in Microelectronics*, Reston VA: ISHM Press, 1989.

15   "Guidelines for Chip-on-Board Technology Implementations, ANSI/IPC-SM-784, Lincolnwood, IL: IPC, 1990.

16   D. G. Grabbe, H. Merkelo, "High Density Electronic Connector for High Speed Digital Application," *AMP J. of Techn.*, pp. 800-90, Nov. 1991.

17   C. W. Pike, R. Hassan, "Wire Button Contacts as a Connection Alternative: Design Opportunities and Challenges," *Proc. IEPS Conf.*, (Marlborough MA), p/ 944, Sept. 1990.

18   CINCH Connectors, 1500 Morse Avenue, Elk Grove Village, IL 60007.

19   H. W. Markstein, "Applications Widen for Elastomeric Connector," *Electr. Packaging and Production*, vol. 32, no. 5, pp. 29-32, May 1992.

20   W. R. Lambert, W. H. Knausenberger, "Elastomeric Connections-Attributes, Comparisons and Potential," *Proc NEPCON West*, (Anaheim CA), pp. 1512-1516, 1991.

21   C. A. Haque, ""Characterization of the Metal-in-Elastomer Contact Material," *Proc. 35th Holm Conf. Elec. Contacts*, Chicago, IL, pp. 117-120, 1989.

22   A. Strange, L. S. Buchoff, S. Ross, "Elastomeric Connectors Meet Critical Aerospace Requirements," *Proc. NEPCON West*, (Anaheim CA), pp. 651-656, Feb. 1992.

23   L. S. Buchoff, "Elastomeric Sockets for Chip Carriers and MCMs," *Proc. 42nd Electronic Components and Techn. Conf.*, (San Diego CA) pp. 316-320, May 1992.

24   M. Yamada, *et al.*, "Packaging Technology for the NEC ACOS System 3900," *Proc. 42nd Electr. Components and Techn. Conf.* (San Diego CA), pp. 745-751, May 1992.

11

ELECTRICAL DESIGN
OF DIGITAL
MULTICHIP MODULES

Paul D. Franzon

11.1 INTRODUCTION

The aims in package electrical design are to maximize the performance of the system, as limited by the interconnect (here "interconnect" refers to interconnections and connections) delay, while minimizing the possibility of false operation in the field, due to electrical noise, and minimizing the cost. To this end, electrical design of digital systems consists of the following:

1. Selecting the appropriate packaging and semiconductor technology mixture, and the appropriate partitioning of functions between chips and packages, so that the system design is likely to meet its cost and performance aims (see Chapter 3).

2. Generating the logical design (gate level design), and determining the logic families to be used.

3. Generating the timing design (when signal events will take place) and the noise budget (the required signal quality or signal integrity).

4. Determining the appropriate models and selecting the appropriate

simulation tools that allow interconnect signals and timing to be accurately predicted from the physical design.

5. Generating the physical design. This includes a placement, which describes where the chips and other components are located, and a layout, which describes where conductors run.

Electrical design is very important because a large fraction of the clock cycle time of a computer or any other digital system can be attributed to the delay involved in getting a signal off the chip and transmitting it between chips. A good electrical design maximizes performance by controlling these delays. In today's high performance system designs, the speed of ICs already strains the ability of conventional PWB technologies to provide comparably fast interconnections. The use of MCM technology allows clock speed increases of between 30% and 100%. A good electrical design also minimizes the possibility of noise induced errors occurring during operation. Good design practices are particularly relevant to MCM design because it is much more difficult and costly to diagnose and fix an electrical design problem on a fine pitch MCM prototype than it is on a PWB prototype. Nevertheless, in both technologies it is highly desirable not to have to iterate the design once the prototype is constructed.

This chapter starts by defining delay and noise and how they relate to digital interconnection design. It then describes the primary delay and noise phenomena important to MCM interconnect design (many of which are common to conventional package design). Finally the activities that take place during electrical design are discussed.

11.2 DELAY AND NOISE IN DIGITAL DESIGN

In any digital design there is some critical path whose delay limits the maximum possible speed of the system. For example, imagine that the performance of a system is limited by the delay, t_{total}, between the two flip-flops or latches shown in Figure 11-1. (A flip-flop or latch samples the voltage at the input, pin D, such as D1 in Figure 11-1, and transfers the same logical level, 0 or 1, to the output, pin Q, such as Q1, whenever a 0 to 1 edge occurs at the clock pin, Ck, such as Ck1.) A buffer is placed at the output of the first flip-flop for the purpose of driving the package interconnect structures. The buffer might also be called a driver or just an output. The gate input at the end of the interconnection is referred to as an input, a receiver or just a load.

The total delay, t_{total}, is the sum of the following components:

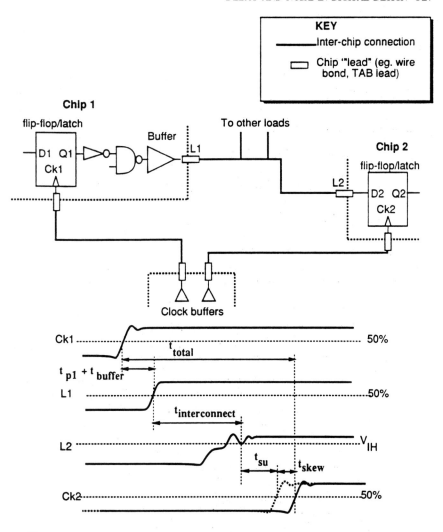

Figure 11-1 An example showing path delay components in a synchronous digital system. (A "synchronous" system is one where all signal events are defined with respect to clocks. Most digital systems are synchronous.)

- The maximum expected internal delay inside latch 1 and the logic gates that come after it, t_{p1}.
- The internal delay within the buffer, t_{buffer}.
- The delay introduced by the interconnect, $t_{interconnect}$.

- The setup time required by the receiver latch t_{su}. (This is the minimum time for the signal at D2 to be stable before being sampled by a clock edge at Ck2).
- The maximum possible uncertainty in the exact position of the edge on Ck2 with respect to the edge on Ck1. This is referred to as clock skew, t_{skew} when the uncertainty is constant. Each individual clock also experiences a small amount of jitter, t_{jitter} (uncertainty when each edge will occur as compared with when its arrival is expected), which should also be added in.

The maximum expected result for the delay, t_{total}, determines the minimum timing between the clock events Ck1 and Ck2. If Ck1 and Ck2 come from the same clock source, as is usually the case, then t_{total} would be the clock period, T_{clock}, of the system and its inverse would be the clock frequency, $f_{clock} = 1/T_{clock}$. The words "maximum expected result" refer to the fact that manufacturing and process variations in the circuits and packages result in different actual delays in different produced systems. The clock period must be selected so that it is greater than the delay expected in more than 999 out of 1000 of the manufactured systems. If nominal values of delay are used to select the clock period, then only 50% of the manufactured systems will work.

Of main interest here, is the package interconnect delay, $t_{interconnect}$. This is shown in Figure 11-2 as consisting of two delays. First there is the time, t_{prop} between when the output L1 reaches the 50% point of its logic swing and when the input L2 reaches a similar point. The 50% point is used as a reference as it

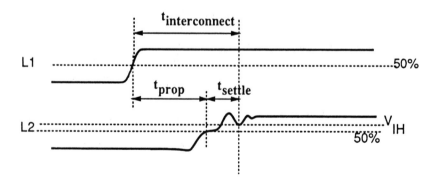

Figure 11-2 Data signal delay, defined as the delay between the 50% points in the signal with noise settling delay added in.

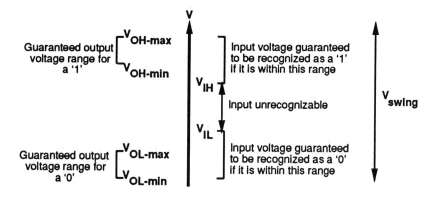

Figure 11-3 Definitions of DC voltage properties for a digital gate.

is the standard point used to define delay in digital circuits. Second there is the time, t_{settle}, that is introduced by the requirement that any electrical noise on the signal must settle before the signal can be safely sampled by the flip flop and determined to be a logic-0 or logic-1. In Figure 11-2, the signal must settle to a value greater than V_{IH} in order to be considered a valid logic-1. The reason for this relates to the DC properties of the circuit (Figure 11-3):

- Maximum and minimum output low voltage, $V_{OL\text{-}max}$ and $V_{OL\text{-}min}$, represent respectively the highest and lowest possible output voltage which corresponds to the logic-0 state. $V_{OL\text{-}min}$ is also the nominal output low voltage (such as 0 V in CMOS and TTL circuits).

- Maximum input low voltage, V_{IL}, is the highest possible input voltage that the circuit recognizes as a logic-0.

- Maximum and minimum output high voltage, $V_{OH\text{-}max}$ and $V_{OH\text{-}min}$, represent respectively the highest and lowest possible output voltage which corresponds to the logic-1 state. $V_{OH\text{-}max}$ is the nominal output high voltage (such as 5 V in a CMOS circuit).

- Minimum input high voltage, V_{IH}, is the highest possible input voltage that the circuit recognizes as a logic-1. (This is V_{IH} as used above.)

- The nominal voltage swing V_{swing}.

Any voltage between V_{IL} and V_{IH} is not classifiable as a logic-0 or logic-1 and if sampled could possibly be interpreted as either, resulting in a logic error.

Thus, if the output of a gate is at a logic-0 level, the DC properties of the gate guarantee that as long as the magnitude of any noise voltage added to this signal is less than V_{IL} - V_{OL-max}, then the signal is always recognized as a logic-0 at the input of any similar gate. This difference is thus defined as the low DC noise margin,

$$NM_L = V_{IL} - V_{OL-max} \qquad (11\text{-}1)$$

of the logic gate. The high DC noise margin is similarly defined as

$$NM_H = V_{OH-min} - V_{IH} \qquad (11\text{-}2)$$

Logic gates are usually designed so that these properties remain constant for most of the products within one circuit family for any one manufacturer. Typical DC voltage properties for different families are given in Table 11-1. (Here, and from now on the -max and -min suffixes will be dropped, with $V_{OL} = V_{OL-max}$ and $V_{OH} = V_{OH-min}$.)

Table 11-1 Typical DC Parameters and Noise Margins for Different Logic Families.

FAMILY	V_{swing}	V_{OL}	V_{IL}	NM_L	V_{OH}	V_{IH}	NM_H
Advanced CMOS	5	0.1	1.65	1.55	4.9	3.85	1.05
Advanced LS TTL	3.4	0.5	0.8	0.3	2.7	2.0	0.7
10K ECL	0.8	-1.63	-1.48	0.16	-0.96	-1.11	0.15

It has been recognized that digital gates can safely withstand short AC noise pulses of magnitude greater than the DC noise margin. A plot of the maximum safe noise voltage versus its 50% width is referred to as the AC noise immunity curve for a particular logic family, an example of which is given in Figure 11-4. Virtually no manufacturers guarantee values for AC noise immunity as they guarantee values for DC noise margins. It must be characterized by the designer.

The noise control requirements on digital signals are referred to collectively as "signal integrity" requirements. Note that signal integrity requirements are

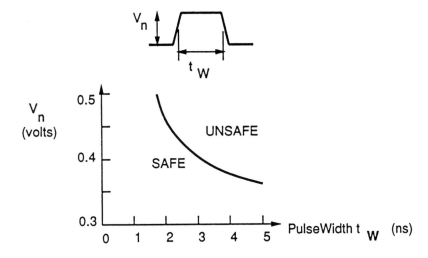

Figure 11-4 A typical AC noise immunity curve for ECL logic. If the gate input pulse height and width fall in the "safe" region, it will not result in a potential logic error [3].

more severe on clock signals than on data signals. On the latter, extra delay time can sometimes be included to allow noise to settle. On the former, excessive noise might be incorrectly recognized as a clock edge, leading to the latching of an incorrect data value. The techniques used to establish the noise margin and noise immunity of digital circuits are discussed in a number of sources: [1] - [5].

The remainder of this chapter discusses the basics of delay and noise control for MCM interconnections, including how they affect decision-making in the technology selection process and during the design process. It starts by discussing the main phenomena of interest: delay, reflection noise, crosstalk noise and simultaneous switching noise and then describes the design process.

11.3 PROPAGATION DELAY AND REFLECTION NOISE

In order to understand propagation delay and reflection noise it is first necessary to determine if an interconnecting segment (signal line or signal connector) should be treated as a transmission line or as a lumped circuit modeled by discrete capacitors, inductors and resistors. This determination is related to the frequency spectrum of the signal. A digital signal is actually communicated over

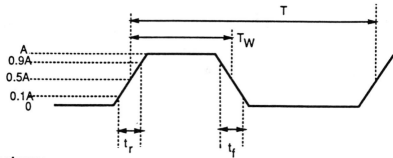

Spectrum:

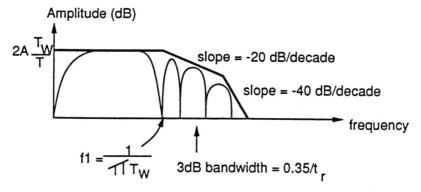

$$f1 = \frac{1}{\pi T_W}$$

3dB bandwidth = 0.35/t_r

EX: 100 MHz signal, rise time = 1 ns, 3dB bandwidth = 350 MHz.

Figure 11-5 Frequency spectrum of a digital signal.

a wide range of electromagnetic frequencies as shown in Figure 11-5. The bandwidth, BW, of this signal is given by

$$BW = 0.365 / t_r \tag{11-3}$$

where t_r is the rise time of the signal or the time taken for the signal to transit between 10% and 90% of the voltage swing. For example, the bandwidth of a signal with a 1 ns rise time is 350 MHz. For this signal, we define a "pulse design wavelength" λ,

$$\lambda = \frac{c / \sqrt{\varepsilon_r \mu_r}}{N \times BW} \tag{11-4}$$

where $c = 3.0 \times 10^8$ m/s is the speed of light in a vacuum, ε_r is the relative dielectric constant, μ_r is the relative permeability constant, normally 1, (and thus $c / \sqrt{\varepsilon_r}$ is the speed of light in that medium) and N is an integer that determines the quality of the signal. If the length of a structure exceeds $\lambda/8$ then a transmission line treatment is in order. Otherwise a lumped circuit treatment suffices. With a polyimide dielectric constant, $\varepsilon_r = 3.5$, and a rise time, $t_r = 1$ ns, then with N $= 4$, $\lambda / 8 = 1.4$ cm suggests that even 2 cm long interconnect structures should be treated as transmission lines.

If the length is shorter than $\lambda / 8$ then a lumped circuit, as shown in Figure 11-6, is used to model a point-to-point connection. A lower bound estimate on the propagation delay of this line is the RC delay given by

$$t_{prop} = t_{50\%}$$

$$t_{50\%} \approx 0.7 \left(R_{out} (C_1 + C_{line} + C_2 + C_{in}) + \frac{1}{2} R_{line} C_{line} + R_{line} (C_2 + C_{in}) \right)$$

$$(11\text{-}5)$$

where R_{out} is the equivalent resistance of the driver, R_{line} is the total line resistance, C_{line} is the total line capacitance, C_1 and C_2 are the lead (chip connection) capacitances, and C_{in} is the input capacitance of the die (typically about 2 pF for CMOS). The effect of any line inductance, L_{line}, is to increase this delay by 10 - 30% or more.

Figure 11-7 shows a point-to-point connection and simple equivalent circuit model for a line that must be treated as a transmission line. Note that because the chip connection leads are shorter than $\lambda / 8$, they are treated as lumped circuit elements. If a short via or connector were placed in the center of this line, it would be modeled by a lumped circuit equivalent and referred to as a discontinuity. Typical values for lead inductance and capacitance are given in Table 11-2.

Figure 11-7 also shows what happens to a signal produced by the output buffer. In Figure 11-7, the buffer is modeled by its Thévenin equivalent circuit, that is a voltage source and an equivalent output resistor. The input or receiver is replaced by its equivalent capacitance. The combined effect of the output circuits and the load presented to the signal at the output is that the signal will have a certain rise time, t_{rise}, which contributes to the 50% delay. The rise time is determined in part by these lumped circuit parasitics. The faster the circuit family, the faster the rise time. For example, ECL rise times are often well under a nanosecond. The buffer design determines t_{buffer} and t_{rise}, both of which decrease with reduced load capacitance. For example, a CMOS buffer that might

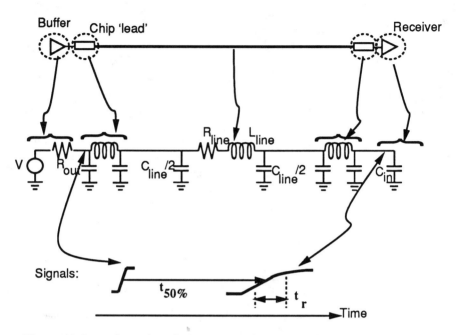

Figure 11-6 A lumped equivalent model for an electrically short point-to-point interconnect line.

have a delay t_{buffer} = 10 ns and a rise time, t_{rise} ≈ 1.5 ns when mounted in a single chip package, finds these times reduced to t_{buffer} = 7 ns [6] and has a rise time, t_{rise} ≈ 800 ps when mounted on a MCM. This results from the lower capacitance of the MCM interconnection and represents significant advantages for MCM technology.

When the signal leaves the output buffer it travels down the transmission line at the speed of light in that medium. At the receiver, the rise time of the signal is increased further by the inductive and capacitive parasitics of the receiver chip lead. All these effects combine to give the total delay of the signal as shown at the bottom of Figure 11-7. If first incidence switching is achieved (see Section 11.3.3), the total propagation delay is given by

$$t_{prop} = \frac{l}{c/\sqrt{\varepsilon_r}} + t_{rise-time-degradation} \qquad (11\text{-}6)$$

Table 11-2 Typical Values for Lead Inductance and Capacitance.

Lead Type	Capacitance (pF)	Inductance (nH)
SMT Package	1	1 - 12
PGA	1	2
Wire Bond	0.5	1 - 2
TAB	0.6	1 - 6
Solder Bump	0.1	0.01

Note:
• Inductance increases linearly with lead length unless a ground plane is provided (hence the reduced inductance of a PGA when compared with an SMT package).
• Lead capacitance is split evenly between the two capacitors shown in the chip lead model.

where l is the length of the interconnection, c is the velocity of light in a vacuum, ε_r is the relative dielectric constant of the dielectric and $t_{\text{rise-time-degradation}}$ is the delay effect of the increase in rise time between the end and start of the line (no simple equation is available for estimating the size of this delay contribution). The propagation velocity is

$$v_{prop} = c / \sqrt{\varepsilon_r} \qquad (11\text{-}7)$$

the speed of light in that medium and $l / \left(c / \sqrt{\varepsilon_r} \right)$ is referred to as the time-of-flight, t_{flight}. Note that the propagation delay decreases with decreasing dielectric constant.

A transmission line allows the electromagnetic waves to propagate uniformly, an example of which is given in Figure 11-8. These electromagnetic waves are associated with a voltage wave and current wave as shown in Figure 11-9. The return current flowing in the reference plane in Figure 11-9 has the same magnitude as the current on the signal line. The characteristic impedance Z_o of the transmission line is defined as the ratio of the voltage and current waves traveling down the line and thus has the units of ohms [7] - [9].

The signal line shown in Figure 11-8 must maintain a constant characteristic impedance over any length greater than $\lambda / 8$. When it does so it is referred to as a controlled impedance line. It can be shown that for a lossless line, the

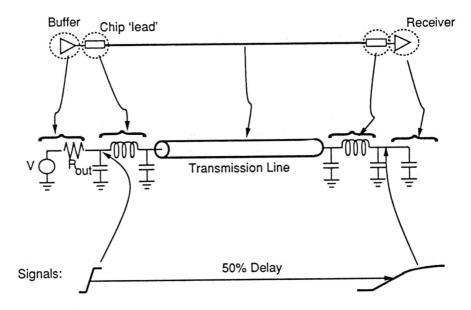

Figure 11-7 A transmission line model for an interconnection showing how propagation time and increased rise time due to discrete capacitive and inductive loads all contribute to delay.

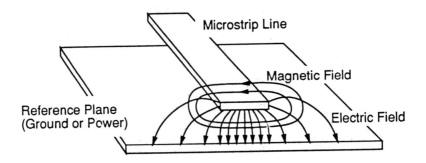

Figure 11-8 In a transmission line an electromagnetic wave is propagated along the line at the speed of light in that dielectric.

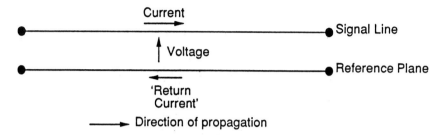

Figure 11-9 Voltage and current associated with the electromagnetic wave being propagated along the line.

characteristic impedance is expressed in terms of the inductance per unit length, L_o, and the capacitance per unit length, C_o.

$$Z_o = \sqrt{\frac{L_o}{C_o}} \ \Omega. \qquad (11\text{-}8)$$

One way to maintain a constant impedance is to maintain a constant geometrical relationship between the signal conductor and the reference plane. If the line narrows, for example, the capacitance decreases and the impedance increases.

The three arrangements that are most commonly used in MCMs and PWBs to provide controlled impedance interconnections are shown in Figure 11-10. While all three arrangements are common in PWBs and laminate MCMs, only the last two are generally used in thin film and cofired ceramic MCMs (though a multilayer ceramic might use surface microstrips).

An empirical (and approximate) equation for Z_o for a surface microstrip transmission line is

$$Z_o = \frac{87}{\sqrt{\varepsilon_r + 1.41}} \ \ln \frac{5.98h}{0.8w + t}, \qquad (11\text{-}9)$$

ε_r is the dielectric constant of the dielectric and the dimensions h, w and t are described in Figure 11-10.

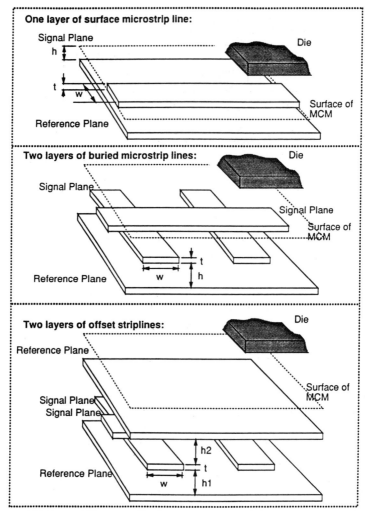

A 'Reference' Plane can either be a power or ground plane.

Figure 11-10 The three most common techniques for layering signal and reference (ground/power) planes in a multichip module.

Note that in order to obtain a higher value for Z_0 you either have to reduce the width, w, reduce the thickness, t, increase the height, h, and/or reduce ε_r. (All of these reduce the capacitance of the line and the first three also increase the inductance of the line.) The time-of-flight delay for a signal traveling down a surface microstrip (in nanoseconds per meter) is

$$t_{flight} = 3.337 \sqrt{0.475\varepsilon_r + 0.67}. \qquad (11\text{-}10)$$

Similar equations for a buried microstrip are

$$Z_o = \frac{60}{\sqrt{\varepsilon_r + 1.41}} \ln \frac{5.98h}{0.8w + t}, \qquad (11\text{-}11)$$

and

$$t_{flight} = 3.337 \sqrt{\varepsilon_r}. \qquad (11\text{-}12)$$

Note that signals propagate fastest on surface microstrips. Also note that a buried microstrip line must be narrower or thinner than a surface microstrip to achieve the same value of Z_o if h and ε_r are the same. The reason is that some of the signal is propagating in the air layer ($\varepsilon_r = 1$) above the surface microstrip, reducing the capacitance. While there is no simple empirical relationship for the characteristic impedance of offset striplines, the time-of-flight delay is given in Equation 11-12. Note that as an offset stripline has two reference planes, its capacitance is higher (inductance is lower), and thus, its impedance is lower than the same sized microstrip line with the same dielectric thickness.

The reference planes (either power or ground) tend to be meshed in thin film and many cofired ceramic MCM structures to promote adhesion between layers. (Each plane is actually a solid sheet with square holes in it.) By breaking up the reference planes, the meshing can affect the characteristic impedance and propagation delay of the signal lines. It also increases the DC resistance of the power and ground circuits. If the holes are large compared with the line pitch, then considerable care must be taken to ensure that the signal lines are routed only over the solid parts of the ground plane. Unrestricted routing might result in characteristic impedance variations of 30% or more. On the other hand, if the mesh pitch is comparable to the conductor pitch, the effect on conductor characteristics is small [10]. If the holes cannot be made this small, then this effect can be reduced partially by running the lines diagonally across the grid.

11.3.1 Reflections

Whenever a change in characteristic impedance occurs, part of the incident electromagnetic wave is reflected, just like part of a light beam is reflected upon striking a sheet of glass.

Characteristic impedance changes occur whenever the line branches (the waves "see" two lines in parallel) or the line ends. The portion of the incident electromagnetic wave voltage that is reflected at a change of characteristic impedance is given as the reflection coefficient, Γ

$$\Gamma = \frac{Z_{\text{load}} - Z_{\text{o}}}{Z_{\text{load}} + Z_{\text{o}}}. \qquad (11\text{-}13)$$

Z_{o} is the characteristic impedance of the line that the incident wave is traveling on and Z_{load} is the impedance of the load being seen by the line. If the load is an open circuit, such as a gate input, then $Z_{\text{load}} = 0$ and $\Gamma = 1$ and the reflected wave has the same voltage as the incident wave. (This reflected wave also causes the voltage at the end of the line to be doubled, the reflected traveling wave voltage to be added to the voltage provide by the previous wave.) If the load is a short circuit, then $Z_{\text{load}} = 0$ and $\Gamma = -1$ and the reflected wave is inverted with respect to the incident wave. On the other hand, if the load is matched to the line impedance, $Z_{\text{load}} = Z_{\text{o}}$, then no reflection occurs.

The successive partial reflections of the wave from each end creates a damped ringing signal as shown in Figure 11-11. The ringing signal can be

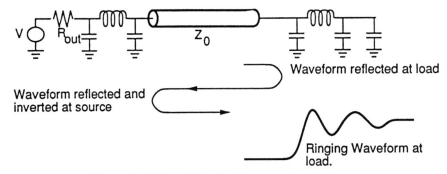

Figure 11-11 Multiple reflections act to cause ringing.

analyzed quantitatively using a lattice diagram [11] and [7] or a Bergeron diagram [12].

Ringing is a potential problem if the time it takes for the wave to travel down the line is longer than one-quarter of the rise time,

$$t_{prop} > t_{rise}/4. \qquad (11\text{-}14)$$

Waiting for the ringing to settle down can take up to an additional $4 \times t_{prop}$. The reason why a short line controls ringing noise is that the ringing settles down before the end of the rise time. Equations 11-14 and 11-6 indicate that for a 30 cm long PWB line ($\varepsilon_r = 4.5$), ringing could be a problem for signals with rise times less than 8 ns. For a 10 cm long cofired MCM line ($\varepsilon_r = 10$), on the other hand, ringing becomes a potential problem with rise times less than 4 ns. Rise times less than 1 ns - 2 ns are not uncommon in high speed circuits.

If ringing is a potential problem, then one or sometimes both ends of the line must be terminated with a matched resistive load. The four most common techniques for providing a matched impedance load are shown in Figure 11-12. (The series termination shown in Figure 11-12 prevents noise by not allowing the wave reflected from the load to be re-reflected.) Note that circuits operating at CMOS voltage levels do not typically use the Thévenin equivalent style of termination because the DC voltage at the join of the two resistors usually falls between the V_{IL} and V_{IH} when there are no other signals on the line. The AC termination style is used to prevent a DC current flowing through the termination, reducing power consumption [31].

Reflections also occur whenever the line branches, for example whenever a short section of line called a stub is used to attach the main line to an intermediate load. In this case, a signal traveling up the line sees a load of Z_0 in parallel with itself, $Z_0 / 2$, and a reflection occurs. The effect of the reflection is small, however, if the propagation delay along the length of the stub length is kept small when compared with the signal rise time.

Small reflections also occur when the uniform transmission line structure is changed even for a distance of less than $\lambda / 8$ such as at vias, bends and input gates placed midway along the line.

11.3.2 Line Losses

There are two main sources of losses in transmission lines:

1. Resistive losses experienced by currents traveling in the signal conductor and reference plane return path.

Parallel termination:

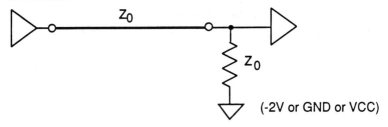

Thévenin equivalent parallel termination:

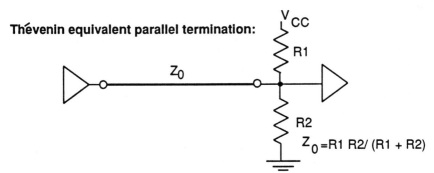

AC termination:

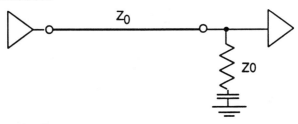

Series termination:

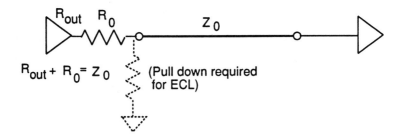

Figure 11-12 The four most common matching termination styles.

2. Dielectric losses due to some of the electromagnetic wave being absorbed in the dielectric material.

If the dielectric materials are chosen correctly, dielectric losses are negligible at most frequencies of interest. This is discussed in Chapters 5 through 8. However resistive losses can be significant, particularly for the very thin conductors used in thin film MCMs. The net effect of line losses is to attenuate the signal voltage and to increase its rise time as shown in Figure 11-13.

The signal voltage, V, placed initially on the line is attenuated to a voltage,

$$V_{end-of-line} = V e^{-Rl/2Z_o} \qquad (11-15)$$

as it travels down the line. Here, R is the resistance per unit length and l is the line length. As shown in Figure 11-13, this attenuated signal has a slower rise time because of the line losses, and is followed by a slowly rising signal that eventually brings the voltage at the end of the line to the output voltage of the driver.

The line losses are dominated by the resistance of the signal line, the resistance of the reference return path usually being small (but not negligible at high speeds). This resistance depends on the material being used, the cross sectional geometry, and the pulse design bandwidth.

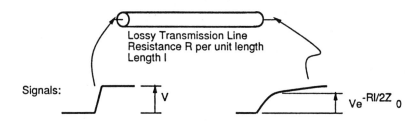

Figure 11-13 Effect of line losses on signal propagation.

Table 11-3 Resistivities of Common Conductor Materials.

MATERIAL	RESISTIVITY (10^{-8} Ω-m)	TYPICAL APPLICATIONS
Molybdenum	5.7	Cofired MCMs
Tungsten	5.7	Cofired MCMs
Copper	1.67	Thin film MCMs LTCCs PWBs
Aluminum	2.8	Thin film MCMs

Typical metal resistivities are given in Table 11-3. At DC, this translates into a resistance per unit length of

$$R_{DC} = \frac{\rho}{W \times T} \qquad (11\text{-}16)$$

where ρ is the resistivity, W is the line width and T is the thickness. For conductors typically found on PWBs, laminate MCMs and cofired ceramic MCMs (typically W = 100 μm and T = 30 μm) this translates into a resistance of less than 10 - 20 Ω/m. For a 30 cm long 50 Ω characteristic impedance line, less than 5 - 10% of the signal would be lost due to this resistance. On the other hand, the conductors in thin film MCMs are very small, typically 2.5 - 10 μm in thickness and 10 - 25 μm wide. A typical 2.5 μm $\times$ 15 μm aluminum conductor has a resistance of 747 Ω/m while a 10 μm $\times$ 15 μm copper conductor has a resistance of 110 Ω/m. For a 10 cm long, 50 Ω line, these resistances result in losses equal to 67% and 10% respectively. This is a major disadvantage in using aluminum conductors.

At frequencies above DC, however, the current concentrates in the skin of the conductor, as shown in Figure 11-14, with the current density decreasing with distance from the conductor edge. The current density distribution is characterized by a skin depth, δ_S, which is the depth at which the current density is the fraction $1/e$ of the density at the conductor surface. This depth is given by

$$\delta_s = \sqrt{\frac{\rho}{\pi \mu f}} \qquad (11\text{-}17)$$

where ρ is the resistivity of the metal (see Table 11-3), μ is the magnetic permeability $\mu \approx \mu_0 = 4\pi \times 10^{-7}$ H/m and f is the frequency. The skin effect causes a significant increase in resistance at frequencies above those for which the skin depth becomes half the thickness of the conductor. This occurs at a frequency of 64 MHz for a 30 μm thick cofired ceramic MCM molybdenum conductor, for example, and above 670 MHz for a 5 μm thick thin film MCM copper conductor. If a significant portion of the pulse design bandwidth is much greater than this frequency then the skin effect increases signal attenuation. Furthermore, in thin film MCMs this causes the current to be concentrated in the higher resistivity nickel or chromium barrier metal [14]. Above these frequencies, the only way to decrease the resistance of the conductor is to make

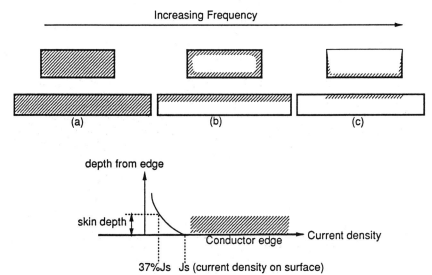

Figure 11-14 The skin effect. As the frequency increases, the current becomes more concentrated a → b → c.

it wider, the resistance being independent of thickness. Skin effect is only of concern for very fast digital systems operating at clock frequencies above several hundred MHz [15].

Note that as line losses attenuate reflected signals, the reflection noise in a lossy line is less than the noise in a lossless line. Sometimes it is possible to rely on line losses, rather than matching terminations, to control reflection noise, at the expense of increased propagation delay.

11.3.3 First Incidence Switching

To get the fastest propagation delay on a controlled impedance line, the first signal to arrive at the end of the line must have sufficient voltage to switch the receiver. (It should exceed V_{IH} on a $0 \rightarrow 1$ transition and be below V_{IL} on a $1 \rightarrow 0$ transition.) When this is achieved, the situation is called "first incidence switching." First incidence switching also requires that reflection noise be controlled well enough for the noise settling delay t_{settle} to be close to zero. If first incident switching is not achieved, then the propagation delay might be as much as five times the time-of-flight delay.

The first incidence voltage at the end of a matched terminated line is calculated by

$$ V_{first} = V_{swing} \frac{Z_o}{R_{out} + Z_o} e^{-Rl/2Z_o} \qquad (11\text{-}18) $$

where V_{swing} is the open circuit output voltage swing of the gate, l is the length of the line, R is the resistance of the line per unit length and R_{out} is the output impedance of the buffer circuit. In Equation 11-18, the $Z_o / (R_{out} + Z_o)$ factor arises from the potential division between the gate output and the line impedance. The Thévenin equivalent output resistance, R_{out}, is nonlinear and may have a value anywhere from 5 - 7 Ω for ECL, fast TTL and GaAs outputs, to 10 - 20 Ω or more for high speed CMOS and regular TTL outputs and to over 100 Ω for low speed outputs. Thus, in order to achieve first incidence switching, line losses must be low and the characteristic impedance of the line must be significantly greater than the Thévenin equivalent output resistance of the driver. For this reason, in high speed systems, the line characteristic impedance tends to fall in the range between 40 Ω and 75 Ω, higher values being preferred, particularly for CMOS drivers with their high values for R_{out}. The tradeoff is, however, that a higher value for Z_O requires either a narrower line or a thicker dielectric, both of which increase the wiring pitch. (See the crosstalk discussion below.)

Because the wire cross sections can be so thin in MCM-D technology, controlling line losses to achieve first incidence switching sometimes requires careful design. For example, consider a CMOS circuit. As $NM_H < NM_L$ (see Table 11-1), the $0 \rightarrow 1$ transition requires the largest V_{first} magnitude. Assuming $R_{out} = 15\ \Omega$ in the logic-1 state, $Z_0 = 65\ \Omega$, $V_{swing} = 5$ V, and $V_{first} = V_{IH} = 3.85$ V, then using Equation 11-18, first incidence switching requires that $R \times l < 7\ \Omega$. For a thin film 2.5 μm × 15 μm aluminum line $R_{DC} = 747\ \Omega/m$ and the longest line would be 9 mm, which is too short to be useful. For a 5 μm × 20 μm thin film copper conductor $R_{DC} = 165\ \Omega/m$ and the longest line would be 4.2 cm. For a cofired tungsten line (dimensions 30 μm × 100 μm, $R_{DC} = 19\ \Omega/m$), so first incidence switching at low frequencies, is not a problem for any practical line length. If no terminations are used then V_{first} is increased by a factor of (1 + Γ) and as long as the line resistance sufficiently attenuates the reflection signal, then first incident switching can sometimes be achieved with longer lossy lines. However, the resistance increases with frequency. This changes the shape of the signal [15], increases the rise time and makes first incidence switching more difficult for long lossy lines for design bandwidths above 200 - 500 MHz. The smaller output impedances of TTL and ECL circuits make first incidence switching easier to achieve than in CMOS circuits.

If first incidence switching is achieved, then delay primarily depends on the values of dielectric constant and the capacitive load, as shown in Figure 11-15. Reducing delay is the main reason to reduce dielectric constant. Delay is also controlled by limiting the number of loads on a line, preferably to three or four. Propagation delay is affected, to some extent, by the chip connection choice because the chip connection technology is included in the load capacitance. Flip chip solder bump connections add the smallest excess capacitance and single chip packaging the most capacitance.

If first incidence switching is not achieved, then the delay is increased. The value of the extra delay depends on the reason for the non-first incident switching. If excessive line losses are the cause, then the extra delay is only an increase in $t_{rise-time-degradation}$, which might be small or large, depending on the magnitude of the losses. No simple analytic formula exists to predict this. If the cause is a high output impedance, $R_{out} > Z_0$, then the extra delay is large, several time-of-flight delays, t_{flight}. (The reflection coefficient at both ends is positive and the successive reflections build the voltage up to the required value.) A similarly sized extra delay is expected if excessive ringing noise is the cause.

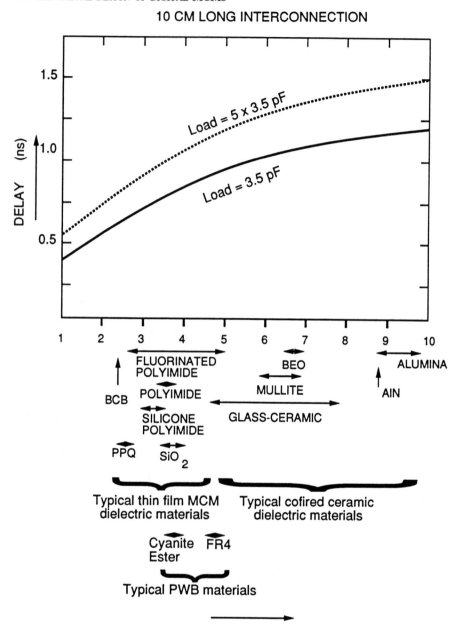

Figure 11-15 Effect of dielectric constant and capacitive loading on delay [16].

11.3.4 Net Topology

A net refers to the network of wires that join a set of transistor circuits. In this case, digital drivers and receivers. The topology refers to the net shape, as viewed from above.

If the timing design requires that ringing noise be controlled, and the net length is long ($t_{prop} > t_{rise} / 4$), then a matching termination, either series or parallel, is needed and the stub lengths must be limited. Controlling the stub lengths means controlling the net topology. Acceptable net topologies for controlling reflection noise are shown in Figure 11-16. Note that a ring topology

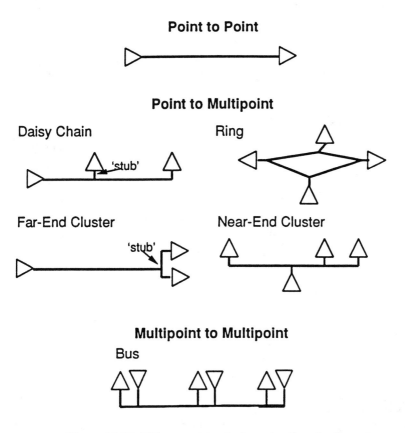

Figure 11-16 Different net topologies to handle reflection noise.

does not require a matching termination, at the expense of increased wiring, and the bus topology can be treated as either a daisy chain or near end cluster, depending on which driver is in use. The most common point to multipoint topology is the daisy chain. The acceptable stub length depends on how much ringing noise is acceptable and how fast the rise time is. The faster the rise time, the shorter the acceptable stub length.

11.3.5 Effect of Loading

If the loads (receivers) are closely spaced along a net, then their capacitance effectively adds to the self-capacitance of the line. This is referred to as distributed loading along the line. By closely spaced, a spacing, l_s, between the loads such that $l_s / v_{prop} < t_r / 2$ is meant. The capacitive loading might be in the form of the capacitance of the attached drivers and receivers, mutual capacitance from crossing lines or mutual capacitance with surface pads. If the total distributed load capacitance is C_L, then the effect of this loading is to slow the propagation velocity to

$$v_{prop} = \frac{c / \sqrt{\varepsilon_r}}{\sqrt{1 + C_L / C_0 \, l}} \qquad (11\text{-}19)$$

where C_0 is the per unit length self capacitance of the line, and l is the length of line over which the load C_L can be considered to be distributed. As the load capacitance has a large effect on delay, critical nets are limited to a maximum of three to four loads if possible.

The characteristic impedance Z_o is reduced to

$$Z_L = \frac{Z_o}{\sqrt{1 + C_L / C_0 l}} \qquad (11\text{-}20)$$

With heavily loaded lines, this reduction might be 50% or more. This can be compensated by increasing Z_o (this is often done in backplanes, where Z_o might be as high as 120 Ω). If Z_L becomes too low then first incidence switching cannot be achieved and the increase in delay is even more than the decrease in v_{prop} would indicate.

The effects of loading also are compensated by reducing R_{out}. This is done by inserting a special driver chip after the outputs of the ASIC. Manufacturers produce drivers that guarantee first incidence switching for loaded characteristic impedances as low as 30 Ω. The internal delay of the driver must be accounted for in the total delay.

11.4 CROSSTALK NOISE

Mutual inductance and capacitance between different electrical signal paths contribute to unwanted electrical coupling known as crosstalk noise. Whenever a signal edge travels down a signal wire, chip attach lead or connector lead, both forward and backward crosstalk noise pulses are induced in the neighboring two wires, as shown in Figure 11-17. Capacitive coupling, $K_C = C_m / C_0$, and inductive coupling, $K_L = L_m / L_0$, between adjacent lines add at the near end of the quiet line and subtract at the far end. The maximum noise voltage at the near end can be approximated by:

$$V_n \approx K_B \left(\frac{2}{v_{prop}} \right) \left(\frac{V_8}{T_1} \right) l \text{ if } l < \frac{v_{prop} T_1}{2} \qquad (11\text{-}21)$$

$$V_n \approx K_B V_s \text{ if } l > \frac{v_{prop} T_1}{2} \qquad (11\text{-}22)$$

where $K_B = (K_C + K_L) / 4$ is the coupling coefficient, C_m is the mutual capacitance between the lines per unit length, L_m is the mutual inductance per unit length, C_0 is the self capacitance per unit length of each line, L_0 is the self inductance per unit length, V_s is the voltage swing on the active line, $v_{prop} = c/\sqrt{\varepsilon_r}$ is the propagation velocity of electromagnetic waves in the dielectric, T_1 is 0% - 100% rise time of the signal, and l is the coupled line length. When the crosstalk stops increasing with coupled length (see Equation 11-24) is referred to as saturated crosstalk and the length $l = v_{prop} T_1 / 2$ is referred to as the saturated line length. More accurate, but more complex, analytic expressions for crosstalk can be found in reference [17].

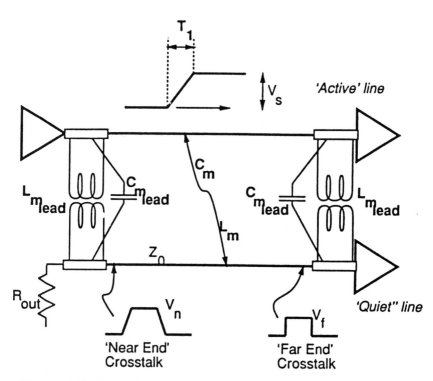

Figure 11-17 Crosstalk noise arises from the effects of mutual inductance and capacitance. This diagram show how a signal travelling down an "active" line causes crosstalk signals at both ends of a quiet line.

The maximum noise voltage at the far end is given approximately as,

$$V_f \approx K_F \left(\frac{2}{v_{prop}} \right) \left(\frac{V_s}{T_1} \right) l. \qquad (11\text{-}23)$$

where $K_F = (K_C - K_L) / 4$ is the coupling coefficient. If the medium in which the line is buried contains no other materials beside the dielectric then the far end crosstalk is zero because $C_m / C_0 = L_m / L_0$. This is only perfectly achieved within homogeneous dielectrics stripline conductors, and if no other conductors are present. Even with other conductors present, and for buried microstrips, however, K_F is usually small.

If the neighboring wires do not have matching terminating resistors then the noise pulses are reflected from each end. As a result, part of the near end noise shown in Figure 11-17 will still arrive at the receiver and matching terminations can be used to help reduce the effects of crosstalk noise.

Plots showing near end crosstalk versus length for an alumina cofired MCM with a 7.5 mil line separation, taken from Sons *et al*, [18], are shown in Figure 11-18. For the plot labeled $T_1 = 1$ ns it can be seen that the saturated line length is about 2.5" (6.35 cm).

Crosstalk requirements can determine the required spacing between the lines and thus the signal line pitch (line width + line spacing). Often the required spacing is at least twice the minimum spacing that the technology would allow. As the spacing increases, C_m and L_m decrease and crosstalk noise decreases. The required line pitch to control crosstalk also must increase with characteristic impedance Z_o. When Z_o is increased by making the lines narrower, the mutual capacitance and inductance increases if the spacing remains the same. To keep the crosstalk noise voltage constant the line spacing must be increased. As the increase in spacing is greater than the decrease in width, the line pitch increases. The characteristic impedance value chosen might actually be a compromise between the desire for first incidence switching, particularly when the line is loaded, and the desire to maximize interconnect density [1].

Choosing a material with a lower dielectric constant has two positive effects on crosstalk. First, by allowing the lines to be brought closer to the reference planes for the same impedance Z_o the interline spacing can be reduced with the coupling coefficients K_B and K_F remaining unchanged. Second, it reduces v_{prop} and increases the saturated length, effectively moving the plots in Figure 11-18 to the right.

Choosing a stripline over a microstrip configuration also allows a reduced spacing for the same impedance Z_o, as the presence of two reference planes in the former reduces the ratio C_m / C_0 (therefore, K_B and K_F). In either case, the greater the distance from the ground plane, the larger both the line width and spacing must be for the same Z_o and crosstalk.

The placement of a ground line between adjacent microstrip or stripline signal lines can be used to reduce crosstalk by almost 50%. This is only beneficial if the spacing already is large enough to consider adding a ground line. The ground line cannot be placed too close to the signal lines or it changes the characteristic impedance of the signal line [19].

Sometimes, crosstalk also may be reduced by rerouting parallel lines so that the length of the parallelism is minimized. In thin film MCMs this can be done with little penalty because of the MCM's large routing capacity.

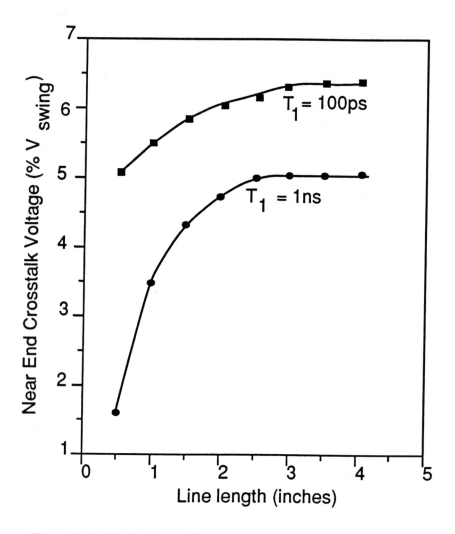

Figure 11-18 Simulated crosstalk versus length for an alumina cofired ceramic substrate [18].

Crosstalk is greatly affected by the capacitive loading represented by the crossing of lines in the adjacent signal layer. The reason why the lines are crossing in the first place is to prevent the high amounts of crosstalk that occurs when lines in different layers run parallel on top of each other for any significant length. This orthogonality is enforced by designating alternate layers as being mainly X running or Y running layers. The effect of multiple orthogonally crossing lines, however, is to increase C_0 and to decrease C_m. Deutsch *et al*, [20] report an example, where the capacitive coupling K_C reduces by about 40% when the crossing lines are pitched at minimum pitch and are continuously distributed along the signal line. As a result, the near end noise decreases and the far end noise increases. This effect is greater for two crossing microstrip layers than for two crossing offset stripline layers. The effect disappears when symmetric stripline layers are used, that is one signal line only centered between every pair of reference planes.

Another source of crosstalk is between chip connection bonding leads. Inter-lead crosstalk gets worse as the leads become longer or become more tightly spaced, and also as the rise time of the signal becomes shorter. For wire bonds and TAB leads of equal length and pitch, the crosstalk contribution on the bonding leads is nearly the same (with $C_m \approx 1$ pF and $L_m \approx 1$ nH). With high speed systems ($f_{clock} > 50 - 75$ MHz), wire bond and TAB leads must be kept short and/or consideration be given to incorporating a ground plane beneath them (multimetal TAB [20]). The crosstalk between flip chip bonds is very small due to the short lead lengths. Similar considerations apply in single chip packages. For example, most PGAs contain extensive ground planes, while most surface mount packages do not. As a result, lead crosstalk usually is smaller in a PGA package.

Crosstalk between MCM-to-board connector leads must also be considered for signal lines that leave the MCM. Controlling this noise at high speeds might require that every other pin be assigned to power or ground.

11.5 SIMULTANEOUS SWITCHING NOISE

When a number of off-chip loads are switched simultaneously in a digital system, a current change is produced in the power and ground supply network. For example, consider the CMOS circuits at the top of Figure 11-19. Whenever a $1 \rightarrow 0$ occurs at the outputs of a buffer (driver), the capacitive loads connected to that driver are discharged through the ground, producing a current spike through the ground. For example, consider a 5 V, 32-bit driver chip with a rise time of 2 ns driving a load of 320 pF (10 pF/gate). This corresponds to a di/dt $= C\Delta V / \Delta t = 0.8$ A/s. Switching noise is also a problem in TTL and ECL logic families though it tends to be worse in CMOS logic. For example, in an ECL logic state change, an 0.8 V swing into a load with an impedance of 50 Ω might occur in 700 ps. This corresponds to a rate of change of current di/dt of 16 mA/700 ps $= 0.02$ A/s. In either case, when this transient current passes through the inductive power distribution network, a noise voltage is produced,

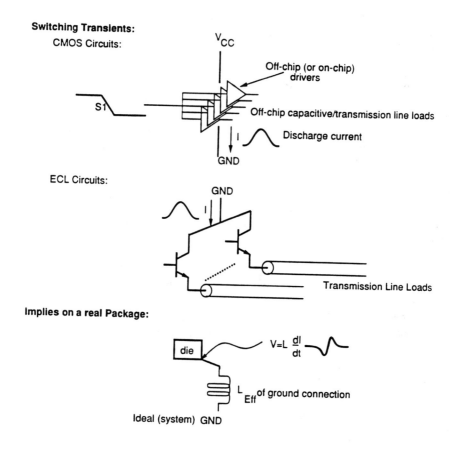

Figure 11-19 Simultaneous switching noise in CMOS and ECL circuits.

to as simultaneous switching or delta-I noise and shown in Figure 11-19.

Note that noise can occur in a similar fashion on power rails, such as V_{CC} and V_{EE}. However, this noise is less than the noise on the ground connections mainly due to the properties of the circuits. For this reason, simultaneous switching noise is sometimes referred to solely as "ground bounce." (The noise is likened to the ground voltage bouncing.) The transmission line return current transients (see Figure 11-9) flowing on the ground and power reference planes make a small contribution to the ground and power supply noise at each chip.

Switching noise can result in a number of problems if not handled correctly [21]. From top to bottom in Figure 11-20, the following effects occur:

Implies for non-switching drivers:

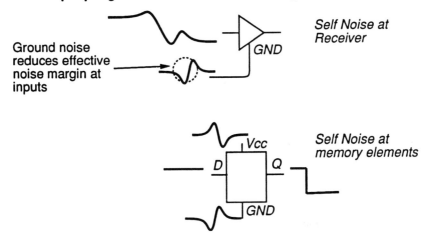

Transmitted Noise

Low input

GND

And for switching drivers:

Delay Increase + over/undershoot

And for chip input gates connected to the same ground connection:

Ground noise reduces effective noise margin at inputs

GND

Self Noise at Receiver

Vcc

D   Q

GND

Self Noise at memory elements

Figure 11-20 Effects of simultaneous switching noise.

1. Noise appears at the output of what were intended to be quiet off-chip drivers. This noise appears at the inputs of connected receivers.

2. The changes in internal chip supply voltage make the circuits operate more slowly and, thus, increase the delay in the switching drivers. The delay increase might be up to 2 - 3 ns or more for CMOS and TTL circuits, depending on the circuit details and the number of switching drivers. Overshoots and undershoots might also appear in these drivers.

3. For on-chip circuits acting as input gates, simultaneous switching noise acts to reduce the effective noise margin at the inputs.

4. For on-chip memory devices, such as latches, large amounts of ground-rail and power-rail noise might cause false changes in state (a logic-1 becoming a logic-0).

To a first order, the noise generated by the simultaneous switching of N output drivers can be estimated as

$$\Delta V = N L_{eff} \frac{di}{dt} \qquad (11\text{-}24)$$

where L_{eff} is the effective inductance of the power and ground connections and di/dt is the peak rate of change of current. This rate is often approximated as $\Delta i / \Delta t$ where Δi is the current demand of each driver during the switching event, and Δt is the rise (fall) time of the signal. It is only a first order equation because it contains the assumption that di/dt is independent of N and L_{eff}. In fact ΔV does not increase linearly with L_{eff} or N because of a feedback effect between ΔV and di/dt. Any increase in ΔV due to an increase in L_{eff} or N tends to result in a decrease in di/dt as the reduced voltage slows down the circuits. Thus use of Equation 11-24 often leads to an overestimate of expected noise. Improved expressions accounting for this effect in CMOS drivers are presented in references [23] and [24]. Even these expressions will tend to overestimate ΔV, however.

The effective inductance, L_{eff}, is primarily a function of the package design. In contrast, N is dependent on the logic design and di/dt on the circuit design. Reducing L_{eff} requires minimizing the inductances of the power and ground distribution networks and also making use of bypass capacitors. Bypass capacitors placed between the power and ground pins of each chip can act as a local source of charge during switching events so that not all of the switching current has to be supplied from the system ground, minimizing the local change in voltage. The equivalent circuit model for a CMOS output driving a capacitance is shown in Figure 11-21. In this model, $L_{ground\text{-}lead}$ is the inductance of the ground lead in the chip attach, L_{gnd}, is the inductance of the ground plane or wiring between the chip attach and the bypass capacitor, R_0 and L_0 are the resistance and inductance between the bypass capacitor and the power supply. Also shown is the parasitic inductance and capacitance associated with the bypass capacitor. Thus, the factors through which the package contributes to L_{eff} are as follows:

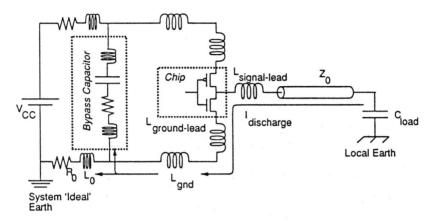

Figure 11-21 CMOS circuit model for simultaneous switching noise. With correct choice of bypass capacitor, most of the discharge current is supplied by the charge stored on this capacitor.

- **Choice of chip connection technology** - typical lead self inductances for different technologies are given in Table 11-2. Basically, inductance increases with the length of any round or rectangular lead. Of the MCM connection techniques, solder bump technology has the least inductance while fanout TAB and wire bonds have the most. The ground lead inductance, however, is less than the self inductance due to the effect of mutual inductances [25]:

$$L_{eff} = \frac{L_{self} \, I_{ground} - \Sigma \, L_{mutual-i} \, I_{signal-i}}{I_{ground}} \qquad (11\text{-}25)$$

where L_{self} is the self inductance of a ground lead, $L_{mutual-i}$ is the mutual inductance of the ground lead with signal lead i, $I_{signal-i}$ is the current flowing in signal lead i and I_{ground} is the current flowing in the ground lead. Note that the ratio of $I_{signal-i}$ to I_{ground} is the same as the signal to ground lead count ratio. The more ground leads provided for the same number of signal lines, the less the switching noise (actually this is mainly due to the decrease in N in Equation 11-24). The effective inductance is also substantially reduced if ground and power planes are used instead of leads, as in multimetal TAB and PGAs.

- **Choice of bypass capacitor** - at high frequencies it is important to minimize the parasitic inductance of the bypass capacitance. One technique to do this in a thin film MCM is to closely space the power and ground planes and use this as a low inductance bypass capacitor as done in the DEC VAX-9000 MCM (see Chapter 17).

- **Location of bypass capacitor** - on PWBs, bypass capacitors are located right next to the chips. On MCMs this might be undesirable due to their size (about 1 mm × 3 mm). One solution is to use the power and ground planes in a thin film MCM. Another is to build the capacitors into the MCM in a cofired or multilayer thick film ceramic MCM. Alternatively, the capacitors could be placed at the edge of the MCM, or off the MCM entirely. Both of the last two approaches lead to an increase in L_{gnd} (shown in Figure 11-21) and L_{eff}. If the capacitor is placed off the MCM, the MCM-PWB connector inductance must be included in L_{eff} and N must be evaluated MCM-wide rather than for each chip [26] and [22] - [23].

- **Choice of ground and power network components** - to minimize L_{gnd} and L_0, ground and power planes should be used, the connectors should be selected for minimum inductance and multiple ground and power pins should be distributed evenly through all connectors.

Techniques to minimize switching noise and its effects are discussed in a number of sources [11], [12], [22] and [27] - [28].

11.6 OTHER SOURCES OF NOISE

There are several minor sources of noise that must be considered. First, the resistive voltage drop caused by the DC supply currents in the power and ground planes must be kept to less than 1% of the power supply voltage. This is a potential problem in high power thin film MCMs. If so, solutions include using multiple power supply planes (such as the VAX-9000, see Chapter 17) or using a ceramic substrate base with thick film ground and power planes within it.

Second, if a parallel termination style is used, the resistance of the signal lines prevents the DC line voltage from settling at the nominal voltages for logic-0 or logic-1, thus consuming part of the noise margin. The driver output resistance, the line resistance and the parallel termination form a potential divider circuit that reduces the voltage across the latter and across the chip input.

Third, there might be small amounts of electromagnetic interference noise (EMI) produced in the circuit from outside sources of electromagnetic radiation, such as electric motors. This effect is small in MCMs because of their small size (the noise depends on the size of the antennas formed by the circuit interconnections) and can be neglected.

Fourth, some of the internal chip noise might appear at the outputs of the chip.

Finally, there is so called thermal noise caused by different operating temperatures for connected chips. The operating temperature of the chip has an effect on DC parameters, such as V_{IH} and V_{OH}, particularly for ECL circuits. Thus, the expected temperature difference between chips reduces the noise margin by a small amount.

11.7 THE ELECTRICAL DESIGN PROCESS

11.7.1 Technology Selection and System Planning

This process is summarized in Chapter 3. From an electrical design point of view, a packaging technology must be selected so that, with a suitable choice for partitioning and floorplan, delay and noise aims are met.

Noise control has the greatest number of requirements. As the circuits in the system get faster, rise times become faster, and more stringent requirements on the technology arise. In particular, controlled impedance interconnection is required, and chip connection and connector styles with lower inductances and less mutual coupling are preferred. For example, at signal frequencies above 50 MHz, long leaded wire bonds and TAB frames become undesirable. Either short lead connection methods must be used or reference planes are needed beneath the leads. (One way to provide this reference plane for wire bonds is to use multitier chip connection structures, examples of which are given in Chapter 6. The lower tier forms a ground plane over which the signal wire bonds run.) As the rise time increases, striplines are preferred over microstrips so that the mutual coupling between crossing lines in adjacent signal layers is reduced. Also the design of bypass capacitors becomes more critical. One advantage of MCM technology over single chip technology is that it makes it easier to meet these requirements.

The desire to produce fast first incidence switching creates requirements on line impedance, dielectric constant and line losses. The latter become particularly difficult to handle with thin film interconnect if aluminum lines are

Table 11-4 Noise Budget for the ECL VAX-8600 [29].

SOURCE	BUDGET (mV)
Load Reflections	100
Interconnect Impedance Mismatch	100
Crosstalk	100
Simultaneous Switching Noise	150
-2.0 V AC Noise	25
Signal IR Drop	25
V_{CC} IR Drop	14
Internal Chip Noise	50
Temperature	6
Sum	570
RSS	237

used or if the design bandwidth is so high that the skin effect becomes important. Guidelines for thin film technology selection accounting for this and other effects, are given by Gilbert and Walters [27].

The desire for a high line impedance must be balanced against the desire to minimize the line pitch, as determined by crosstalk requirements. If the line pitch is large, then extra layers might have to be added to create more interconnect capacity. An impedance of 50 Ω is commonly used but higher impedances are sometimes used for CMOS systems and heavily loaded lines. In any case, as the rise time gets faster, the line spacing has to increase. For longer lines, the minimum manufacturable spacing is unlikely to be acceptable. Tighter spacings can be used in stripline than in microstrip layers. This consideration is particularly important in laminate and ceramic technologies as one wishes to avoid the extra cost of adding layers. The wiring capacity provided by two layers of interconnect in a thin film MCM is sufficient for most systems today.

Noise Budgeting

The relative effort that goes into controlling each noise source, reflection noise, crosstalk noise and simultaneous switching noise, depends on the noise budgeted for each. Determining the noise budget is part of the system planning process. An example of a noise budget, used in the design of the VAX-8600 [29], an ECL system, is given in Table 11-4. This noise budget accounts separately for two different types of reflection noise, reflections from loads and reflections due to mismatches between different transmission lines. The crosstalk noise is given a budget of 100 mV. The simultaneous switching noise refers to the noise placed at the outputs of quiet drivers when grounded. Simultaneous switching noise on the -2.0 V power rail, resistive voltage drop noise, internal chip noise and thermal noise also are accounted for. Another good example of a noise budget is given in reference [30].

The total DC noise margin for ECL circuits is given in Table 11-1 as 150 mV. A typical ECL AC noise immunity curve is given in Figure 11-4. It can be seen that the sum of the different noise sources exceeds both the DC noise margin and the highest point on the AC noise immunity curve. This worst case, when pulses produced by the various noise sources arrive simultaneously at the receiver, is unlikely. If this assumption is used to determine the noise budget, the allowed noises would be very small and the system very difficult to design. Instead it is assumed that the noise sources arrive at the receiver at random times. This is far more reasonable and is sufficiently safe if a margin of safety is included. With this assumption, it can be shown that the total noise does exceed the Root Sum of Squares (RSS) of the different noise voltages for over 99.7% of the time. In this case, the RSS Noise voltage is

$$V_{RSS} = \left(V^2_{load-reflection} + V^2_{mismatch-reflection} + V^2_{crosstalk} \right.$$
$$\left. + V^2_{SSN} + V^2_{AC} + V^2_{IR-sig} + V^2_{IR-V_{CC}} + V^2_{thermal} \right)^{\frac{1}{2}}$$

(11-26)

In the case of the VAX-8600, the designer chose a V_{RSS} that was significantly less than the noise immunity for a suitably long pulse. (It appears to be about 150 mV less.)

The relative weight given to the different noise sources depends on how difficult it is to control each source. In the noise budget in Table 11-4, it is recognized that the simultaneous switching noise is the most difficult to reduce further and, therefore, is given the largest weighting. For example, if the simultaneous switching noise could be reduced, by improved technology, it might be possible to increase the crosstalk budget and thus allow narrower line spacings.

The noise budget for Table 11-4 is for a data connection. As discussed earlier, noise control on clock connections is more important. Noise budgets for clock signals tend to be far more conservative.

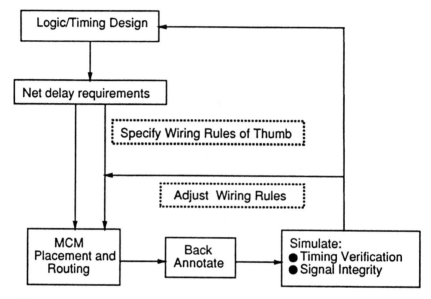

Figure 11-22 The steps in producing an MCM layout from an electrical design point of view.

11.7.2 Modeling, Simulation, and MCM Layout

The final objective of the electrical design of the MCM is to produce a layout which is a description of the artwork used to make the masks to be applied in MCM production.

The process to produce this layout, as supported by today's computer aided engineering (CAE) tools is described in Figure 11-22. From the timing design, delay requirements for each net are produced by subtracting the worst case delays of the active components between each pair of latches (see Figure 11-1) from the clock period. (Note: Any effects of simultaneous switching noise on this delay must be included. Data books rarely include this in their delay specifications. Typically for CMOS drivers, an extra 250 ps of extra delay is required for each simultaneously switching driver.) From these delays an estimate of the wiring rules is produced. The wiring rules specify which nets need to use controlled topologies and matching terminations, the limits on the lengths to each receiver and the stub length limit for those nets, as well as the spacing requirements between nets. Not all nets require that the conditions for first incident switching be satisfied. The rules for these nets do not have to be as stringent. See references [12] and [1] for more details. Care must be taken in using Equation 11-19 to estimate delay because the propagation delay is only a part of the total delay. This is particularly true in thin film MCMs where line losses can significantly increase $t_{\text{rise-time-degradation}}$ [31]-[32].

The wiring rules are passed to the placement and routing CAD tools which produce the layout. It is necessary to verify that delay requirements and noise budget requirements (signal integrity requirements) are met. This requires that electrical models be obtained and the simulations be undertaken and studied. The results of these simulations are compared with the requirements. If they predict that the layout will not produce a working design, then the wiring rules are adjusted and the process iterated. For example, if the simulated delay is too long, the maximum allowed length is reduced. Unfortunately, many iterations might be required. The need for iterations might be avoided by taking more care in producing the initial rules [1]. Computer aided techniques to automate and improve the determination of initial wiring rules are a current area of investigation.

Models must be obtained for the drivers, receivers, transmission lines, and line discontinuities including vias, chip connection leads and connectors. The driver and receiver models are either Thévenin equivalents or, for more accuracy, full circuit models (typically use of full circuit models improve delay prediction accuracy by at least 300 ps). Take care, however, as many of the circuit models provided by parts vendors are approximate models, not the real ones they use in design. They might need to be verified and qualified through measurements. There are a variety of modeling approaches for the packaging structures including using empirical equations (for example, Equation 11-11 for Z_o of a microstrip), various numerical solution techniques implemented in a range of CAE tools and measurement-based models.

A variety of simulation techniques have been implemented in a number of computer-based simulators. It is important to note that some interconnection parameters have frequency dependent behavior, particularly the skin effect resistance. If this frequency dependence is strong within the design bandwidth then specific modeling and simulation techniques must be used [33].

11.8 SUMMARY

Electrical design is concerned with delay and noise control, the control limits being determined by the timing design and the noise budget. To some degree, over-the-budget noise on data signal lines can be tolerated by allowing some time for the noise to settle. Excess noise on clock lines cannot be tolerated. The delay of the data paths depends on the output buffer characteristics, the characteristic impedance of the line, the length of the line, the dielectric constant of the dielectric, the line losses and the load capacitance of the line and the amount of noise that must settle out. Thus, delay is improved by ensuring that the paths are short, the dielectric constant is low, the line resistance is low, the chip leads have low capacitance and inductance and that noise is controlled (high levels of signal integrity). Simultaneous switching noise is controlled through careful system design including the use of bypass capacitors and other techniques. The control of reflection noise and crosstalk noise involves careful design of the interconnections and connections themselves. Many of the design issues for CMOS and TTL systems are covered further by Buchanan [12], for ECL systems by Blood [3] and for GaAs systems by Long and Butner [5].

The packaging technology choice has a very strong effect on delay and noise. By using MCMs, particularly thin film MCMs, delay and noise are substantially reduced. Most of this improvement comes from the removal of the single chip package whose leads introduce considerable capacitive and inductive parasitics and whose size often increases path lengths substantially. Even after the basic technology is chosen, however, there are still many decisions to make. For example, the interconnect material affects line resistance and delay. The chip connection technology affects both crosstalk noise and simultaneous switching noise. Fanout TAB, with its long leads, is the worst in both regards unless a ground plane is added. Short lead TAB and short wire bonds usually are acceptable except for the fastest systems. Flip chip connection provides the most superior electrical connection. Further technology factor guidelines for thin film MCMs, categorized by desired clock frequency of operation, have been formulated by Gilbert and Walters [27].

The use of MCM technology has one disadvantage, however. The MCM has to be carefully designed as prototyping is expensive, debugging difficult and redesign undesirable. The appropriate CAE tools must be used in high speed designs. Nevertheless, the resulting performance advantage is a significant factor in driving many system users to consider MCM technology.

Acknowledgments

The author sincerely thanks Hassam Hashemi and Peter Sandborn for their extensive suggestions and feedback on earlier portions of this draft. He would also like to thank Steve Lipa, Sharad Mehrotra, Slobodan Simovich, Michael Steer, Douglas Thomae, and the anonymous reviewers for their comments and suggestions.

References

1   E. E. Davidson, G. A. Katopis, "Package Electrical Design," R. R. Tummala, E. J. Rymaszewski, eds., *Microelectronics Packaing Handbook*, New York: Van Nostrand Reinhold, 1989, Chapter 3.
2   C. F. Hill, "Noise Margins and Noise Immunity in Logic Circuits," *Microelectronics*, vol. 1, pp. 16-21, April 1968.
3   W. R. Blood Jr., *MECL System Design Handbook*, Phoenix AZ: Motorola Semiconductor Products, Inc., 1983.
4   J. Lohstroh, "Static and Dynamic Noise Margins in Logic Circuits," *IEEE J. of Solid State Circuits*, vol. 14, no. 3, pp. 591-598, June 1979.
5   S. I. Long, S. E. Butner, *Gallium Arsenide Digital Integrated Circuit Design*, New York: McGraw Hill, 1990.
6   J. Shiao, D. Nguyen, "Performance Modeling of a Cache System with Three Interconnect Technologies: Cyanate Ester PCB, Chip-on-Board and CU/PI MCM," *Proc. IEEE MCM Conf.* (Santa Cruz CA), pp. 134-137, March 1992.
7   R. E. Matick, *Transmission Lines for Digital and Communcations Networks*, New York: McGraw Hill, 1969.
8   E. H. Fooks, R. A. Zakarevicius, *Microwave Engineering Using Microstrip Circuits*, New York: Prentice Hall, 1990.
9   T. Itoh, *Planar Transmission Line Structures*, New York: IEEE Press, 1987.
10   L. Smith, "High Density Copper/Polyimide Interconnect," *3rd Internat. SAMPE Electr. Conf.*, pp. 939-947, 1989.
11   H. B. Bakoglu, *Circuits, Interconnections and Packaging for VLSI*, New York: Addision Wesley, 1990.
12   J. E. Buchanan, *BiCMOS/CMOS Systems Design*, New York: McGraw Hill, 1990.
13   E. Burton, "Transmission Line Methods Aid Memory-Board Design," *Electr. Design*, vol. 28, pp. 87-91, Dec. 1988.
14   M. S. Lin, A. H. Engvik, J. S. Loos, "Measurements of Transient Response on Lossy Microstrips with Small Dimensions," *IEEE Trans. Circuits and Systems*, vol. 37, no. 11, pp. 1383-1393, Nov. 1990.
15   L. T. Hwang, *et al.*, "Thin Film Pulse Propagation Analysis Using Frequency Techniques," *IEEE Trans. CHMT*, vol. 14, no. 1, pp. 192-198, March 1991
16   M. W. Hartnett, P. D. Franzon, M. B. Steer, E. J. Vardaman, "Worldwide Status and Trends in Multichip Module Packaging," Austin TX: Techsearch International, 1990.
17   H. You, M Soma, "Crosstalk Analysis of Interconnection Lines and Packages in High Speed Integrated Circuits," *IEEE Trans. Circuits and Systems*, vol. 37, no. 8, pp. 1019-1026, Aug. 1990.
18   T. Sons, Y. Wen, A Agrawal, "Electrical Considerations for Multichip Module Design," *Proc. ISHM Conf.*, (Orlando FL), pp. 287-291, 1991.
19   C. S. Chang, "Electrical Design of Signal Lines for Multilayer Printed Circuit Boards," *IBM J. of Res. and Devel.*, vol. 32, no. 5, pp. 647-657, Sept. 1988.
20   A. Deutsch, *et al.*, "Electrical Charactristrics of Lossy Transmission Lines for High Performance Computer Applications," *Internat. Symp. on Advances in*

Interconnections and Packaging, SPIE (Boston MA), vol. 1389, pp. 161-186, Nov. 1990.

21 R. T. Smith, H. Hashemi, "Electrical Analysis of TAB Tape in Bonding High I/O Chips to High Density Boards," *Proc. Internat. Electr. Packaging Conf.*, (San Diego CA), Nov. 1988.

22 G. A. Katopis, "Delta-I Noise Specification for a High Performance Computing Machine," *Proc. IEEE*, vol. 73, no. 9, pp. 1405-1415, Sept. 1985.

23 P. A. Sandborn, H. Hashemi, B. Weigler, "Switching Noise in a Medium Film Copper/Polyimide Multichip Module," *Internat. Symp. on Advances in Interconnection and Packaging*, SPIE, vol. 1389, pp. 177-186, Nov. 1990.

24 R. Senthinathan, J. L. Prince, "Simultaneous Switching Ground Noise Calculation for Packaged CMOS Devices," *IEEE JSSC*, vol. 26, no. 11, pp. 1724-1728, 1989.

25 R. Kaw, R. Liu, R. Crawford, "Effective Inductance for Switching Noise in Single Chip Packages," *Proc. Internat. Electr. Packaging Conf.*, (Marlborough MA) pp. 756-761, Sept. 1990.

26 H. Hashemi, *et al.*, "Analytical and Simulation Study of Switching Noise in CMOS Circuits," *Proc Internat. Electr. Packaging Conf.*, (Marlborough MA) pp. 762-774, Sept. 1990.

27 B. K. Gilbert, W. L. Walters, "Design Options for Digital Multichip Modules Operating at High System Clock Rates," *Proc Internat. Conf. on Multichip Modules*, ISHM, (Denver CO), pp. 167-173, 1992.

28 A. J. Rainal, "Computing Inductive Noise of Chip Packages," *AT&T Bell Laboratories Technical J.*, vol. 63, no. 1, pp. 177-195, Jan. 1984.

29 H. Hackenburg, "Signal Integrity in the VAX-8600 System," *Digital Technical Review*, vol. 1, no. 1, pp. 43-65, Aug. 1985.

30 National Semiconductor, *FAST Applications Handbook*, South Portland, MN: National Semiconductor Corporation, 1987.

31 P. D. Franzon, *et al.*, "Tools to Aid in Wiring Rule Generation for High Speed Interconnects," *Proc Design Automation Conf.*, IEEE and ACM, (Anaheim CA), pp. 466-471, June 1992.

32 T. Mikazuki, N. Matsui, "Statistical Design Techniques for High-Speed Circuit Boards," *Proc. IEEE CHMT '90 IEMT Symp.* (Baltimore MD), pp. 185-191,

33 M. S. Basel, M. B. Steer, P. D. Franzon, "Simulation of High Speed Digital Interconnection with Nonlinear Terminations," submitted for publication, available from author.

12

THERMAL DESIGN CONSIDERATIONS FOR MULTICHIP MODULE APPLICATIONS

Kaveh Azar

12.1 INTRODUCTION

Operational integrity and longevity of electronic components are directly affected by their operating temperature. Thus the object in thermal design is to control the temperature sufficiently to meet the reliability requirements of the system. Achieving this goal requires the cooperation of many engineers on a design team. It also requires close attention of the thermal engineer. However, the relative ease or difficulty of achieving thermal aims depends a lot on design decisions made by the systems engineer, the logic and circuit engineers and the layout engineer. They also need to understand thermal design principles and processes.

The thermal design of multichip modules (MCMs) is both more critical and more complex than for single chip modules (SCMs). It is more critical because a single MCM generates a lot more heat in a space comparable to a large SCM. Removing this additional heat requires larger cooling capacity. It is more complex for a number of reasons. First, an MCM contains several components whose temperatures must be controlled. Second, it contains several heat sources, all of which might dissipate different powers. Third, an MCM contains multiple materials and materials interfaces and is asymmetric. In comparison, a SCM usually contains only two materials (a metal and plastic or ceramic) and has several degrees of symmetry, both of which simplify modeling.

This chapter emphasizes a systems approach to module cooling and also modeling and evaluation techniques used in the design process. It starts with a description of the objectives in thermal design and a brief overview of the thermal design process. Following that, the thermal phenomena that a reader must understand in order to conduct design are discussed. This leads into a description of thermal management alternatives for MCMs. Finally, the analytical, computational and experimental methods used to conduct thermal design are discussed.

12.2 THERMAL MANAGEMENT

12.2.1 Objectives in Thermal Management

Temperature is the key player in reliable operations of MCMs and other electronic components. Though it is difficult to define the failure properties of a composite structure with one or a few parameters, transistor junction temperature reduction is viewed as the primary goal toward reliability enhancement. Having such a single goal simplifies the entire design process. Thermal management then embraces efforts to reduce or maintain junction temperature, T_j, within the design specifications. Although there is no industry set standard, T_j design limits vary from 80°C - 180°C. The industry norm for most components appears to be 125°C.

The reason for using a single temperature metric becomes more evident if we look at the equation used for reliability calculation of electronic components. Consider the ratio A_T of the time to failure at a temperature T_2 relative to that at T_1 [1]:

$$A_T = \exp\left[\frac{E_a}{k_B} \left(\frac{1}{T_1} - \frac{1}{T_2} \right) \right] \qquad (12\text{-}1)$$

Here, E_a is a characteristic activation energy, and k_B is the Boltzmann constant. By reducing the operating temperature, T_2, the failure rate is reduced exponentially.

Also, as the temperature increases in the MCM the materials within it expand. Unfortunately, the rates of expansion of the different materials are often different and thermal stresses arise. If the thermally induced stresses exceed the elastic limit, the materials fail by coming apart. This might happen during manufacturing with a poorly designed process. If the composite structure experiences many temperature variations, such as power on, power off cycles, then fatigue field failure might eventually result.

12.2.2 Thermal Paths

Thermal design of MCMs, similarly to SCMs, can be divided into two areas: internal and external. For the sake of discussion, we refer to these as paths. The internal path is the one that directly deals with the structure of the component (module). Hence, it varies from component to component and differs significantly from the SCMs. In the external path, heat that has come to the component surface from the internal path is taken away either by gases or liquids. This part of the design does not have a strong dependency on the internal design of the MCM; it is more a function of the circuit board and system configuration.

Thermal design of MCMs, specifically with respect to their internal paths, typically takes two forms. These are experimental or computational simulations. The latter is the first chosen form for two reasons. First, it is much quicker to develop a model based on finite element or finite difference methods. Second, the computational models allow parametric simulations. This means that key parameters influencing design are varied to gauge their effects on junction temperature. The experimental simulation often follows the computational modeling. Some computational simulations of SCMs are described in references [2] - [4]; an example of the more involved modeling required for MCMs is given in [5].

12.3 THERMAL PHENOMENA IN ELECTRONIC ENCLOSURES

Thermal phenomena govern the removal of heat from components. A thermal process is defined as the merger of heat transfer and fluid flow to transport energy.

In this section, we begin by defining the basic principles of heat transfer. Then, heat transfer in electronic components is discussed. The concept of thermal resistance is presented with its uses. We also discuss why it should not be used to uniquely characterize SCM and MCM thermal properties. Since circuit boards contain MCMs and SCMs and play an important role in the thermal response of the MCM, thermal transports in circuit boards are discussed. The last two parts of this section discuss the thermal coupling (communication) between elements that form an electronic enclosure (system).

12.3.1 Heat Transfer Mechanisms

There are three modes of heat transfer: conduction, convection and radiation. Conduction heat transfer is when the heat is transferred by molecular vibration - solids or stagnant fluids. An example of a solid is the molding compound or the substrate in an MCM. An example of a stagnant fluid is the air trapped between the MCM and circuit board. The conduction heat transfer through a block of material is governed by the Fourier cooling law defined as:

$$Q = \frac{kA}{L} (T_h - T_c). \qquad (12\text{-}2)$$

where Q is the heat flow in units of power (watts) and k is the thermal conductivity of the material. The other terms are defined in Figure 12-1. T_h and T_c are the temperatures on opposite sides of the block. L is the length of the heat path and A is the cross sectional area. Table 12-1 shows the thermal conductivity of typical materials used in MCMs.

Convection heat transfer is when the transport of heat takes place by fluid motion. Three types of convection heat transfer are recognized: natural (free), forced and mixed. Natural convection occurs as a result of fluid (air) being in contact with a heated surface. The density of the fluid decreases causing it to rise, thus creating a natural circulation.

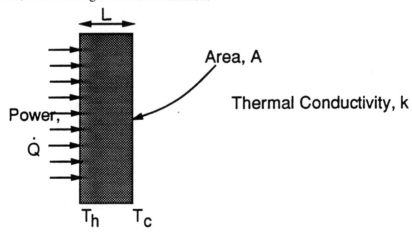

Figure 12-1 Conduction heat transfer in a solid.

Table 12-1 Typical Materials Used in MCMs.

MATERIAL	MAX TEMP (°C)	(kW/cm°C)	APPLICATION
METALS			
Aluminum	660	2.1	Chip conductor and wire bonds
Gold	1,063	3.4	Hybrid conductor and wire bonds
Copper	1,083	3.8	Lead frame and hybrid, PWB, conductor
Lead	327	0.3	Solder attach
Molybdenum	2,610	1.3	Cofired on ceramic conductor
Tungsten	3,380	1.5	Cofired on ceramic conductor
ORGANICS			
Epoxy (70% SiO$_2$)	170	0.002	Packaging
Epoxy glass (FR-4)	120	0.02	Multilayer board substrate
Adv. epoxy (resin only)	180	0.02	Multilayer board substrate
Triazine	250	0.002	Hybrid dielectric
BT resin (laminate)	290	0.005	Flexible substrate
Polyimide	400	0.0007	Flexible substrate
Polyimide	310	0.0007	Interlayer dielectric
INORGANICS			
Alumina (ceramic)	1,600	0.3	Hybrid substrates, chip carriers
Silica (fused)	1,100	0.02	Filler for molding epoxies
Silicon nitride	2,000	0.3	Candidate substrates
Aluminum nitride	1,800	3.2	Candidate substrates
Silicon carbide	2,100	2.7	Candidate substrates
Silicon	1,400	1.5	Candidate substrates
Diamond	>3,500	20	Candidate encapsulation
Glass-ceramic	>1,000	0.05	Candidate substrates
Beryllia	1,500	2.6	Chip carriers

Forced convection occurs when the fluid motion is induced by external sources. These include fans, pumps and blowers. Mixed convection occurs when natural and forced convection both are present. This is typically observed in low velocity flows at the presence of high powered components.

Convection heat transfer is governed by Newton's cooling law:

$$Q = hA(T_s - T_f),$$ (12-3)

where h is the heat transfer coefficient, A is the surface area exposed to the fluid, T_s is the temperature of that surface and T_f is the temperature of the fluid. The value of a specially constructed heatsink in thermal design is that it greatly increases the surface area.

Regardless of the type of convective heat transfer, Equation 12-3 is used for its solution. What sets the three types of convective heat transfer apart is h (heat transfer coefficient), which is obtained from empirical data. Figure 12-2 gives ranges of h encountered in the cooling of electronic components [6] - [10].

A word of caution seems merited at this point with respect to h and its correlations with fluid parameters. The data is typically reported in a form of correlation that relates the Nusselt number (Nu) to the Reynolds number (Re). When using a given correlation, its constraints must match the specific problem. Otherwise, that particular correlation is unsuitable for your analysis.

Radiative heat transfer occurs when heat is transported by photons or electromagnetic waves. What sets radiation heat transfer apart from conduction and convection is the medium for transport. Radiation heat transfer does not require a medium to transport energy. Radiation heat transfer is always present and its magnitude, similar to other modes of heat transfer, is a function of temperature difference. The belief that radiation heat transfer can be ignored is in error. Especially in natural convection problems, an excess of 20% of heat transfer is attributed to radiation.

Radiation heat transfer is governed by Equation 12-4,

$$Q = \sigma \varepsilon F_{hc} A (T_h^4 - T_c^4).$$ (12-4)

Here σ is the Stefan-Boltzmann constant, and ε is the emissivity of the radiating surface of area A. The temperatures of the (hot) radiating surface and of (cold) neighboring bodies are T_h and T_c, respectively, in degrees Kelvin. Values for F_{hc} (view or shape factor) are tabulated in heat transfer texts [11]-[12].

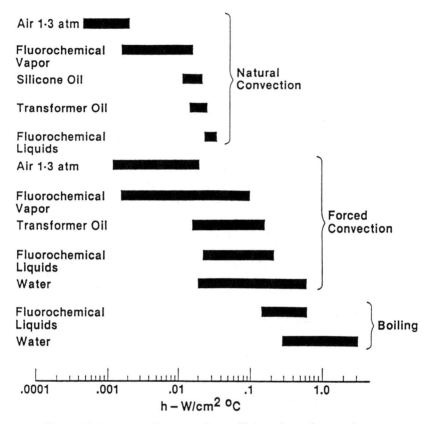

Figure 12-2 Range of heat transfer coefficients for various coolants.

12.3.2 Heat Transfer in Electronic Components (Modules)

This section discusses generation, spreading and eventual departure of heat from a component, in accordance with the three heat transfer mechanisms described above. Electronics components (modules) are made of an aggregate of materials with different physical geometries.

Consider a typical low performance plastic packaged MCM residing on circuit board in an air cooled system as shown in Figure 12-3. The electrical signals are brought to MCMs via the leads and then to the chips (typically) through the wire bonds. The leads are either press fitted, brazed or soldered to the substrate. The component is connected to the board through its leads either by surface mounting or through hole methods. This creates a package with

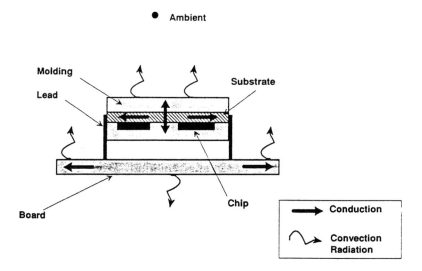

Figure 12-3 Heat transfer in a low performance multichip module.

potentially multiple heat sources in thermal communication with each other via the substrate and its ambient.

The heat generated at the chips seeks the most conductive path to reach the sinks. The sinks to which the heat is eventually transferred are the cooling fluid and the boards. The paths available for heat flow are through the substrate then the molding material and the leads. The flow of heat is impeded by each material, regardless of its thickness, as it travels from the sources to the sink. As the heat reaches the leads, part of it is conducted to the board, and the rest is either radiated or convected to the ambient.

The flow of heat spreads, seeking the greatest ratio of A to L by Equation 12-2 to maximize the conductive heat transfer rate. As a side effect of this heat spreading, each chip has a higher temperature due to the presence of its neighbors. The closer the chips are spaced, the greater will be this effect.

However, heat spreading also can be beneficial because of the increase in heat transfer rate. In plastic SCMs, this is done by inserting an aluminum plate, called a "heat spreader," into the package body. In some MCMs this is done with copper plates. As long as these copper plates conduct much more heat to the heatsinks than to each other, the net result is beneficial.

A similar process occurs as the heat reaches the physical boundaries of the component. Figure 12-3 gives a schematic depiction of different modes of heat transfer and heat flow process in an MCM. Combination of multiple heat sources and different possible avenues for heat flow have created a rather complex and nonuniform temperature field.

12.3.3 The Concept of Thermal Resistance

The concept of thermal resistance is associated with impeding the flow of heat through a medium. Similar to electrical resistance, if resistance is decreased, less voltage is required to pass the current through the wire. Current is similar to heat flow, and voltage to temperature. Hence, if thermal resistance is decreased, smaller temperature differences across a medium result. Thus, it becomes intuitive that if thermal resistance is minimized, internally and externally, the junction temperature is reduced.

The package thermal response can be viewed by two resistances, external and internal. The internal resistance, Θ_{jc} (junction to case resistance), addresses heat transfer within the package, from the chip to the surface of the package. The external resistance Θ_{ca} (case to ambient resistance) is a measure of thermal transport occurring between the package surface and the ambient.

Θ_{jc} and Θ_{ca} are defined by the following equations,

$$\Theta_{jc} = \frac{T_j - T_c}{P} \qquad (12\text{-}5)$$

and

$$\Theta_{ca} = \frac{T_c - T_a}{P}. \qquad (12\text{-}6)$$

In Equations 12-5 and 12-6, subscripts a, c and j refer to ambient, case and junction, respectively. The denominator, P, is the total power dissipation in the component. The units of these resistances are °C/W. Figure 12-4 shows a schematic representation of these resistances.

There are three problems, however, with this use of the concept of thermal resistance. First, it is difficult to calculate a single number for thermal resistance with the presence of multiple heat paths. Second, even if you did calculate it, you could not reuse the resistance in other designs. Consider what would happen to the heat flow pictured in Figure 12-4 if another MCM were placed on the bottom of the board. Less heat would flow in this direction due to the smaller ΔT. The total heat flow would change and the thermal resistance would increase. These problems are common to MCMs and SCMs [13]. In fact, a common error in SCM design is to assume that the data sheet value for thermal resistance

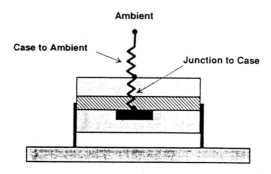

Figure 12-4   Thermal resistance representation in an electronic component.

applies to a crowded two-sided board. It often has to be increased by 50% or more to compensate for the ideal conditions under which the manufacturer measured it. The third problem is unique with MCMs. With multiple chips, each with a different power, P, each chip will have a different temperature, T_j, and thus the choice of values to be used in Equations 12-5 and 12-6 are arbitrary and somewhat meaningless.

Thermal resistance is a widely used concept. The preceding discussion has shown the weaknesses associated with using Equations 12-5 and 12-6 even for SCMs. These equations contain even higher error levels when applied to low performance MCMs. In the case of high performance MCMs, thermal resistance in this form is completely useless. Although thermal resistances for these MCMs still are reported, it typically is for the chip and not the entire module.

Nevertheless, thermal resistance is a very valuable concept for qualitatively understanding the thermal effects of different materials and cooling approaches. Equations 12-2 and 12-3 show conduction and convection heat transfer. Conductive and convective thermal resistances are defined as

$$R_k = \frac{L}{kA} \qquad (12\text{-}7)$$

$$R_h = \frac{1}{hA} \qquad (12\text{-}8)$$

which refer to internal and external resistances. By definition, Equations 12-7 and 12-8 are identical to equations 12-5 and 12-6. We see that R_k is inversely proportional to k and A (area normal to the direction of heat flow). R_h is inversely proportional to h and A (convective surface area). To reduce these resistances, the denominator would have to increase.

Consider an MCM where the chips are epoxied to the substrate. The thermal resistance between the chip and the substrate (where the heat is conducted away) is the one imposed by the epoxy. The area, A, associated with the epoxy is constrained to the size of the chip. Often times in the application of epoxy, air gaps are created adding to overall chip to substrate resistance. The resistance is reduced by removing the air gaps or using an epoxy with a higher thermal conductivity or another bonding technique. The Hitachi Silicon Carbide (SiC) RAM [14] is an example of such practice where 52 solder bumps are used for heat transfer purposes only (Figure 12-5). Of course, as we change process or epoxy, we have to be concerned with material compatibility to avoid uneven expansion. Otherwise, stresses induced as the result of uneven expansion may

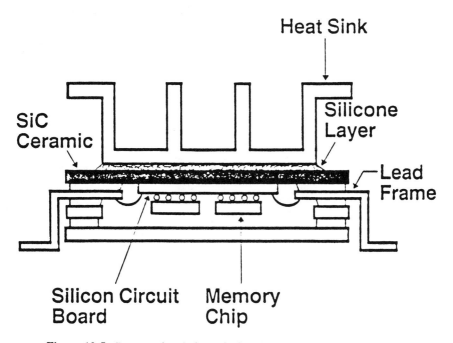

Figure 12-5 Cross sectional view of Hitachi air-cooled multichip module.

result in component failure. For this reason, thicker epoxies typically are used in laminate MCMs.

Similarly, if we look at R_h, it can also be reduced by increasing h or A. The heat transfer coefficient, h, is increased by going to higher velocity flows (fans or jet impingement) or changing the fluid (gas to liquid). The surface area is increased by adding heatsinks to the component.

Based on the above discussion, thermal resistance plays a pivotal role in the magnitude of junction temperature. Reducing thermal resistance either internal or external to the MCM package impacts thermal control positively. In the design of MCMs, it is important to locate the heat sources and the thermal resistances on their paths. The resistances should be reduced so that minimum spreading takes place, and the path from the chip to the sink has the least thermal resistance.

12.3.4 Heat Transfer On a Board

For thermal design purposes, each component cannot be considered in isolation. The heat being produced by one component is transferred amongst the others. This is referred to as thermal coupling and is discussed here at the board level, and in the next section at the system level. As discussed above, thermal coupling can be very strong within an MCM.

Consider the forced air cooled board shown in Figure 12-6. There are two mechanisms leading to thermal coupling. First, the MCMs and SCMs heat the air as it passes over them. The downstream parts will experience a hotter fluid

Figure 12-6 Flow over a circuit board containing SCMs and MCMs.

temperature, T_f and, according to Equation 12-3, the component is hotter, by the same amount, to dissipate the generated heat. Second, some of the heat produced by each component is passed to the board through leads, conduction and radiation across the air gap. Though part of this heat is convected away from the board, part of it is conducted to the other components on the board, raising their temperature. While the board dielectric is a poor conductor of heat, the copper layers within it are excellent heat conductors. An eight layer board exhibits a stronger conductive thermal coupling between components than does a four layer board.

The net effect of this is that the chip with the highest T_j might not be the chip that produces the most heat. The critical chip is the one that has a T_j closest to or over the specified limit. For example in Figure 12-6, the high power CPU core MCM is placed near the fan while the low power main memories are placed at the other end. Often this arrangement requires that special memories be purchased with an aluminum heat spreader within them.

12.3.5 Thermal Coupling in Electronic Enclosures

To appreciate the impact of the system (enclosure) on the thermal performance of MCMs, it is necessary to review the thermal phenomenon in an enclosure. (see Figure 12-7 for an example of an enclosure.)

The shelf or card holder (cage) is where a circuit board resides in the system. Boards are normally inserted into the shelves through card guides. Except in some specialized cases where a latching mechanism is used to rigidly attach the board to the shelf, the boards are loosely fitted inside the shelf. Therefore, the necessary contact to facilitate conduction heat transfer from the board to the shelf does not usually exist.

The backplane, or motherboard, in a PC is another avenue for the heat to be transported to the ambient or the shelf. If the thermal conductivity of the board is very large, that is multilayered boards with several layers of copper, conduction heat transfer through the backplane can be significant. However, the thermal coupling by convection and radiation heat transfers is significantly larger than conduction heat transfer.

Frames or enclosures that house single or multiple shelves generally are designed to be isolated from the shelves. Thus, the heat generated within the system normally is convected through the vent holes. Although this constitutes the bulk of heat flow, there exists significant thermal coupling between the boards (and shelves) and the frame. The thermal coupling, in the order of significance, is by radiation, convection and conduction heat transfer. Since the frame is in contact with the system ambient, it acts as a sink and source of heat for the system.

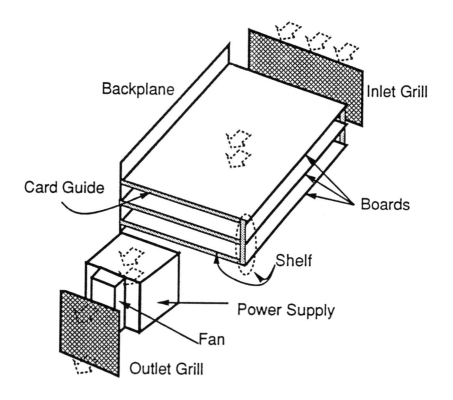

Figure 12-7 Schematic view of electronic system configuration.

The magnitude of conduction heat transfer is very system design dependent. The radiation heat transfer, however, is generally the predominant mode of thermal coupling between the shelves and the frame. The radiation heat transfer tends to be even more significant if the system is cooled by natural convection.

The frame is coupled to the surrounding ambient via radiation and convection heat transfers. The system ambient also can act as a source and a sink. The magnitude of these heat transfers varies significantly with the changes in the system surroundings.

The thermal transport process in electronic systems is quite involved and can become very complex. Because of many different thermal processes and strong coupling at various system levels, thermal bookkeeping is necessary for accurate design. In addition, it should be clear that we cannot only focus on a component (module) without considering the system, environment and other parameters affecting thermal design.

12.4 THERMAL MANAGEMENT OF MCMs

Power dissipation levels exceeding 4 W/cm^2 and specific system performance requirements have forced design of highly customized cooling systems for some MCMs. MCMs typically have higher power dissipations than SCMs and generally are placed on circuit boards or substrates that contain other potentially high power components. The propensity for thermal spreading within the circuit board or the substrate has sometimes led the designers to build individual cooling elements for every chip on the substrate. This combination at times has created a challenging problem for thermal management of MCMs. The challenge has embraced packaging of the MCM itself and integration of its cooling system into the overall frame.

In low performance systems, thermal management of the MCM is typically an after thought and is constrained by the application. In high performance systems, the cooling system design is an integral part of the design cycle. Because of the customized nature of cooling systems, their spatial restrictions in terms of system compactness or electrical distance between high speed components has been an added challenge to thermal management.

The constraints imposed by temperature limits and system compactness have produced innovative packaging and thermal control methods. In this section, some of these techniques, with emphasis on cooling method, are presented. The objectives of thermal management, with respect to design and manufacturing constraints, are discussed. Then, the cooling approaches and order of application are reviewed.

12.4.1 Alternate Thermal Control Methods for MCMs

There are alternate thermal control methods for both the internal and external paths. The choice of the primary internal path is closely related to the choice of chip attach. The primary internal path alternatives are (Figure 12-8) through the substrate, through the substrate with thermal vias or thermal cutouts, and chip backside.

With through the substrate cooling, the main heat dissipation surface in contact with the fluid is beneath the chip. The primary internal heat path is through the substrate. There might be considerable heat spreading and the thermal resistance might be high, particularly as the thermal conductivity of the most common substrate materials is low. It can be reduced by using different materials, such as aluminum nitride instead of alumina. However, highly thermally conductive alternatives to polyimide and laminate MCM materials are not available.

Through-the-substrate: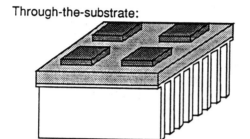

Chip Backside: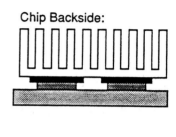

Figure 12-8 Through-the-substrate and chip backside technology for MCMs.

If the material properties cannot be improved, then the heat spreading and the thermal resistance can be reduced, at the expense of wiring capacity, by placing copper vias in the substrate or by sinking the chip into the substrate. Either of these techniques can be used with any chip connection technique. If wire bond or TAB leads are used, the chip can be attached with a thermal epoxy. Part of the heat is transferred through the thermal epoxy and part through the metal leads. With solder bump attachment, the heat is transferred through the bumps and the air gap. Often extra bumps are added to reduce thermal resistance.

With chip backside cooling, the main heat dissipation surface is above the chip and is attached with a metal mount, a high thermal conductivity solder, a thermal epoxy or a thermal grease. Only flip techniques (flip TAB or solder bumps) can be used because of potential damage to the surface of the chips. Chip backside cooling generally has the smallest internal thermal resistance.

In either thermal path, careful attention must be given to the interfaces. For example, if the epoxy chip interface has many air bubbles in it, the thermal resistance increases substantially.

In general, the external thermal control methods can be categorized as follows:

- Natural convection
- Forced convection
- Conduction or radiation cooling
- Liquid immersion
- Phase change (boiling)

Table 12-2 Thermal Control Methods.

Primary Cooling Mechanism	Typical HTC (W/m²K)	Relative Effectiveness	Achievable Density	Complexity
Natural convection (air)	10	0.1	Low	Very low
2orced convection (air)	100	1.0	Medium	Low
Natural convection (liquid)	100	1.0	Medium	Medium
Forced convection (liquid)	1000	10.0	High	High
Phase change (liquid)	5000	50.0	High	High

Natural convection cooling is the case when no fluid movers are used in circulating the fluid in the system. Cooling by forced convection utilizes a fluid mover to circulate the fluid. Conduction or radiation cooling is when a cold plate [15] or radiation plate is used to remove heat. The application of the latter is seen in military and space electronics. Liquid immersion is when a component or system is immersed in a liquid. The liquid can be fluoroinerts or others such as liquid nitrogen. With boiling, the fluid boils at the MCM contact surface. General features of the convection-based cooling modes are given in Table 12-2 [16]. In that Table, HTC is the heat transfer coefficient h (see Equation 12-3).

A designer often is confronted with the decision of selecting a cooling method, keeping in mind manufacturing issues and end use application constraints. Selecting a thermal control method is a function of component temperature rating and the heat removal capacity of a specific design. The coolant fluid, gas or liquid, typically sets the capability of these cooling methods apart. This is made more clear by looking at Figure 12-9 [17]. The figure merges the thermal control methods with power dissipation (heat flux) and temperature rise. It shows a representation of the expected temperature rise over ambient for different cooling methods. It also hints to a potential thermal control method as a function of component power density.

Figure 12-9 does not suggest or provide an absolute case for a thermal control method. It suggests the expected range or type necessary for heat removal to ensure the junction temperature meets its constraint. For example, for an MCM with 10 W/cm² and a temperature rating, T_j, of 85°C, Figure 12-9 suggests some sort of liquid cooling. However, the same component can be

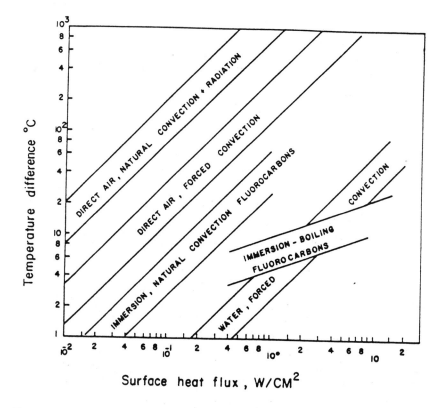

Figure 12-9 Temperature differences attainable as a function of heat flux for various heat transfer modes and various coolant fluids.

effectively cooled with a high level air system (jet impingement, where the air is blown directly onto each heatsink [18]) and may not require exotic cooling. The high end air-based thermal control methods tend to be not as expensive as liquid based ones, yet they yield junction temperatures within the rated limits. This has sparked much interest in the community to explore and expand air cooling limits.

To provide an overall view of cooling techniques practiced in the industry, it is worthwhile to review some of the designs and highlight their salient thermal management features. References [19] and [20] provide an excellent overview of this subject and excerpts from these and other references are used in the forthcoming discussion. Figure 12-5 shows Hitachi's SiC RAM representing low performing systems. The module has six 1 W chips that provide 1 kbit of memory. The cooling system is an 8 mm high 4 fin simple heatsink with air as the coolant fluid. Using a finned heatsink increases the area in contact with the coolant and thus reduces T_j. The chips contain 77 solder bumps; 52 of them are there purely for thermal reasons. The bumps act as thermal paths, carrying the dissipated heat from the chips to the substrate. The substrate's thermal conductivity is approximately 14 times that of alumina, thus it is very effective for spreading the heat [20]-[21].

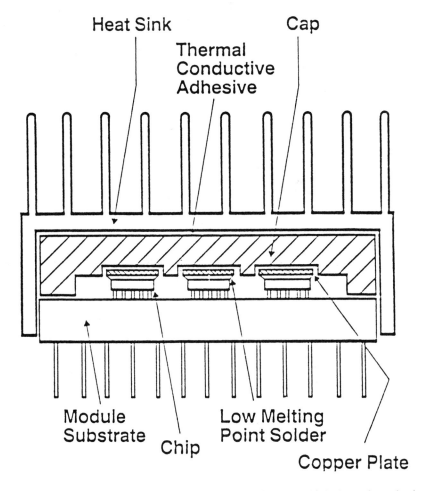

Figure 12-10 Cross sectional view of Mitsubishi air-cooled high thermal conduction module.

Figure 12-10 shows another example of air cooled modules designed by Mitsubishi known as the High Thermal Conduction Modules. The component's total power dissipation is 36 W generated uniformly in nine 3K gate ECL chips. The internal thermal paths consist of both the top and bottom of the chips. The chips are placed on bumps to accommodate heat transfer from the bottom side to the substrate. The combination of a copper plate heat spreader placed on the top of the chips (Figure 12-10), and a cap heatsink assembly provide thermal paths from the top side of the chips. The heat generated by the chips is dissipated through the heatsink and convected away by air with a velocity of 6 m/s.

References [19] and [22] discuss the details of heat transfer analysis and subsequent thermal resistances for this module. A note worthy point that relates to our earlier discussion is the variations in heat flow paths and thermal resistance. Reference [22] reports that in the absence of heat transfer from the pins to the air, the central chip conducts 13% of its heat through the solder bumps versus 18% for the peripheral chips. The central chip has 3°C/W and peripheral chips have 2.5°C/W chip to heatsink thermal resistance, resulting in 0.5°C/W difference in thermal resistance for these chips. The combination of internal multiple heat paths (stemming from heat spreading) and the nature of the air flow through the heatsink accounts for this difference. The analysis presented in the noted references has ignored the heat path to the PCB via the module's pins. Inclusion of this heat path may have led to an even higher difference between these chips.

Although the chip power dissipation is uniform, the boundaries of the peripheral chips are different from the central one. The central chip is surrounded, on all sides, by 4 W chips. The peripheral ones have at least one side facing the periphery of the module. Therefore, as a result of thermal coupling through the substrate and the surrounding gas, the central chip is expected to operate at a higher temperature. As a side note, this example typifies the difficulty of reporting a single thermal resistance value for MCMs.

The IBM 4381 module, Figure 12-11 [19], is an example of a high performance air cooled MCM that uses higher level air cooling (impingement). The module is dimensioned 64 mm × 64 mm × 40 mm and contains 36 chip sites. The chips are solder bumped on a multilayered ceramic (MLC) substrate and separated from the ceramic cap by a layer of thermal paste designed by IBM. The power dissipation per module varies from 36 W - 90 W resulting in a typical circuit board power dissipation of 1.3 kW. Reliable system operation requires that the chip temperature not exceed 90°C.

The chips are closely spaced to minimize delay. The power dissipation per chip is high (3.8 W maximum). For this power dissipation level and temperature rating constraint, each chip requires individual thermal management and control.

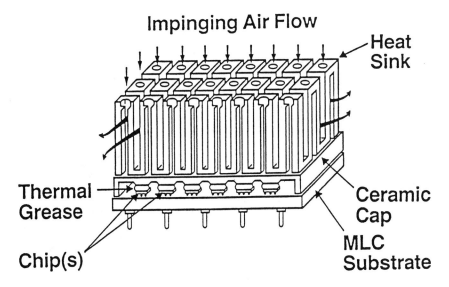

Figure 12-11 Impingement air-cooled MCM used in IBM 4381 processor.

The heatsink, air impingement combination provides an adequate level of thermal control at the chip level to ensure chip temperature below 90°C.

Perhaps the most talked about example of liquid cooling is the IBM Thermal Conduction Module (TCM), used in the 3081 processor, Figure 12-12 [19]. The design requirement specified an 85°C temperature limit for the chip and achieved 69°C. The system performance required that up to nine TCMs be mounted on a single PCB. Since each TCM can dissipate up to 300 W, the total power on the PCB is 2700 W. Stringent chip temperature limit and high system performance was the driving force behind development of this MCM. The thermal design objective was reached by removing heat from the chips as directly as possible and minimizing thermal spreading. The heat dissipated by each chip is conducted via the spring loaded piston in a helium atmosphere to the water-cooled heat exchanger, Figure 12-12.

Fujitsu's FACOM M-780 is a water cooled MCM that departs markedly from the IBM and other similar water cooled modules (Figure 12-13) [19] and [23]. In this design, the thermal control unit consists of bellows and water jets packaged in a closed system. The tip of the bellows is in contact with chip surface through a compliant material to ensure adequate thermal contact.

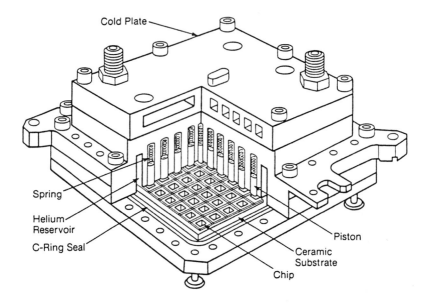

Figure 12-12 IBM thermal conduction module (TCM) with water cooled cold plate.

The FACOM M-780 has 336 single chip modules mounted on both sides of 540 mm × 488 mm PCB. The maximum chip power is 9.5 W and the board dissipates 3,000 W [19]. The cold plate is introduced to the section of the PCB containing the single chip modules. The cold plate is factory assembled and cannot be separated for field repairs.

The next level of cooling that a few computer companies have gravitated toward is liquid immersion. The same criteria drive the selection of the thermal control method: system performance and temperature limit. One of the advantages of immersion cooling is eliminating interface resistances seen in the cold plate thermal control methods. By immersing the MCM or the entire circuit board in a Fluorocarbon (FC-72, FC-77), immersion cooling is attained. The most noted forced immersion cooled system is the CRAY-2 supercomputer.

The immersion cooled portion of the CRAY-2 consists of SCMs mounted on eight PCBs, dissipating a total of 600 - 700 W for a heat density of 0.21 W/cm^2. Though this is within air cooling limits, the large air flow rate required would have been impractical to design. Hence the CRAY-2 uses FC-77 fluorocarbon cooling forced horizontally over the surface of the PCBs with 2.5 cm/sec. velocity.

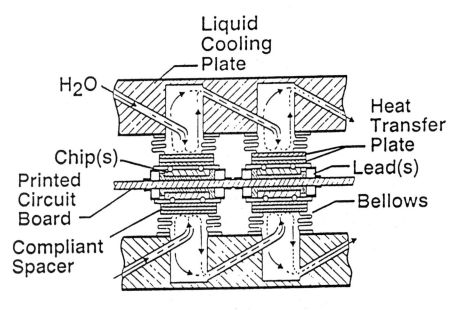

Figure 12-13 Cross sectional view of Fujitsu water-cooled bellows cold plate cooling system.

Before closing this section, it is important to revisit a few points regarding the thermal control techniques used for MCMs. We looked at many examples from the simple application of a flat finned heatsink to immersion cooling of MCMs. Two issues should be evident by now. First, thermal control techniques, beyond the use of a heatsink, are system dependent, or customized. This dependency stems from system level packaging and performance requirements. For example, CRAY-2 designers determined that with 0.21 W/cm^2 heat flux, air cooling was possible. Volumetric air flow requirements, however, were not practical and a liquid cooling system was designed. Therefore, customization of thermal control methods suggests that it is not safe to define general design rules.

The second point is thermal resistance minimization. In all the examples reviewed here, and in many more, the designers have gone through much effort to ensure that the internal and external resistances are as small as possible. An example to be cited is the IBM TCM, in which helium is used in place of air to improve conduction heat transfer through the gas. By replacing air with helium inside the TCM, the internal thermal resistance was reduced from 25°C/W to 8.08°C/W [24]. The importance of this point cannot be emphasized enough. Use of different chip mounting technologies and materials (discussed in other chapters) for internal chip design can have a major impact in thermal performance of the MCM. Hence, examination of alternatives, with regard to thermal tradeoffs, should be a routine exercise for an MCM designer.

12.4.2 Cooling Methods - Cost Impact of Thermal Management Techniques

The electronics industry can be segmented into four categories:

- Computer
- Military and Space
- Telecommunications
- Consumer products

MCMs have been and will continue to be used in products produced by these industries. The thermal control technique is a function of system application and, therefore, varies between each industry. The high end computer industry has led the way in cooling system design and can afford to use exotic, although not desirable, cooling methods. The military has close tolerance requirements, resulting in unique and system specific cooling systems. Telecommunications tends to gravitate toward lower level cooling methods. Consumer electronics seeks passive cooling techniques because of their application. With the increase in processing speed, all these industries seek higher order cooling methods. This trend will continue until significant changes in the packaging of electronics components are introduced.

The power density, system packaging and junction temperature specifications set the foundation for selection of a cooling method. Additionally, this selection is constrained by manufacturability and cost. The hierarchy of the cooling methods are:

- Air in natural convection, with or without heatsink
- Air in forced convection, with or without heatsink and other flow enhancement methods
- Air jet impingement
- Radiation
- Liquid cooling:
  Natural convection
  Forced convection
  Jet impingement
  Immersion
- Immersion by boiling
- Cryogenics

The above list is ordered in the ease of implementation, but in a reverse order of cooling capacity (see Figure 12-9). Natural convection, with air as the working fluid, is most desirable since no cooling system design is required. This does not imply that thermal analysis is not needed. It suggests that no cooling system (fluid movers, heat exchanger etc.) is required for thermal management of the system. Forced air convection is the next most desirable mode. Use of heatsinks with air cooling can further increase the heat removal capacity of forced air convection. Some industry segments (telecommunications and consumer products) tend to shy away from fans (fluid movers) because of reliability issues and noise. Fluid movers are unavoidable if power dissipation and temperature specifications are at such a level that fans are required.

From jet impingement to higher cooling methods, heat removal capability goes up significantly. But implementation in a system from cost, manufacturing and user impression becomes complex. For example, jet impingement requires a compressor and placement of jet nozzles throughout the circuit board, creating a reliability and physical design dilemma. Additionally, there is a whistling noise as the air expands and leaves the nozzle. Imagine your PC or workstation whistling continuously - fan noise is uncomfortable enough.

Liquid cooling is a very attractive proposition for high powered components and systems. But implementation, cost, maintenance and use is difficult and only seen in a small number of systems. This difficulty has invigorated the search for air cooling methods for high powered systems. Reference [25] gives an excellent review of liquid cooling in microelectronics components. Reference [26] provides a general review of thermal management of electronic equipment.

12.4.3 Parameters Impacting MCM Thermal Performance

We have discussed thermal phenomena at length in MCM and SCMs. The intent of this section is not to revisit them, but to give a list of parameters impacting junction temperature.

External
External parameters are outside the package (module). These include the following:

- Environment where the system resides

- Method of thermal control

- Coolant fluid temperature, velocity and flow regime - laminar, transition or turbulent

- Neighboring circuit board power, locational proximity and surface emissivity

- Circuit board or substrate thermal conductivity

- Neighboring component's power dissipation, size and locational proximity to the MCM

- The size of the gap between the component and the circuit board. In the case of military applications, this gap is typically filled with a conductive material

- Method of lead attachment to the circuit board

Internal

The internal parameters are specific to inside the package (module). These include the following:

- Dimensions - exact dimensions of component including leads, wire bond, chips, via type and density, etc.

- Method of chip attachment to the substrate (bonded, epoxy) with material dimension and thermal properties

- Method of lead attachment to the substrate

- Molding compound material property and dimension

- Power dissipation of each chip on the substrate

- Metallization - material and dimensions

- Material and dimensions of the substrate

- Property and the dimension of interface material used between the chip and the cooling tower

The response to the above list provides the information necessary for thermal analysis or experimental simulation of MCMs. In the case of analytical simulation, the impact of various parameters can easily be highlighted. For example, one can calculate the junction temperature as a function of chip attachment method and molding compound thermal conductivity. Various what-if cases allow for selection of more appropriate material to help reduce junction temperature. Similarly, if a system level analytical model is developed, the what-if cases provide the designer with insight into alternative options in system design and cooling method implementation.

12.5 TOOLS FOR THERMAL DESIGN

12.5.1 Overview of Design Analysis Tools

The tools used for thermal analysis do not discriminate between SCMs and MCMs. The tools are of generic nature and apply to any thermal problem. In this section, the use and the domain of applicability of these tools are presented. The discussion is followed by two examples to demonstrate the application and utility of these tools.

The principal tools in thermal design are categorized (Table 12-3) into three areas:

- Integral (Analytical)
- Numerical (Computational)
- Experimental

The integral method is the first and most essential method for forming the solution. Because of its analytical nature, it typically is referred to as a first order solution. The numerical solution is a second order solution with limited domain of application, mainly component and circuit board. The experimental method is the highest order solution utilized when the other two methods are not suitable for the specific problem. The important point is that the solution techniques are interdependent. The integral method often forms the foundation or the starting point of the numerical solution procedure.

The requirements of the two higher order solutions show an interdependency with the integral method. The numerical method needs boundary or initial conditions to initiate a solution. The conditions are obtained either by direct measurement or by application of an integral method. When experimentation is deemed necessary, the physical trend and pertinent parameters should be identified before measurement is done. These and the premises for experimentation are obtained by forming the solution by the integral method. Since analytical methods play such a pivotal role in the analysis and design of electronic components and systems, it is worth the time to further develop this base.

Table 12-3 Comparison and Application of Solution Techniques in Electronics Cooling.

	INTEGRAL	NUMERICAL	EXPERIMENTAL
Order of Solution	First	Second	Third
Accuracy	Moderate	High	Best
Effort	Small	Small-High	Difficult
Domain of Application	All problem domains	Component and circuit pack	All program domains
Cost	Small	Medium-High	Very high
Expertise	Introductory	Introductory-Specialized	Specialized
Tools	PC	High speed computers and software	Laboratory facility

12.5.2 Analysis Tools

Thermal problems are a synthesis of heat transfer, fluid mechanics and thermodynamics. The analytical tools are based on the conservation laws and equation of state. These laws are:

- Conservation of mass
- Conservation of momentum
- Conservation of energy
- Equation of state

Thermal problems can be formulated by the general application of these laws to a finite or infinitesimal region of the problem domain, called a control volume (CV). To accurately formulate the problem, it is necessary to understand the utility and the domain of the application of each law. Therefore, we start by defining each equation and briefly discuss its utility when applied to a thermal design problem.

Conservation of mass simply states that the mass in a given thermal process is conserved. The equation is formed from the cross sectional area, velocity and density in two regions of the problem. The equation can be expressed as follows:

Rate of mass flow into CV - Rate of mass flow out of CV =
Rate of accumulation of mass inside the CV,   (12-9)

where CV denotes control volume. The conservation of mass is a very useful tool, and its utility or application is often overlooked in thermal analysis problems.

The flow field in thermal problems is resolved by the application of the conservation of momentum (Newton's second law). This equation relates the forces that govern the flow to the rate of the fluid's change of momentum. Once applied to a fluid enclosed by a CV, the equation can be expressed by the following:

      Rate of momentum accumulating in the CV
=   Rate of momentum into the CV
-   Rate of momentum out of the CV
+   Sum of the forces acting on the CV.   (12-10)

Application of the conservation of momentum yields information regarding velocity and pressure distribution within the system. The equation is nonlinear in nature and inherently unstable. Hence, its solution can become quite involved and may require higher order solution techniques such as numerical computation or experimentation.

The last equation in the conservation laws is the energy equation. Conservation of energy, or the first law of thermodynamics, simply states that within a thermal process, the energy is conserved. In general, the energy can have many facets, from viscous heating to heat released by a chemical reaction. Therefore, to include all forms of energy, let us state the conservation of energy as applied to a CV by the following:

      Rate of energy entering the CV
+   Rate of generation of energy in CV
-   Rate of energy leaving CV
=   Rate of accumulation of energy in CV.   (12-11)

The solution of the conservation of energy yields the temperature and heat transfer of the thermal process. If fluid motion also is associated with the problem, then the conservation of energy and momentum equations are coupled. The energy equation, in this case, includes the velocity terms. Their magnitude must be known before a solution can be obtained. Furthermore, if the fluid properties are temperature dependent the conservation of energy and momentum equations also must be solved simultaneously. Otherwise, each equation can be solved independently, significantly simplifying the solution process.

All of these laws contain thermodynamic properties describing the fluid at a given state. If these properties become non-constant and vary from state to state (for example at different temperatures), their values must be known. Therefore, an equation of state, such as the ideal gas law, is used for determination of the properties.

12.5.3 Solution Procedure

Complexities in electronic cooling problems stem from the intercoupling that occurs among various elements of thermal transport. To develop a solution, it is necessary to establish a methodical approach or engineering bookkeeping. Hence, the general approach to solution of thermal problems is first outlined. We then focus more specifically on the electronic cooling.

The general approach for problem solving, independent of electronic cooling, is shown in a flow chart in Figure 12-14 [27]. The approach consists of four parts: problem preparation, solution preparation, solution procedure and solution verification. The approach provides a method for accurately making assumptions and keeping track of the solution process. Additionally, the critical issue of what method is necessary for the solution procedure can be addressed early on. This also provides a process for generating the necessary information to select a solution technique, for example, boundary conditions for numerical simulation. The general procedure for electronic cooling consists of fluid flow and heat transfer analyses, with the outcome focused on the reliability predictions of the components (Figure 12-15).

12.5.4 Analytical Modeling - Integral Method

The thermal model of an electronic system is developed by applying Equations 12-9 through 12-11 to a CV embracing the region of interest. The model yields an analytical expression describing the desired variable in terms of other parameters governing the problem. The process of modeling, however, is not confined to a specific problem domain. The procedure is best described by the following example:

Example 1: Governing equations for a single component residing in a circuit board channel

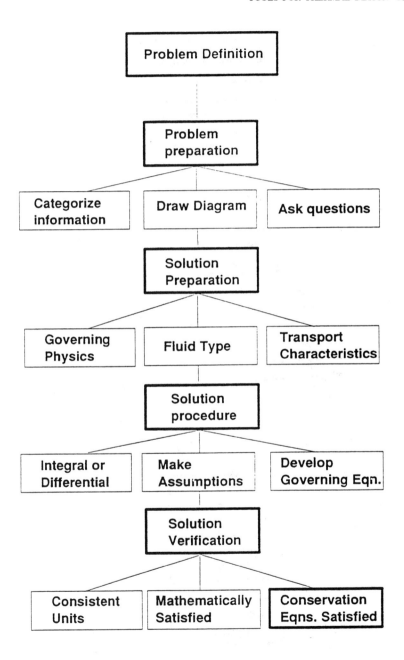

Figure 12-14 Flow chart for general problem solving approach.

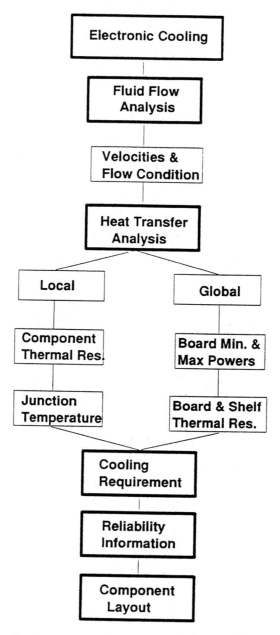

Figure 12-15 Flow chart for problem solving approach in electronic cooling.

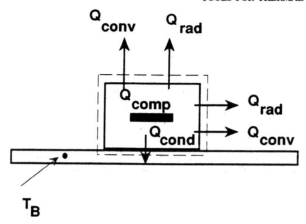

Figure 12-16 Single SCM with control volume around component.

Consider an air cooled SCM residing on a board. The objective is to obtain the junction temperature of the component as a function of SCM board and coolant parameters. This will be done by constructing the equations applicable to each CV.

The following assumptions are made within the CV:

- The problem is at steady state.
- The fluid behaves like an ideal gas, and is incompressible.
- In the designated CVs, the fluid temperature is uniform. Note that since the first order solution method is selected, the temperatures and velocity are averaged over an area of interest, for example, an area perpendicular to the direction of the fluid flow.
- The board temperature is uniform over the component's footprint. This area is equal to twice the planar area of the component.
- The convective heat transfer coefficient is also uniform within the CV.

Component
The information generally available for the component is the geometry and an estimate of the power dissipation. However, on may occasions, the power dissipation is not known exactly. This is not an obstacle in the analytical model since the power dissipation can be treated as a variable. Applying Equation 12-11 to the CV shown in Figure 12-16 we get:

$$Q_{comp} = Q_{cond} + (Q_{conv} + Q_{rad})_{top} + (Q_{conv} + Q_{rad})_{bottom}$$

$$(12\text{-}12)$$

Replacing the heat flows, Q, in Equations 12-2 through 12-4 with their respective temperature definitions and linearizing the radiation heat transfer terms, we find:

$$Q_{comp} = \frac{kA}{L}(T_1 - T_B) + hA_{top}(T_1 - T_f)$$
$$+ 4\sigma\varepsilon F_{1,top} A_{top} T_m^3 (T_1 - T_{board,top})$$
$$+ hA_{side}(T_1 - T_f) + 4\sigma\varepsilon F_{1,side} A_{side} T_m^3 (T_1 - T_{board,top})$$

$$(12\text{-}13)$$

where T_1 is the component case temperature and T_B is the board temperature in the footprint of the component. A, L and k are the lead total cross section, length and thermal conductivity, respectively. A_{top} is the area of the top of the component, and A_{side} is the total area of the component sides. T_f is the mean fluid temperature $T_f = (T_{f,i} + T_{f,o})/2$, where $T_{f,i}$ and $T_{f,o}$ are the fluid inlet and outlet temperatures, respectively. $T_{board,top}$ is the temperature of the board top away from the component. T_m is the mean board temperature.

Board
The energy balance for the board with heat flows shown on Figure 12-17 becomes:

$$Q_{cond} = Q_{cond,B} + (Q_{rad} + Q_{conv})_{top} + (Q_{rad} + Q_{conv})_{bottom}$$
$$(12\text{-}14)$$

Substituting the temperatures from Equation 12-2 through 12-4 and linearizing the radiation heat transfer:

$$\frac{kA}{L}(T_1 - T_B) = k_B \frac{t_B W_B}{L_B}(T_B - T_{amb})$$
$$+ 4\sigma\varepsilon A_{f,p} T_m^3 (T_B - T_{board,top})$$
$$+ h(A_{f,p} - A_{top})(T_B - T_f)$$
$$+ 4\sigma\varepsilon A_{f,p} T_m^3 (T_B - T_{board,bottom})$$
$$+ hA_{f,p}(T_B - T_{f,bottom})$$

$$(12\text{-}15)$$

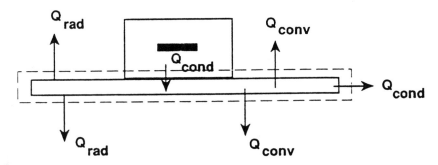

Figure 12-17 Single SCM with control volume around the circuit board.

where k_B is the thermal conductivity of the board and t_B, W_B and L_B are board thickness, width and the distance from the component to the thermal connection with the frame, respectively. T_{amb} is the ambient temperature of the frame. $A_{f,p}$ = $2A_{bottom}$ is the area of the board assumed to be at temperature T_B (component footprint), $(A_{f,p} - A_{top})$ is the area not covered by the component, $T_{board, bottom}$ is the temperature of the board away from the component and $T_{f,bottom}$ is the mean fluid temperature on the bottom of the board. The radiation view factor is ignored since it approximately is equal to 1.

Fluid
Conservation of energy for the fluid (Figure 12-18) yields the following:

$$\dot{m} C_P (T_{f,o} - T_{f,i}) = h(A_{top} + A_{side})(T_1 - T_f) \\ + h(A_{f,p} - A_{top})(T_B - T_f), \quad (12\text{-}16)$$

where $\dot{m}$ is the mass flow rate of the coolant and C_P is its specific heat. Conservation of mass becomes

$$V_1 A_1 = V_2 A_2, \quad (12\text{-}17)$$

where A_1 is the inlet temperature, A_2 is the outlet temperature, V_1 is the inlet velocity and V_2 is the outlet velocity. If V_1 is given, then V_2 and $V = (V_1 + V_2)/2$ can be calculated and h determined.

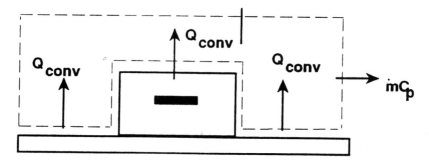

Figure 12-18 Single SCM with control volume around fluid.

Component Interior
The equation applicable to the CV inside the component body is

$$Q_{comp} = \frac{k_{eff} A_{eff}}{L_{eff}} (T_j - T_1) \qquad (12\text{-}18)$$

where k_{eff} is the effective conductivity, A_{eff} is the effective area and L_{eff} the effective length of the SCM internal path.

Thus, if Q_{comp}, $T_{f,\,i}$, V_1, $T_{f,\,bottom}$, $\dot{m}$, $T_{board,top}$, $T_{board,\,bottom}$ and T_{amb} are given, then the designer can solve for T_j, T_1, T_B and $T_{f,o}$.

12.5.5 Computer Based Tools - Numerical Method

The numerical approach to the solution of thermal design problems is based on the differential form of the conservation laws, [28]. The differential model yields the distribution of the desired parameters: velocity and temperature. The solution can be obtained by either finite difference method (FDM), finite element method (FEM) or spectral method (SM). Since the differential model tends to be highly nonlinear and inherently unstable, the numerical approach to these problems must be done with much care.

The numerical solution can be categorized into three areas: solid, fluid or a combination of the two. For solids (component or board), the modeler must have a good insight for fluid to solid (heat transfer coefficient) and solid to solid (board where component resides) boundary conditions. The fluid case yields velocity and pressure distributions in the flow field. If the boundary conditions are not available at the interface of the solid and fluid, the solution becomes a simultaneous simulation of the two.

The accuracy of the numerical simulation is a function of boundary or initial conditions. In addition, the solution is dependent on the mesh or grid size, resulting in an iterative process to verify the dependency of the solution on the numerical mesh. Since the differential models are developed on the basis of infinitesimal CVs, their application is suitable indeed for localized analyses. This implies that higher accuracy with reasonable effort can be obtained when the domain of analysis is comparatively small, such as with components and circuit boards. Therefore, to obtain the highest accuracy, some rules of thumb can be developed when using numerical techniques.

1. Do not model the entire system or the shelf unless you are seeking some very qualitative data.

2. Increase the mesh or gird density at the points of concentration. These points can be categorized as areas such as heat sources and points of separation in the flow field. Many tools do this automatically.

3. Avoid idealization of the boundary conditions. That is, if the boundary condition is a gradient, do not assume it is uniform unless there is a sound engineering reason for making such as assumption.

4. Do not automatically assume that the answer obtained from the numerical simulation is correct. The answer from numerical simulation must always be verified via integral methods or experimentation.

5. Do not to analyze data by just looking at the color graphics output. In most electronic cooling problems, the geometry is small. Therefore, spatial variation can be significant. Color graphics outputs, even with 16 colors, do not have the necessary resolution to highlight the spatial variations. The color graphics, therefore, can be very misleading and should be used only for obtaining insight into the trend of the problem. The XY plots or numeric outputs are the best way to see the variations in the area of interest.

There are three numerical methods for obtaining the solution to thermal design problems, FDM, FEM and SM. They might be used in two dimensional form (usually FDM) for obtaining quick low order solutions or in three dimensional form for high order solutions. The two dimensional forms typically use analytic expressions (such as the ones above) to handle the effects of the third dimension. There are many good references that discuss these methods in detail [29]. It is beyond the scope of this text to discuss these methods in detail. Instead this section attempts to answer a common question that confronts a practitioner - "which method is most suited for my application?"

Let us start by first giving a general description of each method. The finite difference method uses a Taylor series expansion to discretize the entire problem domain and converts the differential equation into a series of analytic expressions. The finite element method uses approximate functions as a local solution and also converts the differential equation into a set of analytical expressions. The spectral method can be categorized as an extension of finite element methods in which the spatial frequency domain also is utilized. From this simple description of each method, one can conclude that the FDM is perhaps the most simple to model and the SM the most complex.

To determine which method is most suited for thermal design of MCMs, or other electronic components, let us divide the domain of application into the fluid and solid solvers. The FDM and FEM are most suited as solid solvers, that is, component and board. The suitability for solid modeling is basically a function of geometry. The FDM, which converges much faster than FEM, is ideal for regular geometry, such as rectangular. Irregular geometries also are handled effectively if the FDM utilizes the Body Fitted (or fixed) Coordinate (BFC) methodology. The FEM addresses the irregular geometry much more accurately. The SM method appears to be an over description, with a very slow rate of convergence, as a solid modeler. Therefore, FDM and FEM are the techniques of choice for obtaining temperature and stress distribution in a solid.

Four points are of interest in selecting the fluid solver: geometry, condition of the boundaries, transient/separated flows, and rate of convergence. The FDM method is most suited for regular geometry and the FEM for irregular geometry. If the boundary conditions are non-stationary, FEM is more suitable for modeling than FDM. If the flow field contains separation or transience, that is, flow past a component, SM appears to be most capable to address the physics associated with the flow regime. One should add that all these methods are capable of handling transient problems. But much more effort may be required to obtain a solution with one method versus another. The last point of concern is the rate of convergence. Computational fluid dynamics tend to be very computer time intensive. The author's comparison of all three methods applied to the same problem on similar computers has resulted in the following order (where least computer time is first): FDM, FEM, and SM. It is important to point out that the numerical methods are not black boxes. These are tools that must fit the problem. Therefore, the selection of a technique should be based on the type of problem and the capabilities of the code.

To see a typical application of the computational fluid dynamics method, the following example is presented.

Example 2: Temperature and air flow distribution in a horizontal channel containing an MCM.

Consider a single MCM placed inside a horizontal channel. The total power dissipation of the module is 5 W. The center chip has 4 W and the side chips have 0.5 W of power dissipation each. The MCM also is surrounded by four SCMs each dissipating 0.5 W. The components are placed on a four layer board and cooled with natural convection. We have two objectives in this case. First, we would like to see how the module functions in natural convection - does it meet the 125°C temperature specification? Second, what if a flow of 200 ft/min (forced convection) is introduced from left to right - what is T_j in this situation?

A commercially available computational fluid dynamics (CFD) software, FLOTHERM™ [30], was used for the simulation. Figures 12-19 and 12-20 show the results of such analyses. These results show the temperature distribution in the circuit board and the velocity distribution in vectorial form, with arrow length proportional to air flow velocity. The results show that natural convection is not adequate for cooling of this MCM since junction temperature is very close to the design limit, Figure 12-19. The margin is not sufficient to approve this design for natural convection cooling.

In the case of forced convection (Figure 12-20), although the fins are misaligned with the flow direction, we see that the junction temperature is considerably lower. The temperature would be lower yet if the fins were aligned with the air flow. This maximizes the average flow rate. Fin misalignment is a common error (and done here solely for illustrative purposes). The temperature with forced air cooling provides a sufficient margin in T_j making it the recommended solution. Note the thermal coupling evident in Figure 12-20. The component downstream of the MCM resides in a slower moving fluid at an elevated temperature. Thus, its T_j is hotter.

12.5.6 Experimentation - Why, When and How

The highest order solution in a thermal design process is experimentation. As the ranking of the solution techniques show, it is last on the list. Although we tend to believe experimental data over analytical ones, and rightfully so, it is

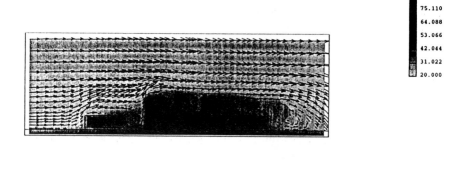

Vector   Fill
Temperature

Zero Range

97.154
86.132
75.110
64.088
53.066
42.044
31.022
20.000

z=2.5000E-02 z=2.5000E-02

| 0.50
Ref Vector

Figure 12-19 Flow distribution around a 5W MCM cooled by natural convection.

normally avoided because of its cost and duration. Therefore, this section highlights when an experiment is necessary and what parameters should be measured. It is certainly not the scope of this section to show methodologies for fluid flow and heat transfer measurements. For a detailed description of these types of measurements, references [31] and [32] are recommended.

As discussed earlier, because of thermal intercoupling that exists inside the electronic circuit boards, analytical and numerical predictions are difficult. The level or magnitude of intercoupling indicates when an experiment should be conducted. When the boundary conditions are not clearly defined, an experiment closely simulating the actual case must be conducted. Another case when experimentation is deemed necessary is when the problem is not phenomenologically understood. These are cases when the problem at hand is new and has not been investigated before. These cases typically are encountered in new technologies or processes. A good experiment is one that closely resembles the actual process. An experimenter should understand the details of the problem and design the experiment accordingly. Often because of the nature of the problem, such as chip size or limitation in the test facility, scaling is required [33]. In this case, the dimensionless fluid dynamics numbers such as Reynolds (Re), Grashof (Gr) and Prandtl (Pr) [11], must be identical for the model and actual process. Another important aspect of a good experimental design is the uncertainty analysis [34].

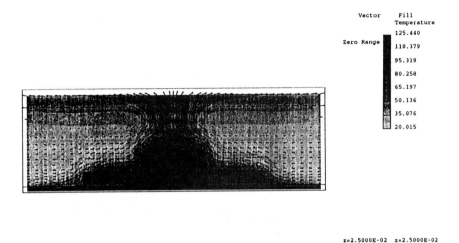

Figure 12-20 Flow distribution around a 5W MCM cooled by moderate forced convection.

The selection of parameters that best describe a process is always a difficult task. This difficulty is even more profound in electronic cooling since the problem can be a very dynamic one. For example, if the case temperature is to be measured, where should the temperature sensor be placed? What follows are some general guidelines for measurement in electronic cooling.

Let us start by stating that pressure (velocity) and temperature are the two parameters that readily can be measured in laboratory. The parameters and points of measurements should be the ones that best describe the process. Two parameters, component temperature and temperature rise in the system, can be quantified as the most important for temperature measurement. The component temperatures are junction, case on top of the hottest chip, and case at the edge. The two case temperatures provide the limits, minimum and maximum, reached by the case under test conditions.

Use of thermal resistance in MCMs should be avoided. If one insists on reporting Θ, chip junction to ambient should be used. Here, the ambient is defined as the reference temperature or the fluid temperature in the vicinity of the component. For conventionally cooled MCMs (low performance systems), the latter is calculated by an energy balance performed on the components, preceding the component of interest. For liquid cooled MCMs, fluid temperature at the source is used as the reference.

Temperature rise in the system is defined as the increase of cooling fluid's temperature over an established reference. The reference temperature in the case of air cooled systems is the environment's temperature, and for liquid cooled systems, it is the supply temperature.

Likewise, fluid velocity and pressure drop would have to be measured to accurately quantify the component thermal performance and the system cooling mechanism. Fluid velocity and temperature measurements at the local level are labor intensive and do not yield much practical information. For air cooled systems, the measurements are at the boundaries, that is, inlet and outlet.

12.6 SUMMARY

In this chapter a description of thermal design and management of MCMs is presented. The chapter describes thermal phenomena in MCMs, thermal control methods and different methodologies for analysis of MCMs. There are some salient points that merit reiteration. These are:

- Principles of thermal management and design have not changed for MCMs from what was practiced with SCMs.

- The challenge or complexity in thermal control of MCMs versus SCMs stems from multiple chips on the substrate. Air cooling of MCMs is the most desirable method of cooling. This is based on the ease of use, abundance of air and lesser manufacturing and maintenance costs.

- A logical progression for thermal control of MCMs is to first exhaust all air cooling options before considering any form of liquid cooling.

- Liquid cooling tends to be highly customized for most MCM designs. Prepackaged cooling systems with liquid cooling capability typically are not found.

MCM thermal performance is increased by minimizing internal and external resistances. The internal resistance can be reduced by selecting:

- Materials with higher thermal conductivity
- Minimizing thermal spreading near the multiple chips and
- Considering alternate chip mounting techniques.

The external resistance is reduced for:

- Air cooled MCMs by delivering air directly to the MCM and minimizing flow resistances.
- Liquid cooled MCMs (excluding immersion) by reducing interfacial resistances between the cold plate and the chip.

An important point to note for low performance MCMs is the impact of board layout on thermal response. Judicious placement of components on the board significantly improves component thermal response. Often, a combination of thermally appropriate board layout and minimization of internal resistance can avoid higher order cooling methods.

MCM thermal analysis for either characterization or verification are done by three methods: integral, computational and experimental. Each method's domain of utility is specified. An important point is that the integral method forms the foundation for any analysis initiative. Computational and experimental studies, although very useful, are boundary condition dependent. A noteworthy point is the utility of integral and computational methods for parametric studies. The what-if scenarios performed early in the design cycle play a pivotal role in the robust design of MCMs. Additionally, the design duration and its associated development cost can be reduced if these analyses are performed at the onset of system conceptualization.

Acknowledgments

The author would like to thank the Editors for their very significant technical and organizational help in defining the structure of the chapter. He also would like to thank the reviewers for their concrete and substantial suggestions.

References

1   D. J. Klinger, Y. Nakada, M. A. Menendez, *AT&T Reliability Manual*, New York: Van Nostrand Reinhold, 1990.

2   M. Bonnifiat, M. Cader, "Thermal Simulation for Electronic Components using Finite Elements and Nodal Networks," *ASME HTD*, vol. 57, pp. 183-188, 1986.

3   H. Hardisty, J. Abboud, "Thermal Analysis of a Dual-in-Line Package Using Finite Element Method," *IEE Proc.*, (UK), vol. 134, Part 1, pp. 23-31, 1987.

4   M. Aghazadeh, D. Mallik, "Thermal Characteristics of Single and Multi-Layer High Performance PQFP Packages," *Proc. 6th Annual Semi. Thermal and Temperature Measurement Symp.*, (Phoenix AZ), pp. 33-39, 1990.

5   M. A. Zimmerman, K. Azar, C. D. Mandrone, "Thermal Characteristics of Molded

Multi-Chip Modules," *ASME Winter Annual Mtg.*, (Dallas TX), 90-WA/EEP-21, 1990

6   W. M. Kays, M. E. Crawford, "Convective Heat and Mass Transfer," 2nd Edition, New York: McGraw Hill, 1980, pp. 139.

7   R. A. Wirtz, W. McAuliffe, "Experimental Modeling of Convective Downstream from an Electronic Package Row," *ASME J. of Electr. Packaging*, vol. 11, pp. 207-212, 1989.

8   R. A. Wirtz, P. Dykshoorn, "Heat Transfer from Arrays of Flat Packs in a Channel Flow," *Proc. Inst. of Electr. Packaging Symp.*, pp. 318-326, 1984.

9   E. M. Sparrow, J. E. Neithammer, A. Chaboki, "Heat Transfer and Pressure Drop Characteristics of Arrays of Tectangular Modules Encountered in Electronic Equipment," *Internat. J. of Heat and Mass Transfer*, vol. 25, no. 7, pp. 961-973, 1982.

10   S. Sridhar, *et al.*, "Heat Transfer Behavior Including Thermal Wake Effects in Forced Air Cooling of Rectangular Blocks," *ASME HTD*, WAM, (Dallas TX), vol. 153, pp. 15-26, 1990.

11   F. P. Incropera, D. P. Dewitt, *Introduction to Heat Transfer*, New York: Wiley and Sons, 1985.

12   F. M. Shite, *Heat Transfer*, New York: Addison-Wesley Publishing Company, 1984.

13   V. P. Manno, K. Azar, "The Effect of Neighboring Components on Thermal Performance of Air-Cooled Circuit Packs," *ASME J. of Electr. Packaging*, vol. 113, pp. 50-57, March 1992.

14   K. Okutani, *et al.*, "Packaging Design of SiC Ceramic Multi-Chip RAM Module," *Proc. of Internat. Electr. Packaging Soc.*, pp. 299-304, 1984.

15   K. Azar, C. D. Mandrone, "Effect of Pin Fin Density on Thermal Performance of Un-Shrouded Pin Fin Heat Sinks," *Proc. of Electro/92 Conf.*, (Boston MA), 1992.

16   A. D. Kraus, A. Bar-Cohen, *Thermal Analysis and Control of Electronic Equipment*, Hemisphere Publishing, 1983, pp. 593-596.

17   R. Hannemann, *Thermal Control for Mini and Microcomputers: The Limits of Air Cooling,*" Heat Transfer in Electronic and Microelectronic Equipment, Hemisphere Publishing, 1990, pp. 61-84.

18   K. Azar, J. R. Benson, V. P. Manno, "An Experimental Investigation of Microjet Impingement Cooling," *27th Nat. Heat Transfer Conf.*, July 1991.

19   A. Bar-Cohen, "Thermal Management of Air and Liquid-Cooled Multichip Modules," *IEEE Trans. CHMT*, vol. CHMT-10, no. 2, pp. 159-175, 1987.

20   R. C. Chu, R. E. Simons, "Recent Developments in Thermal Technology For Electronics Packaging," *Electr. Packaging Forum*, vol 2, New York: Van Nostrand Reinhold, 1991.

21   K. Otsuka, *et al.*, "Considerations of VLSI Chip Interconnection Methods," *IEEE Computer Society Spring Workshop*, (Palm Desert CA), 1985.

22   M. Kohara, *et al.*, "High Thermal Conduction Package For Flip Chip Devices," *IEEE Trans. of Components, Hybrids, Manuf. Techn.*, vol. CHMT-6, pp. 267-271, 1983.

23   H. Yamaoto, Y. Udagawa, M. Suzuki, "Cooling System for FACOM M-780 Large Scale Computer," *Proc. of Internat. Symp. on Cooling Techn. for Electr. Equipment*, (Hawaii), pp. 110-125, 1987.

24 Personal communication with R.E. Simons of IBM.

25 A. E. Bergles, A. Bar-Cohen "Direct Liquid Cooling of Microelectronic Components," *Advances in Thermal Modeling of Electr. Components and System*, ASME Press, vol. 2, pp. 233-342, 1990.

26 W. Nakayama, "Thermal Management of Electronic Equipment: A Review of Technology and Research Topics," *Appl. Mechanics Review*, vol. 39, no 12, 1986.

27 K. Azar, "Problem Solving Techniques in Electronic Cooling," *Proc. 5th Annual Semi. Thermal and Temperature Measurement Symp.*, (San Diego CA), pp. 37-44, 1989.

28 B. R. Bird, W. E. Stewart, E. N. Lightfoot, "Transport Phenomenon," New York: John Wiley and Sons, 1976, pp. 83-85.

29 C. A. J. Fletcher, "Computational Techniques for Fluid Dynamics," Springer Verlag, 1988.

30 FLOTHERM, Flomerics Corp., Westborough, MA 01581.

31 R. J. Goldstein, "Fluid Mechanics Measurements," Hemisphere Publishing, 1984.

32 E. R. Eckert, R. J. Goldstein, *Measurements in Heat Transfer*, Hemisphere Publishing, 2nd ed., 1976.

33 G. Murphy, *Similitude in Engineering*, Ronald Press Co. 1950.

34 H. W. Coleman, W. G. Steele, "Experimentation and Uncertainty Analysis for Engineers," New York: John Wiley and Sons, 1989.

13

ELECTRICAL TESTING OF MULTICHIP MODULES

Thomas C. Russell and Yenting Wen

13.1 INTRODUCTION

Electrical testing is used throughout the multichip module fabrication process to verify the quality of each processing step and component which goes into the module. Once module assembly is complete, a final test is performed to make sure that the module functions to its specifications. It is desirable to locate any faults as early in the manufacturing process as possible since this results in a lower overall cost. Testing can be divided into three basic areas: substrate test, integrated circuit (IC) test and module test. Substrates are tested during fabrication and prior to component attach. ICs and other components are tested prior to mounting on the substrate. The assembled module is tested prior to final sealing to permit repair of faulty components.

Multichip modules (MCMs) provide an increased level of performance over printed circuit boards by virtue of their dense packaging. It is this same dense packaging that complicates the testing process. The small size of the substrate and component pads makes access by electrical probes difficult and creates a need for test techniques which are unique to MCMs. The ICs mounted on the substrate are unpackaged and typically run at high speeds. A bare IC is often called a die and requires special handling and test to meet MCM needs. Finally, the completed module itself is a complex electrical system which must be tested as a single unit. Simulation and specialized diagnostic techniques are required to test fully the operation and to locate any faults.

MCMs provide considerable testing challenges which will be addressed in this chapter. While not all of the problems associated with MCMs test have been solved as of the writing of this book, each of the testing issues is presented here along with the current state of the art. Substrate testing is discussed first since this forms the foundation for the module. Testing of ICs and other components is then presented, followed by the simulation and testing of assembled modules. Figure 13-1 presents a flow chart for the entire MCM testing process.

13.2 SUBSTRATE TEST

13.2.1 Introduction

Electrical testing of the MCM substrate is an integral part of the overall MCM fabrication process. The substrate contains the wiring used to interconnect all of the components on the completed MCM. Testing is used to verify the connectivity of the substrate and to monitor the fabrication process for quality control. A final electrical test is performed typically at or near the end of the substrate fabrication. In some cases, electrical testing is performed in-process to monitor the yield after various steps.

MCM substrates contain a number of electrical networks (nets) which are terminated in pads on the surface. These pads are used to bring the signals in and out of the substrate. Some pads connect to the chips which are mounted on the substrate while others connect to off module components. Electrical testing is performed by bringing one or more probes into contact with the pads on the substrate and making measurements. Each net on the substrate must be tested to ensure that there are no open circuits within the net and also that there are no short circuits to any other net. Nets also can be tested to make sure that the resistance from one pad to another meets design criteria. Additional testing can include measurements of impedance, propagation delay, crosstalk and high voltage leakage. Typically, every substrate receives testing for opens and shorts. Additional measurements are made as required to meet specifications and to maintain quality control.

MCMs fabricated using thin film technology on silicon or ceramic substrates are built up in successive layers using lithographic techniques. Pads occur on only one side of the substrate. These substrates are characterized by pad size $\leq$ 100 μm, pad spacing $\leq$ 200 μm and pad densities > 100 pads per square cm. The layers are defined using a lithographic process which results in very accurate

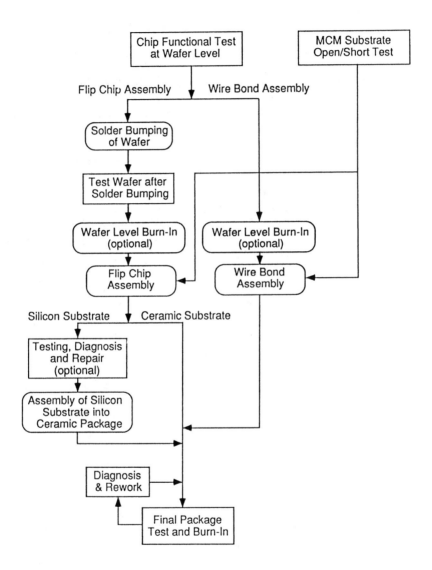

Figure 13-1 Flow chart of testing procedure for multichip modules.

placement of the pads across the entire substrate. Pads are aluminum, nickel, copper or gold providing good contact for probes, but they can be damaged by excessive probe pressure. A single substrate may contain several modules which range in size from 10 - 15 cm across, with larger sizes anticipated in the future. The thermal mismatch between the substrate and the thin film layers can cause camber (warpage) of the wafer, as discussed in Chapter 7. This camber does not usually cause a problem during interconnect testing since the substrate is held flat on a vacuum chuck. Automatic handling equipment is readily available and the modules are cut apart after interconnect testing.

MCMs fabricated using cofired ceramic technology contain layers of ceramic material which are fired at high temperature. The firing process causes the substrate to shrink, introducing distortion in the placement of pads across the substrate. This distortion complicates the alignment of the probes used with testing systems. The pad size is typically ≥ 100 μm, the spacing ≥ 200 μm and the densities typically lower than those found on thin film substrates. Cofired ceramic MCM substrates often contain pads on both sides and testing must include verification of the interconnection from top to bottom side. In addition to shrinkage, cofired ceramic parts are subject to camber resulting in variations in the height of the test pads and require adjustments in the probing level. The metallization is tungsten followed by layers of nickel and gold. Interconnect testing of cofired parts is done after firing, when the part is at or near its final size. In some instances, the circuitry can come very close to the edge of the part which complicates the fixturing. Automatic handling of ceramic parts tends to be more complex than that for thin film parts since the cofired part is formed in a variety of sizes and shapes.

MCMs fabricated using an overlay technology as described in Chapter 7 start with the bare die mounted on a ceramic substrate and deposit interconnect layers over the die to complete the circuit. Since there is no interconnect substrate distinct from the completed module, testing is performed as part of the module test described in Section 13.4.

No matter what the fabrication technology, MCMs are essentially miniature printed wiring boards (PWBs). Some testing techniques from PWBs can be employed in testing MCMs, but the small size and high density of pads creates a formidable testing challenge. Access to the individual pads is more difficult than on a PWB, so there is a temptation to wait and test the functionality of the assembled MCM from its input/output (I/O) pins. Unfortunately, this is not often viable economically since repairing the module at that point would require removal of all of the ICs and then mounting them on a fresh substrate or simply discarding the entire assembly. The rule of tens holds in this case - it is ten times as expensive to diagnose and correct a defect at the next level of integration. It is therefore wise to locate the fault as early in the process as possible. Adding this fact to the substantial benefit gained by using electrical measurements to monitor process quality provides an overwhelming motivation for full electrical test of the substrate.

The test methods described in this section are used to test for the most common types of electrical failure modes: opens, high net resistance and shorts. The selection of any test method will depend on the level of testing required and the volume of parts to be tested. A testing program should be developed as an integral part of the design and manufacture process. Proper design of the substrate can greatly reduce the testing time and decrease the testing cost. Design for testability applies equally to the substrate as well as to the whole system. It no longer is possible to create designs without considering the test issues. Careful consideration of testability results in a final product that is not only less expensive to produce, but also higher in quality.

13.2.2 Fixed Probe Array Testing

A fixed probe array tester (often called a bed-of-nails tester) uses a head with many probes arrayed to contact all of the pads on the substrate at the same time. Each probe is connected to one channel of an analog multiplexer. Each channel of the multiplexer can connect to either side of the measurement unit (typically an ohmmeter). This allows the meter to be connected between any pair of pins or between two groups of pins. A schematic diagram of a $2 \times n$ switch matrix is given in Figure 13-2. In actual practice, the switch channels are often configured as a $4 \times n$ matrix to permit 4-wire Kelvin type measurements to be made which negate the effect of the resistance in the matrix and the wiring.

Opens and high resistance within a net can be located by measuring the resistance between all the pads on the net. The total number of tests required for this is calculated using:

$$\text{number of open tests} = \sum_{i=1}^{n} (p_i - 1) \tag{13-1}$$

where n = number of nets and p_i = number of pads in net i. Shorts between nets are located by measuring from each net to every other net. This is a simple combination of n objects taken two at a time; the number of tests required is figured as:

$$\text{number of short tests} = \frac{n(n-1)}{2}. \qquad (13\text{-}2)$$

For a circuit with 100 nets and an average of four pads per net, the total number of tests is:

$$(100)(4-1) + \frac{(100)(99)}{2} = 5{,}250 \text{ tests.} \qquad (13\text{-}3)$$

An analog multiplexer uses solid state relays and can make 2400 measurements per second [1] so the total test time is still less than 3 seconds.

If the number of nets increases to 1000, then the number of tests required, and consequently the test time, increases dramatically:

$$(1000)(4-1) + \frac{(1000)(999)}{2} = 502{,}500 \text{ tests.} \qquad (13\text{-}4)$$

Fortunately, if the multiplexer allows more than one probe to be connected to the measuring unit at the same time, the number of tests required to locate shorts can be reduced radically. One method is to check continuity from each net in turn to all the other nets which have been tied together. If the part is good, then only n measurements will be needed. For each shorted net identified, additional (n-1) measurements will be needed to isolate the short. Other methods employing binary search techniques can reduce the number of measurements still further.

Bed-of-nails testers are commonly employed to test PWBs. Test heads for PWBs are typically a regular array of probes since many boards are designed on fixed centers (either 100 or 50 mils). Since MCMs contain bare die, with no pads on a grid, a custom probe array is required for each new substrate design. Various types of probes are employed to build the test head including spring loaded pogo probes, buckling beam probes and cantilever probes. When the head is contacted to the board, each probe aligns with a single pad on the board. As the size and density of the pads increase, the probe head becomes increasingly difficult to fabricate. While there is no absolute limit to the complexity of the test head, the viability of the method diminishes due to the high per pin cost of the test head and the fact that a head is useable for only one design. In addition, the reliability of individual probes tends to go down with their size. A single bent or nonfunctional probe will disable the entire tester.

For silicon substrates, the pad density is often too high to use a fixed probe array approach. If the pad density is low except for one or two small areas of

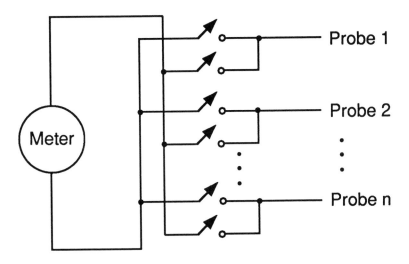

Figure 13-2 Schematic diagram of fixed probe array tester.

high density, it is often possible to build up a test head out of varying probe technologies. Usually, cofired ceramic MCM parts cannot be tested using a fixed probe array due to shrinkage variations across the substrate and from part to part. Cofired parts typically have a size specification of ±0.5 percent across the entire substrate [2]. For a part measuring 4" on a side, this translates into a pad placement error of ±20 mil, which is often larger than the pad to be probed. A fixed probe array only can be used if the pads are larger than the errors in pad placement. One way to reduce the pad placement error is to apply the top layer metallization after the part is fired. This layer contacts pads on the fired layer below and provides probing pads with much less distortion across the substrate.

Fixed probe array testing provides very high throughput since all the pads are contacted at once and only electronic switching is required to complete the test. It is employed most favorably for high volume testing since the tooling cost is high. It also adds considerable lead time to the test setup. Although it cannot be used for cofired parts with small pads or with very high density parts, it can be used with many lower density designs. For this reason, it may be prudent to design the substrate to accommodate this type of testing if high volume is anticipated.

13.2.3 Single Probe Testing

Single probe testers build on an idea originally patented by Honeywell [3] and made popular by Teledyne/TAC [4]. Each net in an MCM substrate is isolated

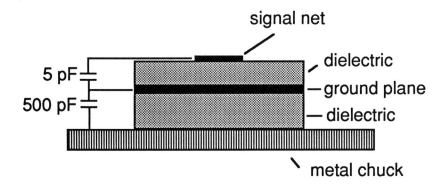

Figure 13-3 Capacitance relationship of a signal net to the ground plane on a substrate.

from all other nets by a dielectric material (polymer, ceramic etc.). Due to this relationship, there is a certain capacitance between each net and a nearby ground plane based solely on the geometry of conductors in the net, the separation between the net and the ground plane and the dielectric constant. The ground plane itself can be an internal plane inside of the part or it can be the chuck that the part rests on during the measurement process. The tester operates by connecting one side of a capacitance meter to the ground plane and connecting the other side to a single probe which is brought into contact with each pad on the substrate. While the theoretical capacitance of the net can be determined from its design, measurement of this capacitance on a known good part (also called a golden part) to create a reference is more common. If a golden part is not available, capacitance is measured on several parts and the median value for each net is computed to create a statistically derived reference. During testing, if the capacitance of any net has increased above the reference value, then its size must have increased, and it is likely therefore that it is shorted to another net. If the capacitance has decreased, then the net probably contains an open circuit.

The chuck used as the ground plane often has a thin dielectric coating over it to prevent shorting of any exposed circuitry on the substrate. Use of an external ground plane is the most common method since only one probe needs to contact the substrate, simplifying the mechanical system of the tester. A ground plane inside the substrate also can be used, but this requires the addition of a fixed probe to contact a pad connected to the plane. If an external plane is used, the measured capacitance of the net is dominated by the capacitance between the net and any internal ground plane. This is illustrated in the Figure 13-3. Assume the capacitance between the signal net and the ground plane is 5 pF and the capacitance between the ground plane and the external ground of the chuck is 500 pF. Since these two capacitances are effectively in series, the total capacitance is calculated using:

$$\frac{1}{C_T} = \frac{1}{C_1} + \frac{1}{C_2} = \frac{1}{5} + \frac{1}{500} \qquad (13\text{-}5)$$
$$C_T = 4.95\,pF$$

If the capacitance between the internal ground plane and the chuck varies due to changes in substrate thickness or to mounting on the chuck, the effect on the measured capacitance is very small.

Figure 13-4 shows two nets labeled N1 and N2 on a substrate. If net N1 has a nominal capacitance of 5 pF and net N2 has a nominal value of 8 pF, then a short between the two would result in a capacitance of approximately 13 pF. The total might be slightly more or less than the sum of the two net capacitances due to the additional metal area of the short itself and the possibility that there are some overlapping areas in the nets. A fault condition would be indicated if the capacitance of either net were significantly greater than its nominal value. The tolerance setting is somewhat dependent on the substrate technology and on the degree of feature size control, but a typical value is 25%.

Using this method, a substrate with n nets will require exactly n measurements to verify that there are no shorts. It should be pointed out that not all shorting nets can be identified specifically using this method. If a 20 pF net shorts to a 1 pF net, then the total capacitance will be about 21 pF. This is much larger than the expected value of 1 pF, so that net will be flagged as a fault. However, since it is too close to the expected value for the 20 pF net, no fault will be indicated on that net. For this reason, the capacitance method is used primarily as a go/no go test. Further testing of the part is required usually to isolate the fault.

Opens within a net are indicated by a reduction in capacitance. Figure 13-5 shows a typical net with two possible break sites labeled X1 and X2. If the capacitance test is made at pad A, then a break at X1 will cause a loss of about half of the capacitance. A break at X2 will cause only a very small change in capacitance since only one pad was lost from the net and might not be detected as a fault. If, however, an additional capacitance measurement were made at pad D, then the capacitance measured would be much lower than the expected value since pad D is very near break X2. In order to make sure that opens are

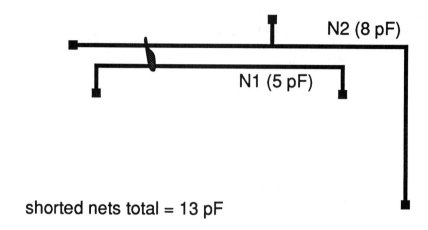

Figure 13-4 Capacitance of two nets shorted together will be approximately equal to the sum of the capacitance of each net.

identified regardless of their location in the net, a capacitance measurement is required for each pad on the net. The relative capacitance values measured at each pad on the faulty net can be used to help locate the fault within the net.

Single probe testing can be used to provide opens and shorts testing on a wide variety of MCM substrates. There is no inherent limitation on pad size or density since only a single probe is involved. As pad sizes decrease, the probe can be made smaller and the motion system more accurate. Since there is no fixed array of probes, distortion from one part to the next, as is common for co-fired ceramic parts, can be handled by an alignment procedure which maps out the distortion. The simple mechanical system results in a small equipment footprint with no need for the bulky and complex switch matrix required by a bed-of-nails tester. New parts can be set up in a matter of hours since there is no fixture required. The pad locations for the new part are simply programmed and the capacitance values learned from a sample of parts.

Single probe testing does have its limitations. The capacitance method can locate opens and shorts but it cannot find high resistance within a net. To the capacitance meter, any resistance less than a few megohms looks like a short circuit. This may not be adequate for MCM parts, particularly in the early prototyping stage, where high resistance faults due to line width and thickness variations may be common. Since the capacitance reference values are learned

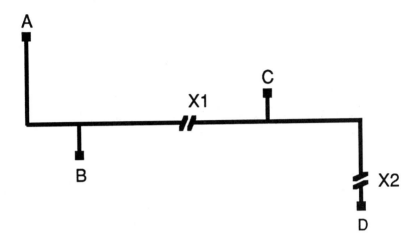

Figure 13-5 Net showing two possible break locations, X1 and X2.

from actual parts, a design error may go undetected. If the substrate has circuitry on both sides, then it must be turned over to complete the test. The testing speed of this tester is on the order of five measurements per second and is a strong function of the distance from one pad to the next. While the throughput is much lower than that for the bed-of-nails tester, the setup cost for a new product is much less and there is an inherent improvement in reliability realized by the use of a single probe instead of thousands of probes.

13.2.4 Two Probe Testing

Two probe testers provide the benefits of single probe testers with the added ability to make resistance measurements. In a two probe tester, the substrate is held stationary on a vacuum chuck or other fixture and the two probes move over the substrate making measurements at the required pad pairs. The measurement can be the resistance from one pad to another within a net or the leakage current between two nets. The simplest two probe testers have only resistance capability. These can perform measurements similar to those of a bed-of-nails tester by moving the probes from pad to pad. This is adequate for opens and high resistance testing within a net, but too slow for shorts testing. As was discussed in Section 13.2.2, the number of tests required to detect all shorts increases geometrically with the number of nets. To locate shorts on a substrate with 1000 nets would require over 500,000 measurements. A high performance two probe tester can make about ten measurements per second so it would take nearly 14 hours to test a single substrate. This could be reduced somewhat by eliminating tests between nets that could not possibly be shorted due to their routing, but in many substrates this would not be adequate.

Two probe testers typically add capacitance test capability to their resistance testing. In this way, they can perform with all of the advantages of a single probe tester for shorts identification but with the added ability to isolate shorts to specific nets. A resistance measurement is made between any net with excessive capacitance and all other nets with similar capacitance to determine which pair or pairs of nets is shorted. Integri-Test® and Bath Scientific Limited (BSL) both offer testers for the MCM market utilizing this combined measurement capability. Integri-Test® uses a patented method to combine the tests for resistance and capacitance and thus provide a higher throughput [5].

By making use of the resistance measurement feature of a two probe tester, a design verification of a new product can be performed. A design verification checks that the net list of the physical substrate matches the design. During verification, each net is checked for opens and then each net is checked to every other net for shorts. The shorts verification can be made faster by using measurements of capacitance to limit the number of tests, checking the resistance only between nets with similar capacitance. Also, two probe testers can make net to net leakage measurements using a high voltage source and an electrometer to measure the current.

Dual sided two probe testers, such as the BSL PRECISIONPROBE™ pictured in Figure 13-6, can test both sides of an MCM. The part is mounted in a fixture so as to expose both sides at the same time. The motion system is duplicated above and below the part to permit probing of connections on each side of the part and also from one side to the other. When a part is mounted for probing of both sides, there is no external ground plane provided by a mounting chuck. In this case, one of the probes is brought in contact with a pad connected to an internal power or ground plane to provide the electrical reference needed for capacitance measurement. If the part does not have internal metal planes, then it cannot be tested using the capacitance method unless it is mounted on a conductive chuck to provide the plane. Unfortunately, this restricts access to one side of the part at a time. An MCM with pads on both sides should include an internal metal plane in its design if the part is to be tested on a dual sided two probe tester.

The testing speed of both single and two probe testers is limited by the mechanical motion of the probes. Increases in the positioning speed of the probes translate directly into decreases in test time. The positioning speed

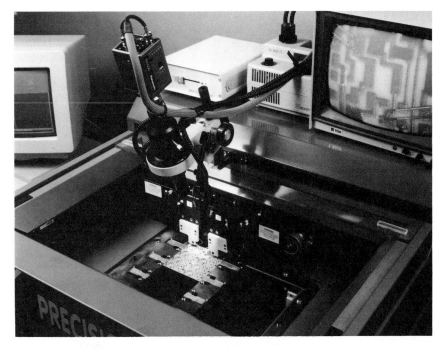

Figure 13-6 BSL PRECISIONPROBE™ two sided tester: interior view (courtesy of Bath Scientific Limited).

improves as the distance covered from one measurement to the next decreases, so moving probe testers operate faster on smaller substrates. A proprietary two probe tester designed and built by the author at AT&T Bell Laboratories averaged 12 measurements per second while testing thin film on silicon substrates.

One method to improve the testing speed is to reduce the total distance that the probes need to travel to make all measurements. For single probe testers, this is equivalent to a popular mathematical problem known as the Traveling Salesman Problem. This problem postulates a salesman who needs to visit a number of cities and would like to determine the best order in which to make the visits covering the least possible total distance. Solutions of this problem fall into a class of problems which are intractable; that is as the number of cities gets larger, the calculation time required to solve the problem grows explosively [6]-[7]. Exact solutions to the problem are feasible for a few hundred cities at most. There are many solutions to the problem which yield a good but not optimum route. A single probe moving around an MCM tester can be considered as a kind of traveling salesman.

Two probe testing somewhat complicates the problem since now we can consider the probes as two salesman which must visit different cities at the same time. Integri-Test® uses a combination of serpentine motion of one probe with the other probe moving as required to contact the remaining points on the net [8]. The two probe tester developed at AT&T Bell Labs used a scheduling algorithm which considered all measurements for all nets at once and then sorted them to obtain an optimal route [9]. The original sorting method used at Bell Labs was a Greedy Algorithm solution to the traveling salesman problem modified to handle the notion of two probes. In the greedy algorithm approach, the next probe position selected is the one which requires the least possible movement from the current position. Improvements in the route are achieved using more sophisticated algorithms and additional computation [10].

Two probe testers provide great flexibility in MCM substrate testing. Opens, shorts and high resistance faults can all be identified. Other two terminal measurements can be made by simply connecting an external meter to the probes. Two probe testers can accommodate parts with pad sizes down to 2 mils and even smaller with improvements in the positioning systems. Substrate distortion found on cofired MCMs can be handled using distortion mapping just like in single probe testers. Parts with die cavities present no problem by using a mechanism with programmable probe actuators. Probing from one side of a part to the other side also is possible using a tester configured for this mode of test. The primary disadvantage of two probe testing is the testing speed. Typical speeds are three to six tests per second on commercial machines, but this is improving with new models. As the testing speed increases however, it becomes increasingly difficult to bring the probes to a vibration free stop due to the high acceleration and deceleration required.

13.2.5 Electron Beam Probing

All of the probing methods described above use a physical probe which is brought into contact with the test pads and which can leave a mark on the pad.

In a properly designed test system, this mark usually does not cause any problems with wire bonding or with flip chip attachment. There is a practical minimum size to the pads which can be physically probed. Mechanical probing can usually handle any interconnect pad on the surface of the substrate since these must be large enough to connect to a component such as a die. In some instances, however, it may be desirable to probe a substrate during the fabrication process. Thin film MCMs are built up one layer at a time and the interconnect pads may not exist until the final layer is applied. Before application of the final layer, only lines and vias may be available for probing with probing areas well below 20 μm across. In this case, automated mechanical probing is not feasible and electron beam probing provides an alternative.

Electron beam probing operates on the same principle as a scanning electron microscope. An electron beam is directed toward test points on the surface of the substrate. The beam charges the selected test point to a preset voltage. A secondary electron energy analyzer is used then to locate other places on the substrate which also have been charged. Opens can be detected by charging one pad on a net and then checking for charge on every other pad. Shorts can be detected by charging one net and then checking that no other net has been charged. An electron beam flood gun is used to discharge or charge all nets at once. Electron beam testers are not commercially available yet. Significant development efforts have been made both at the IBM Research Center [11] and at Microelectronics and Computer Technology Corporation (MCC) [12]. The system developed at IBM has the additional capability of an electron beam flood gun mounted below the substrate. This is useful for testing continuity through a substrate by flooding the bottom surface with electrons and then locating all pads on the top surface which become charged.

The electron beam is deflected across the substrate circuit in the same way that a television paints a picture on the screen. Since there are no moving parts, the testing speed can be very fast. Speeds of 100 tests per second have been achieved and greater speeds are possible [12]. Electron beam testers are not without their disadvantages, however. They require that the substrate be placed in a high vacuum chamber that takes several minutes to pump down. Systems can be equipped with load air locks to speed this process, but there will always be a throughput penalty. There is also a tradeoff between maximum substrate size and resolution. As the substrate becomes larger, the resolution of the beam placement decreases. Errors on the order of 100 ppm are achievable, [12] so for a 10 cm substrate the placement accuracy can be held to 10 μm. While this is adequate to probe pads on the substrate, it may not be enough to probe vias or lines. As substrates grow larger, mechanical movement of the substrate under the beam may be required which would diminish the speed advantages. Electron beam testers also cannot check for high net resistance. The net charging is

Table 13-1 Comparison of MCM Substrate Test Methods.

Test Method	Typical Tests/sec	Fixturing Required	Minimum Feature Size	Distortion Compen- sation	Double Sided
Fixed Probe Array	2400	yes	4 mils	no	yes
Single Probe	3 - 10	no	2 mils	yes	no
Two Probes	3 - 10	no	2 mils	yes	yes
Electron Beam	100+	no	<1 mil	yes	limited

similar to the capacitance method used by a single probe tester and is insensitive to high resistance within the net. Resistances of less than several megohms all appear as shorts to an electron beam system. Finally, electron beam testers are inherently expensive due to the vacuum chamber and sophisticated electronics required for their operation. Despite these limitations, electron beam probing may provide high throughput production capability for MCM testing in the future as the technology is developed further.

13.2.6 Developing a Substrate Test Strategy

The development of a substrate test strategy requires careful consideration of a number of factors including volume of parts, technology of substrate, failure modes, level of testing required and where in the process testing will occur. For high volume parts and low pad density, a fixed probe array tester probably provides the best solution. If the pad density is high and the process is well developed, then single probe testing may be adequate. If the resistance of each net must be tested, then a two probe tester is a good choice. If probe marks cannot be tolerated on the finished component, then testing the substrate prior to final metallization will be required. Above all, consideration of these factors should feed back to the design area to provide a design which is easy to test. Table 13-1 summarizes some of the features of each of the available test methods.

13.2.7 Summary

Electrical testing of MCM substrates is essential to ensure a quality product and commercial equipment is readily available. Fixed probe array testers provide high speed but are limited in the range of pad sizes and densities that can be handled. Single probe testers provide a simple go/no-go test for a variety of substrates at limited speed. Two probe testers provide excellent test flexibility with limited speed, but may provide a cost advantage compared to fixed probe testing when used for low volume production runs [8]. Electron beam probers provide very high speed and can handle a wide range of MCM substrates but are not yet commercially viable. For many MCM substrates, including high density interconnects on silicon and ceramic, the only testing method which can provide a range of test capabilities is the two probe tester. This technology is fairly mature, but further increased in testing speed can be expected as the motion systems are improved. While all of the substrate test technologies described above have been demonstrated for pilot line volumes, further development will be required to handle the large volume of MCM production expected in the coming years. Substrate test will be a significant factor in the determination of the cost effectiveness of MCMs.

13.3 IC TEST

13.3.1 Requirements for MCM Modules

By definition, MCMs contain two or more bare ICs mounted to a common substrate. These ICs can belong to any logic family (CMOS, ECL, etc.) and may be analog as well as digital. The dividing line between hybrids and MCMs is indistinct, but hybrids usually contain a variety of components including lower lead count analog and digital ICs, while MCMs primarily contain high lead count application specific integrated circuits (ASICs), microprocessors and memory ICs. Since the vast majority of current MCM designs do not contain any analog ICs, we will focus our attention on the testing of digital components. While IC testing is a mature technology, MCMs present some unique challenges. Most ICs today are not tested fully and burned-in until they have been mounted into their packages since it is much easier to perform the final testing of the chip in its packaged form. ICs which are destined for MCM use must be tested fully as bare die since it may be difficult to perform a complete test once the die are mounted to the MCM substrate.

MCMs are utilized typically for high speed applications and require ICs and substrates capable of operating at high speed. Flip chip bonding is likely to be the method of choice for chip-to-substrate connection as MCM technology matures since it provides the shortest electrical path from the die to substrate and permits the die to be packed very close together. Some ICs designed for MCM use may have very small output drivers since they will only need to connect to other devices on the same MCM through low capacitance paths. Testing of these devices will require special electrical test equipment designed to present a very small load on the device.

The quality of die used in an MCM is a key factor to producing a finished part with high yield. To illustrate this point, consider an MCM with 10 die. Even if there is no yield loss from the assembly process, but 5% of the die fail during module burn-in, then the yield will be $(0.95)^{10}$ or only 60%. This means that 40% of the MCMs will require some sort of rework to obtain a fully functional module. This rework is usually costly since the die are bonded to the substrate and not placed in sockets like on a PWB. From this example, it is clear that a die which is to be used on an MCM should be as free of defects as possible prior to installation on the substrate.

While it very desirable to obtain known good die, there are still some technical and infrastructure obstacles which need to be overcome. It is difficult to perform high speed functional testing of ICs in wafer form due to parasitics introduced by the wafer probes. There is also no satisfactory method for burning in ICs either in wafer form or as individual die. From a business standpoint, most companies are not interested in selling bare die since part of their profit is derived from the package and it is difficult to guarantee the quality of bare die due to the technical problems mentioned above. In addition, IC manufacturers often shrink the size of their die without telling the customers since the die are fully packaged anyway. The success of MCMs hinges on the availability of known good die. It is for this reason that many companies are now participating in programs to create standards for bare die [13].

13.3.2 Introduction to IC Test

A complete test of a digital IC includes testing of the functionality of the device as well as its quality. IC testing is broken down into two basic areas: functional and parametrics. In functional testing, the IC is subjected to a sequence of input states known as vectors, and is monitored for the correct sequence of output states. Parametric testing provides a measure of how quickly the IC responds and the electrical specifications of the inputs and outputs. As the complexity of ICs grows, it becomes increasingly difficult to test completely all the functions from the I/O pads. This has given rise to a built-in self test (BIST) method which uses a set of structures designed into the IC to exercise some of the internal functioning of the IC. The self test process is controlled from the I/O pads and usually complements the traditional functional test. In addition to BIST, boundary and internal scan have become important design for testability features. Boundary scan provides access to all of the I/Os on a IC through four special purpose pads added to the IC. Internal scan provides a similar function to access internal nodes of the IC. Boundary scan is of great benefit to MCM test since access to all of the device I/O pads may be impossible.

13.3.3 Test Generation

Most of the ICs incorporated into MCMs consist of microprocessors, memory controllers, digital signal processors, memories and ASICs. Functional testing for all of these IC types is similar. Prior to any testing, the IC is modeled on a digital simulator and the functionality of the IC represented by software models. In some simulations, each gate on the device is modeled, while in others, the functionality of groups of gates or cells is modeled. Once the model has been created, a set of bit patterns is applied to the model inputs and the corresponding outputs are determined by simulation.

These inputs and outputs form the basis for a functional test of the IC. Creating the model is a fairly straightforward task since it is a representation of known physical elements. Generating the set of inputs to exercise the model is quite another matter. A thorough knowledge of the expected use of the device is required to create the input patterns. Creating the input patterns for a simple combinatorial logic circuit is easy since the device does not have any internal memory. This is not true for a microprocessor where the output state depends not only on the current input state, but also on one or more previous states. For this reason, it can be very difficult to create a set of input patterns to generate completely all possible output states.

In addition to covering all possible output states, the input patterns should be able to identify all possible faults inside of the IC. Faults inside the IC can take many forms, but the most common are stuck-at faults and shorts and opens. Stuck-at faults are gates which will not change state regardless of their inputs. Shorts and opens are caused by defects resulting in improper interconnection of gates. An internal fault can be detected only if the application of input patterns causes an unexpected set of output patterns. The ability of a set of input patterns to reveal a fault is measured as the test or fault coverage. For complex circuits, it is usually not practical to provide 100% fault coverage. The fault coverage typically increases with the length of the test pattern, so coverage usually can be increased by making the test longer. Microprocessors often require millions of input test patterns to obtain adequate fault coverage. Clearly, patterns of this length cannot be generated by hand. One method to create the test patterns for a microprocessor is to write a program using the appropriate assembler language which makes use of all of the microprocessor functions. The compiled test program is captured in its binary form and applied as inputs to the simulation program of the microprocessor. Input test patterns for memory devices can be created by simple algorithms such as walking bit and checkerboard. In ASICs, the generation of test patterns is aided by creating the device from a set of standard cells.

13.3.4 Boundary Scan and Built-In Self Test

A new standard has emerged recently to improve the testability of both ICs and completed MCMs. The standard, IEEE 1149.1 [14], commonly referred to as boundary scan, provides for the addition of a shift register at each of the IC I/Os. All of the shift registers are connected together in a serial chain with each end of the chain connected to an added I/O pad. Two additional pads are added to the IC to provide clocking and control of the modes of operation of this boundary chain. The boundary scan controller or test access port (TAP) is accessible through these four new pads. A detailed description of the TAP modes can be found in the data sheets for the boundary scan equipped IC or the IEEE 1149.1 standard. A single boundary scan register is shown in Figure 13-7. When the IC operates normally, the internal logic is connected to the I/O pads. In this mode, the boundary is transparent, although there may be some delay penalty due to the addition of the extra gates. In the test mode, the IC logic is disconnected first from the I/O pads. A sequence of bits then is shifted into the

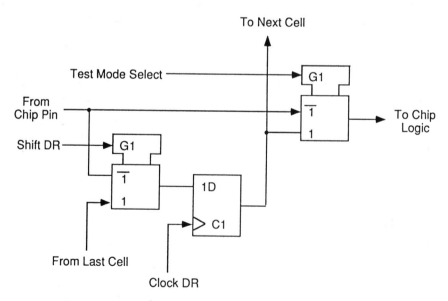

Figure 13-7 Basic boundary scan cell configured for an IC input pin.

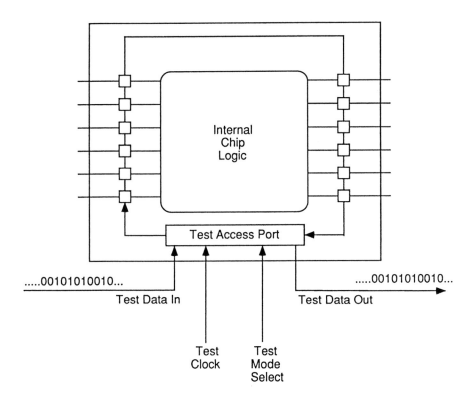

Figure 13-8 IC showing boundary scan cells connected together in a serial chain with TAP controller.

boundary so that each register contains the required logic level for each input on the device. The boundary control then connects the internal logic to the boundary registers and clocks the data on the boundary into the IC logic. Next, the state of the outputs is transferred to their boundary registers. The IC logic is disconnected from the boundary and all of the boundary registers are shifted out to read the results. Figure 13-8 shows an IC with boundary scan registers (or cells) and a TAP controller.

By using boundary scan, it is possible to access any I/O on the IC through connection to the TAP only. Internal scan builds on the boundary scan concept by adding additional serial chain paths inside of the IC as illustrated in Figure 13-9. These paths allow internal nodes of the IC to be accessed and thus tested.

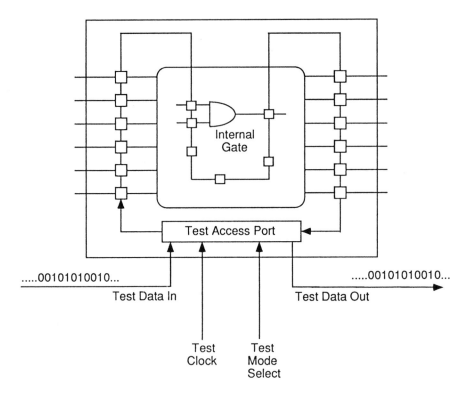

Figure 13-9 IC showing boundary scan and internal scan cells connected together in a serial chain with TAP controller.

In fact, if all gates in the device were accessible through internal scan paths, it would be possible to test fully the device by individually testing all of its gates and interconnections without an understanding of its function at all. Of course, this would not be practical as it would add greatly to the size of the IC and also would decrease its speed.

Another form of advanced IC testing is BIST. The concept of BIST is to provide additional logic on the IC itself which is designed to exercise some of the functions of the IC. There are many forms which BIST can take, but in general, an external tester provides clocking and control signals through IC pads to the BIST structure. After the test runs, the BIST reports its results to the external tester. This result is often in the form of a result code or check sum which is then compared to the expected results in the external tester. High speed

ICs can benefit greatly from BIST by qualifying the operation of the device at-speed [15] without the need for a costly tester with very high speed drivers and receivers. BIST devices require only that the tester provide a high speed clock signal and then wait for the final results. The use of BIST greatly simplifies the testing of complex ICs, but adds to the number of gates and, as a result, increases the die size. BIST can also be initiated from boundary or internal scan paths to reduce the need for extra I/O pads. Several articles and books have been published describing the use of BIST [16]-[18].

The most important function of boundary scan is to simplify testing of ICs once they have been mounted on a PWB or MCM. It is possible to connect the boundary scan paths of each IC together to form one long chain. This provides access to all of the chip I/Os through a single TAP. This feature is important particularly to MCMs and will be discussed in more detail in the section on module test. Boundary scan can be used in IC testing to perform functional testing without connection of all of the I/O pads. Devices with a large number of I/Os can be tested by converting all of the functional test vectors into serial chains and then loading them through the TAP. This does not provide a complete test, however, since there is no way to be sure that the IC I/O pads are connected to the IC logic since no signal is applied to the pads. In addition, an IC cannot be tested at full operating speed since all of the inputs and outputs must be accessed by loading in a serial fashion instead of in parallel as in normal operation.

13.3.5 Parametric Testing

Testing of ICs does not end with simply verifying the functionality of the device. Parametric testing is performed also to make sure that the device meets its performance specifications. Parametrics include both AC and DC measurements. Some of the more commonly measured AC parameters include propagation delay from an input to an output (t_{pd}), rise and fall time of output signals (t_r and t_f), the time that an input must be present before a clock transition (set up time, t_{su}), and the time the input must be maintained after a clock transition (hold time, t_{hd}). AC parametric testing is especially critical in ICs destined for MCMs due to their high operating speeds. DC parametrics include the voltage levels associated with an input high and low (V_{ih} and V_{il}), the voltage levels produced by an output high and low (V_{oh} and V_{ol}) and the quiescent and dynamic current required by the device (I_{DDQ} and I_{DD}) [19]. The specifications are determined by the designer in order to meet the requirements for devices in that logic family. An IC may be operating perfectly according to its functional test, but may be unable to meet its parametric specifications. In this case, the IC will be unable to interact properly with other components in the circuit.

13.3.6 Wafer Probing

During the fabrication of the wafers each of which contain many ICs, several tests are performed to maintain the quality of the process. The first functional tests of the ICs are made when the wafer is complete and the devices have not been separated into individual die. ICs which fail this test are discarded and the rest are mounted into packages. Typically a more complete test is made once the package has been sealed.

Functional and parametric testing is performed using automatic test equipment (ATE). A functional tester contains a number of driver/receiver channels which are wired to a fixture which makes contact with the pads of the device under test (DUT). Each signal pad of the DUT is connected to one channel of the tester. Power is also supplied to the DUT from power supplies built into the tester. The tester presents each of the input patterns which have been stored in a special pattern memory to the DUT in sequence and collects the results from the output pads. Discrepancies between the collected and expected results are reported as failures. Functional testers can operate over a wide range of clock speeds and provide the levels needed to drive a variety of logic families. The same testers usually include parametric measurement units to permit these measurements on the selected pads.

The wafer form provides an easy vehicle for handling, and permits testing of many parts without operator intervention. Functional testers configured for wafer testing use a wafer probing attachment equipped with a probe card as shown in Figure 13-10. The wafer is translated under the probe card so that each IC can be aligned and contacted by the probes. A low speed functional test typically is performed on each IC in turn. While a high speed test is possible at this point, it is complicated by the parasitic capacitance and inductance inherent in the probe card. High speed functional testing can be performed through the use of specially designed probe cards and connections to the functional tester [20]-[21]. The selection of tests to be performed at wafer level is determined by trading off the increased cost of additional testing and the expense of packaging a defective die. In the case of ICs to be used in MCMs, it is very desirable to test the part as completely as possible in wafer form since it will not be mounted in an individual package prior to installation on the MCM.

13.3.7 Die Carriers and Packages

Once the wafers have been probed, they are sawn apart to produce the individual die which are mounted then in single chip or multichip packages. In single chip

Figure 13-10 Typical probe card for use in testing IC wafers. (Courtesy of Micro-Probe, Inc.)

package applications, the IC undergoes a final test and burn-in once it is sealed inside the package. In the case of die to be used in MCMs, usually it is desirable to perform a complete test and burn-in prior to mounting on the MCM, as will be discussed later. Unfortunately, it is very difficult to handle bare die and there are no commercially available sockets which can hold a single die. One solution to this problem is the use of a carrier to hold the die during test and burn-in. The die may be removed from the carrier after test or it can be mounted directly on the MCM with the carrier.

One common carrier form is tape automated bonding (TAB) [22]. TAB is suitable for very high production since the carrier takes the form of a continuous tape onto which are mounted one die after another. Each TAB carrier consists of a lead fanout which is connected electrically to the pads on the die. There are many types of TAB carriers, but to be useful for electrical test and burn-in, the lead fanout must be supported with a nonconductive tape using a two or three layer construction. It is possible to test the ICs while still in tape form or to cut the tape up into individual carriers and test each part in much the same way as packaged parts are tested. Once the IC has been tested and burned-in, the TAB lead frame is trimmed to its final size and the IC is mounted directly to the package.

There are many issues surrounding the use of TAB carriers. The custom TAB design required for each new die can be costly and add to the manufacturing lead time. The testing socket must provide adequate electrical and thermal performance without harming the carrier. Handling can be difficult due to the fine lead pitch. For very high speed ICs, additional parasitic capacitance and inductance in the untrimmed TAB leads may make it impossible to test the part at full speed. Ideally, for MCM applications, an IC manufacturer would supply die mounted on TAB which have been tested fully and burned-in. While TAB is popular in Japan, it is not in widespread use in the United States, making it difficult to obtain die in this form.

Another carrier form is a simplified plastic or ceramic sacrificial package. This carrier can take many forms including a simple fanout pattern on a substrate or an actual pin grid array (PGA) package. The die is mounted to the carrier and removed after testing is complete. If the die is wire bonded to the carrier, then these bonds must be broken to remove the die. This can be a difficult and costly process which has its own yield loss. If removal of the die from the carrier is too difficult, then the die with carrier can be mounted on the MCM [23]. In this case, the carrier is designed to be as small as possible and to have minimal impact on the electrical performance. If the die is designed for flip chip solder attachment, then a carrier can be designed to permit easy mounting and removal of the die with no damage. The carrier can be designed so that the die can be separated mechanically or reheated to melt the solder bump pads. After separation, the die is cleaned and additional solder added as required.

13.3.8 Final Test

Final testing is usually performed on the IC mounted into its package or carrier. Fixturing of the part to the ATE depends on the type of package. For those parts such as PGAs which are designed to be inserted into a socket, a matching easy insertion socket is provided on a special board which mounts on the tester. This performance board is customized to each IC design, connecting signal and power to the proper channels on the tester. A special socket is required to hold surface mount packages or carriers during testing without lead damage. The electronics and wiring of the tester is designed to mimic the actual operating conditions of the IC and thus determine its true performance. A photo of a typical functional

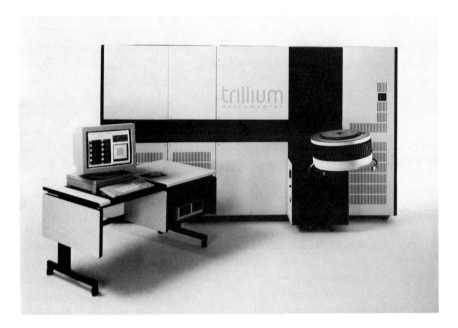

Figure 13-11 LTX/Trillium® Deltamaster™ functional tester capable of 256 pins with data rates of 160 MHz and pattern depth up to 128 million words (Courtesy of LTX Corp.)

tester configured to test packaged parts is shown in Figure 13-11. If no carrier or package is to be used, then the final testing must be performed on the IC in wafer form.

Final testing includes a full complement of tests. The part is first tested functionally at its full rated clock speed. If it does not pass this test, the clock speed is typically reduced and the test repeated. The parts are sorted according to their maximum clock speed to provide varying levels of performance. A full parametric test is performed then, and again parts may be sorted according to the results. Rejected parts often go on to diagnostic stations to determine the exact cause of the defect. The defect data can be transferred back to the manufacturing and design area to help improve the yield.

13.3.9 Burn-In

Burn-in is the process where the ICs are subject to accelerated aging conditions for many hours. A certain percentage of the ICs will fail during burn-in due to infant mortality (a term used in testing for early operational failure). These failures result from manufacturing defects, including gate oxide pinholes in CMOS parts, for example or photoresist or etching defects resulting in poor geometry and contamination [24]. The need for burn-in is related to the maturity of the IC fabrication process. New IC designs which utilize smaller design rules or new process steps are more likely to fail during burn-in. As the process is refined, the failure modes are analyzed and process steps corrected, reducing the number of failures at burn-in. In some cases, the failures at burn-in can drop low enough to warrant eliminating the burn-in step. In fact, one recent study has shown that in some cases burn-in creates more failures than the actual defects that it detects [25].

Burn-in is performed inside of an environmental chamber. The ambient temperature in the chamber is elevated to provide accelerated life testing. The IC can be operated in either a static or dynamic mode. In static burn-in, the IC is powered and static loads placed on the outputs. After a period typically ranging from 48 hours for some commercial parts to hundreds of hours to meet military specifications [26], the IC is removed from the chamber and again subjected to a full functional and parametric test. The power supply current to each IC is monitored during burn-in and if it changes radically, the IC has failed and no further testing is needed. In dynamic burn-in, the IC is exercised by supplying a limited number of input vectors and monitoring of the output states. If the outputs or current consumption changes from the expected results, the IC has failed. In both static and dynamic burn-in, the power supply voltage may be increased above nominal to accelerate the aging of the device.

Burn-in systems resemble large ovens with racks for several burn-in boards. Each burn-in board holds sockets for a number of individual IC packages. These sockets are similar to the sockets used in a functional tester. They must be rugged enough to tolerate the high temperature environment in the chamber and permit easy insertion and removal of the part without pin damage. Burn-in boards are customized for each product type since the pin-out and burn-in requirements vary. Depending on the burn-in test to be performed, the ICs are wired in parallel with some monitoring lines from each IC brought out to the edge of the burn-in board. Since many parts are burned-in at one time, the burn-in system has large power supplies and fuses to isolate each burn-in board or individual IC. If one device shorts during burn-in, its fuse will blow and the rest of the parts can continue to operate.

Burn-in is perhaps the most troublesome problem for ICs designed for use in MCMs. While it is possible to perform a full functional test on the IC in wafer form if required, wafer level burn-in or the equivalent is not yet a reality. Instead, if burn-in of the IC is required, it must be done in the bare die form using a carrier. In prototype MCM designs, the solution has been simply to eliminate the burn-in step for the die and substitute a burn-in (and possible subsequent repair) of the assembled module. While this may be acceptable for prototyping, it is unlikely to be viable economically for full scale production. While the subject of bare die versus module burn-in is still being debated, there are several arguments that suggest bare die burn-in will be required.

Perhaps the most persuasive argument for bare die burn-in relates to the nature of functional testing to be performed on the MCM. In many MCM designs, the final testing will verify only that it has been assembled correctly and operates at its rated speed. It is unlikely that a complete test of each die will be performed after MCM assembly since the access to each device I/O is limited. For comparison, in PWBs the most common final test is an in-circuit test which confirms that each IC is present and connected correctly. A functional test may also be performed from the edge connector of the board, but this test does not exercise each IC fully. It is extremely difficult to obtain high fault coverage for every IC from the edge of the module. Since the burn-in process stresses the internal logic of the ICs, it is essential that a full functional test be performed on each IC after burn-in (and before installation on the module) to ensure that no faults have developed.

In addition to the limitation imposed by functional test requirements, MCMs, by their very nature, integrate a range of components. An MCM may contain some CMOS devices as well as some bipolar or ECL devices. Each of these technologies will have it own burn-in requirements. Burn-in of the full module may not meet the needs of all of the die on the module. Even within a single logic family, individual part types may have differing burn-in needs related to the maturity of the product. A complex microprocessor may need a longer burn-in than a simpler part with more relaxed design rules. Finally, the ICs that find their way into MCMs tend to be the newest and highest performance products. These devices often use new technologies that usually have a higher fallout rate than for more mature technologies. As a result, burn-in is even more critical for these ICs than for lower performance devices found in single chip packages

The most promising solutions to bare die burn-in are wafer level reliability and wafer level burn-in. Wafer level reliability subjects the die and test structures on the wafer to a range of tests designed to provide stress aimed at likely failure modes. Typical tests include elevated temperature, voltage and output currents. Increasing the temperature and current will cause shorts or opens in improperly formed structures. High voltage will cause electromigration and breakdown in gate oxides and junctions. Wafer level reliability is really an extension of standard quality control procedures, stressing the ICs in the most revealing way possible. Wafer level burn-in, on the other hand, seeks to move the burn-in process back to the wafer. The entire wafer is burned-in, eliminating the need to handle and to socket bare die. The individual ICs are connected together on the wafer using additional layers of metallization. The wafer is then contacted using some sort of socket or probe card and placed in the burn-in system. After burn-in this metallization is either stripped off or disconnected in some way. This is not yet a commercially available process and those IC manufacturers pursuing wafer burn-in employ their own proprietary methodology.

Despite the technical problems, burned-in bare die are available from some IC manufacturers. The methods used are proprietary, but utilize wafer level reliability techniques and in some cases, wafer burn-in. Burned-in bare die are also available from vendors who mount the die on carriers to perform the burn-in. The die is supplied either on the carrier or it is unmounted if required. This is done most commonly for the military market where the cost of the finished MCM is high and the volume is small. At the current time, the supply of burned-in bare die is very limited and the price is much higher than that of fully packaged die. As MCM volume increases, it will become essential that known good bare die be available at a reasonable cost. It is likely that this goal will be met only by a combination of quality control and wafer level testing which can guarantee defect-free die without the need for further test or burn-in.

13.3.10 Summary

MCMs present some additional requirements to the standard methods of test and acquisition of ICs. The use of bare die on the MCM necessitates that full functional and parametric tests be performed on the die in unpackaged form. While this is not the current practice, it is possible technically and only needs the demand from MCM fabricators to make it happen. Burn-in, however, presents some additional difficulties. There is currently no widely accepted method of performing burn-in on bare die, and this will continue to have a significant depressing effect on the introduction of multichip modules. Once a technology is in place to produce known good bare die at a reasonable cost, the yield of MCMs will increase and their ultimate cost will be reduced. In the interim, die carriers can be used to permit fixturing of the die during burn-in and thus enable the MCM fabricator to obtain reliable functional parts. Users of ICs for MCMs also are more likely to require test facilities such as boundary scan and BIST which make it easier to test completed modules. These points will be discussed in the next section.

13.4 ASSEMBLED MODULE TEST

13.4.1 Introduction

After the die and other components have been mounted onto the MCM substrate, the completed module must be tested fully to ensure that it meets the design specifications. This testing includes both functional testing and measurement of AC and DC parametrics, similar to that performed on each individual die mounted onto the substrate. The module may also be subjected to a burn-in procedure to locate any defects in the assembly process or in the substrate itself.

MCMs can be viewed as either large, complex components or as small, high performance subsystems. The testing process used is influenced strongly by the way that the MCM is considered. Since most MCMs are destined to be plugged in or mounted to a PWB, the system designer will view the MCM as a component. As a component, the defect rate must be as low as that of other semiconductor components on the board. On the other hand, the MCM contains a number of individual components and may be repaired by diagnosing the fault and replacing the offending component. MCMs thus have a *split personality*: while they are being fabricated, it is possible and perhaps even essential that individual components be replaced as needed. Once the MCM is complete and sealed, the end user has no way to repair it. This leads to a testing methodology which fits into the dual nature of the MCM.

Fabrication of MCMs requires known good components. As previously mentioned, diagnosing faults as early in a process as possible invariably leads to the least cost. The assembled MCM, while still at the manufacturer, needs to be treated as a very small and complex PWB. Test equipment, diagnostic techniques and repair methods will be based on this requirement and focus on functionality. Prior to shipment to a system house where the MCM will be used, it will require testing to verify its quality as a component. This will include high speed functional testing, parametrics and burn-in.

This section reviews the methods that can be employed to test assembled MCMs along with a guideline to design for testability. The module test deployment sequence is covered from test generation through bring-up to production test. A reference covering design for test, boundary scan, built-in self test, modeling and fault simulation is available [18]. While this reference does not cover MCMs specifically, its concepts provide an excellent introduction.

13.4.2 Testing Strategies

There are a number of different test strategies that can be employed favorably to test the assembled MCM. The selection of the testing method depends on a number of factors, including quality of the individual die and substrate, acceptable failure rate at the customer's site, ease of fault diagnosis and cost of repair. The most aggressive testing program subjects the module to a full test of the internal logic of each individual die on the module. A greatly reduced level of testing just verifies that the components have been assembled properly onto the substrate. A review of these testing options will enable the MCM fabricator to select the most desirable methodology.

Full Functional Testing

From a fault coverage standpoint, the ideal module test would combine all of the tests for the individual components into one large test. Running this test would guarantee the functionality and performance of all of the internal logic in the module. The problem with this testing concept is that it may be extremely difficult and time consuming to generate a test which can exercise fully all of the internal logic from the module I/O pins. The test itself could have an enormous number of test vectors which would lengthen the test time and add to the testing cost. Despite these problems, there are some MCM designs that lend themselves well to this testing methodology. An MCM consisting of a microprocessor and SRAM in a bus arrangement can be tested by combining the tests of the microprocessor and SRAM together as long as the SRAM is accessible from the module I/O pins. If the SRAM is connected only to an internal bus, then the microprocessor can be used to test the SRAM, although this requires creating a new combined test. A full functional test of all components in a module has the advantage that it can detect logic defects that might have been created during the assembly or burn-in process.

Limited Functional Testing

If the assumption is made that the primary defects that show up during module test are assembly errors, (an assumption often made in the PWB industry) then a simpler testing method can be employed. Simulation of the MCM can be used to generate a test that provides a high degree of fault coverage for faults related to the individual chip I/Os. This limited functional test is much reduced in scope from the full functional test, but require much less effort to create it. In some cases, adequate coverage can be obtained by modeling the way the ICs interact without considering their internal behavior or gate level logic. Prior to limited functional test, passive testing can be used to make sure that no module I/O pins are shorted together.

Boundary Scan and BIST

If the chips on the module incorporate boundary scan features, then it is possible to create a test that detects all opens and shorts at the chip I/Os without the need to understand the functionality of the ICs [24]. Boundary scan is useful particularly in MCMs since it permits access to the chip I/Os with the addition of just four pins on the module. Figure 13-12 shows the boundary interconnection of several ICs on a module. Opens and shorts testing can be

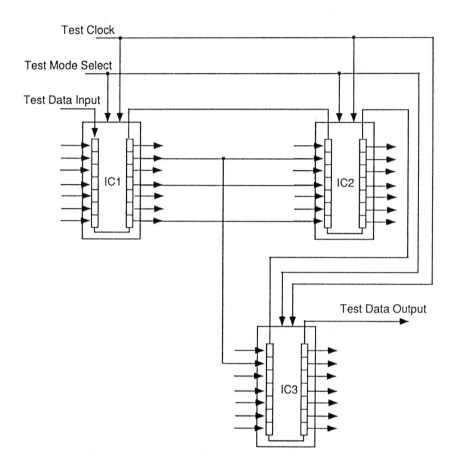

Figure 13-12 Boundary scan interconnection of several ICs on a module.

accomplished by applying an input test pattern to the TAP and by using the boundary scan chain to shift a known state to a single chip output [29]. The state at the output is transferred to all of the inputs on the same net through the substrate interconnect. The boundary scan chain is shifted out then, and examined to see if the pattern matches the expected result. This process is repeated for each chip output on the MCM, creating a walking bit type test. This test can be quite lengthy since the number of input bits = $(n_o)(n_t)$, where n_o is the number of chip outputs and n_t is the total length of the boundary scan chain.

The test time can be reduced by utilizing algorithms which use an adaptive technique to reduce the number of tests [30]. The connections from the module pins to the chip I/Os can be tested by applying vectors at the module pins, clocking the states that appear at the chip I/Os into the boundary scan registers, and then serial shifting the results out the TAP. In the case where an MCM contains some ICs with boundary scan and some without, a combination of boundary scan and limited functional test is required to test the interconnects fully.

The boundary cells are located near the edge of the die so the cells themselves are particularly vulnerable to mishandling of the die. Just the fact that the boundary scan path is intact provides some measure of the quality of the assembly. Boundary scan can be used also in a module to perform an IC level functional test. The test vectors which have been designed for parallel application are converted to serial form, shifted through the boundary to the correct chip and then clocked into the internal logic of the IC. Then, results from the chip outputs are shifted out in the same manner. This is useful for detecting any failures in the IC level logic which may have occurred as a result of assembly errors. Internal testing of the logic can be enhanced if the chips contain BIST features. In this case, the boundary scan path can be used to initiate and further to read the results of BIST [31].

Final Performance Test

MCMs are often considered for those applications requiring very high speed operation due to the short distances between die and the low substrate parasitics. If the substrate has been fully tested and characterized and the MCM assembled from die which meet the required high performance specifications, then no at speed testing of the MCM should be required. Unfortunately, there are many factors which can contribute to an MCM not running at its rated speed. Timing skew between chips, ground bounce, crosstalk between lines, variations in loading on the IC outputs and errors in the assembly process can combine to reduce the performance of the MCM. It is for this reason that a final test of the module at rated speed is required. The AC and DC parametrics are tested also to ensure that the module performs up to its expectations.

A final performance test of the MCM is similar to the limited functional test used to diagnose assembly defects. In the performance test, the vectors selected are designed to make sure that all ICs on the substrate can communicate together properly and that the MCM can communicate at its rated speed with external components. An MCM may pass a limited functional test and fail its at-speed or parametric test. Similarly, the at-speed test may not provide full coverage of all open, short and stuck-at faults for the internal chip I/Os.

The proper choice of module tests will depend strongly on the quality of both the incoming components and the assembly yield. In keeping with the dual nature of the MCM, a limited functional test or boundary scan test will be performed after assembly. Any defects in assembly are repaired and then the module will undergo a high speed functional and parametric test as a final qualification. While these two tests could be combined into one single test, it may be more appropriate to utilize a PWB type tester for the assembly test and an IC type functional tester for the final test. Board testers have many features designed to aid in isolating a fault to an individual component, as will be discussed later. Table 13-2 presents a comparison of the testing strategies available for module test.

13.4.3 Design for Testability

It is important to consider the testability of any product designed for volume production. This is particularly true of MCMs due to their complexity and small size which limits access to their internal components. Changes in design can have significant impact on the ease of testing. The designer should incorporate as many testability features as possible while still providing the required performance. A list of design for test features for MCMs is given below in the approximate order of desirability:

1. Design the circuit with modularity in mind. Breaking the circuit down into functional units at design time will lead to more modular test programs and simpler fault isolation.

2. Bring as many internal chip I/Os as possible to the module pins, even if not needed to operate the module. This reduces the problem of limited access since the internal nodes become external nodes. The selection of which I/Os to bring out is made by evaluating any improvements in fault coverage, and the ease in test generation and fault diagnosis. This simple concept is overlooked often in the rush to make the MCM design as elegant as possible, although many designs can tolerate the additional trace length and extra module pins.

3. Use ICs that have outputs which can be placed in a high impedance state. The use of ICs with tristate outputs permits one chip to be isolated from all the others by simply turning off the other ICs. This is a particularly good design for test feature when combined with Item 2 above, since it makes it possible to test a chip with its standard IC test vectors isolated from the other ICs.

Table 13-2 Comparison of Test Strategies for MCMs.

Test type	Test IC Logic	Test Interconnects	Test At-speed	Test Module Pins	AC Parametric	Fault Diagnosis Level	Physical Access
Full Functional	yes	yes	yes	yes	yes	logic level	module pins and test pads
Limited Functional	some	yes	yes	yes	no	die level	module pins
Final Performance	some	no	yes	yes	yes	die level	module pins
Boundary Scan Interconnect	no	within module	no	no	no	interconnect	module TAP
Boundary Scan + Module Pins	no	yes	no	yes	no	interconnect	module pins
Boundary Scan + Internal Scan	some	within module	no	no	no	partial logic level	module TAP
Boundary Scan + Chip BIST	some	within module	within chip	no	no	partial logic level	module TAP
Module Level BIST	some	some	within module	some	no	die level	module pins

4. Provide test access pads where possible for internal chip I/Os. This is similar to Item 2 above, but requires only that accessible test pads be added to the substrate instead of using more module I/O pins. The test pads can be accessed using a separate probe card or a manual probe. The test pads can be used for diagnostics or they can be part of the standard testing sequence. Once the module is complete, the test pads may be hidden by the package lid or heat sink, if desired.

5. Use ICs equipped with boundary scan and BIST. Boundary scan is the next best thing to having physical access to the internal nodes. Assembly defects which cause opens and shorts can be detected easily and the boundary can be used for diagnostic probing of an internal node in the event of a failure in the functional test. BIST allows high performance testing of IC level logic and can be used to make sure that the chip still operates at its rated speed.

6. Incorporate a boundary scan controller on the module where appropriate [32]. Additional chips can be placed on the MCM to control testing and to aid in fault diagnosis. In addition to running boundary scan tests, a module level BIST controller can perform a variety of test functions. If a microprocessor is present on the MCM it may take the place of a dedicated BIST controller.

13.4.4 Test Generation

Simulation
The first step in generating a test for a MCM is to create a computer simulation of the circuit. The simulation is used to model the behavior of the circuit in response to input patterns. The simulator calculates the output states of the MCM which form the basis for a functional test. AC and DC parametric tests are generated by combining the specifications for each IC used in the module to calculate expected values such as propagation delay through the MCM or quiescent current. It should be noted that the test generation process is accomplished best by the designer of the MCM who possesses an intimate working knowledge of its functions and its intended uses. The MCM fabricator needs to rely on the designer to lead the test generation process. The only exception to this is the generation of an assembly test for MCM parts designed with ICs utilizing boundary scan. These MCMs can be tested for interconnect opens and shorts without any knowledge of the IC functionality. Of course, a final test of the MCM functionality will still require design knowledge.

Simulation programs typically run on workstations and may require hours of computation time to provide an acceptable level of coverage. Common simulators include VERILOG™ by Cadence Design Systems, Inc. and QUICKSIM™ by Mentor Graphics, Inc. These same simulators are used to model the performance of individual ICs. Simulators typically represent an IC by modeling either its logic elements (AND, OR etc.), its overall behavior including timing or just its bus level interaction. In a logic level model, the response of each gate on the IC is modeled. This provides the most detailed simulation, but it is not required usually for MCM design. A behavioral model calculates the response of the IC to input patterns using an algorithm appropriate for the device. For example, it is quite straightforward to write a small program to model the behavior of a memory IC since it acts the same way as an array in a high level computer language. Models for more complex devices can be much more difficult, but they still only require the data book for the device, since this book fully describes the device behavior. The simplest models are known as bus level models. These models are created for ICs which operate on a bus and deal only with the communication to and from the IC. This can be quite useful for MCM test generation if the primary goal is to make sure that the ICs can run at full speed. It does not, however, provide very good fault coverage for opens and shorts at the chip I/Os. Both behavioral and bus models are offered by Logic Modeling Corp., for a wide range of ICs. In some cases, the IC manufacturer may be willing to supply the needed models.

Hardware Modelers

If a software model is not available, a hardware modeler can be used in its place. Instead of relying on software, the hardware modeler uses an actual IC to determine the next state. Most simulators have interfacing available to connect to a hardware modeler. Input patterns from the simulator are sent to the hardware modeler which contains a packaged IC mounted on a board inside of the unit. The modeler can hold several ICs at the same time, each on a separate board in the unit. The IC processes the input vectors and returns to the simulator the output vectors while operating at the clock speed of the modeler. Hardware modelers are an excellent method of running simulations when the software model is not available yet. One disadvantage to using this method is that the response returned from the device is not the typical response, but is specific to that copy of the part. One could create a simulation based on hardware modeling and then find that it does not work correctly due to manufacturing tolerances of the ICs. In addition, some ICs designed specifically for MCM use may not be available in a single chip package.

Timing

Timing information for the ICs must be included also in the simulation. This is available from the data book for each IC or may be part of a purchased hardware or software model. Since MCMs are designed for high performance, the timing information is critical to obtaining a reliable simulation. Timing information allows the simulation to check that input and output waveforms are present when expected. The propagation delay of a signal traveling through the substrate interconnect can be added to the model to provide a more accurate timing simulation.

Input Vectors

Perhaps the most difficult simulation task for the MCM designer is the creation of a set of input patterns for the MCM model. Just creating a model of the MCM does not provide test vectors. A set of input conditions is required also. The goal is to create a set of input patterns that will exercise the ICs on the MCM to obtain adequate fault coverage. The determination of adequate coverage is made by the system designer, but corresponds to the most likely set of failure modes. In many instances, the MCM contains a microprocessor, and responds to programmed instructions. In this case, input vectors can be created by writing programs designed for the MCM and capturing the binary form of the compiled program. If the MCM performs a simple function such as static memory or digital switching, an algorithmic approach can be used to create a program for generation of the inputs. A complex MCM will require the combination of several different approaches to create the input vectors. A sound philosophy to this process is to break up the functions of the MCM and then generate vectors designed to exercise each of these functions. This process is facilitated greatly by employing design for test strategies as outlined in the previous section.

Fault Simulation and Coverage

The test coverage level provided by the input patterns is obtained by observing the operation of the simulation in the presence of faults introduced into the model. For a fault to be observable, it must be propagated to an MCM output pin where it changes the expected output state. Functional tests designed to uncover assembly defects use a set of inputs that causes each internal node on the MCM to change state and propagate this change to the MCM pins during the test. An at-speed test might have less node coverage, instead emphasizing the interchip communications. A full functional test that tests the logic of each individual chip will require enough input vectors to observe any internal fault from the MCM pins. This can be extremely difficult and it is not expected to be used as a general method.

The level of fault coverage can be determined during the simulation by inducing faults into the internal nodes of the MCM and then by observing if the output vectors have been changed. Opens, shorts and stuck-at faults can be added easily to the simulation, and the effect of each noted. Fault simulation can, however, be an extremely time-consuming process. As an example, if an MCM has 500 nets, then the simulation will need to be run 500 times to introduce a single stuck-at fault into each net. If multiple faults are to be simulated, the number of simulations required increases dramatically. To simulate a stuck-at fault in just two nets will require (500 (499) / 2 = 124,750 simulations. The computation time for this would be excessive. The designer must rely on knowledge of the likely failure modes to choose an acceptable level of fault simulation.

In addition to determining the level of fault coverage, a fault simulation also can be used to create a fault dictionary. The changes in the output patterns are recorded as a function of the induced fault. This dictionary can be used by the tester to isolate a fault by comparing the fault dictionary to the actual vectors measured on the MCM under test. Unfortunately, if an MCM has more than one fault, the test vectors collected may not match any single entry in the fault dictionary. In this case, some testers employ a best fit approach to selecting the most likely cause of failure based on the test results.

Once the simulation has been run and fault coverage determined to be adequate, a test program specific to the test equipment must be generated. This can be done with the assistance of translator programs which convert the output of popular simulators into the specific format needed by a wide range of test equipment.

13.4.5 Test Equipment

Selection of test equipment to be used for the module test is dependent on the nature of the test to be performed. In the past, a variety of test equipment has been used to perform the functional test. Functional testers can be divided into two broad categories: board testers and IC testers. Board testers have high pin counts and diagnostic features including the ability to handle fault dictionaries and provide a trace back method for locating defects on a PWB using a manual probe. IC testers are configured to test single devices and can run at much higher speed with greater timing accuracy, but their per channel cost is much higher than a board tester. IC testers usually can handle test programs with long sets of test vectors, often one million vectors or more. Since performance has been a prime motivation in MCMs, IC testers have been used often for functional testing. In addition, early MCM designs often provided access to all I/Os of all chips through pins on the module, permitting each chip to be tested the same way as was done for single chip packaging. As the industry develops, it is expected that MCM test will require a two step approach. A low cost board type tester will be used to locate assembly defects using either limited functional testing or boundary scan. An IC type tester will be used then to check the at-speed performance and measure the parametrics.

Fixturing of the MCM to the ATE can be more difficult than for a single chip package or PWB. MCMs tend to have high pin counts and unusual lead arrangements. If the MCM is designed for solder installation, then a special test socket will be needed to hold the MCM without damage to the leads. The high speed of many MCMs also presents some challenges to fixturing. It may be necessary to bring controlled impedance lines to the MCM to terminate its I/O pins properly. This is practiced already for ultra high speed ICs and is adaptable readily to MCMs. As MCMs gain in popularity, standard packages will emerge making fixturing much easier. For thin film MCMs, it may be desirable to test the module after die attach, but before assembly into the final package. In this case, the modules will need to be placed in carriers and tested using a wafer probe as is done for ICs.

13.4.6 Bring-Up

Bring-up is the process whereby a new MCM design is tested and debugged. The test program generated from the design simulation is installed in the ATE and an appropriate socket wired on the tester adapter board. The tester is then configured for the part by setting up the mapping of test channels to package pins, by setting of the timing parameters and by configuring the power and ground. As with any complex system, there is a good chance that the module will not pass the test program the first time. Problems can include wrong channel assignments in the tester, wrong timing for the module, errors in the simulation, errors in the models and finally errors in the design itself. In addition, the module may contain assembly errors or defective die presenting a very difficult diagnostic problem. Once again we turn to design for testability as the savior of this situation. Strict adherence to modular design, along with features designed to assist in the diagnostic process, can provide a great deal of assistance. Steps to diagnosing a failed module at bring-up include:

1. Make sure that the module pins are mapped correctly to the tester channel pins and that the continuity at the pins of the module is as expected.

2. Run the boundary scan test to make sure that the parts are installed and wired as expected. This can go a long way to revealing errors since both errors in the substrate interconnect and in improper descriptions of the boundary will be revealed from this test.

3. Run a limited functional test which has been organized by the functionality of the module. The first section of the test might just reset all the ICs and check the output states. Diagnose any timing errors at this time.

4. If all seems well in the functional test, but the module does not produce the expected response, go back to the simulation with information on which vectors have failed and locate any errors in simulation.

5. Re-check the performance of relevant software models and make sure the hardware modeler is producing the expected responses.

6. If all else fails, go back to the original substrate and circuit design and make sure that the module will function as intended.

One simple method to verify that an MCM is functioning properly is to plug it into its intended final circuit and see if it works. This is a useful procedure in the bring-up stage, but does not provide adequate fault coverage in production testing. In addition, if the MCM does not work, it is difficult to diagnose the fault. An advantage to this type of test during bring-up is that once the MCM is found to be functioning in a real circuit, it is easier to debug the test program without the uncertainty of a faulty part.

If there is no test program available for the MCM because no simulation was performed, it is possible to capture the test vectors from the MCM operating in a real circuit. The clocking between the circuit and the tester must be synchronized and then the vectors can be captured. This method is not recommended. It is imprudent to design and fabricate an MCM without performing a simulation. The functional tester is operated in a vector capture mode and connected to the circuit in parallel with the MCM. This method has an additional disadvantage. It provides no way to diagnose faults since the intent of each vector in the test is unknown.

13.4.7 Production Test

Production testing of MCMs, like virtually any other manufactured item, will always be a tradeoff of cost versus yield. The more time spent on the final test, the more reliable the final product will be and the higher the yield will be after it is shipped. Testing does have a cost in time and equipment, so the desired quality level must be determined before any production test program can be implemented. A production test will include an initial test for assembly defects, a test of MCM performance as a component and an optional burn-in step.

Since each MCM product will be different in the quality of its components, the level of its performance and the desired final quality, it is useful to provide an ideal case as an example. Assume that the substrate has been tested fully for opens and shorts and that the ICs utilize boundary scan and have been tested fully at-speed and burned-in, if required. After assembly, the module will be tested on a board type tester by testing first for shorts with no power applied. Power is applied and a DC parametric test is performed. A boundary scan test is performed then to confirm that the die are installed correctly and that the substrate interconnect is correct. If the MCM fails at this stage, the test results are used to guide the repair or replacement of a die or in the case of overlay interconnect technology, the replacement of the interconnect itself. This cycle is repeated until the MCM has passed its assembly test. A high speed tester then is employed to perform a limited functional test of the MCM at its rated speed. The AC and DC parametrics also are measured at this time. Failures are diagnosed with the aid of a modular test program and the use of a fault dictionary generated during the simulation. Die are replaced as needed and the module starts at the beginning of the test process again.

Once the module is tested fully, it is then subject to a burn-in if required. Burn-in of a module uses the same equipment and methodology as is employed for single chip packages. A module will fail at burn-in due to a number of possible causes including assembly, interconnect, substrate and die defects. Module burn-in is recommended highly for new MCM products which do not have adequate production experience. If the die on the module have been burned-in and the level of defects in assembly and substrate are small, this step may be eliminated. As with IC burn-in, some failures will be detected on the burn-in unit itself by monitoring selected output pins and power supply currents. If no failures are detected on the burn-in unit, the module is retested to make sure no subtle problem has developed. Modules that fail after burn-in can be repaired and retested. In some cases, burned-in parts will not be repaired since it would require sending the entire module through burn-in again which might shorten the life of some of the components.

13.4.8 Summary

The testing of the assembled MCM presents many challenges to the world of semiconductor testing. Design for testability plays a key role in the success of bringing up a new design and in isolating the faults to a single component. Extensive simulation of the circuit is required to assure adequate fault coverage. Boundary scan will be an essential component in nearly every MCM design in the future since it provides the ability to probe internal nodes of the device without the use of extra test pads or pins. Finally, the MCM must be treated as both a large component and a small system. The small system nature assumed by an MCM during fabrication requires diagnosis and repair of defects, while the large component nature requires a very high level of quality for the part as shipped to the customer.

13.5 CONCLUSION

Testing of MCMs requires a broad knowledge of technologies. The substrates must be tested using special equipment which can probe the very small pads. There is a tradeoff in testing speed, flexibility and cost, but commercial equipment is available to achieve the desired goals. A successful MCM program requires a source of known good die. At the current time, fully tested bare die are in very limited supply, but as the demand increases, the supply will expand. Burn-in of these die remains a roadblock to the widespread availability of known good die. In the interim period, MCMs can be fabricated using die placed on carriers which ease the test and burn-in problems. Module test is mostly a case of employing a proper design methodology from the conception of the product through to the creation of the test. Good design for test methods can greatly simplify first time bring-up of the device and isolation of faults. Boundary scan and BIST are new technologies that are here to stay in MCM fabrication and their use cannot be emphasized too much. Simulation of the module as a whole is essential also. It is no longer possible to create a schematic and wire a circuit without the benefit of an extensive simulation. While MCM testing is not yet a mature technology, it offers the excitement of a new area of endeavor and the rewards of ensuring the quality of the finished product.

The testing of MCMs presents one of the greatest challenges to their widespread use. Significant improvements are still required in substrate testing speed, quality of bare die, module level test program generation and fault diagnosis. Once these problems have been brought under control, MCMs will move closer to offering cost and performance improvements over single chip packaging in many applications.

References

1   Kryterion 4000 Specifications, *Everett/Charles Test Equipment, Inc.*, Pomona, CA, Document Number 8803-10M, 1988.

2   "Cofired Ceramic Multichip Module Design Guide," *Alcoa Electronic Packaging, Inc.*, San Diego, CA, Rev. 91-08, Aug. 1991.

3   L. J. Webb, "Non-Contact Coupling Plate for Circuit Board Tester," U.S. Patent Number 3,975,680, Issued Aug. 17, 1976.

4   C. F. Spence, "Capacitive Continuity Testing of Multilayer Hybrid Substrates: Its Capabilities and Limitations," *Proc. of the 1985 Intl. Symp. on Microelect.*, pp. 174-182, 1985.

5   R. P. Burr, *et al.*, "Method and Apparatus for Testing of Electrical Interconnection Networks," U.S. Patent Number 4,565,966, Issued Jan. 21, 1986.

6   S. Lin, "Computer Solutions of the Traveling Salesman Problem," *Bell System Technical J.*, vol. 44, pp. 2245-2269, 1965.

7   Michael Garey, David Johnson, *Computers and Intractability: A Guide to the Theory of NP-Completeness*, New York: Freeman, 1979.

8   J. A. Conti and P. Hohne-Kranz, "Advantages of Moving Probe Testers Compared to Fixtured Systems," *IPC 31st Annual Meeting*, (Lincolnwood, IL), Technical Paper IPC-TP-679, 1988.

9   J. A. Moran, T. C. Russell, "Testing Process for Electronic Devices," U.S. Patent Number 5,023,557, Issued Jun. 11, 1991.

10 S. Z. Yen, *et al.*, "A Multi-Chip Module Substrate Testing Algorithm," *Proc. of the Fourth Annual IEEE Intl. ASIC Conf. and Exhibit*, pp. 9-4.1-4, Sept. 1991.

11 S. D. Golladay, *et al.*, "Electron-Beam Technology for Open/Short Testing of Multi-Chip Substrates," *IBM J. of Res. and Dev.*, vol. 34, no. 2/3, pp. 250-259, Mar./May 1990.

12 O. C. Woodard Sr., *et al.*, "Voltage Contrast Electron Beam Tester for Testing Unpopulated Multichip Modules," *Proc. of the 1990 Intl. Symp. on Microelect.*, pp. 370-377, 1990.

13 Microelectronics and Computer Technology Corp., 12100-A Technology Blvd., Austin, TX, a consortium of high technology companies, has hosted several workshops on the acquisition of known good die.

14 IEEE Standard 1149.1-1990, IEEE Standard Test Access Port and Boundary-Scan Architecture, New York: IEEE Standards Board, May, 1990.

15 L. Whetsel, "Event Qualification: A Gateway to At-Speed System Testing," *Proc. of the 1990 Intl. Test Conf.*, pp. 135-141, 1990.

16 B. Koenemann, G. Zwiehoff, and R. Bosch, "Built-In Testing: State-of-the-Art," *Proc. of the 1983 Curriculum for Test Tech. Conf.*, pp. 83-89, 1983.

17 E. J. McCluskey, "Built-In-Self-Test Techniques," and "Built-In-Self-Test Structures," *IEEE Design & Test*, pp. 21-36, Apr. 1985.

18 M. Abramovici, M. A. Breuer, and A. D. Friedman, *Digital Systems Testing and Testable Design*, New York: W. H. Freeman and Company, 1990.

19 Parametric measurements are defined in most digital data books. See: *Advanced CMOS Logic Data Book*, Texas Instruments, 1990, pp. 1.6-9 as an example.

20 T. Tada, *et al.*, "A Fine Pitch Probe Technology for VLSI Wafer Testing," *Proc. of the 1990 Intl. Test Conf.*, pp. 900-906, 1990.

21 S. P. Atahn, D. C. Keezer and J. McKinley, "High Frequency Wafer Probing and Power Supply Resonance Effects," *Proc. of the 1991 Intl. Test Conf.*, pp. 1069-1072. 1991.

22 T. C. Chung, D. A. Gibson and P. B. Wesling, "TAB Testing and Burn-in," *Handbook of Tape Automated Bonding*, J. H. Lau, ed., New York: Van Nostrand Reinhold, 1992, pp. 243-304.

23 G. Coors and R. Buck, "Pad Array Carriers—The Surface Mount Alternative," *Proc. of the 1991 Intl. Elect. Packaging Conf.*, pp. 373-389, 1991.

24 D. J. Klinger, Y. Nakada and M. A. Menendez, *AT&T Reliability Manual*, New York: Van Nostrand Reinhold, 1990, p. 44.

25 D. Romanchik, "Burn-In: Still a Hot Topic," *Test & Measurement World*, pp. 51-54, Jan. 1992.

26 MIL-STD-883C—Test Methods and Procedures for Microelectronics, Alexandria, VA: National Technical Information Service, 1983.

27 J. Reedholm and T. Turner, "Wafer-Level Reliability," *Microelect. Manuf. Tech.*, pp. 28-32, Apr. 1991.

28 P. T. Wagner, "Interconnect Testing with Boundary Scan," *Proc. of the 1987 Intl. Test Conf.*, pp. 52-57, 1987.

29 C. M. Maunder and R. E. Tulloss, eds., *The Test Access Port and Boundary Scan Architecture*, Los Alamitos, CA: IEEE Computer Society Press, 1990, pp. 11-21.

30 N. Jarwala, C. W. Yau, "A New Framework for Analyzing Test Generation and Diagnosis Algorithms for Board Interconnects," *Proc. of the 1990 Intl. Test Conf.*, pp. 565-571, 1990.

31 P. Hansen, "Taking Advantage of Boundary-Scan in Loaded-Board Testing," *The Test Access Port and Boundary Scan Architecture*, C. M. Maunder and R. E. Tulloss, eds., Los Alamitos, CA: IEEE Computer Society Press, 1990, pp. 81-96.

32 C. W. Yau and N. Jarwala, "The Boundary-Scan Master: Target Applications and Functional Requirements," *Proc. of the 1990 Intl. Test Conf.*, pp. 311-315, 1990.

Part C—Case Studies

Alice went timidly up to the door, and knocked.
"There's no sort of use in knocking," said the Footman,
"and that for two reasons:
First, because I'm on the same side of the door as you are:
secondly, because they're making such a noise inside,
no one could possibly hear you."

"How am I to get in?" Alice asked again in a louder tone.
"Are you to get in at all?" said the Footman.
That's the first question, you know."
"I shall sit here," he said, "on and off, for days and days."

"Oh, there's no use in talking to him," said Alice desperately,
"He's perfectly idiotic!"
And she opened the door and went in.

Alice in Wonderland
by Lewis Carroll

How are all the elements pulled together to create MCM-based products? The intent of the following chapters is to provide reports from some companies with successful multichip products. Some of these products are commercial in that they compete with non-MCM-based products in applications sensitive to cost. In each case, the MCM program was developed internally from existing expertise. In these case studies, the authors deal with issues that created real and imaginary barriers to considerations of MCM technology.

For the first time, some of the leading companies that have produced MCM-based products tell their **own story.** The list of companies included is not meant to be complete; *it is meant to be a sample.* In particular, the excellent MCM-based products developed at AT&T, Fujitsu, Hitachi, IBM, Mitsubishi, NEC, Siemens, to name a few, are missing from this Part. Case studies also could have been written by MCM vendor companies (that make bare and assembled MCM substrates), such as Alcoa, Cypress, Kyocera, Pacific Hybrid Microelectronics, NTK and TI. At the time of writing there are over

30 MCM vendor companies which also offer design services to help their customers realize their firstMCM product. An even larger number of companies are gaining experience by building MCM prototypes.

In this part of the book, several leading companies describe how they designed and implemented complete MCM products. From these chapters, managers will gain the insight needed to make crucial technology decisions and to organize programs for an MCM-based product or series of products. System engineers will learn what the critical product decisions are and how to make them. Design and manufacturing engineers will see how the concepts discussed in the previous two parts of the book are put into practice. All of these people will be able to understand how concurrent engineering is applied to MCM product development. Marketing personnel will use these case studies to relate packaging alternatives to the needs of their customers.

For the first time, this part provides the reader with:

• Detailed descriptions indicating how some systems houses have made critical MCM-related packaging decisions which best meet their cost and performance goals. For example, Chapter 14 discusses how Unisys made critical technology decisions for a number of its computer products. Chapter 17 shows how Digital Equipment Corporation decided on the details of the MCM technology that was used in the VAX-9000 computer.

• Complete presentations of the end-to-end process of conceptualizing, designing and manufacturing MCM-based products. These descriptions put many of the issues, such as yield and testability, that concern potential MCM product developers, in their proper perspective. Authors also discuss unexpected problems that arose and the solutions they found to them.

- An analysis of the packaging needs of products for aerospace and military applications and how MCMs can be applied as a solution.

- An overview of the most popular thin film MCM technology, silicon on silicon, with a description of the tradeoffs and the range of applications for which this silicon-based technology has been used.

- A full presentation of the manufacturing process for a high end, thin film MCM product, tying together many of the MCM manufacturing issues raised in previous chapters.

The main message conveyed by these chapters is that *MCMs are real* and that products can be made. The complete range of potential MCM types are covered in these chapters, from a simple ceramic package to perhaps the most advanced thin film MCM found. These chapters show the reader that not only is the technology accessible and beneficial to many electronics product manufacturers, but also that it is possible to use the high end of the technology to gain significant systems advantage.

14

THE DEVELOPMENT OF UNISYS
MULTICHIP MODULES

John A. Nelson and Randy D. Rhodes

14.1 INTRODUCTION

Do mainframe computers really need multichip modules (MCMs)? Won't advancements in silicon technology alone be sufficient for next generation computers? What are the challenges in designing MCMs? Can we design future computers with current packaging technology? These are some of the questions this chapter will answer.

One way to begin to answer these questions is to look at trends in computer packaging at Unisys. A mainframe computer consists of many components. The major elements are the instruction processor, the input/output (I/O) subsystem and the mass storage subsystem. All three of these elements utilize semiconductor technology. However, the instruction processor is the brain of the computer where the most demanding packaging challenges originate.

The largest Unisys instruction processor of the late 1970s contained nearly 25,000 semiconductor packages residing on 240 printed wiring boards (PWBs). In addition to long electrical signal paths between computing elements, 400,000 solder joints were required. Major advances in silicon technology permitted nearly a seven to one reduction in the packages and boards for the mid-1980s Unisys computer. Specifically, 3,900 packages were contained on 29 PWBs. Silicon was faster and electrical paths shorter, resulting in significantly increased performance. Reliability also increased because the number of solder joints was reduced to 117,000. The 1991 instruction processor in this family utilized silicon advances along with MCMs to achieve an even larger gain. Only 187 packages on four PWBs were required. Performance was over twice that of the previous generation. Big gains were made in reliability because solder joints were cut to approximately 35,000. The latest Unisys instruction processor, introduced in late 1991, utilized double sided multichip packages and leading edge PWB technology to get down to a single board processor.

How can you do better than that? The whole instruction processor on one PWB! What about putting two to four processors on one PWB? The drive to improve performance, reliability and cost continues. This chapter will preview the type of packaging technology that will permit two to four mainframe instruction processors per single PWB.

Multichip technology doesn't come easily! Electrical requirements, high power densities, complex assembly challenges and lack of an industry infrastructure are all issues to be addressed. This chapter will also provide some insight into the decision making process necessary to resolve these issues.

Finally, this chapter will address the significance of these packaging trends for mainframe computer manufacturers. Mainframes are a modest growth segment of the computing market. The most optimistic predictions result in growths of less than 10% per year. Workstation supporters come up with less than that. Modest growth and drastic parts count reductions result in underutilized factories. It is an industry problem. During 1991 we have seen competing companies willing to sell their excess packaging capacity to their competitors. Are silicon and packaging technologists putting themselves out of business? An article in the *Harvard Business Review* suggests that this may be the case [1]-[2]. However, there is a large market segment out there waiting for the right time to apply MCMs to their products. Exciting work in this area will face packaging engineers as this time arrives.

14.2 THE DRIVING FORCES BEHIND MULTICHIP PACKAGING

The development and application of multichip packaging is primarily driven by the interrelated requirements of performance and density [3]. As silicon performance has increased, the delay associated with the propagation of a signal from one silicon die to another has moved from the status of insignificant to become a limiting factor in system performance. In the past decade, this packaging delay moved from being a very small percentage of the clock cycle time to requiring nearly 100% of the clock cycle. The frequency at which this occurs depends on the number of devices, fanout required, etc.

The term packaging delay as used in this chapter warrants clarification. This delay consists of three major parts. The first is the delay of the output buffer that is required to drive the signal leaving the die. This buffer performs no logic function, therefore, its delay is strictly related to the buffering of the signal as required by the packaging environment.

The second part of the packaging delay is the delay that results from the total interconnect path from the output pad of the source die to the input pad of the destination die. This includes the interconnection of the die to the package, the package interconnect, the next level of interconnect (usually a PWB), the delay of the receiving package interconnection and finally the delay of the package to die connection for the receiving die. Note that this is the description of the delay for a single chip packaging environment.

The third part of this delay is that of the input buffer required to receive the incoming signal. Like the output buffer, this input buffer does not perform any logic function. It is necessary to terminate the incoming signal and isolate the internal nets of the receiving die.

In the case where this packaging delay reaches nearly 100% of the clock cycle, it is not possible to perform any logic function and a die crossing in one clock cycle. An entire cycle must be used to propagate the signal from a register in the source die to a register in the receiving die. This entire cycle is used to move information from one die to another. This cycle is a total loss from a system performance point of view.

14.3 ADVANCES IN SILICON DELAY MULTICHIP MODULES

It must be noted that the advances in silicon density also have helped to limit the impact of the packaging delay on system performance. Even though the packaging delay as a percentage of clock cycle had increased, the number of die crossings (or packaging delays) per logic function have been reduced. This is a direct result of the increase in density, allowing more functionality on each die. Today the increased silicon density allows entire mainframe processors or at least entire processor functions, to reside on one die. In the early 1980s, thousands of die were required. This decrease results not only in a reduced number of die crossings but also reduces the delay itself by shortening the length of interconnect required. This increase in density along with architectural innovations has limited the impact of the packaging delay and delayed the widespread use of MCMs.

14.4 SINGLE CHIP PACKAGES HAVE LIMITATIONS

This same increase in density does result in another issue which favors the use of MCMs. While there is always debate on the relationship of gate count to pin count, the pin count has continued to rise as the gate count increases. As the silicon gets increasingly faster and denser, multiple paths on and off the die will be required, even with MCMs, to provide the data fast enough to maximize the die performance. This increase in pin count per die will make single chip packaging impractical. Single chip packaging gets unattractive above approximately 600 pins, and highly impractical as pin counts near 1000. Array I/O technology will provide relief for the I/Os per die limit. However, the issues with package size, attach to the next level interconnect and the cost impact on the next level of interconnect all contribute to the impracticality of single chip packaging.

The requirement for greater packaging density is the second major driver for MCMs. Small size may be driven by sheer space limitations or by cost considerations for the overall system. Some applications have very strong limits on available space. The most recognized example might be a laptop computer. Actually, all levels of computers have some size restrictions. The allowed size for systems such as personal computers or workstations is basically defined by the application environment where these systems are used. The drive to make them more and more powerful within the same size limitations makes packaging density an issue. Another example is the desire to place powerful digital signal processing capability in satellites, missiles and mobile artillery. Using MCMs would eliminate the necessity to transmit data back to large systems for processing and significantly increase performance of such systems.

Even large mainframes are density driven. Floor space costs for a computer room are quite significant. It is recognized that mainframes have more latitude in dealing with thermal management challenges than do desktop personal computers and workstations. However, reductions in mainframe floor space and power requirements translate directly into cost savings for the user.

The drive to increase density is often related to performance. As packaging density is increased, the physical distances across which signals must be propagated are reduced [4]. This can be translated into higher performance at all levels of the system.

14.5 BASIC PACKAGING GOALS

In any discussion of multichip packaging, one should always review the basic goals. The first goal relates to the drive to increase performance. *The goal is to make the die crossing delay equal to the on die delay.* If this goal is achieved then a signal would cross from one die to another with the same delay and electrical requirements as going from one gate to another on the same die. This means that input and output buffers would be eliminated, making packaging a nonfactor in system performance. The system architect could ignore die partitioning in developing the system. The partitioning would then be driven by other cost factors such as optimum die size, pin count, etc.

The next goal relates to the driving requirement of increased density. The ideal packaging design would not add to the size of the die. *The goal is to have a package to chip area ratio of one to one.* This means that the area required by the package is no more than the die itself. Achievement of this goal would benefit both performance and cost at the system level.

Reliability is always a concern with any packaging and must be considered in any MCM. *The goal is to improve the reliability of MCMs over single chip packaging.* The packaging reliability should be insignificant when compared to the silicon itself.

Any packaging must have a cost goal. *The goal for MCMs is to cost less than the closest equivalent single chip packaging.*

In attempting to achieve the above goals, requirements for thermal performance must be addressed. Thermal performance is a major consideration which requires attention from the very start. Any MCM must be designed to provide the appropriate thermal management in a cost effective manner. *The goal is to provide the required junction temperatures for the silicon while meeting cost and density goals.* The junction temperature requirement is dictated by the reliability goals for the silicon and/or the performance goals. (In CMOS applications, performance goals usually dictate the maximum junction temperature.) Thermal management is quite often the limiting factor in MCM development.

14.6 MULTICHIP MODULE DEVELOPMENT PROCESS

14.6.1 Designing Through Technology Change

This section reviews the major decisions to be made when selecting technology for an MCM program. Techniques for evaluating choices are reviewed. Some experiences are shared to illustrate how unanticipated tasks were dealt with on Unisys MCM programs.

Traditionally, for a given silicon technology, transistor content doubles approximately every three years. As a result, the impact of silicon advances on MCM development programs is more significant than for single chip programs.

Single chip development programs can be very complex but seldom as lengthy as a multichip program. This insures that a custom single chip package has a reasonable life before it is made obsolete by advancements in the silicon family for which it was designed.

14.6.2 Develop a Multichip Packaging Strategy First

All multichip programs are designed for a specific silicon family of devices as well. In order to have a useful life, the multichip package should be developed and available well before the replacement silicon family arrives and reduces its cost/performance effectiveness. Generally, the new silicon family does not fit the prior generation MCM because the increased number of transistors are often in a larger die, consuming more power and containing more I/O.

On the front end of a MCM program, it is important to deal with this and decide on a strategy. One strategy is to accept that with each new silicon family a new multichip package development is required. Development time is reduced since extendability is designed in to the new strategy. No time is required to predict the impact of growth on the package features.

The second strategy is to plan for growth and design the multichip package to be able to accommodate larger die, more pins and more power when it becomes available. This approach results in a more complex development program which will require more resources and more time. The benefits are a longer product life, lower cost, fully characterized reliability and greater production efficiencies. The IBM thermal conduction module used in the 3081, 3090 and S390 machines is an example of this approach, where extendability was designed in to permit several generations of silicon to be accommodated [5]-[6]. The recently announced ES9000 system has a new MCM which leverages many of the design features of the earlier TCMs [7]. These systems illustrate the benefits of an extendable technology by accommodating at least four silicon families spanning more than a decade.

The Unisys Single Chip A-series Mainframe Processor (SCAMP), memory array module (MAM) and multiple random access memory (MRAM) MCMs described in this chapter are an example of the first approach. They were designed to serve one generation of silicon. Their development times were relatively short as they were extensions of mature technology.

14.6.3 Technical Decision Making in an MCM Program

There are less than ten major decisions to be made in the development of an MCM [8]. The following ones will occupy a major part of the program.

- How to connect the die to the package: tape automated bonding (TAB), wire bond or flip chip?

- What type of substrate is to be selected: printed wiring board, metallized ceramic, thin film in combination with a base substrate or silicon?

- Is the module to be hermetic or nonhermetic?

- How is heat to be removed: air, liquid or other?

- What is the strategy for test and repair?

- How will the module be connected to the next level of packaging: leaded or leadless, array or peripheral, socket or no socket, other?

- What type of CAD tools are required and how will they be integrated?

There are other decisions but these are more than can be covered in this chapter.

14.6.4 Making Multichip Module Tradeoffs in a Disciplined Way

To develop a multichip package, start by identifying the core technologies available within the industry. A summary of the most common ones is presented in Tables 14-1 through 14-3 titled Multichip Packaging Core Technology. Unisys has applied MCMs in products ranging from desk top to high end mainframes. As a result, Tables 14-1 through 14-3 have three categories of cost and performance for the differing product needs. Tradeoff studies were completed by considering which of the technologies we already had in-house. Other important factors are the performance attributes of the technologies as well as cost, development time, qualification time and risk factor. The amount of support from industry was also heavily considered. It is effective to prepare lists of performance characteristics and limitations to formalize the tradeoff process. Table 14-4 summarizes the limitations of the packaging core technologies from our perspective. These attributes are considered in narrowing the choices.

As a part of the final selection process, a risk analysis needs to be completed. The amount of development risk you are willing to accept is an important factor when opting for new materials and processes in an MCM program. Table 14-5 illustrates a way to look at the hermetic versus nonhermetic decision. During a development process, analysis is performed during the study phase of a new design. Since most MCM packaging concepts are quite complex,

Table 14-1 Multichip Packaging Core Technology Lowest Cost for Performance.

Chip to substrate connection	Wire bond, TAB or COB, low die I/O
Substrate technology	Low technology PWBs with fanout at OLB
Substrate to PWB connection	Solder module directly to PCB, low I/O count <400
Module sealing concept	Nonhermetic plastic encapsulation, usually at chip level
Module cooling concept	Still air or small fan, <5 Watts
Module manufacturing methods	Highest degree of automation due to volume

Table 14-2 Multichip Packaging Core Technology Medium Cost for Performance.

Chip to substrate connection	Wire bond, TAB or COB, higher die I/O counts
Substrate technology	High technology PWB or ceramic or composite, some fanout may be required
Substrate to PWB connection	Solder module directly to PCB or socket up to 700 I/O
Module sealing concept	Nonhermetic plastic encapsulation, usually at chip level
Module cooling concept	Forced air impinge air, 10 Watts or more
Module manufacturing methods	Fairly high degree of automation, modest volumes

Note:
Most of the above building blocks have limitations that are restrictive and limit extendability to higher levels of integration with increasing performance goals. These limits are summarized in Table 14-4.

Table 14-3 Multichip Packaging Core Technology Highest Cost for Performance.

Chip to substrate connection	TAB or flip chip, die I/O too high for repair with wire bond
Substrate technology	Multilayer ceramic of composites, OLB fanout is not required
Substrate to PWB connection	Sockets most common up 2,000 I/O
Module sealing concept	Usually sealed at module level, not always hermetic
Module cooling concept	Liquid impinged air immersion, 100 Watts or more
Module manufacturing methods	Some automation but volumes are not high

Note:
   Most of the above building blocks have limitations that are restrictive and limit extendability
   to higher levels of integration with increasing performance goals. These limits are summarized
   in Table 14-4.

it is impossible to anticipate all of the surprises one experiences during the implementation phase. A redesign or design tweak is considered a cycle of learning. Note that cycle of learning times are shorter for hermetic designs due to the maturity of industry experience. Furthermore, the number of cycles required is less, as is the sample size in testing programs. The nonhermetic MCMs does not have as much industry data available to reinforce a decision. Therefore, the development program will have to accommodate a larger number of cycles in the schedule.

We are not implying that one is preferred over the other. Rather, the example is designed to show that when you make a decision like this, do it in a disciplined way. Also, it should be noted that there is much interest and work in the area of nonhermetic modules now underway. The IEEE Gel Task Force driven by Jack Balde was an example of industry cooperation to complete evaluations of nonhermetic packages in a shorter time and with less resources overall. Another example is the MCM Substrate Size Task Force, where the goal is to encourage infrastructure participation by standardizing. As stated previously, it is very important to leverage the work done in the industry as a whole.

Table 14-4 Multichip Core Technology Limitations.

Core Technology	Density Limitation	Performance Limitation	Thermal Limitation	Manufacturing Issues
Wire Bonding	I/O count due to peripheral lead/spacing	Inductance of long wire and power distribution	Must remove heat through package	Repair of high I/O not feasible
TAB	Similar to wire bonding	Similar to wire bonding	Not an issue, can be done several ways	Unique lead frame and thermode for each die
Flip Chip	I/O not an issue	Inductance and power distribution not an issue	Not an issue	Major development effort on CAD and die process and assembly
Low Technology PWBs	Lines/cm feature geometries	Fanout at OLB reduces device density	Not an issue	Boards cannot handle temperatures needed for chip assembly
Multilayer Ceramic	Lines/cm feature geometries	Fanout at OLB reduces device density, high dielectric constant	Not an issue	Has upper size limits, long lead time, costly tooling
High Technology PWBs	To be studied	To be studied	Not an issue	To be studied
Composite Substrates of Ceramic and Thin Film	Not an issue, lines/cm approaches silicon	Not an issue	Not an issue	Has upper size limits, long lead time, costly tooling

Table 14-4 Multichip Core Technology Limitations (continued).

Core Technology	Density Limitation	Performance Limitation	Thermal Limitation	Manufacturing Issues
Composite Substrates of Thin Film on Silicon	Not as good as ceramic, lines/cm approaches silicon but layer counts limited by strength	It can only provide peripheral I/O connections, thus it is limited	Must be put in another package that reduces thermal performance	Has upper size limits
Module to Board Connection (hard or soft connect)	Pins per unit of area peripheral or array are only options	Connectors add inductance	Not usually an issue	Soldering high pin count modules into a PWB is a significant challenge
Air Cooling, No Fan	Chips have to be spaced apart	Longer signal runs, more chip crossings	< 2 Watts per device, SCAMP @ 4 Watts used a fan	Lowest cost thermal system
Air Cooling, Fan	Chips can be closer	Some improvement, but still medium performance A-15 class	< 8 Watts per device, A-15 was 5 - 6 Watts	Next lowest cost
Air Cooling, Impingement	Chips almost can be brickwalled	Medium to high performance	Up to 25 Watts per device	Getting to be costly
Liquid Cooling	Chips can be brickwalled	High performance	35 - 40 Watts per device	Very costly
Immersion Cooling	Chips can be brickwalled in three dimensions	Cray, as example	High power density	Very costly

Table 14-5 Package Reliability Assessment Planning Table.

ITEM	HERMETIC		NONHERMETIC	
	Conventional Ceramic Package	Advanced MCM	Conventional Plastic	Advanced MCM
Cycle of learning	9 weeks	15 weeks	12 weeks	24 weeks
Number of cycles of learning required	2	5 - 7	3 - 4	5 - 10
Sample size	50	100	100	100 plus
Residual risk	Very low	High	Medium	High

14.6.5 Temperature Hierarchy Management

The construction of any MCM is driven by the temperature hierarchy of the various processes used in sequence to produce it. Temperature hierarchy refers to the processing temperature at each step in the manufacture of an MCM. This hierarchy is managed to insure that processes down the line don't damage earlier process results. Many potentially useful processes are not feasible because they do not fit well with the other processes. We found it very useful to prepare a formal document as reflected on Table 14-6, titled *Package Temperature Hierarchy - Single and Multichip Formats*. Every process used to make and assemble the module along with the temperature and times involved is listed. It is important to list the processes where you have the potential to reduce reliability so they stand out for frequent review. As this list changes with time, as the MCM development program proceeds, it needs to be kept up to date. The importance of temperature hierarchy management cannot be overemphasized. During Unisys MCM program design reviews, the temperature hierarchy chart was a key program management tool. Final process selections and the required reliability verification tests were developed through using it.

14.6.6 Problems Encountered and Solutions Developed

No matter how well you plan your program, there are bound to be some surprises that result in additional tasks and longer development times. Some are worthy of note as they resulted in some significant achievements for us.

Table 14-6 Process Temperature Hierarchy.

PROCESS	PEAK TEMP (°C)	TIME @ PEAK TEMP (minutes)	REMELT/DAMAGE TEMPERATURE (°C)	REMARKS
Package Processing				
Fire Packages	1600		> 1500	
Braze Pins and Heatriser	830		> 780	
Plate Packages	25		> 300 (time dep)	
Precondition Vacuum Bake	150	1440		Nickel migration
Logic Die Assembly				
Die Attach (Ag/Glass)	440	2	> 450 (time dep)	Nickel oxide formation
Plasma Etch	25	5		
Wire Bond (Au Thermosonic)	150	10	> 300 (time dep)	Intermetallics at die
H_2 Reduction	335	5		
Lid Seal (Au/Sn Solder, N_2)	335	5	> 280	
Stabilization Bake	150	1440		
RAM Die Assembly				
Die Attach (Ag/Epoxy)	180	60	>180 (time dep)	Solids loss up to 200°C acceptable
Plasma Etch	25	5	> 300 (time dep)	Intermetallics at die
Wire Bond (Au Thermosonic)	150	10		
RAM Repair (if required)				
Repair	150 (substrate) 250 (die)	5 4		
Die Attach (Ag/Epoxy)	180	60	> 180 (time dep)	
Plasma Etch	25	5		
W/B (Au Thermosonic)	150	10	>300 (time dep)	
Lid Seal (Au/Sn, N_2)	335	5	>280	
Temperature Cycle	-40/+105	10 cycles		Package level temperature cycling
Heatsink Attach				
Attach AlN (Sn/Ag Solder)	250	2	> 221	
Burn-in	170	1×10^4		
Package-to-Board Attach	170	5	> 140	
Package Removal	185	0.5		
Stress Testing	-40/+100	20 cycles		Board level thermal cycling (5°C/min)

Conventional leak testing of hermetic packages involves subjecting the part to pressures from 15 psi to as high as 60 psi. A lid with nearly 4 square inches of surface area will sustain destructive loads under these pressures. Some companies have dealt with this problem by putting a pillar in the cavity to support the lid during pressure bombing. However, we did not have room for this so we had to seek an alternate method. A computer monitored, deflection leak detection system utilizing very low bombing pressures was successfully developed. On the front end of a new program, it is easy for a design team to assume that standard assembly and test processes will continue to work.

Another significant challenge was module sealing process development. Initially, we intended to use a conventional seam welder to hermetically seal our packages. This was consistent with temperature hierarchy management to minimize the temperature rise during sealing. However, with a cavity down design, seam welding processes existing at the time required that some pins be left off to provide room for the electrodes. System designers do not like to lose I/O pins. The solution was to introduce laser welding as a replacement for seam welding. The laser limited cavity internal temperatures to slightly over 100°C, and did not require any pins to be deleted. Implementation proved quite challenging as little laser welding experience was available from industry at large. Laser welding proved to be a very reliable, high yielding sealing method once established.

Other development tasks which consumed more time and resources than planned were custom multichip package test methods, test hardware, burn-in sockets and accessories for internal handling and shipment of the product.

Burn-in strategy and the burn-in hardware is a significant development task for a MCM. High power dissipation and constraints on temperature hierarchy have to be carefully considered early in the program since they can influence numerous design and process decisions.

As an MCM developer, you need to be sure you understand all of the environments the MCM will see in its assembly and during its lifetime of use. One challenging environment is a board level stress test called Environmental Stress Screening or ESS testing. A typical test could consist of 20 temperature cycles between -40°C and 100°C to break marginal solder joints permitting in factory repair rather than a customer field failure. An environment like this may require additional package strength or special pin metallurgy to provide the reliability margin required to survive the test. Identify these items early in the program when they can be accommodated in the design.

14.6.7 Good News for the MCM Developer

One of the concerns frequently discussed in industry circles is final yield for an MCM. The concern stems from the belief that bare die testing and assembly process complexity combine to impair yields. In practice, bare die testing is facilitated by chip design and by very effective bare die test methods. Also, an MCM can be designed so that with proper assembly process selections, it is possible to remove a defective die and replace it. There is considerable technical information available from published articles, consultants and suppliers to the assembly industry. Those of you about to design your first MCM should know that by paying attention to detail, it is possible to achieve final assembly yields in the high 90s. Continuous improvement programs are a very effective way to insure this.

The following items were confirmed during our programs: package suppliers can build very complex packages very competently; the industry infrastructure exists to supply many of the needed materials and piece parts; published works on others' experiences provide much of the guidance needed, and finally, helpful consultants exist if you need outside support.

14.6.8 Verify the Reliability of the Multichip Module

Upon completion of the design phase, every multichip program must go through a rigorous testing and qualification phase. During this time, the performance properties are measured to verify that system goals are met. Environmental testing confirms that the design is rugged enough to function reliably over the range of conditions anticipated during its lifetime. MCMs are more complex and more costly than an individual single chip package. Therefore, the testing phase of the program requires considerable planning so that costs for test specimens and special test methods are reasonable.

14.6.9 Summary of Development Phase Challenges

In summary, an MCM development program needs careful planning and implementation using proven development methods. Frequent design reviews are an excellent tool, along with test vehicles, to get early feedback on the design features which are new to the development team. A strong team approach with all critical skills represented is a must as well. Use the wealth of industry information available on MCM technology. Attend several of the MCM conferences held every year. Join several of the prominent technical societies and participate. These should include the IEEE, ISHM and IEPS. Follow industry standards activities like JEDEC to stay in the mainstream if you are not a large vertically integrated company.

14.7 UNISYS MULTICHIP MODULE IMPLEMENTATIONS

Historically, Unisys computer systems heavily utilize high performance memory devices for critical functions such as caches, control stores and register files. Industry SRAMs are used for these wherever possible to achieve the best cost/performance advantage. This results in a critical function that contains two die crossings/packaging delays. The first is from an application specific integrated circuit (ASIC) device that generates the address, writes data and controls signals to the static random access memory (SRAM). The second is the data from the SRAM to an ASIC which utilizes the data. It was the need to improve the performance and density of this critical area that drove the first Unisys development and application of MCMs.

14.7.1 SCAMP

The first Unisys MCM was driven by both performance and density requirements. This module, known as the single chip A-series mainframe processor (SCAMP), was developed for the MICRO A system [9]-[10]. For this module, a single ASIC was developed that implemented a mainframe processor. The module contained this ASIC and the required SRAM die for the control store. The features of this MCM, shown in Figure 14-1, are:

- 1 ASIC CMOS processor die
- 10 - 32k × 8 CMOS SRAMs
- Cavity up ceramic PGA
- 5.00 cm × 5.84 cm
- 255 pins, 2.54 mm (100 mil) grid
- 10 ceramic layers
- 4 watts total power

The ASIC die required 346 bond wires at 0.13 mm pitch and utilized aluminum wedge bonding. All the die were attached with a silver filled epoxy. The SRAMS were mounted on the same package layer as the package bond pads, requiring the wires to be down bonded with gold ball bonding. (The package wire bond pads were on the same surface as the die attach pads.) The epoxy die attach allowed the SRAM die to be replaced up to twice per site. The ASIC die was not replaceable due to the I/O pitch on the package. The package bond pads, due to the fine pitch, are not large enough to allow for new bond wires after a die replacement.

The temperature hierarchy required the use of a laser lid seal on a large lid of 4.6 cm × 5.2 cm. The epoxy die attach could not easily withstand the

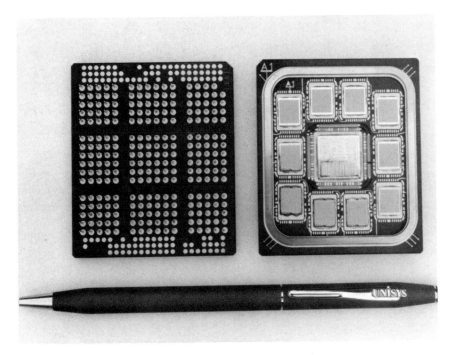

Figure 14-1 Single chip A-series mainframe (SCAMP) multichip module.

temperature of a solder seal operation. The laser allowed sealing without exceeding the temperature, which would damage the epoxy die attach.

The relatively low power of this CMOS module allowed the design to be cavity up. Cavity up is term which means the package cavity is opposite the pin grid array pin field. The resulting thermal performance achieved junction temperatures of 65°C at 0.27 meters per second (MPS) of air flow.

This module achieved the packaging delay reduction necessary for the system performance. The packaging delay for the two chip crossings (to and from the SRAM) would have been 10 ns in the single chip implementation. The MCM reduced this delay to 2 ns.

The MICRO A system required the placement of a mainframe as a coprocessor in a personal computer. This imposed severe space limitations since a single board must contain the processor, its control store, memory control, local main memory and an interface into the personal computer environment. The ratio of package to silicon area for the single chip implementation would have been 12:1. The SCAMP module reduced this ratio to 4:1. This significant improvement allowed the development of the desired system.

14.7.2 A16/A19

The next Unisys module was developed originally for the A16 system which first shipped in June 1990. This large mainframe system required 51 ASICs and 216 SRAMs per processor [10]. For this system the goal was a single board processor.

The MCM developed for the caches, control stores and register files of this system was the result of a highly interactive effort between the system architects and the packaging technologists. This effort resulted in a single ceramic package design used for all the multichip functions in the system by changing only the ASIC die. The resulting processor card contained 27 MCMs and 24 single chip packages.

The basic features of this module, shown in Figure 14-2, are as follows:

- 1 ASIC 10k gate array - emitter coupled logic (ECL)
- 8 - 1k × 4 SRAMs - 5 ns ECL
- 28 watts total power
- Cavity down ceramic PGA
- 4.6 cm × 4.6 cm
- Copper/tungsten die attach slug
  - for thermal performance
- 155 pins, 2.54 mm (100 mil) grid
- 14 ceramic layers
- Copper/Invar/copper convoluted heatsink

The 28 watts of total power made the thermal performance the main challenge in this module. This dictated the cavity down design with the thermal slug that extends through to the back of the package. Cavity down means the cavity is on the same side as the pin grid array pin field. This was necessary to provide as direct a thermal path to the heatsink as possible. The ASIC die was attached directly to this slug with a silver glass die attach. The convoluted heatsink was then soldered to the top of this slug. The result was a junction to ambient for the ASIC of 40°C and 44°C for the SRAMs with 2.9 CFM of impingement air.

The A16 system required only a single cabinet. The floor space required was 80% less than its predecessor. The package to die ratio for the single chip implementation would have been 23:1. The MCM used achieved a ratio of 12:1.

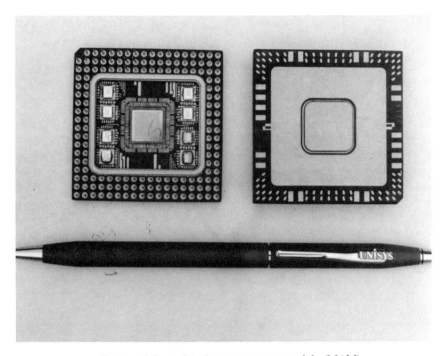

Figure 14-2 A16/A19 memory array module (MAM).

The packaging delay for the SRAM access would have been 6 ns for the single chip implementation. The MCM achieved a 1 ns packaging delay for this critical path. It is interesting to note that the packaging delay at 6 ns for the single chip implementation would have been greater than the 5 ns access time for the SRAM.

The same module is also being used in an even higher performance mainframe, the A19, which was announced in the first quarter of 1991. This system uses a four card processor with a flexible superscalar architecture. A single processor gives a 51 MIPS performance while a 6 processor system yields 240 MIPS performance. The packaging allows this system to require only 186 square feet of floor space, use only 66 KVA of power and use air cooling.

14.7.3 2200/900 Double Sided Multichip Module

The third MCM was developed for a mainframe that required 197 ASICs and

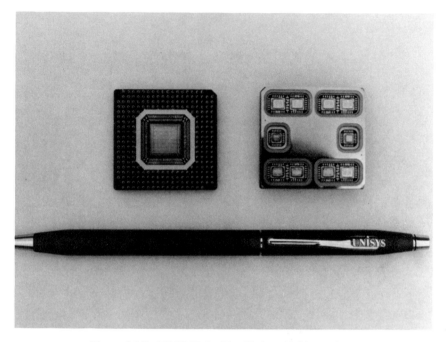

Figure 14-3 2200/900 double sided multichip module.

404 SRAMs per processor [10]. As with the A16, the goal was for a single board processor, but the initial estimates indicated that four cards would be required.

An innovative double sided package was developed to meet the requirements. This package used a cavity down approach for the ASIC and mounted the SRAMs in cavities on the top side above the pins. The basic features of this module, shown in Figure 14-3, are:

- 1 ASIC 9k gate array - ECL
- 10 - 1k by 4 SRAMS - 3 ns ECL
- 34 watts total power
- ASIC cavity down, SRAMs cavities up
- 3.0 cm × 3.2 cm
- Copper/tungsten slug
- Aluminum nitride heat spreader
- 252 pins, 1.52 mm (60 mil) grid
- 14 layers

There were challenges in every area for this package design. The thermal slug extended through the package and covered the entire top side. It contained cavities for the SRAM die that were sealed with individual lids. The aluminum nitride heat spreader was then soldered to the top. An individual liquid cooling jacket was soft attached to each package. The resulting junction to liquid temperature rise for the ASIC is only 18°C and 25°C for the SRAMs.

The single chip package to silicon density ratio for this would have been 10:1. This MCM achieved a ratio of 5:1. The performance improvement was again from a 6 ns packaging delay for the single chip implementation to 1 ns for this package. In this case, the packaging delay would have been twice the access time of the SRAM if single chip packaging were used.

14.7.4 Limits of Cofired MCMs

Each of these first three packages was based on cofired ceramic PGA technology. Each was limited to a single high pin count ASIC for several reasons. The first reason was repairability. The tight bond pad pitch required on the ceramic for the ASIC could not allow room for new wire bonds to be placed reliably if a die were replaced. The spacing on the bonds for the SRAMs allowed for large pads which made die replacement possible. It was found to be acceptable from a cost standpoint to have one die that could not be replaced, rather than multiple ones. This can be understood when one considers that even with a 99% yield on each die, a ten die module would result in 10% of the modules being bad. If the die yield goes to 95%, then over 40% of the modules would be defective without repair.

The high dielectric constant (approximately 10) of the ceramic is also a significant performance limit. For the short lines connecting from the single ASIC to the SRAMs this was not a major factor, but it would be if the general interconnect required for multiple ASICs were implemented. The ability to interconnect multiple ASICs would also be limited severely by the line density limit of approximately 20 lines per centimeter per layer of the cofired ceramic. For modules containing die with I/O counts in the 300 or more range, line densities of 200 to 400 total per centimeter are often required. This would require 10 to 20 ceramic routing layers or more.

The shrinkage factor for cofired ceramic also eliminates effective use of TAB. The final ceramic has a total dimensional uncertainty of up to 1%. For fine pitch TAB, this uncertainty would cause the leads to miss the package pads. Solder bump technology would increase the ability to use such tolerances at low pin counts, but the same problem would occur for high pin count die (along with other stress issues).

Figure 14-4 Unisys thin film multichip module.

14.7.5 Thin Film MCMs

UNISYS Thin Film
The next module developed by Unisys overcomes these limits by utilizing thin film on ceramic. This technology provides 200 lines per centimeter per layer of copper lines with a polyimide dielectric (dielectric constant of 3.5) on a ceramic base. These packages offer higher performance, higher density, higher reliability, full repairability and lower costs. One such module is shown in Figure 14-4.

Many versions of thin film MCMs are being discussed at this time. Some are done on silicon, some on metal and some on ceramic. The interconnect density and packaging features sizes vary significantly. The approach taken by Unisys offers significant advantages when compared to many of these.

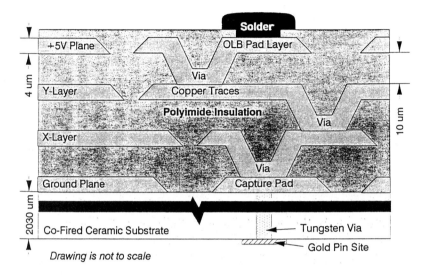

Figure 14-5 Thin film module cross section.

Interconnect Density

The first area of advantage is the interconnect density. The 200 lines per centimeter per layer allows the use of only two signal layers. A total of only four thin film layers are used. A cross section of this is shown in Figure 14-5. The first layer is a combination capture pad and ground plane. The capture pads connect to the vias in the ceramic and are sized to insure connection over the shrinkage range of the ceramic base. The two layers are the x and y signal layers. The top layer is a combination of the attach pads for the die and a five volt plane. This provides a stripline structure for the signals with excellent impedance control. The typical module is designed with the following characteristics:

- Line width = 12 μm,
- Line pitch = 50 μm
- Line impedance = 48 Ω
- Line capacitance = 1.3 pF/cm
- Line resistance = 4.0 Ω/cm

This significantly reduces cost over approaches, that require up to 10 or 12 layers to provide the same level of interconnect. The interconnect is also copper for lower line resistance. This is a major advantage in both AC performance and in the management of DC drops when compared to the aluminum used by many today.

Ceramic Base

The use of a cofired multilayer ceramic base offers several major advantages. First, the high dielectric constant of the ceramic is used to advantage by placing alternating power and ground planes in it. This provides excellent capacitive decoupling. If other substrates are used then this must be accomplished by using additional thin film layers, which is costly, or filter capacitors, which do not provide as good filtering and take up valuable area. The ceramic also provides a strong hermetic base for the module. If silicon is used, it then must be placed in some type of package at additional material and assembly cost.

The ceramic provides for optimum power, ground and signal pin location in a PGA type format. The pins are located under the die and thus provide the best signal and noise performance possible. There are several types of thin film substrates using silicon as the base. There is aluminim/polyimide on silicon, copper/polyimide on silicon and aluminum or copper using silicon oxide as the dielectric. The silicon base does not provide through vias to support an array type I/O design. This requires all power, ground and signals to exit the substrate along an edge. Thus, all interconnection paths to the next level are longer, resulting in a reduction of performance. The location of the pins under the die also results in smaller modules. The periphery connections of the other approaches require area both on the thin film substrate and the next level of packaging. This lowers both performance and increases cost when compared to the ceramic base approach.

Flip TAB

The use of flip TAB (and/or flip chip) is a significant advantage over the other alternatives. No lead forming or die attach is required. Full routing channels are also available under the die, which is not inherent in some of the other approaches. The result is the ability to place the die closer to each other improving the performance (by reducing signal delay) and reducing module size. Again, this results in lower cost and better performance than other methods.

It is recognized that thermal management is always a challenge, regardless of die orientation. With the die mounted to the substrate, the substrate has to be chosen to provide a low resistance thermal path. This usually results in a compromise in reduced routing channels or a high dielectric substrate. With the die backside up, as it is with flip tab, clever heat exchangers have to be employed. The IBM modules previously referenced provide one example as to how this is done. Having accepted this task, the substrate material can now be chosen to optimize electrical properties.

ILB = OLB

The ability to do outer lead bond (OLB) pitch (the attach of the TAB lead to the package) equal to inner lead bond (ILB) pitch (the attach of the TAB lead to the die) is another major performance and density advantage. Improvements of four to one or more over approaches that require the TAB leads to be fanned out to a greater pitch are realized. Once fanout is required the addition of a support ring is often required, which dictates the leads to be even longer. This is illustrated by Figure 14-6.

For example, for an ILB of 0.2 mm, fanning out to 0.5 mm requires leads of 12 mm. With an OLB pitch equal to the ILB pitch the length is less than 1.0 mm. This results in switching noise improvement by a factor of 20 over the 12 mm long leads.

The fine pitch outer lead bonding processes to support this method are challenging. Fortunately, the industry is working in this area. Laser bonding, single point bonding and various combinations of these are emerging as processes to address the challenge.

4 TO 1 DENSITY IMPROVEMENT

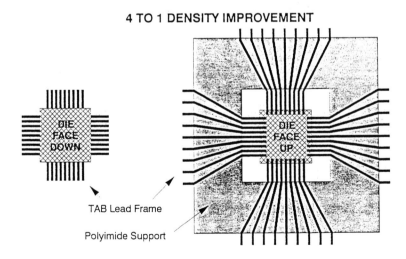

Figure 14-6 TAB lead fanout.

Reliability

All these modules contributed to improved system reliability in their respective applications. The thin film module will offer the greatest gain since it offers the greatest density (a ratio of two for the package to silicon) and the greatest reduction in connections. All examples offered reductions in:

- Interconnects
- PWBs
- Connectors
- Cables
- Backplanes
- Cabinets

These reductions all contribute to improved reliability.

Cost

The thin film MCMs also offer significant cost improvements. First, the packaging material itself is cheaper than the equivalent single chip ceramic PGA if approximately four or more die are placed in a module. The scrap cost also is reduced. Since the thin film module is completely repairable, a defective die is simply replaced. There is no package scrap due to a defective die as is true with ceramic PGAs. The cost of assembly is reduced. Once the die are mounted in the module only a single physical part is handled for test, burn-in, and insertion into the next level of packaging.

While all the above listed cost savings are significant, the greatest savings is at the system level. The savings are the result of:

- Lower PWB cost
  - Most of the interconnect is in the MCM
  - Smaller boards required
- Fewer PWBs
- Fewer connectors
- Fewer and smaller cabinets

14.8 MULTICHIP MODULE INDUSTRY ISSUES

The development and application of multichip packaging has been slower on an industry wide basis than many have predicted [11]. There are many challenges which have faced and continue to face multichip packaging.

Die availability is one of the major issues. Many suppliers of semiconductors are reluctant to sell bare die and some totally refuse to do so. This reluctance exists for several reasons. The bare die does require more careful treatment in handling and storage than a packaged part. Many potential customers are not equipped to meet the requirements of such handling and storage. The supplier is concerned that blame will be misplaced if parts fail in the field due to inadequate handling by the multichip assembler. Ability to burn-in bare die is a complex issue. While TAB claims to facilitate bare die burn-in, it can be a complicated and expensive process for high I/O die. Some suppliers lack the TAB and/or solder bump capability that is desirable for MCMs. While this does not make it impossible for them to supply die, it can be a significant handicap.

The management of die quality is another concern for both supplier and customer. Most suppliers do not have the capability to fully test a die prior to packaging [12]. (TAB, if available, does help this problem in that better testing can be done by more suppliers.) Both the supplier and the customer know how to do business when they both can fully test the component being exchanged (with no potentially damaging processing required by the customer). The issues of how to price, determine initial yields, and determine parts for return are some of the issues.

These are valid concerns that must be addressed. The customers need to develop the expertise to properly handle and store die. Even packaged parts can be damaged by mishandling, but customers long ago developed the expertise to handle them. It will take time for the same to happen for die. Additionally, suppliers will develop better die test methods and also more robust die. Time and the requirements will drive this development.

Semiconductor suppliers eventually will become MCM developers. This will eliminate some of the issues since both the die supplier and customer would then be the same company. However, that same fact will be a limitation. Most modules will require die from multiple suppliers. Semiconductor suppliers will not, in most cases, be able to negotiate for other suppliers' die.

A profitable price structure for the die business must also be developed. Semiconductor suppliers are structured such that packaging is a significant portion of their value added income. Their business is based on the profits generated not only by manufacturing the die, but also from packaging it. Their initial reaction is to view selling die as limiting their profits. This is another serious issue but one which time and volumes should solve.

Another factor which has slowed development is testing. In order to generate a good module test, the simulation models for the die in the module are required. Often suppliers of standard VLSI devices are reluctant to make these simulation models available. The gate level model is basically the design and is considered proprietary by these suppliers. This problem exists even with single chip devices since they also must be tested after board assembly. Application of board level methods and the use of higher level models will solve this problem.

Whereas bare chips for MCM use have not been readily available for the reasons stated above, an interesting phenomenon is occurring in the multichip package/substrate area. Figure 14-7 is a trend plot of the parts count and mainframe growth history discussed in the introduction to this chapter. Vertically integrated mainframe developers have seen parts counts drop, growth almost level and development complexity climb rapidly for each new machine. This trend is predicted to result in excess factory capacities and excess inventory.

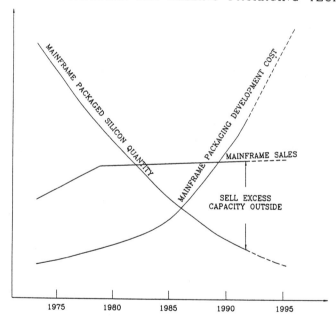

Figure 14-7 Mainframe component trends.

Furthermore, these factories are operated with expensive and complex equipment mandated by the complexity of the products required by mainframes. Although companies will attempt to fill these voids with products from faster growing lines such as workstations, differences in technology will not always make this practical.

One way to deal with this excess capacity and/or inventory is to sell packaging technology/plant capacity to others, including competitors. A number of the Fortune 500 companies publicly announced during 1991 that their packaging technology could be purchased [13], including thin film substrates, advanced ceramic modules, assembly services and so on. This trend should have a positive effect on developing the infrastructure required for smaller companies to develop MCMs. It should also lower the cost of developing modules for the entrants into the marketplace.

MCMs will make it! The development is following a chicken and egg scenario. Today, everyone would use them if the infrastructure were there, and the infrastructure would be there if the volumes warranted it [14]. This is the normal situation for a new technology. Surface mount technology followed a similar, though less complex, development.

Initial uses will be in areas where performance and/or density dictate that no other reasonable alternative exists. These applications will drive the development of the infrastructure required. They will also slowly increase the volume and drive the cost down such that more applications will be addressed.

Acknowledgment

The authors wish to acknowledge the contributions of a large number of people in Mission Viejo, Rancho Bernardo, Roseville and Tredyffrin, who contributed to the developments covered in this chapter.

References

1   A. S. Rappaport and S. Halevi, "The Computerless Computer Company," *Harvard Business Review*, vol. 52, pp. 69-80, July/Aug. 1991.
2   "Should the U.S. Abandon Computer Manufacturing," *Harvard Business Review*, vol. 52, pp. 140-160, Sept./Oct. 1991.
3   E. Davidson, "The Coming of Age For MCM Packaging Technology," *IEEE Computer Packaging Workshop*, (Lake Tahoe, CA), May 1991. No published proceedings.
4   H. W. Markstein, ed., "Multichip Modules Pursue Wafer Scale Performance," *Electronic Packaging & Production*, vol. 31, pp. 40-44, Oct. 1991.
5   A. J. Blodgett and D. R. Barbour, "Thermal Conduction Module: A High Performance Multilayer Ceramic Package," *IBM J. Res. and Devel.*, vol. 26, pp. 30-36, 1982.

6   J. U. Knickerbocker, *et al.*, "IBM System/390 Air-Cooled Alumina Thermal Conduction Module," *IBM J. Res. and Devel.*, vol. 35, pp. 330-348, May 1991.

7   V. L. Gani, *et al.*, "IBM Enterprise System/9000 Type 9121 Model 320 Air Cooled Processor Technology," *IBM J. Res. and Devel.*, vol. 35, pp. 342-351, May 1991.

8   B. Blood, "ASIC Design Methodology For Multichip Modules," *Hybrid Circuit Technology*, vol. 8, pp. 21-27, Dec. 1991.

9   J. Nelson and C. Cheves, "Multichip Module For A Cost Driven Mainframe," *Proc. Internat. Electronic Packaging Soc. Conf.*, IEPS, (San Diego CA), pp. 230-240, Sept. 1989.

10   R. D. Rhodes, "Multichip Packages at Unisys - Cofired Ceramic to Thin Film," *NEPCON Proceedings*, (Anaheim CA), vol. 1, pp. 477-485, Feb. 1991.

11   "Emerging Technologies, What The Experts Say," *Inside ISHM*, vol. 18, pp. 5-12, Nov./Dec. 1991.

12   J. D. Mosley, ed.,"Multichip Modules, Lack of Standards Impedes Design Issues," *EDN*, vol. 21, pp. 35-42, Jan. 1992.

13   K. O'Brien and T. Williams, eds., "A Changing Hybrid Marketplace," *Hybrid Circuit Technology*, vol. 8, pp. 14-18, Dec. 1991.

14   "The MCM Market: It's Anybody's Guess," *Hybrid Circuit Technology*, vol. 9, pp. 6-7, Jan. 1992.

15

HIGH PERFORMANCE AEROSPACE MULTICHIP MODULE TECHNOLOGY DEVELOPMENT AT HUGHES

Robert L. Bone, Dennis F. Elwell, James J. Licari,
Robert S. Miles, Kevin P. Shambrook, Stanley M. Stuhlbarg,
Philip A. Trask, David F. Zarnow

Technology development occurs only when needs are recognized and financial support is applied to focus the necessary personnel and materials. Much recent progress in the development of multichip modules (MCMs) can be attributed to efforts supported by aerospace industries. These industries were among the first to recognize that semiconductor device complexity was beginning to outstrip an engineer's ability to ensure effective intercommunication between chips in a system and among chip clusters.

This chapter discusses the MCM evolution as it parallels the development of military and commercial systems over the past ten years. It stresses the changing role of packaging as a driving force in system definition, describes several key technological developments that are spearheading this packaging drive and outlines the need for a comprehensive computer aided design, manufacturing and testing operation to bring about successful systems implementation.

15.1 MCMs MEET THE NEEDS OF SYSTEMS EVOLUTION

An increase in sophistication and performance levels of military electronic systems, made possible by the rapid development of VHSIC and VLSI chips,

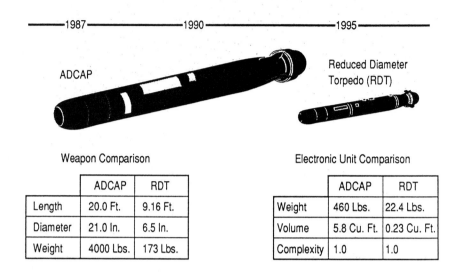

	ADCAP	RDT
Length	20.0 Ft.	9.16 Ft.
Diameter	21.0 In.	6.5 In.
Weight	4000 Lbs.	173 Lbs.

	ADCAP	RDT
Weight	460 Lbs.	22.4 Lbs.
Volume	5.8 Cu. Ft.	0.23 Cu. Ft.
Complexity	1.0	1.0

Figure 15-1 The trend for torpedoes is toward higher performance and smaller electronics. Although the new reduced diameter torpedo has approximately 4% of the volume of the ADCAP torpedo, its electronic unit is of similar complexity.

must be achieved simultaneously with requirements for lighter, more compact microelectronic packaging. This twin challenge - higher performance, but smaller packages - has driven the military electronics industry to seek radical new approaches to the fabrication of MCMs.

The increased complexity of military electronics throughout the aerospace industry parallels a rapid rate of change in the commercial market. In particular, computer manufacturers have developed novel packaging techniques that provide speed and density performance to match the rapid evolution in the speed of memory and logic chips. In military packaging, there are additional requirements of ruggedness and reliability that must be met by manufacturers. Commercial products also are becoming more reliable.

15.1.1 Examples of Trends in Weapons

Examples of military systems that require order-of-magnitude packaging improvements over previous generations can be found in the seas, in the air, and in space. Increasingly sophisticated weapon systems are required to counter improvements in a potential enemy's weapons. Even if the apparent end of the Cold War reduces the need for tactical weapons, a national need is likely to

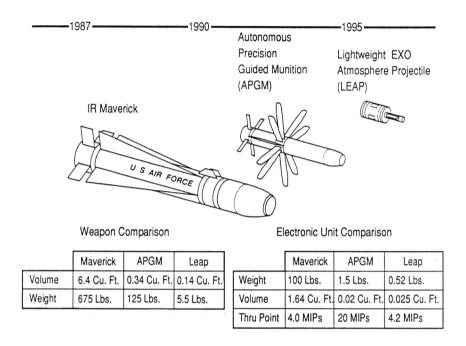

Figure 15-2 Higher performance and smaller electronics is also the trend for missiles. The developmental APGM and LEAP have similar or faster data processing capability than the MAVERICK, but are tiny in comparison.

remain for the most modern surveillance systems and electronic aids to counter terrorism and drug operations. Such equipment generally processes data from various sensors and, in many cases, evaluates the input and provides an automatic response.

A "smarter system" is one with more powerful computers and intelligent sensors. Smarter and smaller weapons lead to an increase in performance and a decrease in cost.

In the Seas

Figure 15-1 shows a size comparison between the Department of Defense (DoD)/Hughes' advanced capability torpedo (ADCAP), developed in the mid to late 1980s, and the projected reduced diameter torpedo, which represents the next generation. The weight of the new torpedo will be approximately 5% of the ADCAP weight. At the same time, the electronic sophistication has increased.

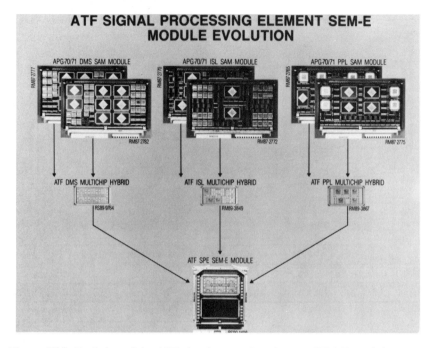

Figure 15-3   Evolution of the ATF signal processing element, SEM-E, module.

In the Air

Successive generations of missiles and projectiles are compared in Figure 15-2. The MAVERICK missile, currently in production, is shown on the same scale as the Autonomous Precision Guided Munition (APGM) and the Lightweight Exo-Atmospheric Projectile (LEAP) missiles. Even though modern, highly integrated semiconductor devices and advanced hybrid technologies are used to control the MAVERICK missile, the electronic units of the APGM and LEAP must be less than 2% and 1%, respectively, of the weight and volume of the MAVERICK electronics unit.

The Advanced Tactical Fighter (ATF) Signal Processing Element (SPE) is undergoing significant packaging evolution. As shown in Figure 15-3, 2" × 4" MCMs have become key elements in this evolutionary change. The modules can be mounted, two per side, on Standard Electronic Modules (SEMs), or three per side on Standard Avionic Modules (SAMs), shown in Figure 15-4.

Figure 15-5 shows the 6" diameter LEAP Guidance Unit electronics implemented in the High Density Multichip Interconnect (HDMI) packaging technology, discussed later in this chapter.

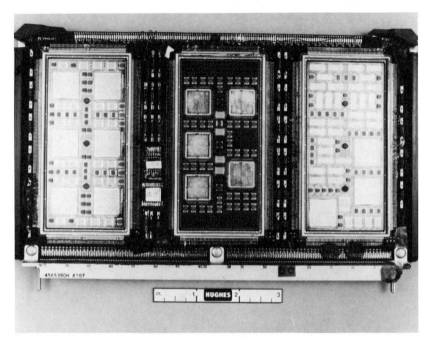

Figure 15-4 MCMs mounted on SAM board.

In Space

The system requirements of the Strategic Defense Initiative (SDI) are making radically new demands on packaging technology. Figure 15-6 is an artist's impression of the Hughes' Boost Surveillance and Tracking System (BSTS), one of the surveillance units to be deployed as part of SDI. According to an article in *Defense Electronics* [1], the BSTS will require analog to digital (A/D) converters with a greater dynamic range than required in any previous system. In this case, military packaging technology is likely to be ahead of that needed in other leading commercial technologies.

The DoD designated microelectronics and packaging as "one of the eight most critical technologies... required to ensure long term qualitative superiority of United States weapons systems" in its 1991 "Critical Technologies Plan" [2]. While there is no ranking order within those eight technologies, improved microelectronics and packaging are the only hardware elements designated as essential to all of the DoD top 12 mission goals. DoD succinctly and

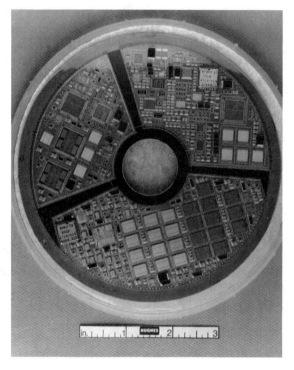

Figure 15-5 Use of HDMI MCM technology in the SDI LEAP program results in a single card electronic guidance unit which weighs only 140 grams.

emphatically states, "... microelectronics technology has a pervasive effect on virtually every U.S. weapon system, current or future" [2]. Order-of-magnitude advancements in microelectronic packaging provides unexcelled leverage in weapon system improvements and are, consequently, essential.

Microelectronic packaging advances are crucial for continued U.S. weapons superiority and essential to the success of the U.S. electronics industry in international commercial competition. Great cost savings can be made by using new miniature, high speed microelectronics to upgrade performance and reliability of existing weapons systems. Order-of-magnitude miniaturization levels and accompanying performance improvements now are needed to create radically new weapons that will fulfill the long term critical mission requirements of the DoD.

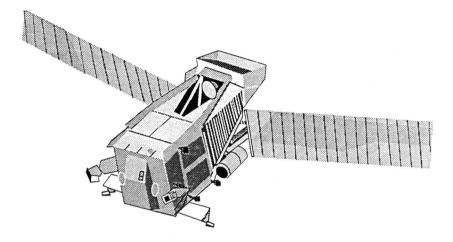

Figure 15-6 An artist's impression of the Boost Surveillance and Tracking System (BSTS), which will be deployed as part of the Strategic Defense Initiative. The BSTS will advance the state of the art for A/D converters.

15.1.2 Measuring the Trends

Recent years have seen progressive increases in the number of input/output (I/O) leads, chip area, power dissipation and speed of semiconductor chips. Future trends in these chip parameters are shown in Table 15-1.

Input/Output Leads

Increase in the complexity of individual semiconductor devices generally requires more I/O pads. The difficulty of packaging chips with large numbers of I/Os is one of the factors that led to MCM technology. One approach toward minimizing I/O connections is wafer scale integration (WSI), in which a 6", or even 8", silicon wafer is filled with functioning, interconnected ICs, including integrated component parts such as resistors and capacitors. Successful WSI usage remains elusive, at least for the present. Multichip microelectronic packaging as a field has maintained a healthy growth rate, because, however complex individual chips have become, circuit designers usually need to mount two or more onto the same substrate. A large number of I/Os on the chips requires a high line density.

Table 15-1 Semiconductor Research Corporation (SRC) Packaging Roadmap - IC Chip Complexity Trends.

	1994	2001
CMOS (Gate Array and MPU) • Size • I/O • Power (watts) • Rise Time (ps)	 1.7 (2.0) 600 (1000) 15 (50) 700 (200)	 2.5 2000 60 150
Switching Lines • Voltage • Number Lines (MPU)	 3.3 (2.1) 128 (256)	 1.6 400
Bipolar (Gate Array) • Size (cm) • I/O • Power (watts) • Rise Time (ps)	 2.2 600 (1000) 60 (100) 100 (75)	 2.5 2000 200 40

Note:
   Average anticipated values are shown.
   Maximum value for small number of devices are shown in parentheses.

Line Density

The width and spacing of conductors used to interconnect circuit devices have been used extensively as a measure of packaging efficiency. Figure 15-7 illustrates the relative rapidity of transition from 0.025" wide conductor lines and 0.025" spacing in 1980 to the present 25 μm line widths with 75 μm spacing. In special cases, these figures can be reduced to 15 μm and 35 μm, respectively. Within the past decade (the 1980s), a notable measure of packaging trends has been the transition from mils to microns as the unit used to describe conductor line widths and spacings. As lines get narrower, they also get shorter due to closer spacing of die, reducing conductor area and capacitance, leading to greater signal throughput.

Signal Throughput

One overall measure of packaging efficiency for military guidance unit processors is signal throughput per unit volume. In Figure 15-8, increased throughput of military signal processors is depicted, using as units millions of

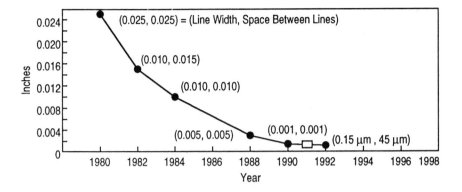

Figure 15-7 Conductor line widths and spaces for typical Hughes interconnect structures, 1980-1992.

instructions per second per cubic inch. From a level of 100 in 1990, this number is expected to increase sharply over the next few years. Higher signal throughput and higher packaging densities lead to more heat dissipation.

Power Density and Power Dissipation

An additional measure of packaging efficiency is power density (watts dissipated per unit volume of package), as shown in Figure 15-9. This value increased from 400 mW per cubic inch for the Hughes' PHOENIX missile in 1980 to 1.2 watts per cubic inch in the currently produced AMRAAM missile. In the LEAP program, this figure will increase to 7.3 watts per cubic inch. As this trend continues, power dissipation will continue to be a challenging factor. A similar problem has been experienced in mainframe computers, where cooling air or water is once again a requirement as it was in the 1950s when computers used vacuum tubes.

Volume Fraction

A common parameter used as a measure of packaging efficiency is the fraction of package surface area occupied by active semiconductor devices. Extending this to the *volume fraction* occupied by active devices is a particularly sensitive parameter and a more useful measure. In Figure 15-10, values of this parameter are plotted for well established technologies and for some developmental and projected approaches. Traditional hybrid microcircuits mounted onto epoxy based printed wiring boards (PWBs), with semiconductor devices mounted on one side

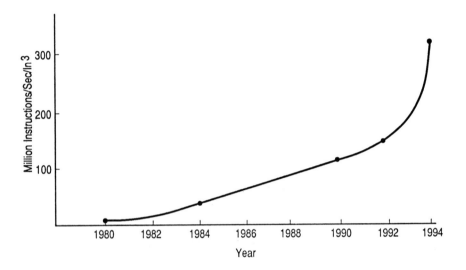

Figure 15-8 Processor throughput per unit volume.

along with typical surface mount passive component parts, have volume packaging efficiencies less than 0.5%. With thick film multilayer ceramic wiring boards (CWBs), developed in the early 1980s, this figure has increased to around 1%.

The earliest attempts at high density MCMs have led to an improvement in volume fraction by a factor of two, while the developmental Hughes' LEAP packaging technology has a volume efficiency around 3.5%. Volumetric efficiency advancements in three dimensional MCM packaging with integrated passive devices is projected to reach efficiencies up to 25%. For monolithic WSI, as yet unrealized on large scale systems, this efficiency figure would become about 40%. Projections for other near term developments are also shown in Figure 15-10. One of the key evolving steps in military packaging is the use of flip chip attachment of the semiconductor die, avoiding leads which extend laterally and occupy considerable surface area. The use of a hermetic sealing coating, eliminating the necessity of a lidded package, would also have a great impact on packaging efficiency.

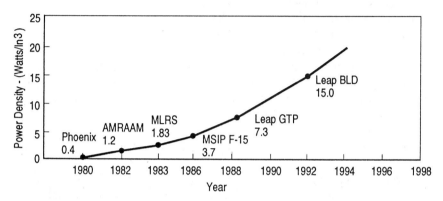

Figure 15-9 Military electronic power density trends, including projections through 1994.

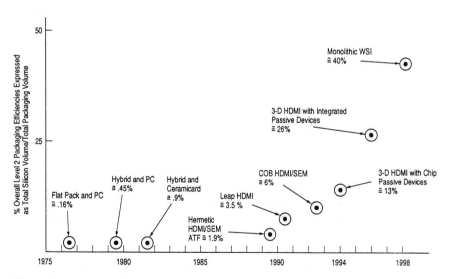

Figure 15-10 While major increases in volumetric packaging efficiency have been made with high density planar MCMs, significant future improvements require three dimensional packaging and chip-on-board approaches.

15.2 NEW PACKAGING ROLES

Within the aerospace industry, progress in the past has been determined mainly by advances in system technology. During the 1970s and 1980s, advances in packaging technology began to accelerate as semiconductors incorporated increasing numbers of electronic functions per unit area of silicon. Various active and passive component parts were assembled within smaller volumes, as building blocks progressed from PWBs with parts assembled through holes, to hybrid microcircuits on CWBs. This acceleration has increased continually to the point where it has impacted the classical systems supported by the packaging scheme described above. Now and in the future, the progress of electronic systems will be determined more and more by advances in semiconductor and packaging technology. Packaging is becoming an increasingly dominant technology.

Two major packaging technologies, described in Section 15.3, are multilayer low temperature cofired ceramic (LTCC) circuits and multilayer thin film substrates. Hughes developed LTCC in the early 1980s for use in microwave and high frequency (over 300 MHz) digital packaging. By the late 1980s, it became clear that multilayer thin film substrates also were required. A new in-house packaging program was set up to address the commercial and aerospace markets.

15.2.1 A High Density Multichip Interconnection Program

In 1987, the High Density Multichip Interconnect (HDMI) program was created to develop advanced MCM packaging required for the DoD/Hughes Advanced Tactical Fighter (ATF), BSTS, LHX, LEAP and ADCAP weapon programs. Progress in the HDMI program led to the creation of an aluminum/polyimide HDMI substrate fabrication facility in Newport Beach CA, with comprehensive high density MCM design, manufacturing and test capabilities. This rapid progress was instrumental in obtaining several key DoD electronic packaging technology contracts, including two Air Force Smart Skins contracts and the Naval Ocean Systems Center (NOSC) VLSIC Packaging Technology (VPT) program [3].

15.2.2 A Company Wide Electronics Packaging Program

Based upon success of the HDMI program and the need to provide timely and unified responses to DoD customers, the Hughes Electronics Packaging Program (HEPP) was established in late 1988. The scope of HEPP was increased to include all high density electronic packaging technologies. Additionally, managers were assigned to major customer electronic packaging categories and to interface with the Microelectronic and Computer Technology Corporation (MCC). To date, the most valuable work of HEPP has been the timely development of high density multichip packaging technologies used to demonstrate packaging viability for programs such as the DoD/Hughes ATF, BSTS, LEAP, LHX, MMIC, and ADCAP and for commercial products.

As known, the increased dominance of packaging technology in overall systems design has modified the classical relationship of the packaging function supporting systems engineering in achieving developmental and production goals. The packaging function now plays a partnership role with systems engineering - an example of concurrent engineering in action.

15.3 DEVELOPING TWO KEY MCM TECHNOLOGIES

MCM development has been driven by system needs, packaging improvements and the increasing complexity and speed of semiconductors, particularly CMOS. Semiconductor interconnection density has been increasing more rapidly than that of PWBs (Figure 15-11), leading to a widening gap in interconnect capability. The first approach toward increasing connectivity was to increase the line density and number of layers in PWBs, leading to development of the MCM-L

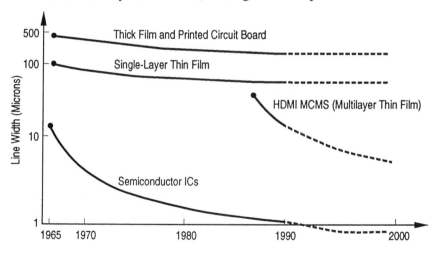

Figure 15-11 HDMI technology reduces this IC versus packaging gap to improve packaging density.

Conventional
Printed Wiring Board

Semiconductor Wafer
HDMI Substrate

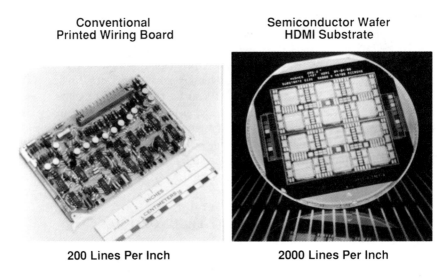

200 Lines Per Inch

2000 Lines Per Inch

Figure 15-12 A 10:1 improvement in wiring density closes the interconnectivity gap.

technology (see Chapter 5). Other principal approaches have been developed to close this gap. One is a multilayer ceramic MCM (MCM-C), which uses LTCC technology, described in Section 15.3.1 and further described in Chapter 6. The other approach is a multilayer thin film (MCM-D), described in 5.3.2 and in Chapter 7. (See Figure 15-12.)

15.3.1 An Evolutionary Approach

An early and important example of MCM-C was the IBM Thermal Conduction Module (TCM), which has die flip chip bonded on a high temperature cofired ceramic (HTCC) interconnect (more than 30 layers). The TCM uses an area array die I/O pattern for two reasons. The desired I/O density would require an extremely fine pitch if it were peripheral I/O. This I/O density cannot be matched by the pad pitch in the ceramic. Making the die larger to achieve a larger I/O pitch creates unacceptable stress problems when using flip chip bonding, as it also reduces die yield. Multilayer ceramic interconnect matches semiconductor area arrays and flip chip bonding very well, but standard die do not have area array interconnects. Aerospace companies generally use relatively small quantities of chips and, therefore, cannot economically justify special area array die configurations. Consequently, their usage has not become a generic aerospace packaging approach. Now, because of the increasing number of I/O contacts required and the development of MCMs, area array patterns are becoming more common.

The traditional MCM was HTCC, which entailed high set up and change costs and excluded the use of gold, silver or copper metals or buried resistors/capacitors. Recent developments have introduced LTCC based on cordierite or glass filled thick film-type materials, which fire around 900°C. IBM is using this LTCC technology as a successor to its TCM. It has been developed at Hughes for both aerospace and commercial applications.

The LTCC approach to MCMs does not address directly the problem of die that achieve high I/O by a fine pitch, about 6 mils in 1990 and reduced to 4 mils in production by 1992. In the early 1980s, Hughes decided it needed a fine line interconnection technology to use most advantageously the fine pitch microprocessor and configurable gate array die projected for the 1990s. Research was done on how best to adapt the technology developed for semiconductors to the die interconnection problem. Traditional PWBs and thick film networks use conductor patterns which are about two orders-of-magnitude greater than the line width of 1 μm being achieved by semiconductors and are limited by their manufacturing process (Figure 15-11). By adopting a photolithographic approach, the line width of the base network is no longer limited by the process but may be optimized for resistance and crosstalk, two parameters which also might limit width and spacing. To match the fine pitch on the die, multilayer fine pitch MCM-D technology was developed. The approach developed is described in detail in Sections 15.2.1 and 15.3.3. The various technologies are compared in the following section.

15.3.2 Technology Comparisons

There are a number of technical choices required in developing an MCM-D technology. Alternative approaches have been adopted by different companies; these are summarized in Table 15-2. Each company optimized different parameters. (Some of these approaches are described in the other case studies.) This is intended not to exhaustively list companies and processes, but to represent the diversity of technology. A primary distinction among the processes is the method of forming vias. Hughes chose a simple approach to achieve high yields. Joining the MCC allowed access to alternative technologies.

Microelectronics and Computer Technology Corporation
The MCC, a consortium of companies, has developed a copper plated thin film technology (Figure 15-13) that plates up the posts that become z-direction interconnects. The polyimide dielectric is then spun on and cured. Before the next plating process, the top surface must be lapped to provide planarization.

Table 15-2 Alternate HDMI Approaches.

FOUNDRY	UNIQUE FEATURE
• MCC	Via: Plated post
• Hughes	Via: Etched, staggered
• ATT*	Via: Filled, stacked
• GE	Overlay: Laser process
• nCHIP	SiO_2 dielectric

This provides a sturdy interconnect using copper (good conductivity) and the number of layers is not limited by planarity. This type of interconnect is also excellent for thermal conduction, which is important since the polyimide dielectric typically used is a thermal barrier.

AT&T
AT&T developed an Advanced VLSI Packaging (AVP) system, in which the via was sputtered and plated in copper and then filled with a nickel plug (see the discussion in Chapter 16 and also Figure 16-9). This allows the stacking of vias as in the MCC approach. AT&T has focused on flip chip bonding and, thus, needed a good thermal path under the solder bumps.

nCHIP
nCHIP departed from the norm by using silicon dioxide as the dielectric rather than polyimide. (See Figure 16-20 as well as the discussions in Chapter 7 pertaining to silicon dioxide as a dielectric material.) Silicon dioxide traditionally has been used in semiconductors for very thin insulation layers, but such thin layers create unacceptable capacitance when extended to the larger length and width of interconnect. nCHIP has developed techniques to create thicker layers of silicon dioxide without cracking. This has certain advantages, since silicon dioxide has good thermal conductivity.

General Electric
The GE overlay process (as described in Figure 7-17) is an innovative approach, requiring that the die be placed in cavities, followed by the interconnect layer on top of the die. This is done by applying a sheet of dielectric material. Vias are created with a laser by aligning the laser to the die pads. A key advantage of this approach is that the interconnect makes a sputtered connection to the die pad, eliminating the need for wire bonding. Another advantage is that the die

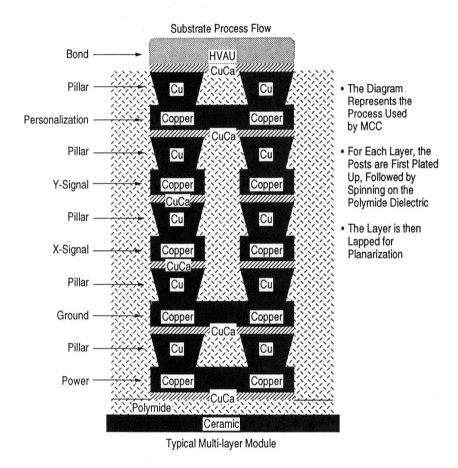

Figure 15-13   The MCC "Copper Post" process.

are placed directly on the base substrate without any intervening multilayer polyimide.   The disadvantage is, that to replace a failed die, the entire interconnect layer must be stripped off and rebuilt.

15.3.3 HDMI Technology

High Density Multichip Interconnect (HDMI), is the term used by Hughes for a high density packaging approach based on techniques that were originally developed to produce semiconductor devices. The base technology involves the use of a spin on polyimide dielectric, with alternate aluminum metallization layers on a variety of substrates, such as silicon, alumina, metal matrix or aluminum nitride [4]-[5].

Many options were considered when confronting the requirements of high interconnect density imposed both by system size and weight on one hand and increasing chip complexity and speed on the other. Two main thrusts were pursued: HDMI for computer-type interconnects, and LTCC for memory, microwave and low cost interconnects. Of course, HDMI can be used for microwave, and LTCC can be used for computers. Even though the technologies are radically different, they both relate to MCM technology.

The most pressing need for HDMI arose from the ATF program at Hughes (Figure 15-3). Initial ATF signal processor speed requirements were below 25 MHz, although evolutionary increases above that figure are expected. The technology has been characterized as high as 350 MHz. It was felt that alumina ceramic substrates could be used initially, and the resulting MCMs could be hermetically sealed within large area alumina ceramic packages. The use of silicon substrates offered the possibility of increased thermal conductivity and processing flatness, plus the possibility of incorporating active devices within the surface area. Thermal mismatch between silicon substrates and alumina ceramic packages created a potential problem, which could be solved through use of an aluminum nitride ceramic package. A family of such packages is being developed in the NOSC program described later in this section.

HDMI-1 describes a five layer baseline process developed primarily for low to moderate digital clock speeds. Substrates with additional layers (HDMI-1A, -1B, etc.) have been developed to extend the range of circuit applications.

Process Description of HDMI

The simplest way to describe this MCM-D process is to look at a cross section of a substrate (Figure 15-14). The interconnection is built on a base substrate, typically ceramic, by depositing a layer of metallization, followed by a layer of polyimide dielectric. Vias are created by etching through the dielectric and sputtering them to the next interconnection layer in the manner used for IC processing. The vias are typically staggered (not placed directly on top of a previous via) to avoid build up of nonplanarity, particularly in that they are not filled. These vias also can be stacked as discussed in Chapter 7.

HDMI technology uses multilayer thin film polyimide and aluminum applied in a Class 100 cleanroom. The substrate is a 150 mm diameter silicon or ceramic wafer. Silicon is available in well polished flat wafers for the semiconductor industry, but this application required new standards for flatness in the ceramic substrate. Fine ceramic dust particles are a hazard, since they may be introduced by the ceramic wafer into an otherwise ultraclean chamber.

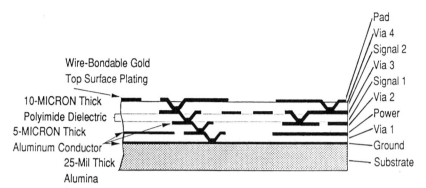

Figure 15-14 Cross section of Hughes HDMI-1 substrate with five aluminum/polyimide interconnection layers.

Polyimide is spun onto the wafers to a typical thickness of 5 μm. It is cured and then metallized by sputtering aluminum to form the ground plane. A second layer of polyimide is then applied and contacts to the ground plane are formed at selected locations, as described above. Via formation occurs by plasma etching after standard lithography is used to define the pattern. The vias are normally filled by aluminum, although alternatives are possible. Sputtering of the next aluminum layer then contacts the vias to form interconnects normal to the substrate. Photolithography and etching are then used to pattern the first level of signal interconnects with 100 μm spacings. (See Figure 15-15 for HDMI-1 baseline process flow.)

This polyimide spin on, via formation and conductor deposition process is repeated to form additional interconnection layers to complete fabrication of the substrate. The VLSI chips are then attached to the surface, with capacitors and other surface mounted component parts as necessary. The high density signal layers yield circuits of enormous complexity with just five conductor layers, one layer being the power plane. The HDMI base process was developed rapidly, because the basic components, using advanced sputtering and photolithographic techniques, were already available from MOS technology. High processing yields are achievable with this base process.

The key issue in production is yield. The substrate is best seen as a large (4" × 4") IC. One defect per wafer in a 150 mm process is 0.005 defects per square centimeter! This low defect density in a simplistic model yields 4" × 4" substrates at 0%; 2" × 4" substrates at 50%: and 2" × 2" substrates at 75%. Note that defect size is different from that of semiconductors: the typical defect is 10 μm or more - a boulder in normal semiconductor terms. However, the defect sources are numerous: the polyimide, the substrate and, especially, the

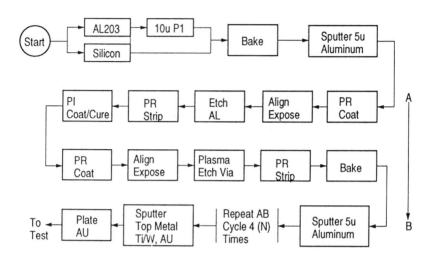

Figure 15-15 HDMI-1 substrate fabrication process flow.

equipment. This is one reason for using standard semiconductor equipment designed especially to eliminate production of any particles. It also relates to the problems of developing new equipment especially suited to substrates. Yield has two components: line yield - the percent of wafers that successfully complete all the process steps within specification, dependent on process maturity and control, and electrical yield - the percent of completed substrates that are good, dependent on defect density and substrate size.

As a process reaches maturity in the semiconductor industry, improvement and modification are needed to meet new demands. One demand was the addition of analog circuitry requiring three additional metal layers. Additional circuit layers may be cofired onto the alumina substrate prior to the first spun on polyimide layer or added as polyimide/conductor layers. This is indicated in Figure 15-16, which also shows the integration of capacitance into the silicon substrates by oxidization forming MOS capacitors. The need for some surface mounted devices is removed. The HDMI MCM performance can be extended to very high speed digital clock rates (2 - 5 GHz), by designing a configuration with signal and power planes closely matched in impedance.

In addition to the above developments, the pattern of history shows the need for even finer lines and spaces. It is possible to achieve 50 μm geometry, with 15 μm wide lines separated by 35 μm spaces. The resulting increase in the

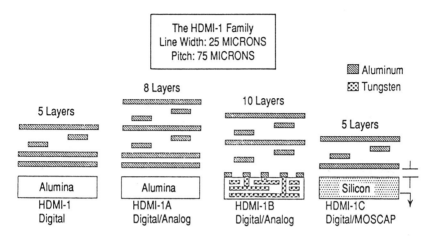

Figure 15-16 Schematic cross section views of progressively more complex substrate configurations. (Beginning with the baseline five layer digital system and extending to integration of capacitive functions for high speed digital processing.

conductor line resistance can be compensated by switching from aluminum to copper, since the conductivity of copper is about 60% higher. Copper does present new challenges; typically it does not adhere well to polyimide and also normally requires a migration barrier. Titanium/tungsten and chrome, respectively, are used as adhesion and diffusion barrier layers.

Table 15-3 shows key physical parameters and design features of the HDMI-1 substrates. This is a production process. Other processes are being developed continuously, but it is essential for production to standardize so that yields can be optimized. Table 15-4 shows electrical performance values for baseline the HDMI-1 substrates. Note that the two signal layers are not symmetrical (SIG1 is near the ground plane) and, hence, have different impedance values. This is not a problem, since most long interconnects contain several SIG1 and SIG2 segments, as illustrated in Figure 15-14, and the average effect predominates. For short interconnects, the impedance characteristics are not critical.

Related MCM-D Work
Hughes' participation in the NOSC VLSIC Packaging Technology (VPT) program included development of a 4.23" × 4.23" aluminum nitride package [6]. Successful performance under this program makes it possible to proceed with HDMI development on a dual substrate basis: alumina in an alumina ceramic package, and silicon in an aluminum nitride ceramic package. An example of a NOSC VPT HDMI test circuit module is shown in Figure 15-17.

Table 15-3 Some Design Features of Baseline HDMI-1 Substrates.

	STANDARD	SPECIAL
Substrate • Material • Wafer size • Substrate sizes	Silicon or alumina 6" (150 mm) diameter 1.8" × 1.8" and 1.8" × 3.8"	Aluminum nitride — 3.8" × 3.8"
Conductors • Material • Sheet resistance • Width • Pitch • Number of layers • Top metal	Aluminum 6 mΩ/□ 25 µm 75 - 100 µm 5 layers Titanium-tungsten/gold Titanium-tungsten/nickel/gold	Copper 3 mΩ/□ 15 µm 50 - 60 µm 7 layers Aluminum (for rad hard applications)
Dielectric • Material • Thickness • Dielectric constant • Dissipation factor • CTE • Thermal stability • Outgassing (1000 hours at 150°C) • Via diameter	Polyimide 5 - 10 µm 2.9 at 1 MHz 0.03% at 1 MHz 3 ppm/°C to 500°C < 5,000 ppm water 35 µm	— — — — — — 15 µm

The initial goals in the program applications of MCM-C and MCM-D technologies were to advance MCM-C and MCM-D packaging to achieve order of magnitude improvement in all three DoD weapon system categories:

- Near term weapon programs
- Performance upgrades of existing weapons
- Radically superior future weapons

While much work remains, initial goals have been met. The greatest value of this project, to date, has been the development and building of high efficiency HDMI MCM-D and LTCC MCM-C modules that demonstrate packaging viabilities and that will be used to achieve performance improvements and costs savings in all three weapon system categories.

Table 15-4 Electrical Performance Values for Baseline Hughes HDMI-1 Substrates.

CHARACTERISTIC	SIG1	SIG2
• Capacitance (pF/in)	3.7	2.6
• Inductance (mH/in)	0.0085	0.0105
• Characteristic Impedance (Ω)	50	64
• Resistance (Ω/in)	9.0	8.1
• Crosstalk, near end (%) (50 μm space, 35 ps T_R)	2.6	5.0
• Crosstalk, far end (%)	2.3	3.8

The F-22 (formerly ATF) is one near term new program for which HDMI MCM-D packaging contributed to operational success. Weight and volume savings (96% and 93% respectively) were achieved for equivalent circuit functions when compared to prior F-15/F-15 radar packaging. By using very large scale integrated (VLSI) devices in conjunction with HDMI MCM-D packaging, the F/A-18 was provided with superior target resolution and higher land mapping resolution. This was a threefold packaging efficiency improvement and a 100% increase in radar processor throughput with a 50% cost reduction. These upgrades, brought about by packaging innovation, enable the F/A-18 to perform new reconnaissance missions which previously required both a reconnaissance plane and an escort.

HDMI packaging of LEAP electronics demonstrates two order weight and volume advancements and establishes credibility of lightweight processors which will be required in theater defense weapons: Brilliant Pebbles, Brilliant Eyes, Ground-Based Interceptor (GBI), advanced lightweight satellites and the Advanced Air to Air Missile (AAAM). LTCC MCM-C packaging technology will be used initially to build prototype transmit/receive (T/R) modules for the MMIC program. This will support active array radar systems developments proposed for upgrade applications in the F-15, F-15, and F/A-18 programs, the F-22 and Multirole Fighter (MRF) programs and in ground-based radar systems. Advancements in 3-D MCM chip on board (COB) packaging are continuing, and progressively more advanced versions of planar MCM packaging are being developed to achieve progressively greater performance reliability and cost advantages in military systems. Application of these advanced packaging

Figure 15-17 Photograph of 3.8" × 3.8" eight layer silicon-based hermetic aluminum nitride package developed under the NOSC VPT contract by Hughes, Coors Ceramics and W. R. Grace Company.

concepts to high end workstations, medical electronics, parallel processors and automotive electronics is a new focus in the commercial sector.

15.3.4 Low Temperature Cofired Ceramic (LTCC) Technology

One alternative method being developed at Hughes involves using a "green" ceramic tape to fabricate multilayer packaging. A schematic of a multilayer structure, together with an LTCC structure, shown in Figure 15-18, illustrates the substrate as an integral part of the package.

LTCC has two major advantages in meeting the challenges of both military and commercial packaging requirements. Although it uses thick film technology that does not have fine line definition, LTCC is effective in the use of the third

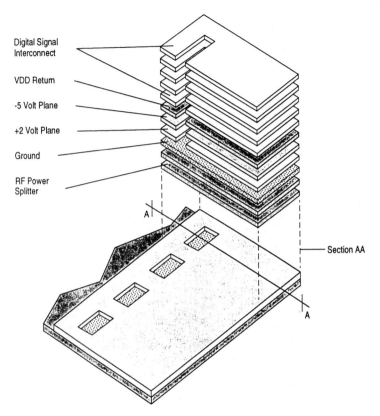

Digital Signal
Interconnect

VDD Return

-5 Volt Plane

+2 Volt Plane

Ground

RF Power
Splitter

A

A

Section AA

Figure 15-18 Artist's conceptual illustration of key features of LTCC interconnect package structure showing ability to install devices within cavities. Substrate integrated passive devices are also common LTCC structures.

or z-dimension. While polyimide-based packaging is currently limited to eight to ten conductor layers because of the stress build up, LTCC packages may have 20 or more layers. Prototype packages of 40 layers have been fabricated.

A second advantage of LTCC is its potential for low cost. In describing present trends in military packaging, the issue of cost was not stressed, but government program managers increasingly emphasize the importance of this issue as well as those of performance and reliability. LTCC has an advantage over conventional thick film technology because it does not require a firing stage after each conductor printing; this saves in processing time. Also, the use of ceramic tape lends itself to reel to reel processing through the lamination step. (LTCC technology is discussed further in Chapter 6.)

An additional advantage of LTCC is that of versatility. Large area packages can be made, limited only by the size of the tape and the maturity of the processing steps, with respect to preserving flatness. The composition of tape materials can be modified for expansion coefficient match with alumina, silicon or gallium arsenide. The dielectric constant may be quite low (4 - 5), but tapes with dielectric constants up to 100 or more can be obtained so that capacitors can be buried within the structure. Similarly, resistors and any other screen printable component parts may be totally embedded within the package, provided that excessive local stress is avoided. Similarly, cavities can be fabricated within the structure to accommodate embedded semiconductor chips as an alternative to surface mounting. The fired LTCC structures are hermetic. Semiconductor devices can be located within a sealed structure without the use of a separate package. Arrangements can be made for I/Os to be brought out to the edges of the LTCC structure for connection to external circuits.

LTCC has been found to be suitable particularly for microwave packages. The requirement here is for impedance matching, achievable through dimensional control of conductor lines and dielectric thickness. Gold is the metallurgy of choice for microwave applications because of its stability and relatively high conductivity. Although most tapes contain glass, the dielectric loss is acceptable for many military applications.

The low thermal conductivity of LTCC normally requires some arrangement to remove heat. The versatility of the structure does permit the insertion of conductive material blocks bondable into cavities formed below the ICs. Arrays of thermal vias also are used to transport heat effectively through the ceramic. The number of vias can be chosen according to the power to be dissipated.

LTCC tapes are now available from DuPont, ElectroScience Labs, Ferro and Rohm and Haas. Of these, DuPont has been most aggressive in developing a tape with low variability; their Green Tape™ Type 851AT has become the industry standard. A detailed review of this tape, with emphasis on economic analysis, has been given by Bender *et al.* [7]. Some early LTCC work at Hughes with this tape has also been reported [8]-[9].

The tape is formed by mixing glass, a ceramic or crystallizable glass powder and a suitable organic vehicle and placing this mixture onto Mylar. In forming a package, each layer of tape is first cut to size with registration marks. Then vias are formed in an appropriate pattern. The tape is patterned using thick film screen printing, including via fill. The individual layers are laminated together by the application of pressure and a small amount of heat to form the green, multilayer package. This then must be fired, with care taken to allow complete burnout of the organic material prior to densification of the glass-ceramic body. The important advance in LTCC over HTCC technology is that the maximum firing temperature is 850°C - 900°C, allowing the internal and external use of gold, silver or copper metallization. High temperature cofiring is restricted to refractory internal conductors such as tungsten, with a resistivity three to four times that of silver.

DuPont has done valuable work in seeking to provide a wide range of conductor and resistor inks and dielectric tapes compatible with their LTCC tape. For military packaging, gold is normally required. Silver is available. A careful study conducted by Gaspar *et al.* [10] demonstrated that silver embedded within an LTCC structure can stand up well to military tests for mechanical integrity and electrical reliability under low DC voltages. Silver can cause warpage if present in high concentrations within a given layer, such as in a ground plane. Similarly, copper metallization is still developmental and suffers from the additional problem of oxidation susceptibility. Firing under a nitrogen atmosphere to prevent this increases production costs. An additional disadvantage is that organics must be evaporated rather than burned out. The concern is that densification will occur before all the organic material has been removed, potentially causing blistering as firing proceeds. Development of a trouble free, low cost silver or copper metallization is an urgent priority for LTCC technology in view of the high cost of gold. Table 15-5 shows key physical properties, dimensional tolerance and electrical properties of LTCC. Processing of LTCC tapes is discussed further in Chapter 6.

15.4 COMPREHENSIVE DESIGN/MANUFACTURING SERVICES

The technology and relationships discussed in Sections 15.1, 15.2 and 15.3 are important to applying successfully MCMs to complex, high performance military and commercial systems. Comprehensive design, fabrication and testing operations also are required to achieve MCM benefits successfully. To minimize cost, scheduling and technological changes in advanced programs, it is vital that comprehensive packaging systems be completed prior to program applications.

Figure 15-19 summarizes essential elements of a comprehensive packaging system. Many process steps are required and a broad scope of manufacturing and testing technologies are applied. Figure 15-20 shows a high performance 2" × 4" HDMI module to be used on the F/A-18 and F/22.

MCM technology is more difficult to apply than other more common PWB techniques. Because many chips are grouped into sealed units, problems associated with dense circuitry, such as testing and thermal dissipation, are amplified. MCM users need a complete design to assembly concurrent engineering process to apply this technology successfully. Such a process is described through a step by step review of Figure 15-19.

Table 15-5 Properties of LTCC.

	LTCC Package	Thick Film on 96% Alumina	High Temperature Cofired Package (92% Alumina)
Physical Properties			
• CTE at 300°C (ppm/°C)	7.9	6.4	6.0
• Density (g/cm^3)	2.9	3.7	3.6
• Camber (mils/in)	1-4[a]	1-2	1-4[a]
• Surface smoothness (microinches)	8.7	14.5	20.0
• Thermal conductivity (W/m-K)	16-20[b]	20[c]	14-18
• Flexural strength (kpsi)	22	40[c]	46
• Thickness/layer after firing (mils)	3.5-10.0	0.5-1.0[d]	5-20
Dimensional Thickness			
• Length and width	±0.2%	N/A	±1.0%
• Thickness	±0.5%	N/A	±5.0%
Electrical Properties			
• Insulation resistance (ohms @ 100 VCS)	> 10^{12}	> 10^{12}	> 10^{12}
• Breakdown field (volts/mil)	> 1,000	> 1,000	> 700
• Dielectric constant (1MHz)	7.1	9.3	8.9
• Dissipation factor (%)	0.3	0.3	0.03

[a] Function of firing setter and part design   [b] With thermal vias
[c] Property of substrate   [d] Property of printed dielectric

15.4.1 Design

The first requisite of a comprehensive MCM packaging system, is that it be developed fully prior to systems applications. It includes design rules, design methods, standards design programs, standard design platforms and training of design personnel. Complete design analysis tools, characterizing electrical performance, mechanical performance, thermal performance, environmental withstanding capability, reliability and producibility must also be available prior to MCM applications in production systems. To support automated low cost MCM manufacture, computer aided testing (CAT) and computer aided manufacturing (CAM) databases, data extraction programs and formatting procedures also are required. New design capabilities had to be created, since MCMs are far more complex in comparison to more conventional hybrid microcircuits.

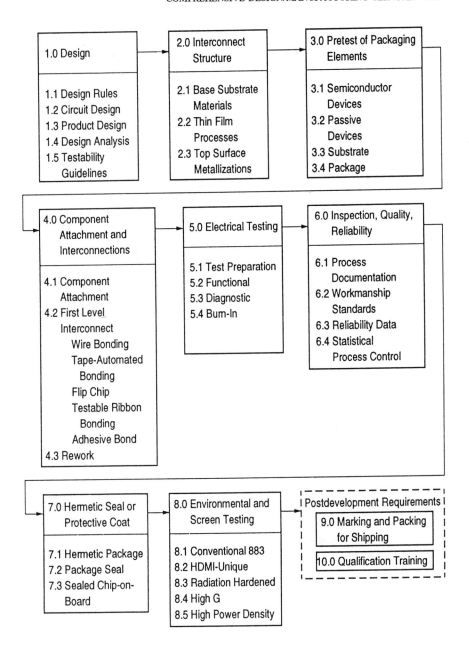

Figure 15-19 Essential elements of a comprehensive packaging system.

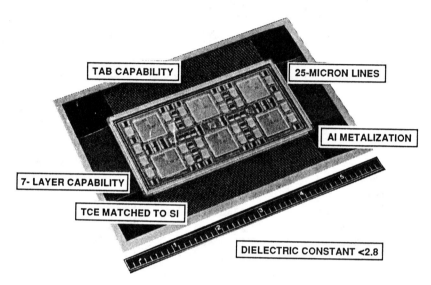

Figure 15-20 Typical 2" × 4" HDMI MCM-D used in F/A-18 and F-22.

Design Automation

Design automation is critical to placing MCM technology into the hands of board designers. A complete MCM CAD strategy must assist at every step, from design entry through verification and manufacturing. Design systems for hybrids usually have been modifications of IC or PWB tools, since hybrids are used much less frequently. The advent of commercial MCM use is changing this situation. A significant amount of commercial software specifically has been created to assist in MCM design.

At Hughes, CAD personnel initiate each customer project with up front strategic advisement. The first step is to identify areas where MCMs can best improve a particular project. Assisting designers to choose the correct MCM technology and set up a development plan is only the beginning. The design and testing of multichip packages requires a system level approach, forcing the integration of system, IC and packaging engineering. Work with customers focuses on understanding and integrating the various disciplines necessary to produce manufacturable, testable MCMs. The ability to integrate all these disciplines in a timely manner may provide the key competitive edge for designers in the next decade. Complete production capability is offered for MCM users. Capabilities exist which include semiconductor processing, hybrid assembly, and experience in bare die purchasing. A need for new materials also can exist, as well as a need for the same processes, discipline and cleanroom facilities used to manufacture semiconductors.

Workstation Role
The most difficult task for MCM users is the design and verification of the circuit. Design automation can simplify much of the task. Mentor Graphics design automation tools were used because of the range of high density MCM technologies on engineering workstations they offer [10].

CAD support at Hughes began in proprietary efforts. Early in the development of its CAD program, it was found that most design automation tools supported either IC or PWB design, but not hybrids or MCMs. Therefore, new hybrid design software was developed on internal DEC VAX computers. This CAD system was based on a variation of an existing PWB router for its thick film processes.

As the company developed the CAD technology for internal HDMI efforts, it became apparent that new resources would be required to keep up with the rapid evolution of the CAD industry. An engineering effort therefore began to evaluate commercial vendors, seeking to adapt existing board or IC design tools to its nascent HDMI design process. To evaluate the capabilities of different vendors, the company used a benchmark circuit of a real processor design used in the ATF project. The benchmark circuit contained six 368 pad ASIC chips and more than a dozen memory chips. Several commercial tools were evaluated for their ability to place and route the circuit, given the special constraints of HDMI technology, such as blind and staggered vias.

Based on this benchmark, the Mentor Graphics program called "Hybrid Station™" was chosen to develop a commercial design capability. In an informal joint effort, Hughes and Mentor Graphics attacked the difficulties of MCM design, including via placement, test point extraction and development of GDS II output for the Hughes verification software. The end result of this effort was a design kit that, when used with the Mentor Hybrid Station™, allows systems designers to create high density digital MCM designs on their engineering workstations. In addition, there is in-house standardization of the internal MCM CAD, from design entry through verification, on Hybrid Station™.

Hybrid Station™ offers design capture, placement, routing and verification for hybrids and MCMs. It also can be used in conjunction with the Mentor Graphics tools for simulation, timing verification, electrical analysis and thermal analysis. The design kit contains the process rules and guidelines for design, placement and routing. With input from up front consulting efforts, the board designer can fully specify an MCM. There is in-house testing of the final design and verification steps (adding test structures) that validate particularly complicated MCM aspects for manufacturing and assembly.

Product Design/Simulation

Having determined the design, testing technology and manufacturing parameters, the designer can begin to design the MCM circuit. At this point, assumed timing delays can suffice for first pass verification. The designer can also implement and verify testability of the circuit and begin to develop test vectors for the MCM.

Given a verified netlist for the design, the designer has the choice of performing physical design on Hybrid Station™ or proceeding directly to complete the MCM. Using Hybrid Station™, the designer can interactively or automatically place components depending on the timing constraints of the design. A critical task for layout is creation of the die outlines; they generally are nonstandard and not normally published by semiconductor manufacturers. The designer must enter the die outlines in parts libraries; Hughes provides template pad outlines in its design kit. With a placement that satisfies density and thermal requirements, the designer can begin routing with preplacement of critical nets. For designs with hundreds or thousands of nets, most of the routing is done automatically. The design kit provides process rules for the automatic router in Hybrid Station™. For very critical nets, the designer can reestimate parasitic effects based on actual routing and resimulate the design.

Design Analysis

Other analyses, such as those conducted on transmission lines, are available through Hybrid Station™ for MCMs, but generally are not necessary for moderate speed digital circuits (up to 100 MHz). The fundamental five layer HDMI modules have been characterized electrically with acceptable performance at clock rates up to 350 MHz. Most designs use simulation with nominal timing information or critical path analysis. After routing is completed, designers can verify that, based upon actual line lengths, specified electrical performance parameters have not been exceeded.

Given these design capabilities, a design process is in place for the commercial development of MCMs. The first step is joint in-house evaluation of the project. This initiates the concurrent engineering process necessary to meet all the criteria of for MCM manufacture, assembly, and test.

Thermal analysis of MCMs is particularly difficult. Most thermal dissipation occurs through conduction through the substrate. PWB thermal analysis tools, however, evaluate thermal convection from component parts to ambient air flowing across the PWB. Designers can use a spreadsheet to program basic thermal equations for accuracy within about 10%. For thorough analysis, system designers can create three dimensional models using general purpose analysis tools such as ANSYS™ (a standard finite element analysis program), or using Mentor Graphics' Autotherm™.

Future Enhancements of MCM Design

The first enhancement will be further characterization of the interconnect systems to provide more parametric data. In addition to capacitance and resistance, inductance will be provided for estimating the impedance of interconnects. This step will become more important as microwave and very high frequency signals are used.

Another evolutionary step will help designers detail specific timing constraints up front. Worst case timing delays and relative timing sensitivity could be noted on schematics. Placement and routing tools then would use this information to preplace certain component parts, preroute critical nets and limit the overall length of timing sensitive lines. This timing driven layout will help to minimize the number of verification and design iterations needed to meet performance goals.

As design automation matures, larger production volumes and varieties of MCM products will be encouraged, tending to decrease manufacturing costs. These convergent trends will help to make MCM technology more accessible for commercial projects that require maximum functional density and performance.

Concurrent engineering of MCM design and manufacturing can identify areas of opportunity in the case of custom devices. For example, I/O pins on die can be arranged to simplify wire bonding and I/O drivers may be downsized according to the parasitic load of the substrate. Similarly, smaller die and I/O drivers can help to reduce the amount of bypass capacitance that may be necessary. Such tradeoffs can improve the operating frequency and yield of the MCM.

15.4.2 Interconnection Structures

The second essential element of a comprehensive MCM packaging system, as indicated in Figure 15-19, is the interconnection structure or substrate. LTCC and HDMI substrates are two possible interconnection structures.

15.4.3 Pretesting of Packaging System Elements

It is particularly important to achieve high initial manufacturing yield, minimizing MCM costs, because high performance MCMs are inherently complex. In early stages of production programs, when the circuit designs and

ASIC semiconductor devices are immature and not fully proven, it is usually advantageous to pretest all critical packaging system elements (Figure 15-19) including substrates, semiconductor devices, passive devices and high pin count hermetic packages. To minimize loss of modules caused by defects during the manufacturing cycle, substrates are tested for both continuity and shorts prior to and after temperature cycling.

The principal source of electrical test failures in the manufacture of complex MCMs is defective semiconductor devices. It is essential that only known good bare semiconductor devices be installed in MCMs. Semiconductor device manufacturers do not usually perform tests that are thorough enough to determine that the bare die satisfy circuit performance speeds and temperatures required of the MCMs. In addition, probe methods used by semiconductor manufacturers frequently cause bare die I/O pad damage, which precludes the multiple probing required for complete die pretesting.

To overcome limitations of conventional current bare die testing, alternative configurations were created, including beam leaded devices and tape automated bonding (TAB) formats. Still, a high proportion of MCMs employ wire bonding to form electrical connections from bare semiconductor device I/O pads to corresponding pads on the interconnection substrates. This involves use of bare die. Consequently, it was necessary to develop methods for performing complete dynamic pretesting of bare die at high, ambient and low temperatures to assure adequate yields of wire bond interconnected MCMs.

Probe testing of bare die minimizes pad damage. It was first used successfully on the LEAP program. This approach was based upon prior work conducted by Hewlett-Packard Corp. and IBM Corp. It utilized carefully balanced probe contacts with soft contact materials, in place of conventional hard tungsten probe tips. This approach was termed the "soft touch" probe card configuration. The soft touch probe cards were highly instrumental in achieving high initial test yield in the manufacture of LEAP MCM sector card assemblies. The extension of conventional probe card technology to high pin count advanced ASIC and microprocessor devices is difficult and costly. Therefore an even less damaging membrane contactor probe system was developed, to pretest in increasingly more complex devices with increasingly fine I/O pitches. Lithographically produced, patented membrane probe contactors show great promise for bare die and semiconductor wafer testing at clock rates as high as 5 GHz (Figure 15-21).

A further issue is pretesting hermetic packages in which valuable MCM assemblies are installed. The hermetic MCM package, shown in Figure 15-22, contains 368 leads on 0.030" centers. The procedure for pretesting this complex package is to install it into the electrical test fixture and perform continuity and short testing in conjunction with temperature cycling.

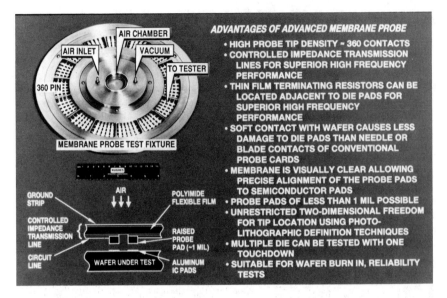

Figure 15-21 Membrane probe technology developed for die pretest.

15.4.4 Component Attachment and Connections

As indicated in Figure 15-19, the next principal packaging system element is attachment of component parts and formation of level 1 (chip level) connections. Manual assembly, while used extensively in less complex hybrid manufacture, is not well suited for production of MCMs. Precision and uniformity required in advanced MCM manufacture require high quality automated assembly equipment. Automated die attach systems are available commercially for high speed, low cost assembly of complex MCMs.

The evaluation and selection of die attach materials is quite demanding for high yield MCM production. For low power MCMs, organic adhesives are used typically for semiconductor and passive device attachment. Extensive tests of both thermoset and thermoplastic adhesives were conducted before adhesives for MCM component attachment were chosen. For MCMs with high power densities, solder attachment of devices to molytab heat spreaders, with subsequent solder attachment of molytabs to metallized pads on the MCM substrate, will result in low thermal resistance paths.

Figure 15-22   HDMI module in carrier and test socket on DUT board (fine pitch, 368 lead electrical test fixture).

First Level Connections

The baseline HDMI-1 and LTCC-1 packaging systems use thermosonic gold wire bonding to form electrical connections from bare semiconductor device I/O pads to corresponding pads on substrate top layer metallized pads.   With recent decreases in I/O pad spacings on complex semiconductor devices below 0.004", new automated wire bond software and greater positioning precision had to be developed.   To accommodate the great variety of MCM types and provide die pretest capability, JEDEC standard TAB technology was developed.   To achieve even greater packaging efficiency, high density TAB technology was developed, which reduced the TAB footprint area by 75%.   TAB nonrecurring tape and tooling costs are high, however, and TAB tape development cycle times are too great for use in early program development stages.   A new connection method, software controlled and using testable ribbons was developed (Figure 15-23). This approach enables electrical test and burn in testing of bare die prior to

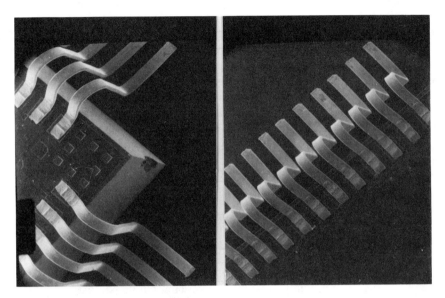

Figure 15-23 Testable ribbon bond (TRB) first level connection. This novel approach provides TAB-like pretest advantages while minimizing nonrecurring cost and cycle times. a) Face up version on left side, b) Flip TAB on right side.

MCM assembly. Moderate pitch flip chip technology is being used to further increase packaging efficiency and to reduce MCM manufacturing cost.

Rework
Rework of defective MCMs with many large, closely spaced devices has been a significant concern. An automated rework station was developed which can remove adjacent semiconductor devices spaced as closely as 0.020" with no resulting damage to neighboring devices or the substrate.

15.4.5 Electrical Testing

Even with extensive use of advanced semiconductor devices and MCM built in test features, functional and diagnostic testing of complex high pin count MCMs frequently exceeds the capabilities of commercially available automated test equipment. MCMs with as many as 1,000 I/O pins are now becoming common. For these complex high performance modules, very precise test fixtures are required, as illustrated in Figure 15-22. It may be necessary to partition tests to account for I/O limitations of commercially available test equipment.

15.4.6 Inspection, Quality and Reliability

Because visual inspection methods are not adequate for large, complex MCMs with more than 100 devices and very small pattern features, automated optical inspection systems are not becoming standard in complex MCM manufacture. In addition, it is necessary to tailor some segments of existing military MCM specifications to accommodate increased package and substrate sizes. Similarly, new reliability models must now be generated to accurately predict overall systems reliability values.

15.4.7 Encapsulation

Hermetic alumina packages, illustrated in Figure 15-20, were initially developed for use in high performance airborne radar processor systems. To provide higher thermal performance and a close coefficient of thermal expansion match with silicon substrates, the team of Hughes, Coors Ceramics and W. R. Grace company developed a family of hermetic aluminum nitride packages under the NOSC VPT program. A 3.8" × 3.8" thick aluminum nitride package with 612 I/O contacts is shown in Figure 15-17.

The weight of a hermetic package into which the bare MCM assembly is installed often is approximately 50 times greater than the weight of the bare MCM assembly. These large hermetic packages also represent a significant portion of overall MCM cost. Accordingly, there are intensive incentives to develop passivation and protective coatings that obviate the need for hermetic packages. Consequently, there is now significant work within the electronics industry to develop and qualify suitable passivation and protective coatings. Several promising passivation and protective coatings, including silicon nitride, epoxy and silicone materials, are being developed and evaluated currently. Coated functional modules have passed Highly Accelerated Screen Tests (HAST). In addition, Hughes is a participant in the Air Force-funded Reliability without Hermeticity (RWoH) program to develop, evaluate and qualify SCOB coating.

Candidate materials are usually silicon dioxide or silicon nitride, alone or in combination [11]. The deposition temperature must be low to avoid damage to the circuitry. Plasma or ion assisted depositions are alternatives to purely chemical deposition methods being evaluated. Diamond-like carbon, a relatively new coating material, is of interest as a possible hermetic coating and will be included in future tests.

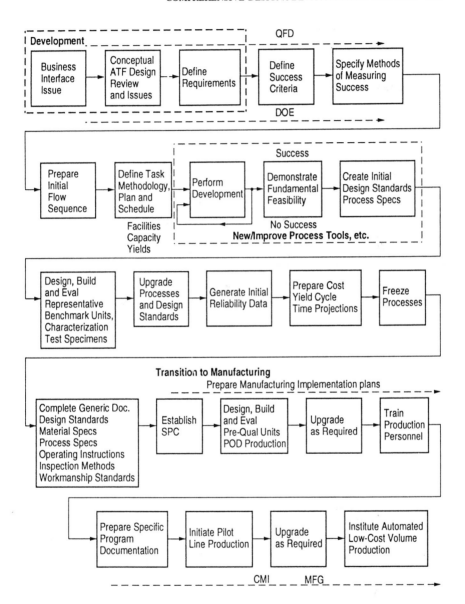

Figure 15-24 Packaging development steps showing transition to manufacturing.

new coating material, is of interest as a possible hermetic coating and will be included in future tests.

15.5 SUMMARY

Like many aerospace manufacturers, Hughes Aircraft Company has recognized the significance of MCMs. As weapon systems technology becomes more complex, signal speeds and the need for high volumetric efficiency increase. Such changes have shifted interrelationships among systems and packaging engineers. Packaging knowledge now is becoming a strong factor in determining the extent and rate of overall technological process to be expected.

At Hughes, progress in two particular areas has determined the degree of overall packaging related advances occurring over the past several years. These two areas, HDMI and LTCC, have been described in this chapter as the core of a foundry which includes a full range of concurrent capabilities from computer aided design through manufacturing and testing. These capabilities are being applied to a broad range of applications. A possible sequence of development and transition to manufacturing is indicated in Figure 15-24.

References

1   F. C. Painter, "Building a Strategic Defense System," *Defense Electronics*, vol 21, p. 58, April 1989.

2   "Critical Technologies Plan for the Committees on Armed Services," *Department of Defense*, United States Congress: Defense Logistics Agency, Defense Technical Information Center, p. ES-1, March 1990.

3   J. J. Licari, "Fabrication and Packaging of High Density Multichip Interconnect Substrates," *Proc. NAECON*, (Dayton, OH), May 1989, pp. 1682-1688.

4   K. P. Shambrook, P. A. Trask, "High Density Multichip Interconnect," *Proc. 39th Electr. Components Conf.*, (Houston, TX), May 1989, pp. 656-662.

5   J. J. Licari, "Hybrid Thin-Film Processing Enters a New Era," *Elect. Packaging & Prod.*, vol. 29, no. 2, pp. 58-63, Sept. 1989.

6   L. E. Dolhert, *et al.*, "Performance and Reliability of Metallized Aluminum Nitride for Multichip Module Applications," *The Internat. J. for Hybrid Microelect.*, vol. 14, no. 4, Dec. 1991.

7   M. F. Bender, *et al.*, "Low-Temperature Cofired Tape Circuits vs. Thick Film: An Economic Analysis," *Proc. ISHM* , (Seattle WA), pp. 12-24, 1988.

8   W. A. Vitriol, J. I. Steinberg, "Development of a Low Temperature Cofired Ceramic Technology," *Proc. ISHM*, (Philadelphia PA), pp. 593-598, 1983.

9   R. G. Pond and W. A. Vitriol, "Custom Packaging in a Thick Film House Using Low Temperature Cofired Multilayer Ceramic Technology," *Proc. ISHM*, (Dallas TX), pp. 268-271, 1984.

10 K. P. Shambrook, E. C. Shi, and J. Isaac, "Bringing MCM Technology to Board Designers," *Printed Circuit Design*, vol. 7, no. 10, pp. 40-49, Oct. 1990.

11 G. Chandra, "Surface Protected Electronic Circuits," *Government Microcircuit Appl. Conf., GOMAC 1987 Digest*, (Orlando FL), vol. 13, pp. 86-88, Oct. 1987.

16

Silicon-Based Multichip Modules

R. Wayne Johnson

16.1 INTRODUCTION

Advances in silicon semiconductor processing and manufacturing technologies are allowing submicrometer scaling of semiconductor features, increasing both the transistor density and switching speed. The need to interconnect ever increasing numbers of transistors on the semiconductor chip has led to significant research and development in thin film multilayer technology. Current semiconductor chips may contain as many as four layers of metal interconnect. The issues addressed in the development of chip level multilayer interconnections include process compatibility, planarization, step coverage, build-in stress, and reliability [1]-[2]. Aluminum and aluminum alloys are used for interconnection metallization. A variety of glasses deposited by chemical vapor deposition (CVD) and spin-on techniques, and polyimides have been used as interlayer dielectrics. Complex processes have been developed to address planarization and step coverage issues.

As transistor switching speeds and semiconductor integration levels increase, the interconnection between chips becomes the limiting factor in realizing high speed system performance. Buschbom [3] has estimated that 55% of the cycle time of a high performance CPU is related to packaging and interconnection delays. Extending semiconductor technology to the level of wafer scale

737

integration (WSI) has been proposed to reduce these delays and to improve performance [4]. While successfully applied to highly repetitive circuits, WSI has been plagued by low yields and the need for elaborate redundancy or discretionary wiring schemes in typical logic systems.

Multichip thin film modules fabricated on silicon substrates provide an alternative to WSI in achieving high density, system level interconnections. The hybrid approach permits pretesting of the substrate and devices prior to assembly, repair of defective modules, flexibility in mixing device technologies and shorter development cycles. The concept of using a silicon substrate is not new. IBM [5] disclosed the concept of an active silicon chip carrier as early as 1972 where field effect transistors were mounted on the surface of a silicon substrate containing monolithic bipolar devices to produce mixed technology circuits.

The fabrication of multilayer thin film substrates is a natural extension of the on chip multilayer interconnection technology. The silicon MCM described by Speilberger [6] in 1984 utilized conventional IC aluminum glass multilayer processing to produce a silicon interconnect substrate. Hitachi also used this technology for a memory module reported in 1985 [7]. The requirements for multichip module (MCM) interconnections are different than for on chip interconnections. Key differences include the average interconnect length (resistive losses, signal attenuation), the module versus chip size (yield) and the need for controlled impedance lines (thicker dielectric layers). Evolution in the materials and processes has occurred in MCM fabrication. Current materials and processes will be discussed in the following sections along with a presentation of multichip applications. The fundamentals of these process steps can be found in Chapter 7. Issues in the selection of dielectric materials are covered in Chapter 8.

16.2 MATERIALS

16.2.1 Silicon Substrates

A number of substrate materials have been used for thin film MCM construction. Table 16-1 presents the properties for the more common substrate materials. In addition, MCMs have been fabricated on metal substrates to improve thermal performance in high power applications [8].

Silicon offers several benefits as a MCM substrate material including:

- Silicon is extremely smooth and flat and dimensionally stable, allowing fine line lithography. Current cleanroom facilities and lithography are used to fabricate semiconductor devices with < 1 μm features at an

Table 16-1 Material Properties for Common Substrate Materials.

	Si	Al_2O_3 99%	Al_2O_3 96%	SiC	AlN	Glass-Ceramic
ELECTRICAL PROPERTIES						
Volume Resistivity (Ω-cm)	2.3×10^5 (10^{-4} to 10^5 doped)	$\geq 10^{14}$	$\geq 10^{14}$	$\geq 10^{11}$	$\geq 10^{14}$	$\geq 10^{13}$
Dielectric Constant at 1 MHz	11.9	9.9	8.9-10.2	40	8.8	5-8
Dielectric Loss at 1 MHz	n/a	≤ 0.0004	0.001	0.05	≤ 0.001	0.002
Dielectric Strength (kV/mm)	n/a	25	23	1.0	14	1.5
THERMAL PROPERTIES						
Thermal Conductivity (W/m-K)	150	25	20	70-270	140-220	1-4
CTE from 25°C-400°C (ppm/°C)	3.5	6.5	7.1	3.8	4.1	3-8
Specific Heat (J/g-K)	0.71	0.88	0.88	0.8	0.7	
MECHANICAL PROPERTIES						
Density (g/cm³)	2.33	3.89	3.75	3.2	3.26	2.9
Grain Size (mm)	Single crystal	1-5	5	2.5	5-10	1-5
Compressive Strength (kpsi)	700	375	340	560	300	
Tensile Strength (kpsi)	12-18	30	25	26	28	10-15
Bending Strength (kg/mm²)	10-50	25-35	25-35	45	40-50	10-20
Young's Modulus (Mpsi)	23.5	40	40	59	40	15-20
Poisson's Ratio	0.28	0.22	0.22	0.15	0.25	0.26
Sintering Temperature (°C)	1412 (mp)	1300	1300	2250	1900	850
Maximum Use Temperature (°C)	1400	1500	1500	1900	1800	500

acceptable yield over areas approaching 1 cm^2. Scaling of multilayer thin film substrate features to 10 - 25 μm should allow acceptable wafer scale interconnection yields.

- Silicon is compatible with thin film multilayer processing.

- The use of a silicon substrate with silicon chips eliminates the coefficient of thermal expansion (CTE) mismatch, between the chip and substrate improving die attach reliability for large area VLSI and VHSIC devices. This is a major concern with flip chip connections since the stress level increases radially outward from the center of the chip. With an alumina substrate, the expansion mismatch is approximately two to one and the maximum chip size is limited for reliable flip chip solder connections. These effects also were discussed in Section 9.6 and are illustrated in Figure 9-42 [9]. Aluminum nitride (AlN) and the cofired glass/ceramic substrate recently developed by IBM for flip chip connections [10] have CTEs which closely match that of silicon. See Chapter 9 for additional discussion of the flip chip solder bump connection technology.

- The thermal conductivity of silicon is six to eight times higher than alumina and is comparable to silicon carbide and aluminum nitride. Since MCMs have higher packaging densities and higher power densities, this is an important consideration if the heat is to be removed through the substrate.

- Active and passive devices can be fabricated in the silicon substrate. Gigabit Logic [11] has fabricated diffused termination resistors and a reversed biased decoupling capacitor in a silicon carrier for a single chip GaAs package. The package is illustrated in Figure 16-1. Potential active circuits include memory arrays, voltage regulators for chips requiring 3.3 volts or other voltages, and off-chip line drivers to reduce chip size and on-chip power dissipation. To maximize substrate yield, the active circuits should be highly repetitive, allowing selective wiring of tested good circuits. MCM examples of active (telecom subscriber line card) and passive (conductive ground plane and decoupling capacitor) circuit elements using the silicon substrate will be discussed in a later section.

- Optical wave guides can be formed on silicon substrates with detectors fabricated in the silicon for electro-optic and photonic systems [12]-[16].

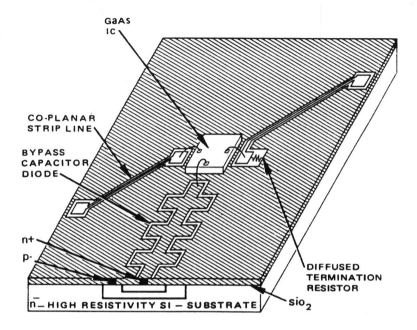

GaAs
IC

CO-PLANAR
STRIP LINE

BYPASS
CAPACITOR
DIODE

n+

p-

n– HIGH RESISTIVITY SI – SUBSTRATE

DIFFUSED
TERMINATION
RESISTOR

SiO$_2$

Figure 16-1 Illustration of silicon substrate carrier for single chip GaAs packaging [11].

- An established base of highly automated fabrication facilities for 4", 5" and 6" wafers exists. Semiconductor manufacturers are integrated vertically to produce silicon MCMs to meet specific application requirements. This will be particularly beneficial in producing MCMs with ASICs.

- Silicon can be micromachined to fabricate sensors [17]-[19] and advanced cooling structures [20]-[21] and to align optical fibers. In sensor applications, it often is necessary to place the signal conditioning electronics as close as possible to the sensor due to the typically low signal level of the sensor. Research is being directed to integrating the sensor and the electronics monolithically in the same piece of silicon. However, the process integration has been difficult, reducing yields and compromising electrical performance. MCM technology would allow interconnection of separately fabricated electronic devices onto a silicon substrate containing the micromachined sensor structure.

Microchannels, re-entrant cavities [20]-[21] and pyramids to initiate nucleated boiling have been fabricated in silicon to improve thermal performance. V-grooves etched in silicon provide a method to align optical fibers for off substrate optical communication.

Silicon does have its limitations as a substrate. Wafer bowing occurs during processing due to stresses arising from the CTE mismatch between the thin film dielectrics and the substrate. The stress in the thin films may be calculated using:

$$\Delta \sigma_f = E_f (\alpha_s - \alpha_f) \Delta T \qquad (16\text{-}1)$$

where: σ_f = film stress,
$(1-\nu_f)$ = biaxial modules of film,
α_s = substrate CTE,
α_f = film CTE and
T = temperature.

The substrate radius of curvature, r, is given by the following equation:

$$r = \frac{E_s D^2}{6 \sigma_f t (1 - \nu)} \qquad (16\text{-}2)$$

where: E_s = Young's modulus of the substrate,
D = thickness of the substrate,
t = the film thickness and
ν = Poisson's ratio.

Young's modulus for silicon is approximately 50% of that for alumina and aluminum nitride. Thus, for equal stress levels and film and substrate thicknesses, the silicon substrate will bow more than the ceramic substrates. Substrate bowing adversely affects automated vacuum handling systems and fine line imaging. Lower stress dielectric films are being developed to address this issue. Thicker wafers also can be used to reduce bowing. Hagge [38] has estimated silicon wafers can be deflected to 0.010 in./in. bow without fracture and can withstand 10^8 bending cycles if the strain range is kept below 0.0005 in./in. For a sample subjected to bending, the strain is calculated as the change in length of the tensile (compressive) surface divided by the length. The strain range is the sum of the compressive and tensile strain.

A second limitation is the need for additional packaging of the silicon substrate for mechanical protection and to provide input/outputs (I/Os) for the next level interface. Various packaging options currently being used will be discussed in a later section. In all cases, the packaging schemes use perimeter I/Os. Internal output signals must travel to the substrate edge before exiting the module, increasing delay time. Power distribution to the substrate interior is also more difficult with perimeter interfaces than with multilayer cofired ceramic substrates using a pin grid array. The number of I/Os is limited by the substrate perimeter and the spatial fanout requirements of the next packaging level. At least one nonperimeter approach has been proposed by Little, *et al.* [22]. In this approach, aluminum is diffused through the silicon substrate to provide an area array of contacts on the substrate backside.

The CTE mismatch between the silicon substrate and the package base must be considered in package selection. The two options are to nearly match the CTE of the silicon substrate (AlN, SiC, mullite packages) or to use a compliant adhesive (silicones, modified epoxies) to attach the substrate into the package. If heat removal from the die is through the substrate and the package base, the thermal conductivity of the substrate attachment material and the package base material must also factor into the decision.

Silicon substrates currently are used in approximately 44% of the thin film MCMs produced worldwide [23]. The choice of substrate material depends on the specific application requirements and the fabrication facilities used. The ideal substrate does not exist. Therefore, substrate choice is based on optimizing the most critical parameters for a particular application. The preceding paragraphs have highlighted the advantages and disadvantages of silicon substrates and have presented comparative properties for other substrate materials. This can serve as a basis for making initial decisions on substrate materials.

16.2.2 Conductors

Aluminum, copper and gold are the three most common interconnection metallizations. Their properties are summarized in Table 16-2.

Early silicon MCMs used aluminum metallization as an extension of semiconductor technology and it is still widely used. The key advantage of aluminum is its ease of processing using sputtering and wet chemical etching. No adhesion or barrier layers are required typically. As operating speeds increase and line widths decrease, the resistivity of aluminum limits its performance. Copper metallization systems have been developed for this reason. Copper requires an adhesion layer typically chromium (Cr) or titanium (Ti). It can be sputter deposited, but is difficult to etch by wet chemical methods.

Table 16-2 Properties of Common Thin Film Metallizations.

	ALUMINUM	COPPER	GOLD
Resistivity	2.66 µΩ·cm	1.67 µΩ·cm	2.35 µΩ·cm
Adhesion layer used	Good adhesion layer	Ti or Cr adhesion used	Cr, TaN or NiCr
Deposition	Sputtering	Sputtering, plating	Sputtering, plating
Corrosion	Corrodes	Corrodes without barrier layer	No

Note: Copper reacts with polyamic acid; requires barrier layer (Ni or Cr).

Liftoff processes have been developed for fine line patterning of sputtered copper [24]. Copper also can be pattern plated to achieve fine lines. In a typical plating process, an adhesion layer followed by a thin copper layer is sputter deposited. Then photoresist is applied and imaged. After electroplating the thicker copper pattern, the photoresist is removed and the background copper and adhesion layer are etched away. When using polyimide dielectrics, the polyamic acid precursor reacts with the copper during the polyimide cure affecting local electrical properties and the polyimide to copper adhesion. Barrier layers such as nickel and chromium are used to protect the copper. These may be deposited by plating or sputtered and patterned by etching techniques. A number of variations exist for depositing and patterning copper metallization but the details will not be discussed here.

Gold metallization also has been used in MCM fabrication [8]. The corrosion resistance of gold is its primary attraction. Gold requires an adhesion layer, but does not react with polyamic acid and no barrier layer is needed. It is typically pattern plated.

Other interconnection systems including superconductors and optical wave guides have been discussed for MCMs and may be used in the future [15]-[16], [25].

16.2.3 Dielectrics

Thin film dielectrics used on silicon substrates include silicon dioxide, polyimide, modified polyimides, benzocyclobutene, and polyphenylquinoxaline. A summary of typical properties is presented in Table 16-3. While this table can be used for the general comparison of dielectrics, caution is advised since the data presented

Table 16-3 Summary of Interlayer Dielectric Properties.

	Standard Polyimide	Fluorinated Polyimides	Silicone Polyimides	Acetylene Terminated Polyimides	Low Stress Polyimides	BCBs	PPQs	SiO$_2$
ELECTRICAL PROPERTIES								
Dielectric Constant (1 kHz)	3.4-3.8	2.7-3.0	3.0-3.5	2.8-3.2	2.9 (z) 3.4 (xy)	2.7	2.7	3.5
Dissipation Factor (1 kHz)	0.002	0.002		0.002	0.002	0.0006	0.0005	0.0001
THERMAL PROPERTIES								
Decomposition Temperature (°C)	520-550	>470	450	500-520	620-650	430	500	>1000
Glass Transition Temperature (°C)	300-320	<300	<300	215-230	>400	350-360	365	
CTE (ppm)	20-40			38	3-6	45-70	55	0.5
Moisture Absorption (% wt.)	1.1-3.5	0.7	0.9	0.8-1.3	0.4-0.5	0.3-0.5	0.9	<0.1
MECHANICAL PROPERTIES								
Young's Modulus (kpsi)	400-1000		50-250	300-450	1280	340		10,000
Stress (10^8 dynes/cm^2)	3	5.3	3	4	0.4-0.5	3.7		±20
PLANARIZATION								
Degree of Planarization (%)	24-34	25-30	22	91-93	10-28	91		0-5
Percent Solids	14-17	15-19	26	35	10-13.5	35-62	35-47	n/a
Viscosity (poise)	11-19	15-25	12	15	20-110	0.15-1.5	11-15	n/a
PROCESSING								
Deposition method	Spin coat, spray	Spin coat, spray	Spin coat, spray	Spin coat, spray	Spin coat, spray	Spin coat, spray	Spin coat, spray	CVD
Curing temperature (°C)	350-400		300-400	300-350	350-400	230-250	400-450	300-400

BCB = Benzocyclobutene   PPQ = Polyphenylquinoxaline   SiO$_2$ = Silicon Dioxide were collected from various references. The dielectric processing and test methods vary and the individual references should be reviewed for specific details. See Chapter 8 and 7 for a more detailed discussion of dielectrics and processing with dielectrics, respectively.

Silicon dioxide, used in early modules, has found limited acceptance. Thick silicon dioxide layers, necessary for controlled impedance transmission lines, have been difficult to deposit crack-free. Horton [39] recently has reported the deposition of silicon dioxide layers greater than 10 μm thick with controllable stress levels by plasma chemical vapor deposition. Examples of this technology will be presented in a later section. Silicon dioxide has the advantage of negligible water uptake, chemical inertness and higher thermal conductivity than organic dielectrics; its dielectric constant is higher than organic dielectrics and it is conformal. The resistance of the dielectric to moisture penetration is a significant advantage for MCMs in nonhermetic packages.

Polyimides are the most widely used interlayer dielectrics. The dielectric typically is applied as a solution containing polyamic acid which is converted by a condensation reaction at high temperature (> 350°C). Standard polyimides are stable at high temperature and compatible with thin film processing and module assembly. Limitations of standard polyimides include high moisture uptake, only moderate planarization and reactivity during cure with copper. Moisture absorption by the cured film effects its electrical properties with the dielectric constant increasing with increasing moisture content. During processing, dehydration bakes are necessary before metal deposition to avoid blistering of the metal layer during subsequent high temperature processing steps. Solid metal power and ground planes must be avoided. Meshed power and ground planes should be used to allow moisture in the polyimide layers below to outdiffuse without blistering the metal planes. The low degree of planarization (20 - 35%) achieved with polyimides results in a loss of planarity as the number of metal and dielectric layers increases. Planarity is important for fine line lithography, step coverage and uniform characteristic impedance. Multiple dielectric coatings improve planarization. Techniques including polishing have been developed to maintain planarity.

Polyimides have been modified to produce materials with lower dielectric constants, lower moisture uptake, lower CTE (low stress) or which are photoimageable. The modifications are typically compromises with other film properties and specific materials should be evaluated against actual dielectric requirements. Compared to standard polyimides, the fluorinated and silicone polyimides have a somewhat lower moisture uptake and the fluorinated polyimides have a slightly lower dielectric constant. Both materials, however, have found limited application to date. The siloxane modified polyimides typically are preimidized and dissolved in suitable solvents, requiring only solvent evaporation during the cure process. Since the material is preimidized, no polyamic acid is present and no water of condensation is released during cure, reducing its reactivity with copper. The acetylene terminated polyimides also are preimidized and crosslink via the acetylene groups located at the ends of the oligomer without releasing volatiles. Fluorinated, acetylene terminated polyimides have been formulated to improve the solvent resistance as compared to earlier fluorinated polyimides.

There is significant interest in the photoimageable materials as a means of simplifying processing for via formation [30]. By exposing the polyimide and developing it by wet chemistry, the process steps of depositing and patterning an etch mask, plasma etching the polyimide and stripping the etch mask are eliminated. Film shrinkage and via distortion during cure (after imaging) have limited the acceptance of these materials in the United States, however there is considerable usage in Japan and development work is continuing.

Benzocyclobutene (BCB) is a highly crosslinked polymer derived by the thermal rearrangement (> 200°C) of bisbenzocyclobutene monomers. No reaction byproducts are produced in the curing process. The prepolymer is B-staged and dissolved in a solvent for application to substrates. BCB has a low dielectric constant, has low moisture absorption, is planarizing, cures at lower temperatures and does not react with copper during curing. Its thermal stability limit is approximately 350°C, which is compatible with most processing and assembly operations, but may be insufficient for certain high temperature metal anneals and soldering operations.

Polyphenylquinoxaline (PPQ) recently has been explored as an interlayer dielectric [26]. This material is supplied as a fully cyclized polymer dissolved in a solvent. Only solvent evaporation is required to cure the dielectric film. The material has a low dielectric constant and a high thermal stability. Additional characterization of this material is required.

Research is continuing to improve existing materials and to develop new ones. Other materials currently being investigated include fluoropolymers and polyquinoline.

16.3 MCM EXAMPLES

Extensive research and development programs in silicon MCMs exist at companies and universities around the world. Many have produced modules. A sampling of these will be presented in this section to highlight the various processing, material and packaging options.

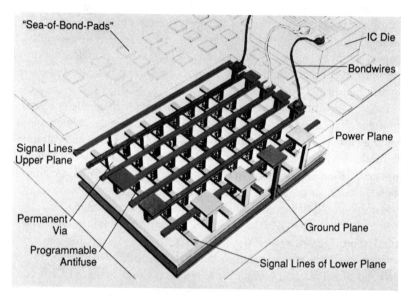

Figure 16-2 Illustration of Mosaic System's programmable silicon substrate. (Courtesy of ERIM.)

16.3.1 Prototype Silicon Substrates

As with any new technology, the development costs for prototypes are high and may be difficult to justify for small volume applications. Development times typically are also long. Mosaic Systems developed an electrically programmable silicon substrate technology to provide rapid prototyping capability [31]. A proprietary antifuse technology, based upon amorphous silicon, is used to form programmable connections between an x and y grid of aluminum lines as shown in Figure 16-2. All x and y lines extend to the perimeter of the substrate for programming. By applying a threshold voltage (approximately 20 V) between appropriate x and y lines, the amorphous silicon crystallizes, forming a conductive path between the two lines. Substrates can be mass produced and programmed directly from a CAD file. Integrated circuit chips are epoxy attached to the silicon substrate and electrically connected by ultrasonic aluminum wire bonding.

The Environmental Research Institute of Michigan (ERIM) has used the Mosaic Systems technology to fabricate the SEM-E memory module shown in

Figure 16-3 SEM-E Memory module. (Courtesy of ERIM).

Figure 16-3 [32]. The module contains 32 32k × 8-bit memory chips. The silicon substrate is mounted on a metal matrix composite (Al graphite) frame. A second example from ERIM is the Symbolics MacIvory processor module shown in Figure 16-4. In this implementation the circuit is partitioned into four smaller silicon substrates which are mounted subsequently on a programmable 4" diameter silicon substrate. The single MCM replaces two 12.5" × 3.9" printed wiring boards (PWBs) and contains 112 components. The thin silicon dioxide layers and the resistance (typically 4 Ω) of the crystallized via connections currently limits this technology to applications below approximately 40 MHz.

A prototype substrate technology based on copper conductors and polyimide dielectrics has been developed by Oerlikon-Contraves AG [33]. In their approach, a grid of x and y conductor lines are separated by a polyimide layer. At crossover points the dielectric is removed between the two conductor traces by plasma etching to form an air bridge. The resulting structure is shown in Figure 16-5. The air gap supports approximately 300 V. To program the substrate, appropriate x and y lines are connected at crossover points by using

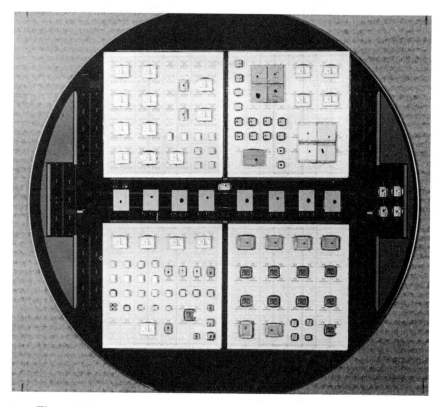

Figure 16-4 Symbolics MacIvory processor module. (Courtesy of ERIM.)

a modified ultrasonic wire bonder to weld the two conductors. A welded contact is shown in Figure 16-6. Laser welding techniques are under development. The laser is being used also to sever unused line segments from the node. These stubs could effect electrical performance at high frequency. Figure 16-7 illustrates a completed assembly.

Another technique, which is similar in philosophy to a gate array design, has been used by MCC (Microelectronics and Computer Technology Corp., Austin TX) for thin film MCM on ceramic substrates, but is also applicable to silicon substrate modules [34]. In their process, the substrate is fabricated using copper and polyimide to produce a standard interconnection pattern with power and

Figure 16-5 Air bridge structure for programmable x-y connections. (Courtesy of Oerlikon-Contraves AG.)

ground planes and short x and y signal line segments. These substrates are placed in inventory. The final interconnection layers are added to connect the appropriate signal line segments providing the necessary signal routing to customize the module.

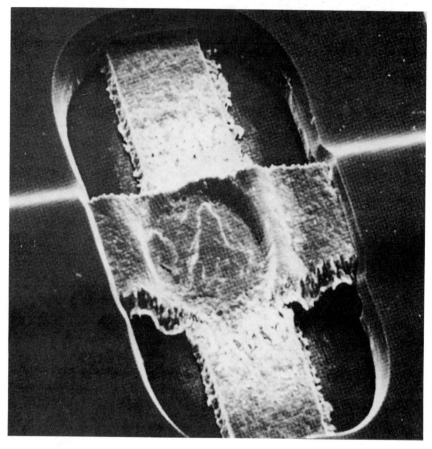

Figure 16-6 Ultrasonically welded connection between x- and y- signal lines. (Courtesy of Oerlikon-Contraves AG.)

16.3.2 Custom Silicon Substrates

As with custom semiconductor chips, custom silicon substrates provide wider latitude in module design. The silicon substrate can be used as an electrically active component of the module. In the case of AT&T, the substrate was heavily

Figure 16-7 12-bit D/A converter. (Courtesy of Oerlikon-Contraves AG.)

doped to create a ground plane. The cross section of their MCM technology is illustrated in Figure 16-8 [35]. A thermally grown oxide and chemical vapor deposited silicon nitride dual dielectric layer forms a decoupling capacitor (25 nF/cm^2) between the power and ground planes. Two micrometer thick copper conductors were electroplated onto a sputter deposited Ti/Cu or Ti/Cr/Cu adhesion layer. Minimum conductor line widths are 10μm wide (line resistance approximately 10 Ω/cm). Characteristic impedances of 50 - 70 Ω for 10 μm wide/20 μm pitch signal lines are achieved with a bottom polyimide layer thickness of 10 μm and a second level polyimide thickness of 6 μm. Vias are filled with electroless Ni to insure electrical continuity through the vias and to allow via stacking. IC chips are bonded to the substrate using flip chip soldering. Figure 16-9 shows a three chip (WE 32100 central processor, memory management unit and math accelerator unit) module fabricated by AT&T. The silicon substrate is bonded to an aluminum heatsink using a thick silicone gel material to provide a compliant bond. The heatsink is assembled to a four layer bismaleimide triazine (BT) epoxy PWB. The PWB contains 160 I/O

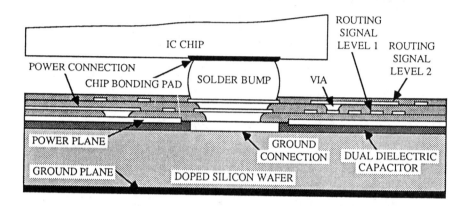

Figure 16-8 Cross section illustrating AT&T MCM technology. (Courtesy of AT&T.)

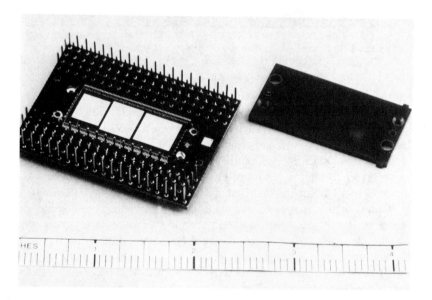

Figure 16-9 Three chip (WE 32100 central processor, memory management unit and math accelerator unit) module. (Courtesy of AT&T.)

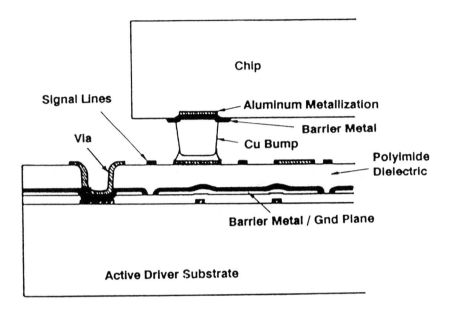

Figure 16-10 Cross section illustrating Texas Instruments' active silicon substrate module. (Courtesy of Texas Instruments.)

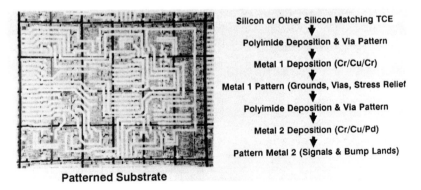

Figure 16-11 Active substrate containing DS3680 line drivers chips. (Courtesy of Texas Instruments.)

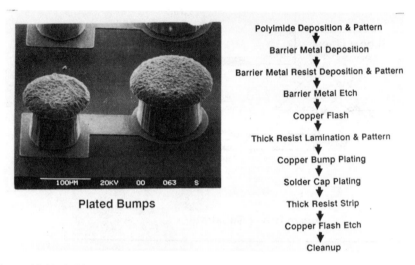

Polyimide Deposition & Pattern
↓
Barrier Metal Deposition
↓
Barrier Metal Resist Deposition & Pattern
↓
Barrier Metal Etch
↓
Copper Flash
↓
Thick Resist Lamination & Pattern
↓
Copper Bump Plating
↓
Solder Cap Plating
↓
Thick Resist Strip
↓
Copper Flash Etch
↓
Cleanup

Plated Bumps

Figure 16-12 Solder capped copper bumps. (Courtesy of Texas Instruments.)

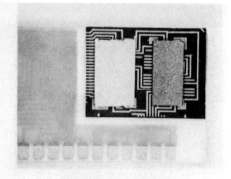

- **Redundant Bumps**
- **Bumps over Active Circuits**
- **Existing Chips with Corrosion Resistant Bond Pads**
- **Plated Copper Bumps with 10/90 Solder Caps**
- **Active Driver in Substrate**
- **Analog & Digital Circuits**
- **Design to Reduce Coupling**
- **Meshed Ground / Stress Relief**

SIP Line Card Module

Figure 16-13 Telecom subscriber line card module. (Courtesy of Texas Instruments.)

Figure 16-14 Prototype automotive MCM module. (Courtesy of Texas Instruments.)

pins in a grid array pattern. Thermosonic gold wire bonding is used to electrically connect the silicon substrate and the PWB. The BT resin system is chosen for the PWB to accommodate the thermosonic wire bonding temperature of 150°C. A plastic cover is attached to provide mechanical protection of the wire bonds.

Texas Instruments [36] has demonstrated a telecom subscriber line card module using an active silicon substrate. A cross section of the assembly is illustrated in Figure 16-10. The circuit uses three chips: a DS3680 line driver, a TCM 4209 SLCC (supervisor control, hybrid and attenuation) and a TCM 2914 COMBO (CODEC and filter). The DS3680 line drivers chips are fabricated monolithically in the MCM substrate as shown in Figure 16-11. The substrate interconnection pattern overlays several DS3680 chips providing redundancy to improve the active substrate yield. The interconnection metallization used is

Cr/Cu/Cr and the interlayer dielectric is polyimide. Palladium metallization is used on the substrate as the solderable pads for the flip chip assembly. Palladium is readily wet by solder, has limited solubility in solder, and is amenable to thin film processing. Chrome is not solderable and copper is chemically reactive and tends to oxidize and corrode prior to soldering. Rather than using solder bumps on the TCM 2914 COMBO and the TCM 4209 SLCC chips, copper bumps with solder caps are plated on the chips. Examples of the copper bumps are shown in Figure 16-12. Compared to solder bumps, the copper bumps provide a lower thermal resistance path from the die to the substrate. Two copper bumps are plated for each I/O on the die to improve thermal performance and also provide redundancy. An assembled module is shown in Figure 16-13.

A surface mount, plastic cavity package for MCMs developed by Texas Instruments is shown in Figure 16-14. The circuit is a prototype for an automotive application. The interconnection substrate has aluminum metallization and polyimide dielectric. The plastic package provides a low-cost packaging option.

Rockwell has designed a number of silicon MCMs [37]-[38]. The avionics processor module shown in Figure 16-15 contains 53 ICs and 40 discrete devices on a 2.2" × 2.2" silicon substrate with aluminum polyimide interconnect. The module is a flight computer system with 64 kbytes of static data memory, 64 kbytes of nonvolatile program storage, system timer, multiple interrupt inputs, discrete output ports and an advanced 16-bit microprocessor operating at 30 MHz. The module is packaged in a Kovar package (2.4" × 2.4") with 180 I/O.

The Rockwell module in Figure 16-16 performs interface and signal processing functions in a Global Positioning System (GPS) receiver. The 1.2" × 1.2" module operates at 88 MHz and uses meshed power and ground planes in a stripline configuration around two signal layers. The aluminum benzocyclobutene multilayer structure has 25 μm wide signal lines.

The Avionics Memory Module shown in Figure 16-17 is packaged in a hermetic 2.2" × 2.2" cofired aluminum nitride package. The module is for storage of flight and operating condition data and contains 32 ICs and 64 chip capacitors and resistors on two silicon substrates. The post in the center of the package is to support the metal lid during the pressure variations encountered in avionics applications. The package contains 136 fine stranded copper leads to provide for ultracompliant surface mount assembly to the next interconnect level. This is necessary to minimize cyclic stress on the surface mount solder joints due to the CTE mismatch between the AlN package and the PWB.

The module in Figure 16-18 contains two 1750A processors with on board memory and selftest fabricated on a 2.5" octagonal substrate. The aluminum

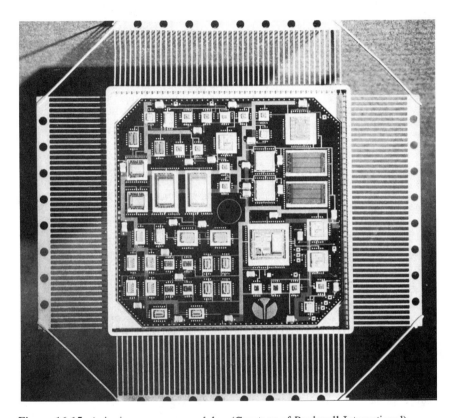

Figure 16-15 Avionics processor module. (Courtesy of Rockwell International).

benzocyclobutene substrate has four conductor layers and interconnects 134 components including 33 ICs. The module operates at 40 MHz and the combined throughput is 5 MIPs. Both hermetic (machineable aluminum nitride) and nonhermetic (graphite aluminum) packages containing an integral plug in connector have been fabricated for the module.

nChip's MCM technology utilizes thick silicon dioxide dielectric layers [39]-[41]. The dielectric is plasma enhanced chemical vapor deposited and undoped. The stress is controllable over the range $\pm 2 \times 10^8$ N/m^2. The three substrate variations illustrated in Figure 16-19 provide increasing performance (lower interconnect resistance, lower power/ground plane resistance, dual power planes, termination resistors and increased decoupling capacitance) with increasing substrate complexity (cost). Target applications for the nC-1000 Series (Figure

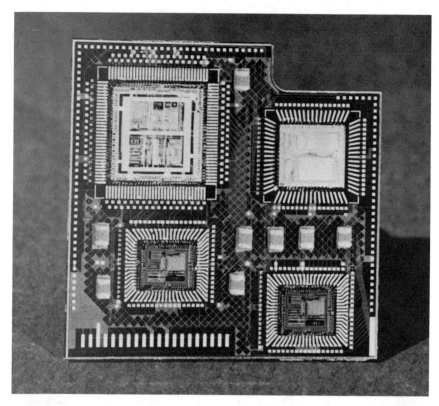

Figure 16-16   Global positioning system (GPS) receiver.   (Courtesy of Rockwell International.)

16-19a) are 40 - 80 MHz CMOS circuits, for the nC-2000 (Figure 16-19b) are > 60 MHz CMOS/BiCMOS circuits, and for the nC-3000 Series (Figure 16-19c) are > 100 MHz GaAs and BiCMOS/ECL circuits. Figure 16-20 shows a five chip, SPARC Central Processor Module packaged in a 224 pin ceramic quad flat pack. The substrate size is 0.9" × 1.2" and replaces a 3.3" × 7.25" PWB.

Three different RISC processor modules fabricated by nChip are shown in Figure 16-21. The five chip module is the SPARC module previously discussed. The other two modules are packaged in pin grid arrays. The eight chip module is CMOS and implemented with nC-1000 substrate technology. The three chip module is ECL and uses nC-2000 substrate technology. This module dissipates 60 watts using conventional air cooling.

Figure 16-17 Avionics memory module. (Courtesy of Rockwell International.)

There continues to be significant development in the use of silicon as a substrate for MCMs and new modules are being announced on a regular basis. Examples of other silicon MCMs may be found in the literature.

16.4 SUMMARY

Continued advances in semiconductor technology, increases in system performance requirements, shorter design windows and short product life cycles will all contribute to the widespread application of MCM technology. Silicon

Figure 16-18 Dual 1750A Processors with on-board memory and self test. (Courtesy of Rockwell International.)

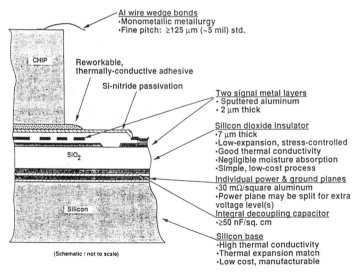

Figure 16-19a Cross section illustration of nC-1000 series silicon circuit board.

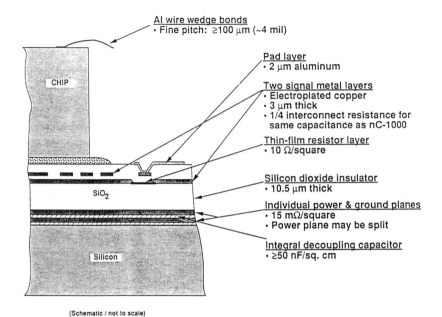

Al wire wedge bonds
• Fine pitch: ≥100 μm (~4 mil)

CHIP

Pad layer
• 2 μm aluminum

Two signal metal layers
• Electroplated copper
• 3 μm thick
• 1/4 interconnect resistance for
  same capacitance as nC-1000

Thin-film resistor layer
• 10 Ω/square

Silicon dioxide insulator
• 10.5 μm thick

SiO₂

Individual power & ground planes
• 15 mΩ/square
• Power plane may be split

Silicon

Integral decoupling capacitor
• ≥50 nF/sq. cm

(Schematic / not to scale)

Figure 16-19b   Cross section illustration of nC-2000 series silicon circuit board.

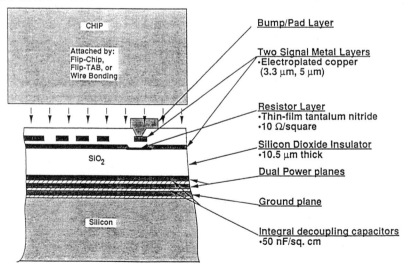

CHIP

Attached by:
Flip-Chip,
Flip-TAB, or
Wire Bonding

SiO₂

Silicon

Bump/Pad Layer

Two Signal Metal Layers
•Electroplated copper
 (3.3 μm, 5 μm)

Resistor Layer
•Thin-film tantalum nitride
•10 Ω/square

Silicon Dioxide Insulator
•10.5 μm thick

Dual Power planes

Ground plane

Integral decoupling capacitors
•50 nF/sq. cm

Figure 16-19c   Cross section illustration of nC-3000 series silicon circuit board.
(Courtesy of nChip.)

Figure 16-20a PWB and silicon MCM versions of the Ross Technology SPARC central processor chip set. (Courtesy of nChip.)

Figure 16-20b Five chip SPARC central processor module. (Courtesy of nChip.)

Figure 16-20c Close up of SPARC module illustrating the fine wire bond pitch. (Courtesy of nChip.)

Figure 16-21 Three RISC processor modules. (Courtesy of nChip.)

provides a number of tradeoffs as a MCM substrate. It will not be universally used; silicon offers unique capabilities which can be exploited in a number of applications.

References

1   E. Ong, H. Chu, and S. Chen, "Metal Planarization with an Excimer Laser," *Solid State Technol.*, vol.34, no.8, pp. 63-68, Aug. 1991.

2   A. Nagy and J. Helbert, "Planarized Inorganic Interlevel Dielectric for Multilevel Metallization—Part I," *Solid State Technol.*, vol. 34, no. 1, pp. 53, Jan. 1991.
      A. Nagy and J. Helbert, "Planarized Inorganic Interlevel Dielectric for Multilevel Metallization—Part II," *Solid State Technol.*, vol. 34. no. 3, pp. 77-80, March 1991.

3   M. L. Buschbom, "Modelling Issues for VLSI Package Design," *Proc. of the Techn. Prog.: NEPCON West*, (Anaheim CA), pp. 686-697, March 1989.

4   S. K. Tewksbury and L. A. Hornak, "Wafer Level System Integration: A Review," *IEEE Circuits and Devices Mag.*, vol. 5, no. 5, pp. 22-30, Sept. 1989.

5   D. J. Bodendorf, K. T. Olsen, J. F. Trinko, J. R. Winnard, "Active Silicon Chip Carrier," *IBM Technical Disclosure Bulletin*, vol. 15, no. 2, pp. 656-657, July 1972.

6   R. K. Spielberger *et al.*, "Silicon-on-Silicon Packaging," *IEEE Trans. on Components, Hybrids, and Manufacturing Techn.*, vol. CHMT-7, no. 2, pp. 193-196, June 1984.

7   T. Yamada *et al.*, "Low Stress Design of Flip Chip Technology for Si on Si Multichip Modules," *Proc. of the Fifth Annual Internat. Elect. Packaging Conf.*, (Orlando FL), pp. 551-557, Oct. 1985.

8   R. L. Hubbard and G. Lehman-Lamer, "Very High Speed Multilayer Interconnect Using Photosensitive Polyimides," *Proc of the 1988 Internat. Symp. on Microelect.*, (Seattle WA), pp. 374-376, Oct. 1988.

9   S. E. Greer, "Low Expansivity Organic Substrate for Flip-Chip Bonding," *Proc. of the 28th Elect. Components Conf.*, (Anaheim CA), pp. 166-171, April 1978.

10   R. R. Tummala, H. R. Potts, and S. Ahmed, "Packaging Technology for IBM's Latest Mainframe Computers (S/390/ES9000)," *Proc. of the 41st Elect. Components and Techn. Conf.*, (Atlanta GA), pp. 682-687, May 1991.

11   T. R. Gheewala, "Packages for Ultra-High Speed GaAs IC's," *Technical Digest of the 1984 GaAs IC Symp.*, (Boston MA), pp. 67-70, Oct. 1984.

12   S. Sriram, "Novel V-Groove Structures in Silicon," *SPIE Vol. 578, Integrated Optical Circuit Engineering II*, pp. 88-94, 1985.

13   R. A. Soref and J. P. Lorenzo, "Silicon Guided-Wave Optics," *Solid State Techn.*, vol. 31, no.11, pp. 895-898, Nov. 1988.

14   J. T. Boyd *et al.*, "Guided Wave Optical Structures Utilizing Silicon," *Optical Engineering*, vol. 24, no. 2, pp. 230-234, March/April 1985.

15   S. E. Schacham *et al.*, "Waveguides as Interconnections for High Performance Packaging," *Proc. of the 9th Annual Internat. Elect. Packaging Conf.*, (San Diego CA), pp. 1003-1013, Sept. 1989.

16   R. Selvaraj, H. T. Lin, and J. F. McDonald, "Integrated Optical Waveguides in Polyimide for Wafer Scale Integration," *J. of Lightwave Techn.*, vol. 6, pp. 1034-1044, 1988.

17 J. B. Angell, S. C. Terry, and P. W. Barth, "Silicon Micromechanical Devices," *Scientific Amer.*, vol. 248, no. 4, pp. 44-55, April 1983.

18 K. E. Bean, "Anisotropic Etching of Silicon," *IEEE Trans. on Electron Devices*, vol. ED-25, no. 10, pp. 1185-1193, Oct. 1978.

19 J. H. Atherton, "Sensor Signal Conditioning: An IC Designer's Perspective," *SENSORS*, vol. 8, no.11, pp. 23-30, Nov. 1991.

20 A. Goyal *et al.*, "Re-entrant Cavity Heat Sink Fabricated by Anisotropic Etching and Silicon Direct Wafer Bonding," *Proc. of the Eighth Annual Semi. Thermal Measurement and Management Symp.*, (Austin TX), pp. 25-29, Feb. 1992.

21 N. K. Phadke *et al.*, "Re-entrant Cavity Surface Enhancements for Immersion Cooling of Silicon Multichip Packages," *Proc. of the Third IEEE/ASME Intersociety Conf. on Thermal Phenomena in Elect. Sys.*, (Austin TX), pp. 59-65, Feb. 1992.

22 M. J. Little *et al.*, "A Three-Dimensional Computer for Image and Signal Processing," *Proc. of the IEEE 1985 Custom Integrated Circuits Conf.*, (Portland OR), pp. 119-123, May 1985.

23 M. W. Hartnett and E. J. Vardaman, "Worldwide MCM Status and Trends: Material Choices," *Proc. of the Technical Program: NEPCON West*, (Anaheim CA), pp. 1111-1120, Feb. 1991.

24 A. Schiltz *et al.*, "Lift Off Techniques of Packaging/ Processing Technologies for Multichip Modules," *Proc. of the Technical Prog.: NEPCON West*, (Anaheim CA), pp. 975-983, March 1990.

25 H. Kroger *et al.*, "Applications of Superconductors to Hybrid Electronics," *Proc. of the 1990 Internat. Symp. on Microelect.*, (Chicago IL), pp. 570-572, Oct. 1990.

26 L. Verdet, J. Reche, and G. Rabilloud, "The P's and Q's of PPQ," *Elect. Packaging and Prod.*, vol. 31, no. 1, pp. 58-61, Jan. 1991.

27 T. G. Tessier, G. M. Adema, and I. Turlik, "Polymer Dielectric Options for Thin Film Packaging Applications," *Proc. of the 39th Elect. Components Conf.*, (Houston TX), pp. 127-134, May 1989.

28 B. T. Merriman *et al.*, "New Low Coefficient of thermal Expansion Polyimide for Inorganic Substrates," *Proc. of the 39th Elect. Components Conf.*, (Houston TX), pp. 155-159, May 1989.

29 D. Burdeaux *et al.*, "Benzocyclobutene (BCB) Dielectrics for the Fabrication of High Density, Thin Film Multichip Modules," *J. of Elect. Materials*, vol. 19, no. 12, pp. 1357-1366, Dec. 1990.

30 T. Ohsaki *et al.*, "A Fine-Line Multilayer Substrate with Photo-Sensitive Polyimide Dielectric and Electroless Copper Plated Conductors," *Proc. of the IEEE Internat. Elect. Manufac. Techn. Symp.*, (Anaheim CA), pp. 178-183, 1987.

31 A. A. Bogdan, "An Electrically Programmable Silicon Circuit Board," *Proc. of BUSCON '87*, pp. 9-14, April 1987.

32 K. L. Drake *et al.*, "Equivalent Hermetic Packaging of High Density Multichip Modules Interconnect Substrate," *Proc. of the Technical Prog.: NEPCON West*, (Anaheim CA), p. 840, Feb. 1991.

33 E. Sutcliffe, "Multichip Modules for High Performance," *Proc. of the Technical Prog.: NEPCON West*, (Anaheim CA), pp. 824-831, Feb. 1991.

34 D. Carey, "A Program to Provide Quick Turnaround Delivery of Multichip

Modules," *Proc. of the Technical Prog.: NEPCON West*, (Anaheim CA), pp. 507-514, Feb. 1991.

35 C. J. Bartlett, J. M. Segelken, and N. A Teneketges, "Multichip Packaging Design for VLSI-Based Systems," *IEEE Trans. on Components, Hybrids, and Manufac. Techn.*, vol. CHMT-12, no. 4, pp. 647-653, Dec. 1987.

36 K. G. Heinen *et al.*, "Multichip Assembly with Flipped Integrated Circuits," *Proc. of the 39th Elect. Components Conf.*, (Houston TX), pp. 672-680, May 1989.

37 J. K. Hagge, "Ultra Reliable HWSI with Aluminum Nitride Packaging," *Proc. of the Technical Prog.: NEPCON West*, (Anaheim CA), pp. 1271-1283, Feb. 1989.

38 J. K. Hagge, "State-of-the-Art Multichip Modules for Avionics," *Internat. Symp. on Advances in Interconnects and Packaging*, (Boston MA), SPIE/OPTCON-90, pp. 175-194, Nov. 1990.
    J. K. Hagge, "Ultra-Reliable Pachaging for Silicon-on-Silicon WSI," *Proc. 38th Electr. Comp. Conf.*, (Los Angeles CA), pp. 282-292, May 1988.

39 T. Horton, "MCM Driving Forces, Applications, and Future Directions," *Proc. of the Technical Prog.: NEPCON West*, (Anaheim CA), pp. 487-494, Feb. 1991.

40 B. McWilliams, "Comparison of Multichip Interconnection Technologies," *Proc. of the 11th Annual Internat. Elect. Packaging Conf.*, (San Diego CA), pp. 63-68, Aug. 1991.

41 J. C. Demmin, "Thermal Modeling of Multi-Chip Modules," *Proc. of the Technical Prog.: NEPCON West*, (Anaheim CA), pp. 1145-1154, Feb. 1991.

17

THE TECHNOLOGY AND MANUFACTURE OF THE VAX-9000 MULTICHIP UNIT

Salma Y. Abbasi, Arun Malhotra, John D. Marshall,
Douglas N. Modlin, Ralph Platz, Hongbee Teoh,
Scott Westbrook, Jimmy Jun-Min Yang

17.1 INTRODUCTION

A multichip module (MCM) is a miniaturized electronic subsystem based on a high density interconnect (HDI) structure. The MCM connects unpackaged components (ICs and/or passive devices) and integrates them with an interconnection system.

Corresponding to the interconnection technology that it employs, the industry has defined MCM-D as having a thin film interconnection structure deposited on a suitable substrate (aluminum, silicon, ceramic). The MCM-D provides the highest density and performance but requires the largest investment in manufacturing technology. This chapter discusses Digital's approach to the development of MCM-D technology and manufacturing processes.

The information presented in Section 17.2 provides a historical perspective on the motivation for the development of MCM technology. Section 17.3 describes development of the Digital's VAX-9000 MCM-D technology as embodied in the multichip unit (MCU). The up front system performance goals are presented, along with the thought process that resulted in the choice of a packaging solution based on MCM-D technology. The VAX-9000 MCU engineering strategy also is described. In addition, the quality strategy used to obtain the very high level of reliability required for a mainframe class machine is presented.

In Section 17.4, the MCU technology is described in detail. Section 17.5 provides a description of the VAX-9000 MCU manufacturing process. Finally, Section 17.6 summarizes the applicability of Digital's MCM-D technology to the development of future products. Examples of applications are presented, along with a strategy for achieving high quality and reliability in the cost sensitive environment in which the industry now finds itself.

17.2 MCM SYSTEM PERSPECTIVE

In order for system designs to reach increasingly higher performance capabilities, designers must continue to develop more effective methods of integrating larger numbers of logic gates and driving them at increasingly faster clock speeds. As clock speeds increase, the intergate signal propagation delays must correspondingly decrease, putting more stringent demands on the design of the physical package and on the physical gate interconnect structures. Shortening the interconnect distances to the minimum by packing the logic into the closest possible physical proximity, optimizing the capacitive, resistive and inductive characteristics of the interconnects, distributing cleanly decoupled power to the circuits and removing the heat are all essential to achieving high clocking speeds. To meet these requirements, two approaches can be used concurrently by the designer: packing the gates densely into ICs and packing the ICs densely onto MCMs.

Integrated circuit technology has been and remains the first order method for achieving dense logic and high performance because constant improvements in chip technology continue to enable higher gate counts, higher gate densities and higher clock frequencies to be realized. However, as the experience of wafer scale integration demonstrates, when the demand for physical density increases by very large factors, numerous problems become exceedingly difficult to resolve.   These problems include managing higher thermal dissipation, implementing the denser and more complex physical interconnect routes, managing clock skews, ensuring signal integrity, providing all the required I/O and power connections and keeping the die and silicon feature sizes within the boundaries where cost effective manufacturing yields are achievable. In addition, some of the performance gain achieved in very large scale ICs is offset by the relatively high inductance and capacitance present in high pin count packages.

To address these issues, designers are turning to MCM technology since it provides alternative methods of achieving performance and density. With MCM technology the system designer can minimize the number of distinct packages and thereby reduce significantly the distances between the ICs. Figure 17-1 illustrates the dramatic size reduction benefits of MCM technology.   The

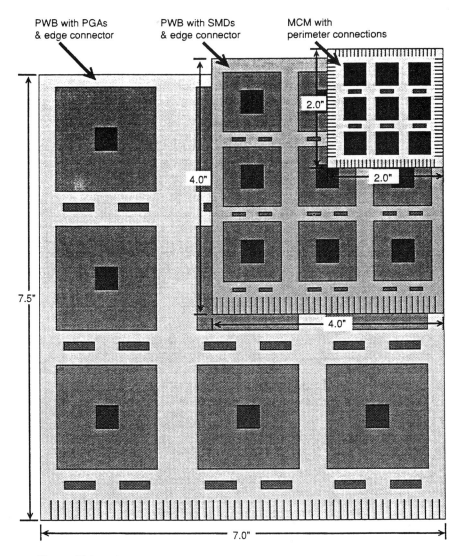

PWB with PGAs
& edge connector

PWB with SMDs
& edge connector

MCM with
perimeter connections

2.0"

4.0"

2.0"

7.5"

4.0"

7.0"

Figure 17-1 Scale perspective of one application implemented in standard through-hole PWB technology, in SMT and in MCM technology.

resultant reduction in interconnect distances, interconnect capacitance and interconnect inductance can reduce signal propagation delays by as much as 300%. Furthermore, MCM technology gives the designer additional options for partitioning the logic in ways that improve its performance, cost or manufacturability.

17.3 VAX-9000 MCU DEVELOPMENT

17.3.1 VAX-9000 Performance Goals

In 1983, the designers of the VAX-9000 set out to design a top of the line high performance mainframe system. The performance goal for the VAX-9000 central processor unit (CPU) was set at 30 VAX units of performance (VPUs). This was to be achieved not only by implementing an innovative CPU architecture that would overlap the execution of CPU instructions and minimize the number of CPU cycles per instruction, but also by deploying a physical technology that would allow the design to be implemented with the fastest available logic gates driven at the highest possible clock rate.

Although system performance was the driving force behind the VAX-9000 MCU technology, aggressive goals were set also for system uptime. High availability was achieved through improvements in the physical machine design as well as through the inherently higher reliability of an MCM package. Specifically, since an entire level of interconnect is eliminated by directly connecting the integrated circuits to the MCM substrate, the projected MCM failure rate is lower than the failure rate projected for the same integrated circuits in conventional packages on a printed wiring board (PWB).

17.3.2 Analysis of Alternative Physical Technologies

To meet such demanding performance requirements, a number of different physical technologies were considered in the beginning of the development program (1983-1984). The evaluation required not only that the technical capabilities of the alternatives be compared, but also that projections be made regarding the feasibility of technically viable options being made available for manufacturing production in the time period of interest (early 1990s).

Against a reference baseline of the VAX-8000 series CPUs (which utilized PGA packaged Motorola MCA-I emitter coupled logic (ECL) gate arrays on large PWBs), the following options were projected and evaluated:

- Chilled CMOS.
- Motorola ECL Mosaic-III gate arrays (MCA-III) PGA packaged on PWBs.
- Motorola ECL Mosaic-III gate arrays (MCA-III) packaged on MCMs.

- GaAs gate arrays packaged on MCMs.
- Wafer scale integration.

To understand the performance differences between these options, a critical path analysis was undertaken for a typical logic path. For each of the several technologies, the signal propagation delay was evaluated through a path that included the arithmetic logic unit (ALU), the translation buffer and the cache. The normalized results of that analysis are illustrated in Figure 17-2.

The results of the evaluation projected the following:

1. Chilled CMOS in 1989 would achieve the same on-chip logic speeds as the earlier ECL logic.

2. Because of the increased level of integration, chilled CMOS would have significantly shorter off-chip delays.

3. The on-chip speeds for Mosaic-III 10,000 ECL gate arrays would be approximately twice as fast as for CMOS.

4. If packaged in PGAs, the performance of the Mosaic-III ECL gate arrays would be dominated by their off-chip delays.

5. The off-chip delays of the Mosaic-III circuits could be reduced by 50% by changing from the PGA package to uncased TAB packaging.

The estimated performance gains from GaAs and wafer scale integration did not appear sufficient to warrant their increased development and time to market risks.

17.3.3 VAX-9000 MCU Strategy

Based on the above analysis, a strategy was established for partitioning the VAX-9000 CPU (greater than 1,000,000 equivalent logic gates and greater than 3,200,000 bits of high speed local storage) into a set of 20 high speed modules which could be implemented with high power (up to 30 watts each), TAB packaged, MCA-III ECL 10,000 gate arrays driven at a clock cycle time of 16 nanoseconds [1]-[3]. To meet these performance demands the set of gate arrays for each module would be interconnected, powered and cooled through use of the proprietary MCM-D technology described in Section 17.4. The modular MCM-D units based on this design were called MCUs. The MCU is comprised of signal and power connectors, integrated circuits and the MCM interconnect is known as the high density signal carrier (HDSC).

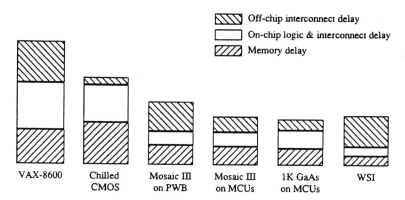

Figure 17-2 Normalized delay of a critical path in an ALU, translation buffer and cache for several technologies.

17.3.4 MCU Manufacturability Goals

To ensure the manufacturability and field serviceability of the VAX-9000 CPU, the MCU had to be designed for testability and diagnosibility. In manufacturing, electrical testing had to be executed progressively through the process at the earliest points of value added. Testability structures and mechanisms had to be established at the IC level, at the interconnect level and at the MCU module level. In the field, the test structures on each MCU had to be directly accessible from the VAX-9000 system console.

To manufacture MCUs in a cost competitive manner, high yielding base process technologies had to be developed and established. These processes had to support the production of 20 distinct MCU design options. In order to establish reliable and cost efficient processes and to meet time-to-market requirements, the processes were designed to use conventional semiconductor and PWB equipment wherever practical. The development of custom equipment was restricted to a selected set of processes (such as HDSC fabrication fixtures, HDSC test, IC TAB excise and lead form, die placement and lead bonding, MCU test) where the payback was high or where there was no other alternative.

17.3.5 Concurrent MCU Engineering and Manufacturing

Because of the long lead time associated with the scale up of an MCM-D process, it was necessary to begin operating the MCU manufacturing line long in advance of the time when the ICs would be available. The concurrent nature of the VAX-9000 engineering and manufacturing efforts necessitated frequent and in depth communication between the respective engineering and manufacturing teams [4]. This was particularly challenging since the engineering group was located in Marlborough MA, and the manufacturing site was located in Cupertino CA. To minimize the impact of distance between the engineering and manufacturing groups, a general purpose networked database system was established in order to allow rapid transfer of design and manufacturing data.

Initial production start up was accomplished by running test vehicles in a small scale process development line. The process was then scaled up in a pilot line and finally transferred to volume production.

17.3.6 MCU Quality Engineering

At the start of the development program, specific MCU and HDSC reliability goals were established to drive the design:

- MCU dead on arrival (DOA) failures - 300 PPM
- MCU mean time between failures (MTBF) - 700,000 hours
- HDSC mean time between failures (MTBF) - 10,000,000 hours

A strong design for reliability (DFR) philosophy then guided the subsequent development of both the HDSC and MCU. The DFR focused specifically on defining the detailed reliability requirements for the design, the material and the manufacturing processes and on establishing methods to verify that the requirements were met. The qualification program encompassed not only internal development but also suppliers' contributions toward product reliability. Working jointly with the suppliers, clear definitions of requirements and responsibilities were negotiated in advance, auditing methods were established and the suppliers' components, subassemblies and subsystems were integrated into the overall verification and qualification processes.

For the HDSC and MCU, the reliability characterizations evaluated the critical areas:

- Process materials
- Contamination
- Solderability
- Die attach integrity
- Encapsulant integrity
- Outer lead bond (OLB) and flex lead bond (FLB) solder joint integrity
- Corrosion

The test and verification methodology was based on applicable military and industry standards:

- Thermal shock, -35°C - 125°C
- Temperature cycling, -65°C - 150°C
- Biased temperature/humidity, 85°C / 85% RH
- High temperature aging, 125°C
- Corrosive gas tests

Preliminary test results showed that some of the early, marginal MCU and HDSC engineering prototype samples had expected failures in PTH integrity and some corrosion was observed on these samples after biased 85/85 tests. Appropriate process changes were then implemented and as a result, no subsequent failures were observed.

The formal MCU/HDSC Design Verification Test (DVT) process was then used to confirm the integrity of the design and to verify that the product met its functional requirements and the requirements imposed by its manufacturing processes, its storage and transport environments and its operating conditions.

Ongoing reliability testing (ORT) was established to ensure that the HDSCs and MCUs produced in manufacturing continued to meet their specifications for infancy and steady state reliability. ORT sample units were randomly drawn from production lots and were stressed thermally and environmentally. The ORT results have reconfirmed the robustness of the HDSC substrate and have demonstrated the high yield and quality of the HDSC manufacturing process.

Finally, field return data is continuously analyzed in order to ensure an accurate empirical understanding of product reliability performance, of failure modes and of failure mechanisms so that corrective actions can be promptly implemented when required. Based on the population of HDSCs in the field over a period of 18 months (since the VAX-9000 new product introduction), an extensive set of data from system installations, customer system run time clocking and field returns has been tracked and analyzed. This data shows only four confirmed HDSC failures to date, with an accumulated total system run time of 37,769,784 hours. This translates to an impressive MTBF of 9,442,446 hours for the HDSC at the point in time at which this is written and this is 94% toward full demonstration of its design goal!

17.4 VAX-9000 MCU TECHNOLOGY

17.4.1 MCU Overview

The MCU serves several functions. It allows for fine pitch bonding so that high density chip placement can be achieved. The characteristic impedance of the

Table 17-1 Summary of MCU Specifications.

Maximum power dissipation	270 watts (air cooled)
Maximum IC junction temperature	85°C @ 25°C room temperature
Maximum number of VLSI chips	72
Minimum chip lead pitch	200 μm
Size: in plane	14.2 cm × 13.21 cm
height	5.44 cm
Minimum pitch on planar module	14.38 cm × 13.46 cm
Weight	1.59 kg (with heatsink)
Clock input frequency	320 MHz - 580 MHz
Signal I/O per MCU	800
Signal rise time	600 ps
Voltage levels	2 plus ground
Maximum current	40 A per voltage level

signal paths between the chips and off the MCU is controlled to within ±5% of the nominal impedance. At the same time, crosstalk is held to less than 5% (see Table 17-3). High pin out connectors increase the system architecture options. The MCU can provide up to 270 W of power with less than 20 mV of resistive voltage drop (IR) between the power supplies and ground. At the same time, the thermal path from semiconductor junction to heatsink is optimized so that no more than a 20°C thermal drop occurs. It is for these reasons that the MCU technology was chosen for packaging the VAX-9000 ECL circuits.

A summary of MCU specifications is given in Table 17-1. To the external world, this device presents four signal connectors capable of connecting to 201 signals each (804 total) and two power connectors capable of distributing two supplies at 40 amps each. Internally, the package can accommodate up to 72 ICs (a maximum of 45 in the actual designs) connected together by a special interconnect known as an HDSC. Variations of the MCU may contain different numbers and designs of ICs and different HDSC designs.

The basic elements of the MCU are shown in Figure 17-3. The lid serves primarily to prevent large objects from inappropriately contacting the HDSC and chip areas. The signal connectors and power connectors are mounted on the housing assembly which supports the HDSC. The pin fin heatsink is bolted to the back of the HDSC.

IC placement on the HDSC is defined by a 3 × 3 matrix of chip placement sites. The middle site is always used by a special clock distribution chip. Any of the other eight sites can be occupied either by one large logic circuit (MCA-III) or by up to nine smaller special RAM circuits.

The VAX-9000 system uses 20 different MCU designs, each utilizing a unique HDSC design. The 20 different MCU options use a total of 79 different IC designs. The common design base for all the MCU options allows for economy in the manufacturing of the device and enables the same manufacturing process to be used for all variations.

The VAX-9000 MCU solution is unique when compared to other solutions for this class of problems such as the IBM multilayer ceramic thermal conduction module (TCM) with solder bumped flipped ICs. There were several differentiating reasons. The investment and effort to follow a similar approach was considered prohibitive. Face up chip attach approaches were considered easier to develop and easier to test, debug and cool. The desire to have the MCU as a field replaceable unit (FRU) drove the power and signal connector to be separable. The signal connector is folded back above the MCU to minimize footprint on the planar board and allow cooling on the opposite side. The separate signal and power cores laminated together was seen to minimize the topography of a single sequential core and the drilled holes replaced the extra via layers of a single sequential core. In summary, the MCU was the point solution to a long series of intersecting constraints.

17.4.2 MCU Integrated Circuits

The MCU uses exclusively ECL based integrated circuits. All of these circuits operate from a -5.2 volt supply, while most also use a -3.4 volt supply. The -3.4 volt supply is used primarily to limit the power dissipation of the off-chip output drivers. There are three basic classes of ECL circuits used: gate array and custom logic circuits, a clock distribution circuit and two types of special RAM devices called "self timed random access memory" (STRAMs). The characteristics of these ICs are summarized in Table 17-2 and Table 17-3.

Gate Array and Custom Logic Circuits
The logic circuit chips all use the Motorola Macrocell Array III (MCA-III) technology. The gate array master slice has 10,000 available gates. The custom circuits were designed for logic functions that could not achieve high efficiency in a gate array implementation. These ICs feature less than 600 ps output rise times.

STRAM

Many of the MCU types require high speed local storage, such as caches, registers and control store. This is provided by 4K × 1 and 4K × 4 STRAMs. The STRAMs are proprietary designs with address latches and internal write pulse generation.

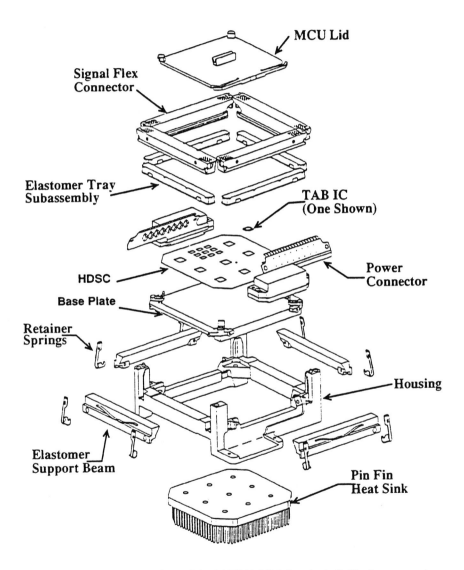

MCU Lid

Signal Flex Connector

Elastomer Tray Subassembly

TAB IC (One Shown)

HDSC

Base Plate

Power Connector

Retainer Springs

Housing

Elastomer Support Beam

Pin Fin Heat Sink

Figure 17-3 An exploded view of the MCU highlighting the individual components.

Table 17-2 Characteristics of MCU Chips.

	MCA-III Gate Array	CDXX (Clock Distribution)	4 K × 1 STRAM	1 K × 1 STRAM
Die Size (mm)	9.8 × 9.8	6.2 × 6.2	6.4 × 4.2	4.9 × 3.6
Signal Pins	256	170	35	33
Total Pins	360	272	48	48
Transistor Count	40.1K	7.2K	103.0K	28.0K
RAM (bits)			16384	4096
Power (watts)	30.0	13.9	2.4	2.4

Clock Distribution Circuit

Central to every MCU is a clock distribution chip (CDXX). This chip has master clock and reference clock inputs, from which de-skewed clocks are generated for every chip within an MCU. This chip also provides scan distribution and scan control for test and diagnostic purposes.

17.4.3 HDSC Interconnect Technology

The heart of MCU technology is the HDSC [5]. The HDSC is built with nine copper conducting layers separated by polyimide dielectric layers. Four metal layers are used for power distribution. Of the remaining five copper layers, two are used for signal interconnects in the x- and y- directions respectively, two act as reference planes for the signal layers and one layer provides surface pads for electrical connections. The HDSC is capable of supporting three to five times the routing channels per unit area available in conventional printed wiring boards. A cross section of the HDSC is shown in Figure 17-4 (not to scale).

Signal Interconnect

The performance of the signal interconnect is determined by the dielectric constant of the insulating medium and the resistance of the metallic interconnect traces. A fairly low dielectric constant (3.5) is achieved by using polyimide as an insulator in the HDSC. The physical dimensions of the interconnects are set so that the line resistance is 1 Ω/cm. The two signal layers in the HDSC are

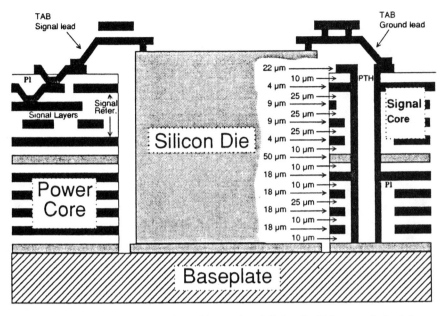

Figure 17-4 HDSC cross section with metal and dielectric thicknesses indicated.

enclosed by upper and lower reference planes to form a stripline. The cross section of this structure is symmetric so both signal layers have the same nominal impedance value of 60 Ω. Crosstalk is held to a minimum by controlling the line resistance and holding the minimum line pitch to 75 μm line pitch. Plasma etched vias create paths for connections between the signal layers and the top metal layer. The pitch of these vias allow for a minimum diagonal placement so that a true 75 μm line pitch can be achieved. Electrical connection to the lower reference plane is made with drilled and plated through-hole technology.

The master and reference clock inputs to the clock distribution chip require 50 Ω impedance, instead of the 60 Ω impedance of the signal interconnects. In this case, the signal traces are made wider (33 μm as opposed to 18 μm) to achieve the desired impedance.

Power Distribution

The HDSC power distribution is designed to minimize IR drops and provide the highest possible level of signal decoupling. Four power planes are used. Two power planes conduct ground or V_{CC} currents, one conducts V_{EEI} (-5.2 V)

Table 17-3 MCA-III Chip Parameters.

Chip Dimensions	385 mils square
Number of Signal Leads	256
Number of Power Leads	104
Total Number of Leads	360
Inner Lead Pitch	4 mils
Output Signal Rise Time	600 ps
Maximum Chip Power	30 watts

currents and the last one conducts V_{EE2} (-3.4 V) currents. Each of the supply planes is paired with a V_{CC} plane separated by polyimide thickness of 10 µm; this produces a distributed capacitance of 0.31 nF/cm^2. The effective impedance across a single power plane is on the order of 20 mΩ. The resistance of each is determined by the low sheet resistance of only 1 mΩ/□.

HDSC Assembly

The HDSC is made up of a top layer of signal interconnect (signal core) laminated to a lower layer of power interconnect (power core) as shown in Figure 17-4. This laminated structure is drilled and plated to connect the power interconnect to the top surface. The top layer of metallization, which later will be bonded to the ICs and the MCU connectors, is also formed in this plating step. Cut outs are excised at the sites where the chips are to be attached and bonded and, finally, the assembly is laminated to a heat spreading and mechanically supporting baseplate.

17.4.4 MCU Structure

The MCU is built in two stages. First the ICs are bonded to the HDSC to constitute the MCU subassembly. Then the MCU subassembly is joined to its housing, connectors, lid and heatsink to form the final MCU assembly.

MCU Subassembly

Prior to subassembly, the integrated circuits have copper clad tape bonded to the gold plated pads on the die [6] (see Figure 17-5). At this point the die are tested at speed and screened before shipment to the MCU assembly plant. The gate arrays and custom logic ICs use a special two metal tape that dedicates one metal layer to a reference ground in order to control impedance and crosstalk. The inner lead pitch is 100 µm for the logic and the clock chips and 450 µm for the STRAMs.

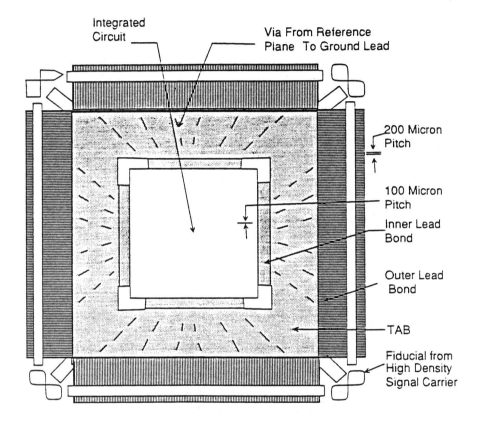

Figure 17-5 MCA-III on TAB frame. Top view of the TABed gate array.

The die are attached to the HDSC baseplate with a diamond filled epoxy. This provides a low thermal resistance path (0.3°C/W) that is electrically isolating (see Section 17.4.6). The electrical isolation is required because the back side of the die is at V_{EE1} potential while the baseplate is required to be at ground potential. During die placement, the outer copper leads of the ICs copper clad tape are formed into the shape required for their subsequent bonding to the solder coated pads on the HDSC. The outer lead pitch of these leads is 200 µm for the logic devices and 450 µm for the memory devices.

MCU Final Assembly
The MCU assembly is complete when it is ready for mounting onto a system's planar board. The electrical connectors are designed to make contact with the

planar board through static pressure (no soldering), so that the MCU can be quickly installed or removed, a fundamental requirement for a FRU. The key parts of the final assembly are the subassembly (the HDSC complete with bonded chips), the power connectors and the signal connectors. There are two identical power connectors bonded to the HDSC on opposite sides and there are four identical signal connectors bonded to the four sides of the HDSC.

MCU Power Connector

Each power connector consists of a semi-flexible tape of three isolated metal planes, eight discrete decoupling capacitors (0.22 µF each) and a spring connector through which it contacts the power bus bars on the planar board. This structure continues the low impedance design implemented in the power planes of the HDSC. Connection to the HDSC is made by soldering 20 metal tabs, each 3.3 mm wide, on a 3.6 mm pitch. Each tab makes connection to the power core of the HDSC through six plated through-holes. The tabs alternate between V_{CC} and either V_{EE1} or V_{EE2}.

MCU Signal Connector

The signal connector is also a flexible tape structure. Its leads are on a 300 µm pitch and are bonded to the HDSC pads through solder reflow. While each connector has 201 signals, there is also a ground return for every three signals, resulting in a total of 267 leads connected to the HDSC. On the side of the signal connector that mates to the planar board, an array of gold bumps on its flexible circuit makes a wiping contact as pressure is applied. Between the gold bumps and the soldered leads the length of the signal conductors is approximately 2.5". The signal conductors reside on one side of the tape and a solid ground plane on the other side. This microstrip structure guarantees controlled impedance signal lines with minimal crosstalk.

17.4.5 MCU Electrical Characteristics

Performance of the MCU is defined by the speed of the internal ICs, by the fidelity of signal transmission through the signal connectors and by suppression of noise in the power distribution system. Overall, these elements are controlled sufficiently to allow for a system cycle time of 16 nsec for the complex instruction set computer (CISC) CPU and for distribution of clocks of up to at least 750 MHz. At the same time, MCU power distribution noise is held within the noise budget of 20 mV.

Signal Integrity

All components of the MCU interconnect - the two layer TAB tape, the HDSC and the signal connector - are designed primarily for a characteristic impedance of 60 Ω and this characteristic impedance is guaranteed by manufacturing tolerances of less than ±5%. In addition, the crosstalk for signals in the HDSC is controlled to less than 5.1%. The resistance of the HDSC signal lines is 1 Ω/cm, enough to help attenuate reflections without adversely affecting delay. Termination of the HDSC transmission lines is accomplished by the use of series terminating resistors of 30 Ω within the ICs.

The gate array ICs required tape to fanout from a 100 μm inner lead pitch to a 200 μm outer lead pitch. The distance required to achieve this fanout is more than 4.0 mm. The two metal tape provides a ground plane that controls the impedance and the inductance. With the plane, the impedance is 60 - 70 Ω and without it, the impedance will exceed 100 Ω.

Several design techniques were used for further improvements in signal integrity. All clocks are delivered in differential form and the clock traces in the HDSC are routed as fixed length, minimum pitch pairs. Restrictions are placed on standard signal traces to limit the length of adjacent traces and crossovers so that crosstalk is minimized beyond the structure specification. Series terminated ECL (STECL) outputs were used to reduce simultaneous switching noise by more than 50% compared to the parallel terminated option.

Power Connections and Distribution

The MCU DC circuit for ground current begins at the pad of the IC, flows through the inner lead bond, through the tape lead to the outer lead bond, through the conductors on the top metal layer of the HDSC and through the top vias to the top reference plane. The top reference plane then spreads the current through all V_{CC} plated through-holes to the power core. The power core shares the ground current between the two ground planes and conducts it to the HDSC perimeter, where additional plated through-holes conduct the current to the power connector bond pads. From the bond pads the current flows through the flexible tape to the spring contacts of the power connector. Similar paths are used for the two power supplies as well.

The resistance, as measured from the outer lead bond pad to the spring contacts of the power connector, is less than 3 mΩ for V_{CC} and less than 30 mΩ for either of the two power supplies. The total distributed capacitance between V_{CC} and either power supply is 30 nF in the HDSC. The distributed nature (ultra low inductance, less than 10 pH/□) of this capacitance provides very effective suppression of high frequency noise [7]-[8]. The supply noise is measured to be less than 30 mV on this power distribution.

17.4.6 MCU Thermal Management

The VAX-9000 was designed to provide an efficient cooling path for high power ECL circuits. This was achieved by conducting the heat generated by the silicon chips through the die attach material, into the baseplate and, finally, into an air stream via a pin fin heatsink [9]-[11].

Since a single ECL chip of the type used in the VAX-9000 system can generate up to 30 watts, a typical MCU generates between 145 - 270 watts of heat. However, because the overall thermal resistance for each MCU chip site is less than 2.0°C/W, the maximum junction temperature will be 85°C for an ambient temperature of 25°C as specified in the design requirements. Achieving this low junction temperature was critical to meeting the reliability goals of the MCU. Consequently, extensive thermal modeling took place in the early design phase to test and verify the feasibility of such a system. A detailed description of the MCU thermal model has been published by Fitch [12].

A significant component of the overall MCU thermal management system is the unique die attach material that was developed specifically for this program. The die attach material ensures maximum heat transfer and high flexibility to accommodate the differing thermal coefficients of expansion between the copper chromium baseplate and the silicon chips. This material was heavily loaded with microscopic (25 μm) diamond particles to provide the highest thermal conductivity possible and yet remain electrically insulating. The design requirement of electrical insulation prohibited the use of conventional die attach materials such as silver filled epoxies. Optimization of the epoxy formulation took several years. In addition, it was necessary to develop concurrently a robust die attach process that would safely allow multiple chip replacements.

The thermal transfer process can be understood in relatively simple terms (see Figure 17-6). The principle is based on the conduction of the heat from the silicon chips through the diamond particles embedded in the epoxy material and into the copper chromium baseplate, from where it is conducted across a dry interface to the base of the aluminum pin fin heatsink. The heatsink is described as "pin fin" because it contains 606 aluminum pins, each 0.078" in diameter, pressed into its base with 0.75" protruding into the air stream. The heat is then removed by air plenums in the VAX-9000 cabinets, which direct approximately 14.6 l/sec of air into each of the MCU heatsinks. In this manner, the IC junction temperature is controlled in the VAX-9000 system.

17.4.7 MCU Test Technology

Before an MCU is installed in a VAX-9000 computer system, each module is tested at least 11 times at various stages in the manufacturing process. The sequence of test operations performed on each MCU is referred to as the test flow. The test flow for the VAX-9000 is shown schematically in Figure 17-7.

The three major test processes performed in the manufacture of each MCU

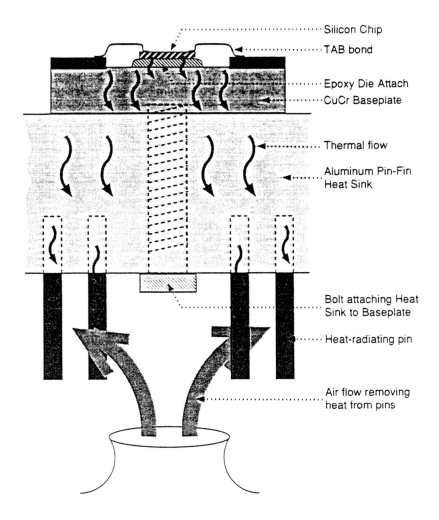

Figure 17-6 The heat conduction and convection paths for the MCU.

are in-process HDSC test, final HDSC test and final MCU test. Each of these three processes can be broken down further into sub-processes and each sub-process can be described as a sequence of test operations. For example, in-process signal core test can be broken down into three successive test operations: surface leakage resistance, via resistance and line resistance measurements. The

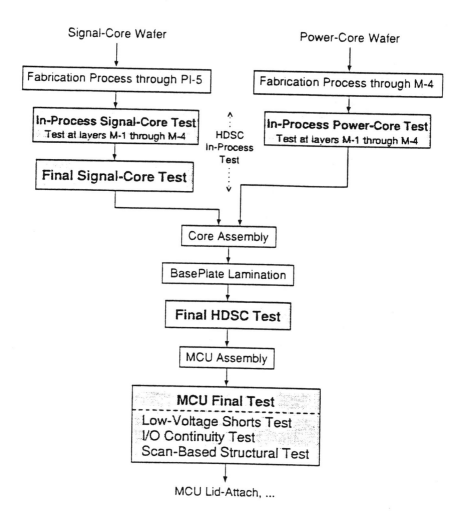

Figure 17-7 This test flow shows that the interconnect is tested several times before MCU assembly and final test.

type of operations and the sequence in which they are performed are chosen to provide visibility into important process parameters that must be monitored and controlled to optimize yield.

HDSC In-Process Test

Both in-process and final signal core tests are performed with a test system consisting of a KLA 1007 wafer prober with auto-alignment and cassette to cassette wafer handling capabilities which is coupled to a semi-custom instrumentation package (Reedholm Instruments Corp. (RI-15a.). An 80386 IBM compatible PC acts as the system controller.

Parametric tests, such as 2-point and 4-point resistance and leakage tests, are carried out with the Reedholm Instruments DC measurement modules. Figure 17-8 illustrates the basic principles underlying the resistance test methods used in in-process testing.

Signal Core and HDSC Final Test

To optimize yield, each net on the signal core must be tested for opens and shorts before lamination to the power core. To optimize cost, test time must be minimized. The most complete test approach would directly test each net segment for DC continuity, measure the DC isolation of each net to every other net and evaluate the performance metric (impedance, crosstalk, propagation delay) of each net. The time and cost of this approach is prohibitive and unnecessary. Parallel probing (such as bed-of-nails) is not practiced due to the small pad size and pitch and high pad count (≈ 4000). Probing each net with one or two probes is acceptable for DC continuity testing but the number of probings required to do the DC isolation test grows as the square of number of nets. To achieve the best optimum of quality assurance and test cost, a capacitance based test approach can be used to verify net continuity and isolation. Since impedance is directly related to the capacitance, the performance of the signal core can be assured. This approach only requires a single probe test setup and only one probing per pad.

Open and shorts testing is carried out by measuring the capacitance at the ends of each net in the interconnect device using a capacitance meter such as the HP 4278A. In Figure 17-9a, it can be seen that for a good net the capacitance measured at either end of the net will have the same value. For an open net, as shown in Figure 17-9b, the capacitance at either end of the net will be lower than the expected net capacitance and the sum of the values at the ends of the net will be equal to the expected value. In the case of a short, as in Figure 17-9c, the capacitance measured at any pad of any shorted net will be equal to the sum of the expected net capacitances for the individual nets.

In practice, subtle processing defects create more complex interconnect faults than the simple cases of Figure 17-9. For example, Figure 17-10a illustrates a resistive open, that is, where a resistive path exists rather than a complete open. This situation, as well as the capacitive open illustrated in Figure 17-10b, can be caused by defective vias between two metal layers.

a) Two-Point Resistance Test (Force current, measure voltage)

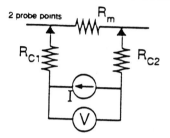

$$R_m + R_{C1} + R_{C2} = \frac{V}{I}$$

Assuming $R_{C1} + R_{C2} \ll R_m$

then $R_m \cong \frac{V}{I}$

b) Leakage Resistance Test (Force voltage, measure current)

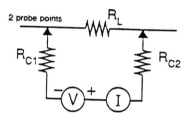

Assuming $R_{C1} \approx R_{C2} \ll R_L$

then $R_L = \frac{V}{I}$

c) 4-Point Resistance Test (Force current, measure voltage)

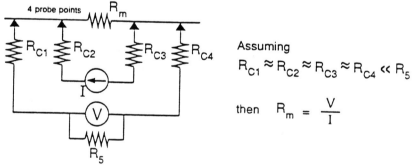

Assuming

$$R_{C1} \approx R_{C2} \approx R_{C3} \approx R_{C4} \ll R_5$$

then $R_m = \frac{V}{I}$

Figure 17-8 Basic resistance test configuration. The probe and meter connections for achieving three types of resistance tests.

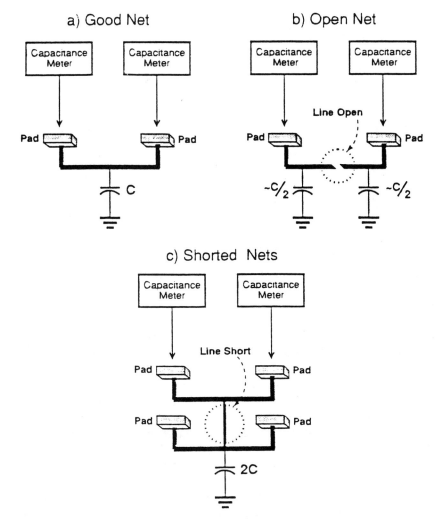

Figure 17-9 Ideal capacitance results for good, open and shorted net conditions.

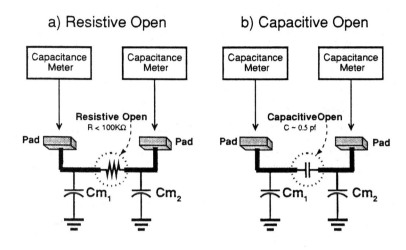

Figure 17-10 Effects of "soft" opens - resistive and capacitive.

In practice, after the capacitance measurements are taken by the tester, the results must be further analyzed to minimize the chance of either false negatives or false positives. False positives can result if a resistive or capacitive open is misinterpreted as a good net. False negatives can be caused by equipment malfunctions or by contamination on either the probe tip or pad.

The expected net capacitance values can be either calculated from design data or derived from measurements of known good devices (commonly referred to as the "learned mode" method). When the expected values are calculated from design data, there are many factors which must be taken into account in order to obtain accurate results. It is straightforward to calculate the capacitance of a parallel plate capacitor if the dielectric constant, the distance between the plates and the area of the plates are known.   However, in controlled impedance interconnect networks, the geometries of the conductors are such that fringe fields become a very important factor in the actual capacitance value. Since more than 20 MCU design options were used in the VAX-9000 system, it proved to be worthwhile to determine the expected capacitance values directly from the design data. There were two compelling reasons for this: automated test file generation capability saved a great deal of time in the debugging phase, when design changes occurred regularly and it was possible to detect and correct masking errors and other sources of repeating defects.

An additive capacitance model was developed to estimate the capacitance of each net. The model used design dependent data such as pad type and pad count, line length and crossover count for input data. Capacitance parameters were determined by two dimensional capacitance modeling and affirmed by a 10% tolerance band was applied to most of the parameters to generate limits that would screen opens and shorts. This tolerance band was also tight enough to screen out parts that would be outside the impedance specification.

Prior to chip connection and TAB bonding, the net capacitance test is repeated to insure that no damage was done during power core and baseplate lamination. In addition, the PTH and power planes are tested.

The PTH are tested using the force current, measure voltage technique described earlier and Figure 17-8c. This measurement allows weeding out parts with suspiciously high PTH resistance. It has been found that PTHs with resistances above several mΩ are subject to reliability problems in the event that the part undergoes extensive thermal cycling. Fortunately, the PTH yield has been found to be very high.

The dielectric integrity of power planes is tested using the force voltage, measure current method described earlier and in Figure 17-8b. A constant voltage (50 volts) is applied to the power planes for one minute, and then the current is measured. In this test, the reject threshold is 100 MΩ.

The final HDSC test suite, which includes resistance and leakage measurements, is performed on a flying probe tester (the STAR tester), designed by Digital especially for the VAX-9000 program. Multichip module substrate testing is covered in more detail in Section 13.2.

MCU Final Test

Because of Digital's experience in testing modular computer systems, the difficulties in testing the VAX-9000 MCU were recognized early in the design process. Given that the module would operate with a clock frequency of 500 MHz and would have over 800 signal inputs and outputs, it was clear that no available commercial tester could provide adequate test capabilities. It was also clear that a viable test strategy and methodology had to be developed (other than using the completed computer system as a tester), since 20 distinct MCU types are required for a VAX-9000 system even in its uniprocessor configuration. It was therefore necessary to undertake the development of a special tester for MCUs; it would be a scan based tester, capable of testing each MCU type individually and independently from the assembled computer system.

One of the important challenges that arises when scan based techniques are used to test a modular system is that even though a module may pass the scan test successfully, it can still fail to function correctly in the final system if the scan test coverage is less than 100%. In practice, test coverage is almost always less than 100%, particularly when a design is newly released to production. Generally a period of maturation is required, during which the set of test vectors for each module is enhanced based on actual experience. Subtle and even not so subtle fault mechanisms are discovered as the number of modules manufactured increases. For this reason, the assembled set of modules must always be tested as a complete system prior to shipment.

Another way of describing the implications of scan based testing is to say that scan based testing provides a validation of an MCU's logical structure but it will not necessarily guarantee its functionality. Scan testing will not detect a logic design error if generated from design data. This error will only appear at the system level.

As was mentioned previously, since no suitable commercial tester was available, it was necessary for Digital to develop its own MCU tester. Digital's MCU tester has now been fully developed and deployed in a production environment for several years. The actual sequence of tests performed are the following:

1. Low voltage pin to pin shorts test (LVS)
2. Input and output pin contact check
3. Scan based structural test

These tests are described in more detail below.

In the LVS test, a small DC voltage (below the level to turn on a PN junction) is applied to the power supply pins, input pins and output pins while the current flow is observed. If all of the currents are below a preset threshold, then the unit passes. If any pins fail, then the MCU is sent for diagnosis and repair.

In the input/output pin contact check, the voltage on each pin is set to a level just high enough to turn on the substrate diode and thus cause a small current to flow. If current flow is sensed at all pins, then the unit passes. If the unit fails, then the problem may be in the connector or the MCU. If the unit does not pass after being re-seated in the test fixture, then the MCU is sent for diagnosis and repair.

Finally, in the scan-based test, an automatically generated set of test vectors (unique to each MCU design option) is scanned into the module through a dedicated scan port. The module is clocked and the data is scanned back out and interpreted by a computer program. If a fault is detected, then the module is sent for diagnosis.

17.5 VAX-9000 MCU Manufacturing

17.5.1 Overview of Manufacturing Process

MCU manufacturing has five major processes: Signal Core Fabrication, Power Core Fabrication, HDSC Assembly, MCU Subassembly and MCU Assembly (Figure 17-11). The signal core and power core processes are similar to IC processes. The HDSC and MCU subassembly processes are like PWB fabrication and IC packaging processes respectively. The core processes are performed in a class 100 cleanroom environment, whereas the assembly processes are executed in a class 10K environment.

The philosophy guiding HDSC/MCU manufacturing has been that of problem prevention and reduction of variability. Up front product specifications were solidified and process capabilities were assessed and improved to ensure a high C_{pk}. Equipment preventive maintenance (PM) and calibration methodologies were developed and implemented. Training modules were developed; it was made mandatory that operators pass certification tests before being allowed to work in the fabrication areas. The fabrication facility was certified and its key parameters were measured and kept in control. Multifunctional process improvement teams were formed and underwent thorough statistical process control (SPC) training. A shop floor control software system was used to manage the work in process (WIP) and to collect and analyze the process data. carefully designed experiments were used to characterize the process and to expedite problem resolution. In process electrical tests and inspections were developed to ensure the quality of the WIP and to provide quick process feedback. As the processes matured and became more stable, sample sizes for test and inspection were reduced to streamline the process without jeopardizing the product quality. An on going Pareto analysis of rejects was maintained and failure analysis tools were used to find and fix the root causes of rejects.

Great effort was put into ensuring the quality of incoming materials by partnering with key suppliers. Ship to stock programs were developed and followed rigorously. Die were fully tested at speed in their TAB frames by the vendors before shipment to the MCU assembly operation. Vendors' process capabilities and SPC programs were assessed to ensure that product specifications would be met consistently and several subassemblies were redesigned in order to match the supplier's capabilities.

In the following sections the core and assembly processes are discussed in detail.

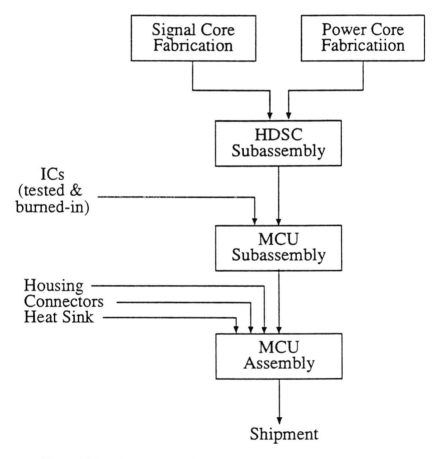

Figure 17-11 High level flow for the complete MCU manufacturing process.

17.5.2 HDSC Signal Core Fabrication Process

The signal core process follows the process flow shown in Figure 17-12. An aluminum wafer of 6" in diameter and 50 mils in thickness is the starting material. The wafer is first solvent cleaned and then is coated on its back side with polyimide to a thickness of 20 µm; this coating isolates the aluminum from the subsequent chemical processing. Photoresist spinners are used for the coating and the soft baking of the polyimide (PI) films. To achieve films of such thicknesses, multiple coatings are used. The polyimide is then cured in a diffusion furnace. Precautions are taken to minimize oxidation in the furnace by controlling the leak back of ambient air and also by cooling the wafers before removing them from the furnace. Another layer of 10 µm thick polyimide is then spun on and cured on the front side of the wafer; this layer protects the first metal layer when the aluminum substrate is later removed.

A Cr/Cu/Cr layer then is sputtered on in a vacuum system. A photoresist layer is spun on and exposed through a mask to define the reference plane. A chemical wet etch process is used to etch off the excess metal. A third layer of polyimide is spun on and cured to separate the reference plane from the signal lines.

A Cr/Cu seed layer is sputtered on and a signal line pattern is defined in the photoresist. Copper is electroplated to form the signal lines. To ensure the uniformity of the thickness, care is taken in pre-cleaning steps and in bath maintenance. The mask layers are thieved to create a uniform pattern and a special wafer plating holder is used. This holder allows for multiple perimeter contacts to the wafer but is sealed from exposure to the bath; the seal prevents plating to the contacts and to the wafer perimeter and ensures repeated uniform contact.

The resist is then stripped, another layer of thicker resist ($\approx$40 µm) is spun on and holes are opened where vias between the two signal layers are required. The holes are then filled with copper by an electroplating process. Because there is minimal redundancy, this is a very critical step; all the holes have to be filled with copper - any small residue left in the hole could prevent or slow down the plating and lead to open vias.

A seed layer is then chemically etched and another thick polyimide layer is spun on and cured. A via mask is then patterned in the thick resist and is plasma hardened so that it can withstand the plasma during the subsequent SF_6 + O_2 plasma etch process that opens the vias. After the via etch and resist strip steps, another Cr/Cu seed layer is sputtered and the process for defining the signal layer and the vias is repeated. A Cr/Cu/Cr layer is sputtered and the second reference plane is defined and etched.

A final polyimide layer deposition and a via etch process complete the signal core fabrication. Figure 17-13 shows the cross section of a finished signal core at various steps of the process.

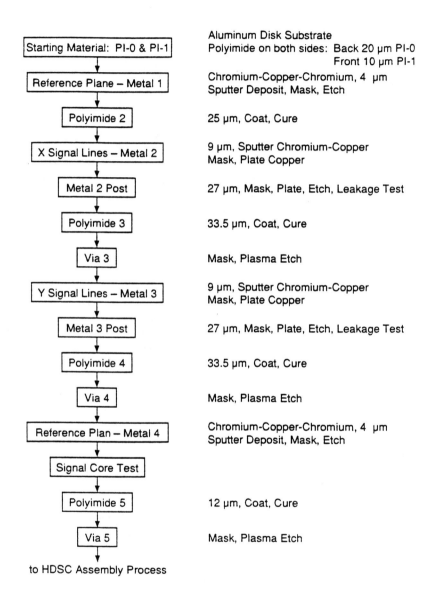

Figure 17-12 Major steps of the HSDC signal core process flow.

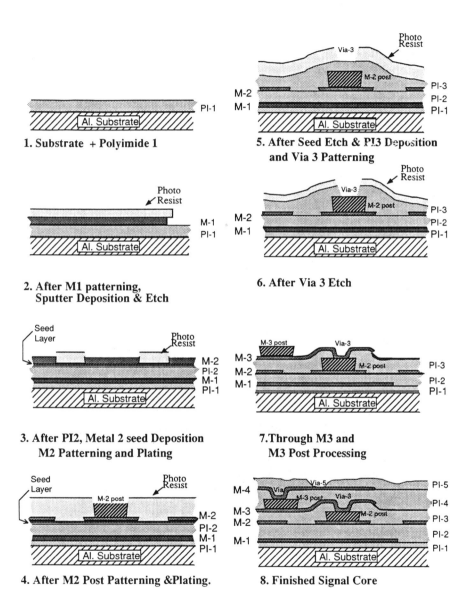

Figure 17-13 Cross sections of the signal core after major steps of the process.

17.5.3 HDSC Power Core Fabrication Process

The starting material for the power core is also a 6" aluminum wafer. The back side and the front side are coated with polyimide through processing steps similar to those used in the signal core process. After the front side polyimide coating, a seed layer of Cr/Cu is sputtered and then built up by use of electroplating. A thin layer of chromium is then sputtered on to ensure good adhesion between the power planes and the polyimide.

Using a photolithographic and a wet etch process, the first power plane is defined. Both interlayer and intralayer tests are performed on all the wafers. Rejected wafers are sent to failure analysis and good wafers are moved forward through the process.

The polyimide coating, metal definition and test steps are then repeated three more times. A top polyimide coating layer then completes the fabrication of the power core. A cross section of a completed power core is diagrammed in Figure 17-14. The detailed process flow is illustrated in Figure 17-15. As can be seen in the figures, there are no vias in the power core; the connections between the power planes and the outside world are made using plated through-holes built during the HDSC assembly process.

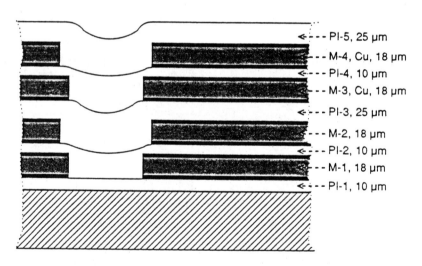

Figure 17-14 Power core cross section with layer thicknesses shown.

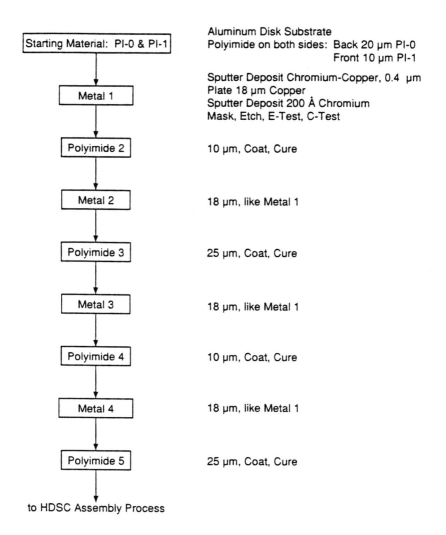

Figure 17-15 Major steps of the power core process flow.

17.5.4 HDSC Assembly Process

In the HDSC assembly process, a signal core and a power core are joined together and then attached to a copper chromium baseplate. The assembly process flow is described in Figure 17-16.

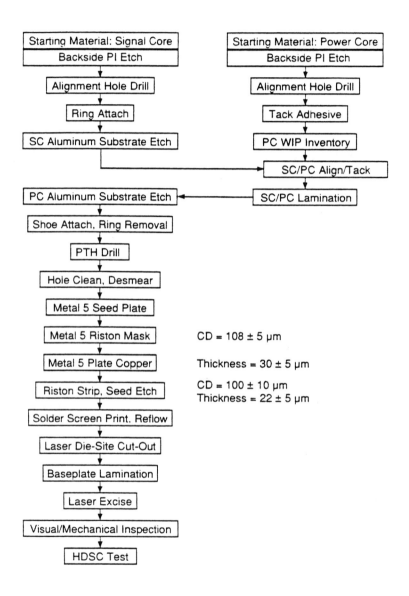

Figure 17-16 Major steps of the HDSC assembly process flow.

The back side polyimide is first plasma etched from both the signal core and the power core. After attaching a support ring to the signal core, its aluminum wafer is etched away, leaving its thin film interconnect on the ring.

An adhesive is tacked onto the power core. The signal core is then aligned to the power core and is tacked to the adhesive on the power core. The whole assembly is then put in a hydrostatic lamination press and the adhesive is cured under pressure.

The aluminum wafer is then etched from the power core to form a composite interconnect decal.

Using a mechanical drill, holes are drilled where connections are required between the power planes and the reference planes. The holes are then cleaned to remove debris and smear, if any, from the metal planes. The debris is removed with an aqueous shear stress clean; smear is removed by plasma etching. This is a very critical step and is controlled by monitoring the amount of polyimide removed from a test wafer. All traces of smear must be removed before any subsequent metal deposition in order to ensure the reliability of the PTH.

A Cr/Cu seed layer is then sputtered and built up using electroless and electroplating processes.

A dry photoresist film is laminated, exposed and developed to define the exterior footprint layer. Copper is electroplated, resist is stripped and the seed layer is wet etched to leave the copper footprint layer on the HDSC surface.

For the subsequent bonding of the TABed ICs, solder is required on the HDSC bonding pads. This is deposited by screening solder paste onto the bonding pads and then reflowing.

Next, the polyimide is excised with a laser to form openings in the decal where the die will be placed so that they can make contact with the baseplate. The decal is then attached to the baseplate with an adhesive. Figure 17-17 shows the cross section of the HDSC after particular process steps.

The finished HDSC then undergoes a complete electrical test (Section 17.4.7) and a thorough visual/mechanical inspection (VMI) and is then transferred to the MCU manufacturing area.

17.5.5 MCU Subassembly Process

An MCU subassembly is built by attaching and bonding the required ICs to the HDSC assembly. See steps 1 through 7 in Figure 17-18.

1. A plasma clean process is used to prepare the surface of the HDSC assembly for the die attach and lead bonding operations.

2. Through a computer controlled system that understands the footprint personality of each distinct MCU type, controls the motion of the HDSC on an x-and y- table and controls the rate of epoxy dispensing, the specified pattern and quantity of epoxy is automatically placed into the die site cut outs on the surface of the HDSC. Unique patterns have been developed for each IC type in order to allow the maximum thermal transfer to occur.

3. Through a similar computer controlled system, a low activation flux is automatically dispensed over those bonding pads on the surface of the HDSC where the TAB tape of the ICs is to be soldered. This activates the surface of the solder so that clean intermetallic bonds occur during the bonding process.

4. With automated bonding equipment, the MCU's ICs, one after another, are excised from their carrier frames and their leads are formed into the required gull wing shape. Each IC is precisely aligned by a vision system to its designated position on the surface of the HDSC. With computer managed control of the time, pressure and temperature parameters, a gang bonding thermode is then applied to the leads of its TAB tape. The solder reflow bonds the ICs leads to its corresponding surface pads on the HDSC.

5. When all ICs have been attached and bonded to the HDSC, the epoxy is cured in a nitrogen purged belt furnace. The bond line thickness, (the thickness of the epoxy material between the die and the HDSC baseplate) is critical for optimum thermal transfer. It is controlled during this process by mechanically applying pressure to the ICs during the curing.

6. To ensure that all soldered leads are bonded reliably, the lead bonds are inspected for possible shorts, opens, and/or weak bonds. Shorts and misalignments are detected by an automated vision system (MRSI 781) that calls failing and marginal conditions to the operator's attention. Opens and weak bonds are detected by an automated system (Vanzetti 6215) which strikes each bond with a pulse of laser energy and measures its thermal decay profile; this system also calls failing and marginal conditions to the attention of the operator. The operator then determines whether touch up corrective action is needed.

7. When touch up is necessary, it consists either of localized removal of shorts or of single point bond reflow.

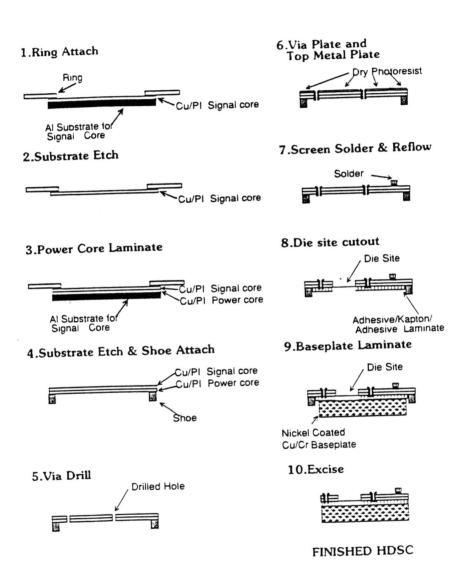

1.Ring Attach

6.Via Plate and Top Metal Plate

2.Substrate Etch

7.Screen Solder & Reflow

3.Power Core Laminate

8.Die site cutout

4.Substrate Etch & Shoe Attach

9.Baseplate Laminate

5.Via Drill

10.Excise

FINISHED HDSC

Figure 17-17 HDSC cross sections after major steps of the process.

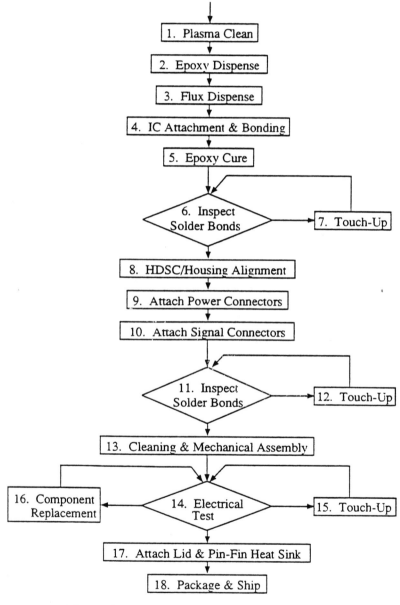

Figure 17-18 Major steps of the MCU assembly process flow.

17.5.6 MCU Assembly Process

In the second phase of assembly, the MCU subassembly is put into a mechanical housing and its remaining hardware (power connectors, signal connectors, lid, pin fin heatsink etc.) is attached to it. See steps 8 through 18 in Figure 17-18.

1. The MCU subassembly (HDSC + ICs) is put into the rigid frame of a mechanical housing and, through use of specialized equipment, is precisely aligned and locked into position with bolts. This alignment ensures that the connector systems, when mounted onto the housing, can readily be aligned with their corresponding bonding pads on the surface of the HDSC.

2. The MCU's two power connectors are bolted onto the housing, their bonding pads are prepared with flux and electrical connections to the HDSC surface pads are established by solder reflow with a single point bonder.

3. The MCU's four flexible signal connectors are attached to the housing, precisely aligned and clamped into position. The bonding pads are prepared with flux and electrical connections to the HDSC surface pads are established by solder reflow effected by series of multi-lead bonder operations.

4. With the same types of systems as mentioned above in Item 6 of Section 17.5.5, the solder bonds of the connectors are examined for shorts, misalignments, opens and weak bonds. The operator is notified if failing or marginal conditions are detected. The operator then determines whether touch up corrective action is needed.

5. When touch up is necessary, it consists either of localized removal of shorts or of single point bond reflow.

6. All flux residues are removed from the MCU through a solvent based cleaning process and solvent residues are driven from the HDSC surface with a baking process. The signal flex connectors are then folded into their final configurations and the final mechanical assembly operations are completed.

7. The completed MCU undergoes a full electrical test. If necessary, appropriate touch up operations are performed or an IC is replaced.

8. When the MCU has satisfactorily completed electrical test, a lid is placed on it to protect the ICs and the pin fin heatsink is bolted to the back side of the HDSC baseplate. The MCU is then ready for packaging and shipment to the VAX-9000 system plant.

17.6 FUTURE EVOLUTION OF MCM TECHNOLOGY

MCMs based on MCM-D technology were originally developed to meet the high performance needs of mainframe computers and military applications since performance, not cost, was the primary driving factor. However, as the evolution of MCM technology has progressed, the potential applications for MCM-D technology have become more numerous and diverse. MCM-D solutions which were once exclusive to high end systems are now being incorporated into a wide range of products in the telecommunications, aerospace, instrumentation and computer workstation market areas.

Lassen [13] has proposed a rationale under which the progression of MCM-D technology from mainframe to lower cost applications can be understood easily. In essence, he argues that as the density of interconnection pads per unit area increases, the cost of a conventional PWB solution begins to climb dramatically due to the need for more and more layers, finer line pitch and smaller PTH. As the pad density moves beyond 250 - 300 pads per square inch, MCM-D technology becomes cost competitive with PWB technology, assuming both must meet the same performance requirements. As the pad density increases beyond 300 - 400 pads per square inch, MCM-D becomes the more cost effective solution.

A case in point is the evolution of the MCM-D technology at Digital Equipment Corporation. The high speed performance of Digital's VAX-9000 mainframe CPU is based primarily on the MCM-D technology developed for and embodied in the VAX-9000 MCU. The MCM-D learning experience that has been gained over the past eight years in the design and manufacture of MCUs is now being extended to the development of more general purpose, low power dissipation MicroModules. When the interconnect pad density exceeds about 300 pads per square inch, Digital's MCM-D can meet the performance/cost requirements of workstations, low end systems, peripheral controllers and similar types of high speed electronic devices.

The two main attributes of MCM packaging from which performance improvements come are the high density device packaging and the elimination of package parasitics. The high density packaging results in time of flight and load capacitance improvements. The elimination of package parasitics improves load capacitance and inductance. The reduction in load capacitance and time of flight reduces signal propagation delay and improves rise and fall times. The reduction of load capacitance also reduces power dissipation. Furthermore, reduced inductance in the signal path and power distribution system reduces noise and ensures reliable high frequency operation. The system speed range at which these attributes will begin to make a difference is from 50 - 100 MHz.

One of the key objectives in the migration from MCU to MicroModule technology is reduced cost. The value added performance of the MCU technology is expected to remain while the base cost is dramatically reduced, allowing for lower cost high performance systems to become practical. In the initial MicroModule applications, cost parity with conventional technology is the goal; the additional performance gained results in an improved price to performance ratio. As systems designers learn to take advantage of this technology, much greater cost benefits will arise. Specifically, if large single chip silicon designs are replaced with equal or better performing multiple chip designs, the expected cost improvement will be dramatic due to the power law governing silicon die yield. Once the potential of low cost MCMs begins to be realized through optimized strategies for design, test and manufacturing, many new application areas are expected to become apparent.

Figure 17-19 illustrates the product benefits of MicroModule technology (reliability, cost and performance) in terms of the characteristic features of the technology. Here, it can be seen how the superior heat removal, high packing density, thin film interconnect and superior power decoupling provided by MicroModule technology are interrelated and can be applied to produce products with superior cost/performance ratios.

For low power applications, it is possible to simplify significantly the processes used for the VAX-9000 MCU and yet retain essentially all of the attractive features of the MCM-D technology. This is so because low power CMOS devices require a much simpler power distribution scheme (V_{CC} and V_{DD} only) and systems with contemporary CMOS ICs consume less power and operate at considerably slower edge rates than the ECL-based system of the VAX-9000.

17.6.1 Strategy for Migration of MCM Technology
from Mainframe to Low Power Applications

A successful technology migration strategy or evolutionary approach seeks to take advantage of the inherent strengths of a technology while correcting its perceived weaknesses. One strength of Digital's mainframe MCU technology is its maturity. The computer aided design (CAD) systems are ready, the module test strategies have been developed and the manufacturing processes are stable

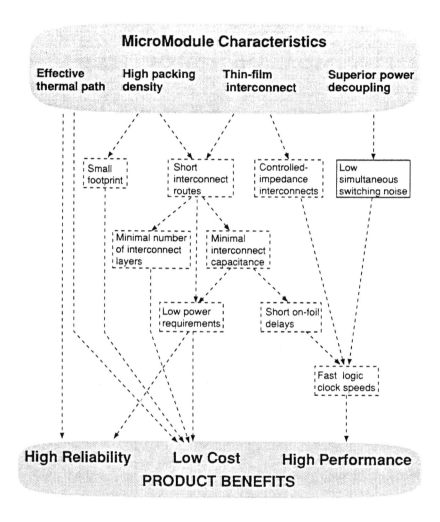

Figure 17-19 How MicroModules create product benefits.

and qualified. The weakness of most concern is cost, a consequence of designing to support high power applications. The key to lowering cost is the exploitation of the lower power requirements of CMOS-based MicroModules.

The MCU technology was developed to handle up to 270 watts. For the applications expected to be addressed with MicroModule technology, the power requirements are expected to be well under 100 watts. Furthermore the greater signal swings of CMOS technologies will tolerate greater noise and/or DC drops. The impact on the HDI (HDSC) design and process is significant. A separate power core is no longer required. Eliminating the power core not only deletes its material contribution to the cost but also eliminates the need to remove the signal core and power core interconnects from their aluminum substrates, the need for any subsequent lamination and the need for the drill and PTH processes. The lower thermal dissipation per IC eliminates the need for the copper baseplate and the excising of die site cut outs and allows for direct die placement. Special copper/polyimide structures can be designed to operate as thermal vias, thereby creating a low thermal resistance path for the hotter chips. In as much as an electrical connection from the top metal to the aluminum substrate is made by a stairway of vias and metal pads, a thermal connection is made as well. Arrays of such structures or thermal vias can achieve less than $0.5°C \ cm^2/W$.

Since it is no longer necessary to remove the interconnect from the substrate, the substrate itself now can be used for one of the power distribution layers. This is true whether the substrate is aluminum or highly doped silicon. While aluminum substrate does not offer the thermal expansion match that a silicon substrate does, it does offer a superior low resistance ($< 0.05 \ m\Omega/\square$); 500 times better than silicon. Aluminum can also serve as a mechanical element of the final package, whereas silicon would require mounting to a plate or a supporting package. Due to the slower CMOS edge rates and the fact that the entire system is contained on one MicroModule, the stripline structure of the MCU (with top and bottom ground planes) can be replaced with a simpler microstrip configuration. Because of the lower power requirement, the ability to take advantage of the substrate for power distribution and the need for only a microstrip configuration, the resulting interconnect needs only four added metal layers as compared with the nine layers of the MCU technology. A cross section of such a low power MicroModule is shown in Figure 17-20.

There are several additional approaches to lowering the cost of the MicroModule. One is to adopt standards so that multiple applications can use the same basic package. For example, a package can be developed to fit a current application but be designed with broader capabilities in mind for the future; an example of such a package is a pin grid array package (PGA). Another approach is to develop a package according to an industry defined standard, such as the 40 mm quad flatpack package (QFP). Note that for the QFP the aluminum substrate not only is electrically active but also serves as a structural element of the package. Other opportunities to reduce cost stem from the fact that the higher levels of integration in CMOS ICs are expected to result in fewer chips per system and therefore to consume less substrate area. This will allow multiple MicroModule substrates to be built on a single wafer.

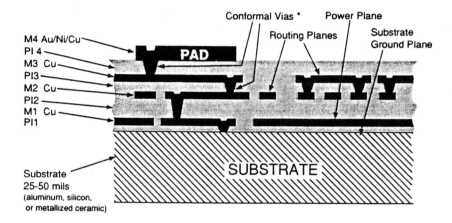

Figure 17-20 Cross section of a low cost microstrip MicroModule interconnection.

Another opportunity to lower system cost is to consider the tradeoffs of partitioning a single large (expensive) chip into smaller, higher yielding (less expensive) chips. The performance capability of the MicroModule substrate is high enough that this can be done not only without sacrificing performance, but also with potential overall system performance improvements. This can lead to greater design flexibility in that silicon performance, for example, can be optimized separately for memory if logic no longer has to be part of the same IC design. More memory could be added to the system since the constraint of the size of single yieldable chip and the size of the stepper field have been removed.

Given the need to accommodate a variety of possible die bonding formats from multiple IC vendors, it is advantageous to provide for maximum flexibility in the assembly process. Although the VAX-9000 MCU assembly was restricted to TAB, the MicroModule process currently supports both TAB and wire bond. In the future, flip chip solder connections will also be supported at the same time, on the same substrate. This flexibility (wire bond, TAB or flip chip capability) minimizes the cost of tooling and shortens the time-to-market.

A key enabler for the wide spread acceptance of MCMs is the availability of so called "known good die." The die requirements for MCM modules are discussed in more detail in Section 13.3. For the MicroModule test flow, known good die are obtained and assembled onto known good substrates.

After assembly, it is necessary to screen out assembly induced defects prior to functional test in order to simplify the diagnostic process on failed units. A suite of *in situ* tests are usually performed; these are tailored to the specifics of the MCM under assembly. For example, open wire bonds or solder joints can be found with visual inspection, boundary scan or other methods. In addition, die attach quality can be checked with ultrasonic, X-ray or thermal imaging. Once the assembly induced defects have been eliminated, the module can be functionally tested with minimal time spent on diagnosis and repair.

It is expected that, in some cases, MicroModules will require burn-in and a second functional test. Burn-in must be carried out prior to lid attach in order to allow for repair, since the lid is not removable in most low cost nonhermetic packaging schemes. After lid attach, the module is given a quick functional test prior to shipment. The issues of module test yield, fault isolation and repair are fundamental to the broad acceptance of MCM technology. This holds true not only for the MCM-D approach outlined, but also for MCM-C and MCM-L technologies. The costs of MCM are dominated by the silicon cost. Therefore the expense of scrapping good silicon due to an unrepairable or unisolated fault is severe.

Acknowledgments

The authors wish to thank the management team of Digital's MicroModules Systems Business Unit for their encouragement and help in our writing of this chapter. We are particularly grateful for the support of Chung Ho, Sharon McAfee-Hunter, James McElroy and Frank Swiatowiec. The authors also thank Jeffrey Slaney for his untiring efforts in researching the material used in this chapter.

References

1   D. E. Marshall, J. B. McElroy Sr., "VAX-9000 Packaging—The Multi Chip Unit," *COMPCON Spring '90: 35th IEEE Comp. Soc.*, (Los Alamitos CA), 1990.

2   P. B. Dunbeck, *et al.*, "HDSC and Multichip Unit Design and Manufacture," *Digital Technical Journal*, vol. 2, no. 4, Fall 1990.

3   T. Fossum, D. B. Fite, Jr., "Designing a VAX for High Performance", *COMPCON Spring 90: 35th Comp. Soc.*, (Los Alamitos CA), 1990.

4   S. G. Baust, R. J. Dischler, S. Westbrook, "Implementing a Packaging Strategy for High-Performance Computers", *High Performance Systems*, pp. 28-31, Jan. 1990, .

5   U. Deshpande, S. Shamouilian, G. Howell, "High Density Interconnect Technology for the VAX-9000 System," *Proc. IEPS Conf.*, (Marlborough MA), Sept. 1990.

6   J. P. Joseph, M. A. Kniffin, "Designing TAB Interconnect for the VAX 9000 Computer," *Proc. IEPS Conf.*, (Marlborough MA), pp. 683-693, Sept. 1990.

7   H. Hashemi, P. Sandborn, "Design and Analysis for a Reduced Parasitic Power Distribution System," unpublished proceedings of the *IEEE VLSI & GaAs Chip Packaging Workshop*, (Westford MA), Sept. 1990.

8   F. C. Chong, *et al.*, "High Density Multichip Module," *WESCON '85*, (San Francisco CA), Nov. 1985.

9   J. M. McPhee, T. S. O'Toole, M. Yedvabny, "Cooling the VAX 9000," *Proc. ELECTRO '90 Elect. Conf.*, (Boston MA), May 1990.

10   S. Heng, J Pei, "Air Impingement Cooled Pin Fin Heatsink for Multi-Chip Unit," *NEPCON West '91 Conf.*, (Anaheim CA), pp. 1155-1166, Feb. 1991.

11   J. Pei, S. F. Heng, "Cooling Components Used in the VAX 9000 Family of Computers," *Proc. IEPS Conf.*, (Marlborough MA), pp. 587-601, Sept. 1990.

12   J. S. Fitch, "A One-Dimensional Thermal Model for the VAX 9000 Multi Chip Units," *Proc. ASME Annual Mtg.*, (Dallas, TX), pp. 64-72, Nov. 1990.

13   C. L. Lassen, "Integrating Multichip Modules into Electronic Equipment," *Proc. IEPS Conf.*, (Marlboro, MA), pp. 3-15, Sept. 1990.

Part D—Closing the Loop

"You're thinking about something my dear,
and that makes you forget to talk.
*I can't tell you now what the **moral** of that is,*
but I shall remember it in a bit."

"Perhaps it hasn't one," Alice ventured to remark.

"Tut, tut child!" said the Duchess.
"Everything's got a moral,
if only you can find it."

Alice in Wonderland
by Lewis Carroll

Over the last 17 chapters we have travelled far and wide in the alternative worlds of packaging! At the start of the book we established a framework, or basis, for understanding the material that followed. This included definitions of the relevant terms and technologies. The main goal in **Part A—Making Decisions: The Big Picture** was to provide a *global picture* of the issues involved in package selection: packaging technologies, materials, manufacturing, systems performance and costs. The chapters in **Part B—The Basics** provided an understanding of the principal technology elements that make up a multichip module and the design sciences that are specific to multichip modules. The **Case Studies** in **Part C** showed examples of how some companies develop MCM products. Now it is time to **Close the Loop**, to tie specific new understanding back to the global picture described in the framework.

The following chapter serves two purposes.

First, it provides *emphasis*. In **Part B—The Basics**, the intent was to provide and describe the set of fundamental technologies needed for MCM package selection. Here the goal is to emphasize which of those technologies are the most important. This emphasis will bemainly given from a systems perspective.

Secondly, it provides a *forward view*. It describes what are the open challenges that need to be solved if future systems are not to be *overconstrained* in performance and cost, due to limitations in the packaging technologies.

These purposes are accomplished through a review of the complementing technologies influential in the package selection process for small and large systems.

18

COMPLEMENTING
TECHNOLOGIES FOR MCM
SUCCESS

Ronald W. Gedney and
Donald P. Seraphim

18.1 INTRODUCTION

In previous discussions in this book, a number of key issues emerge critical to
the future success of MCM-based products and to the future growth of packaging
technology in general. By success, we mean volume manufacturing of MCM-
based products by a range of both large and small companies. By future growth,
we mean those technology elements whose further development are critical if
packaging technology growth is to match the systems requirements and silicon
chip technology density and performance improvements. Often success and
future growth issues overlap.

A major conclusion that the reader has derived by now is that MCMs are
generally more expensive than the same die packaged in single chip packages.
Thus, the use of MCMs must be justified [1]-[3] either by cost savings at higher
levels of the packaging hierarchy (such as printed wiring boards (PWBs),
connectors, etc.) or by performance gains which result in system level
cost/performance improvements. Both of these tie into several of the issues to
be discussed here.

Substantial differences in the complementing technologies used to assemble
large and small systems are influential in the package selection process. The
relevance of these differences to package selection is summarized in the

following list and then are considered in more detail in corresponding sections of this chapter.

- **Separable connectors and the packaging hierarchy.** The connectors between different levels of packaging are the single most important constraint on system wide interconnect capacity.

- **The effect of semiconductor type on package design.** Until recently, most MCMs used bipolar parts. The growing use of CMOS devices creates some unique challenges to success.

- **System performance level (cycle time).** This section reinforces the main advantage gained by using MCMs.

- **Level of chip integration, function size, and chip I/O.** These are the main chip related factors that force the need for high interconnection capacities.

- **Chip joining interface, burn-in, testability and repair.** This is one of the two reasons why using MCMs increases cost. (The other is substrate cost, covered in Chapter 4 and not to be discussed here.) It is a major limitation to success and a major opportunity for future growth.

- **Power density and cooling methods.** MCMs increase power density creating a unique challenge for their use. More care must be taken with cooling on the MCM than is usual in the single chip package.

- **Timeliness of introduction.** It is important that the use of MCMs does not increase time-to-market as discussed in Chapter 3.

Finally, we discuss the advances and limitations of single chip alternatives and the importance of considering the low chip count MCM alternative before concluding.

18.2 SEPARABLE CONNECTORS AND THE PACKAGING HIERARCHY

The previous chapters have shown that performance in the largest systems depends on the footprint and the areal density of the die and circuit functions, as

Figure 18-1 A three dimensional packaging system. A zero insertion force separable connector system along top and bottom edges of the PWBs supports connection of 500 contacts per PWB to two back panel PWBs. Tolerances are in the range of 0.2 mm.

well as on the ability to contain all of those functions within a cycle time. The opportunity does exist to use a variety of packaging architectures to achieve these capabilities. Until the late 70s the dominant approach was to use a three dimensional-type architecture with PWBs plugged edgewise into printed wiring back panels with separable contact arrays (see Figure 18-1). The back panels could interconnect 20 or more PWBs. For larger systems containing hundreds of boards, back panels were interconnected with cables into system configurations. Essentially all packaging of semiconductor main memory still is achieved in this way at extremely low circuit costs and continually improving performance [4]. There appeared to be no real boundary for three dimensional packaging density except for one, all important, limiting factor. This factor, separable contact density, was the key that locked the door on the use of three dimensional packaging for logic in the highest performance systems. Separable contact density is, however, more than adequate for low I/O connection demand of memory and acceptable for small systems and workstations.

Figure 18-2 An area array separable connection system for MCMs. This system is extendable from the 2772 I/O shown here to the range of 8000 I/O.

One significance of the MCM lies in its areal array contacting capability (Figure 18-2) that has grown from 400 I/O (signal and voltage) contacts in the late 70s [3] to the range of 8,000 contacts today [5]-[6]. Even today, there is no qualified competitive approach for contacting large arrays of 2,000 or more separable contacts on the edge of PWBs for back panel connections.

For low end systems (PCs, workstations, servers), and even intermediate systems, a steady but slow stream of innovations in making conventional mechanical pin-in-spring separable contacts allows the continual growth of functional I/O capability of PWBs (Figure 18-3). One approach [7] is to use two opposite edges of the board as shown in Figure 18-1. This separable contact system supplies some 500 contacts. A precision frame with tight tolerances is required with a design to guarantee contact forces in the range of 100 grams per contact, and with a wiping action of the contact of several thousandths of an inch. This is a very reliable system. However, the cost per contact is higher

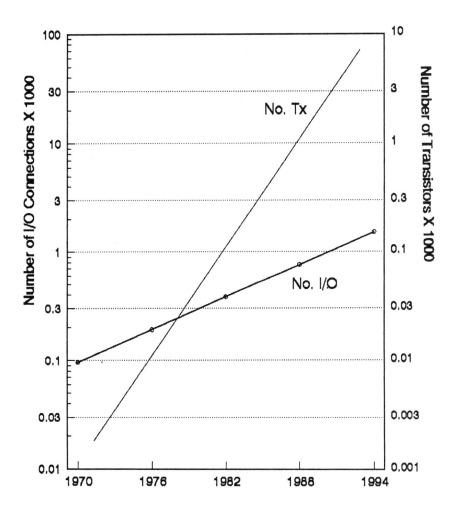

Figure 18-3 Projection and history of semiconductor integration and separable contacts on the edge of PWBs.

than desired by low end system managers. Precision frames are too expensive unless they are amortized over a large number of PWB connectors as applied in intermediate range and larger systems.

The largest board edge (three dimensional) connector systems available, utilizing one edge without precision frame contacts, are now in the range of 800 to 1,400 contacts. One method of extending the contact availability is to extend the edge of the board to an unwieldy, 10 - 15", dimension. Another method is a flex printed circuit to turn the corner into the connector or pin housing from the board edge (see Figure 18-4). This allows sufficient density to place multiple rows of springs beyond the two rows normally used on each side of the PWB. This transforms an edge array on the PWB to an area array at the next package level (PWB backpanel). An MCM pin grid array design accomplished this easily.

An alternative to the flex connection system is elastomeric connections [8]. Either of these technologies are qualified in large arrays. However, there is considerable development activity and it is possible that these two technologies will be developed sufficiently for more general use in the next three to five years. One concern is that the fundamental science base for contacts of these types has not been generated, even though substantial promising empirical data exists. It is essential to provide this science base in the next few years. Clearly, developments in this design space are strategic if PWB alternatives are to be competitive with MCMs and if lower cost connection techniques for MCMs are to be derived from these technologies.

The limitations of the board edge connection systems for application in low end systems are evident in the above discussion. First, the connectors are unwieldy in size requiring board sizes much larger than desired for the component functions and possibly much larger than desired for desktop or portable applications. Second, the conventional mechanical contact systems are reliable only at large forces per contact (60 - 100 grams) and generally with guaranteed wipe. These design constraints impose the use of precision housing and frame tolerances expensive to assemble.

Nevertheless, systems applications growth continues to double in function approximately every three years (see Figure 18-3). The relationship between the growth of circuit functions and increase in I/O is Rent-like, increasing I/Os proportionally to the square root of the number of circuits. This leads to the doubling of I/O and bus widths approximately every six years in the application of microprocessors, as well as in the high end MCM applications [9].

At this juncture, we look for either a new connection technology for PWB edges or for the extended use of MCM arrays and array contact systems to allow functional growth in low end processors.

We conclude that low end systems, with the persistent growth of functions in logic and their persistent pervasive application into all systems use territories, are reaching the same limiting factors faced by high end systems in the late 70s. The question is whether or not the cost learning curves of the high end packaging technologies have been strong enough to penetrate into the low end market. Furthermore, major innovations in the packaging technology have

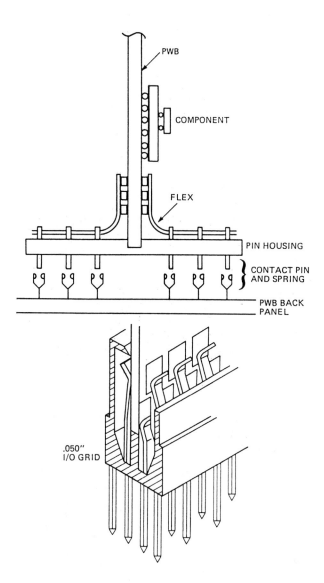

Figure 18-4 Interconnection techniques from PWB edges to PWB back panels. Flexible circuitry is used to extend the number of rows of contacts.

occurred and are developing that may allow alternatives to be applied at the low end. A discussion of the influential technology differences is continued in the sections to follow.

18.3 EFFECT OF SEMICONDUCTOR TYPE ON PACKAGE DESIGN

The discussion on I/O capability should not distract attention from the significance of packaging the die close together. This is an attribute of the MCM. It is desirable to purchase components from a variety of sources at the lowest possible costs. In small systems, there is a tendency for the PWB to contain a variety of components, such as surface mount plastic leaded chip carriers (PLCCs), surface mount small outline package (SOPs), memory boards with edge contacts, dual in line package (DIPs) as well as pin grid arrays (PGAs) and discrete componentry all on the same board surface (see Figure 18-5). At the same time, component sizes are decreasing steadily while device functions are increasing. Capabilities exist to build the processors and the complement of adapters on a single PWB (see Figure 18-6). MCMs, if used at all, are used

Figure 18-5 A typical component menu for a Personal System PWB.

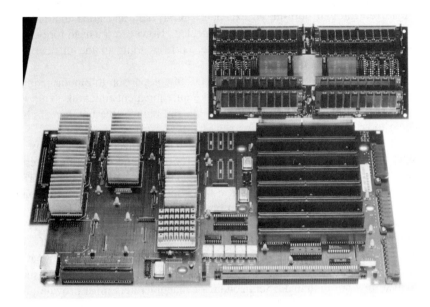

Figure 18-6 A workstation processor board with three dimensional memory board and large SCMs.

sparingly to increase board performance capabilities as a "complement" to the single chip capabilities and to shrink the size of the chip function, allowing it to be packaged on one board.

A substantially larger system with multiple processor logic devices and adapters is more likely to use an MCM. The cost savings, if any, might be minimal. However the initiative to use MCMs in a system of this type is for a cost/performance increase, or to avoid the use of a second PWB and its connections. Another reason to choose MCMs is that printed wiring density, optimized to bring adapters close together for performance increase, can cause the PWB to be very expensive. An MCM allows the use of a much lower cost PWB. The combination of the MCM with a lower cost board meets permissive system costs while providing an important increase in performance.

Logic components used in the low end systems have been primarily CMOS. Substantive improvements have been made in these semiconductors in recent years. However, the very rapid growth of integration, shrink of feature size, decreasing thickness of gate oxide and other factors tend to widen the parametric distribution of CMOS devices. This means that some CMOS parts need to be burned-in and sorted for speed. Once the infant mortality (specification) failures are removed through burn-in, an excellent MCM yield and good thermal and environmental-related reliability can be expected. However, the need for burn-in of logic devices is still a topic of debate, corresponding to the difficulty of accomplishing burn-in on bare die.

The higher level of integration of CMOS in comparison to bipolar requires an order of magnitude fewer chips to make an equivalent function. This is a very positive factor in MCM assembly yields and reduces the number of chips needing to be burned-in to 10 chips per MCM.

It took IBM almost 10 years to develop the manufacturing capability for custom bipolar MCMs. A particular design for a low end multichip processor would require 24 to 36 months to qualify. This is assuming that an appropriate strategy and capability is available for test and burn-in of unpackaged chips, and that the manufacturing capability is in place.

A feature of single chip modules (such as TABs, PGAs, PLCCs, etc.) is the ability to test the function, including its performance. If the chip is faulty, the entire component can be scrapped without concern for its neighbors. For MCM applications, repair of one chip interacts with the other devices and the substrate on which they are mounted. Semiconductor manufacturers must provide good bare die to MCM assemblers as the probability is high that the silicon requirements encompass the product line of more than one manufacturer (see Chapter 1).

18.4 SYSTEM PERFORMANCE LEVEL (CYCLE TIME)

The decision to use an MCM usually is made by a system manager able to assess various alternatives at the complete box level. Since it is very difficult for any MCM to be cheaper than the sum of the single chip carriers it replaces, cost reductions resulting from the use of MCMs are apparent only when the entire system is considered. The system manager also can assess the value of increased performance gained by using a MCM.

Davidson showed that an MCM had the potential for putting 16 or more devices within a 1 ns cycle [10]. If the system cycle time is 20 ns or more, a savings of 1 ns is 5% or less of the cycle; other solutions to meet performance objectives can probably be found. However, when system cycle times are 10 ns or less (100 MHz), a 10% or better improvement can be garnered from packaging alone. Shiao and Nguyen showed that a high performance cache system could see cycle time improvements of up to 25% through use of

Cu/polyamide MCM modules versus single die on board packages when both package and I/O buffer delays were taken into account [11]. Neugebauer *et al.* showed that for a 4.36 MGate array processor system, moving from SCMs to simple 10 chip MCMs would result in a total package footprint reduction of 69%. If 72% of the interconnections are moved from the PWB to the MCM, the power multiplied by the delay product in the average signal path is diminshed by over 75% [12].

Another example of the space and performance advantages of MCMs is the IBM Risc/6000™ system. With slight advances in silicon technology, nine complex single chip modules were condensed into a seven chip MCM, providing a 12× reduction in physical size on the planar performance and a 2× improvement in system performance (see Figure 18-7). IBM also found that the MCM provided a cost reduction at the system level. Krusius *et al.* developed

Figure 18-7 A MCM (SOS) replacement for the processor and cache (nine large single chip modules) shown in Figure 18-6.

computer models for evaluating package performance [13]. Various MCM
alternatives to single chip on PWB designs have been implemented and
substantial performance gains for MCMs have been demonstrated.

It is reasonable to assume that most systems with performances greater than
50 MHz can see measurable cycle time reductions through use of MCMs,
particularly those of the thin film, Cu/polyimide design. While clever use of
high performance single chip carriers, ordered wiring [14] and high performance
surface mount PWBs can achieve the desired system performance, well designed
MCMs may result in better cost/performance tradeoffs at the system level.

18.5 LEVEL OF CHIP INTEGRATION, FUNCTION SIZE AND CHIP I/O

The number of CMOS chips on an MCM has to be small and the quality high
to reduce failures on MCMs during burn-in. Otherwise the test, rework and
scrap costs charged to the MCM will be excessive. Furthermore, a degree of
sophistication in the user's system of assembly and test is required beyond that
necessary for single chip components.

18.5.1 Memory MCMs

An illustration of the impact of test, scrap and rework cost is IBMs original work
with memory MCMs. During the 70s, IBM consistently used multichip memory
modules. These modules, on a 24 mm substrate, would have four DRAM die.
Two substrates could be stacked (pins joined to pinheads with solder) to provide
eight die per module. With the introduction of the 1 Mb DRAM, IBM went
back to a single die module.

Referring to the single chip carrier case in Table 18-1, functional test can
represent as much as 30% of the total package cost. Given the cost of a single
chip small outline integrated circuit (SOIC), it is not cost effective to rework
such a package. Typically, one can expect a complex (not burned-in) CMOS
device (such as a 1 Mb DRAM) to have a chip join to final test yield of 90% to
94%. Assuming a median yield of 92%, approximately 28% of the four chip
modules require repair. If this were a true eight chip substrate, the first pass
yield would be only 51%, but since the individual four chip decks are tested
good before stacking, a very high yield is achieved on the stacked substrates.
Usually there is only one bad die on a substrate, but occasionally there are two
or more. In a production environment, economics would dictate that more than
two bad die would result in scrapping the entire substrate.

Table 18-1 Module Process Steps.

SCM	MCM	
1 Die	4 Die	8 Die
1) chip place and join 2) encapsulate 3) burn-in (optional) 4) functional test 5) ship	1) chip place and join 2) functional test 3) repair and rework 4) functional test 5) encapsulate 6) functional test 7) ship	1) chip place and join 2) functional test 3) repair and rework 4) functional test 5) stack substrates 6) functional test 7) repair and rework 8) functonal test 9) encapsulate 10) functional test 11) ship

Table 18-2 Cost Considerations for MCMs and SCMs.

Assume:	SCM cost (SOIC) = $X Unreworked MCM cost = $X/chip (MCM = SCM cost × Number of SCMs Rework + functional test = $2X (28% of the modules are reworked.) Cost of additional test = $0.3X
Then:	Added module cost over production lot is $0.28 \times 2X = 0.56X$ Added cost per packaged die = $(0.3X + 0.56X)/4 = 0.22X$

Independent of the rework, which is time consuming and expensive, consider the cost of testing. In the case of the four chip module, two complete functional tests on 100% good parts and three on repaired parts have been completed. Keep in mind that, as each die is individually tested, the four chip test is four times as long as the test for a single chip package. (Handling times are usually insignificant compared to test times.) The cost of one of the more expensive parameters in building a module has been doubled. This, coupled with the additional cost of rework and repair (and another test), on 28% of the product means a premium of 22% per chip for the MCM over the SCM cost (see Table 18-2).

This added cost per packaged die might be affordable if cost savings could be made elsewhere in the system (smaller or less dense PWB). However, in the case of memory, the package is almost the same size as the die. With double sided surface mount and stacked SOICs, these savings are not available. This is an example of multichip packaging that did not have any payback at the system level.

We project from this discussion that the number of CMOS chips on an MCM needs to be relatively small and the quality needs to be very high to reduce, to an inconsequential few percent, failures on the MCMs during burn-in. Otherwise the test, rework and scrap costs charged to the MCM will be excessive. To achieve this level of confidence, semiconductor suppliers need to offer tested, burned-in bare die for MCM assemblers.

18.6 CHIP JOINING INTERFACE, BURN-IN, TESTABILITY AND REPAIR

For MCMs to be practical, test and rework processes have to be developed (see Chapter 13). This calls for a chip removal process, a site preparation process where the chip is replaced, a joining process that does not risk the neighboring chips, precision placement capability [15] and a functional test capability with analytical ability built into the design [16] to isolate the fault to a particular chip. All of these factors must be considered during the design stage.

18.6.1 Known Good Die

The next logical question to ask is, "Why not burn-in and test the chip prior to module mounting?" If producers of MCMs could work with known good die, then yields would be far superior and rework and repair minimized, if not eliminated. The problem, of course, is handling, testing and burn-in of devices individually or at the wafer level. Test probes to provide functional test of individual die or at the wafer level are both complex and expensive as discussed in Chapter 13. Connections to bare die for burn-in are also quite difficult, although under active investigation. Our experience to date is that test and burn-in at the die or wafer level leads to device costs equal to or higher than the same device in a single chip carrier. Although we believe this will change in the future as new manufacturing concepts and techniques are devised, there are always some added costs. Given that simple packages cost $0.05 to $0.15 per I/O, it is doubtful that known good die will ever allow MCMs to be cost equivalent at the chip carrier level. The exception to this is the case where, early in a program, a die is committed to an expensive, high I/O, hermetic, multilayer ceramic package and the cost of the MCM interconnection is amortized over a large number of chips.

18.6.2 Chip Joining Interface

Still another MCM design consideration is the physical interface to the semiconductor user. This has been predominantly wire bond technology. Unfortunately, wire bonding is a difficult technology for MCM production. One detriment is the ability to remove a bad die, clean the bond sites and put another die in its place as discussed in Chapter 9, Section 9.3. Furthermore, the inductance and capacitance of the wires are unfavorable (for high performance applications) compared to some of the alternative methods.

We believe that either an assembly process like TAB [17], flip TAB or a process like flip chip (C4), not commonly used by semiconductor manufacturers, has to be introduced as an industry standard for the semiconductor producer and the user of MCMs.

18.6.3 Flip TAB

The flip TAB technology [18] appeared first in the high end systems of NEC [5] where it was used to join chips to a very small multilayer single chip carrier (see discussion in Chapter 9, Section 4). The carrier was interconnected to an MCM with an area array of gold tin bumps. The area array of bumps of the single chip component was on a grid similar to that of the wiring interconnection vias of the MCM. This eliminated the need for fanout interconnection layers, simplifying the MCM design and moving the finest line interconnections to a component package to be tested individually and scrapped, if necessary. This decreased the probability that the MCM had to be reworked after chips were placed and joined because of the simpler, less dense interface and an adequate pretest of the units joined. An advantage of the NEC approach is the versatility in achieving a dense area array grid joined to the surface of the MCM. In fact, the NEC approach is very similar to placing a die directly on a PWB, but it does not eliminate a level of packaging. As pointed out earlier, the dense fanout layers for direct chip joining on an MCM can be equivalent to the fanout interconnections on the single chip component level of packaging.

The single chip carrier may be eliminated in the future by using the flip TAB interconnect directly to the MCM or a printed circuit card. A flip TAB array of this type, recently developed by IBM (see Figure 18-8), could be used to mount chips directly on ceramic or organic packages, eliminating one assembly operation.

18.6.4 Flip Chip

As semiconductor densities increase, the number of I/O, even for low end devices, soon will exceed 600 and push the 1,000 to 1,200 range. Flip TAB on large (> 14 mm) die will handle 600 I/O and might extend to the 800 I/O range.

Figure 18-8 A solder ball array TAB. The ATAB has two metal layers (ground and signal).

After that, flip chip technology (see Chapter 9) seems to be the only fully qualified solution.

Putting solder bumps on the die necessitates additional wafer processing. However, since C4 joints can be placed over active areas (unless there is alpha particle sensitivity), the die does not need a pad for wire bond or TAB mounting outside the active silicon area. Thus, die that are 10% to 30% smaller can be designed with comparable increases in wafer productivity, depending on whether the semiconductor design is circuit or I/O limited. Also, if the flip die bond is designed properly, the via opening in the chip passivation layer is covered and the die is effectively sealed, eliminating the need for hermetic packages.

18.7 POWER DENSITY AND COOLING METHODS

Functions on MCMs for mid-range range systems using bipolar die have run at about 90 watts (Figure 18-9) requiring impingement air cooling on very large heatsink areas [19] and thermally conducting grease between the heatsink and the back of the chip. Power density is about 20 watts per square inch, with air flow up to 1,000 feet per minute. This is not a low end technology. With the advent

Figure 18-9 Impingement air cooled 64 mm MCM (about 90 watts).

of CMOS processors, which run in the range of 3 - 10 watts per chip, a six to nine chip MCM that dissipates about 30 watts is reasonable. This can be cooled with a small fan providing up to 500 feet per minute in a small system environment. For higher powers, it is essential to find efficient methods for heat removal, while decreasing the power levels per circuit. In this case, the cooling cap provides a hermetic or semihermetic seal. This seal eliminates concerns for ionic migration with the high density of interconnection and the resulting high electric field on the top surface of the MCM. CMOS technology is helping by going to lower voltages with resultant lower powers. Devices currently are being announced with voltages of 3.0 - 3.5 volts and expectations that devices with 2.5 volts or less are coming.

Aluminum nitride (AlN) ceramic substrates can be used in future designs since its heat spreading capability is nearly an order of magnitude better than that of alumina. A single C4 joint has a thermal resistance of about 550°C per watt. For high I/O chips, a reasonably good thermal path is provided between the chip and the substrate. (Extra C4 pads can easily be added to assist thermal management.) AlN allows efficient spread of heat through the ceramic without needing a complex process and design.

In the near future (three to five years), we expect 20 - 40 watt CMOS chips still will be used. In this case, the cooling technology will be complex, similar to that discussed for bipolar die and MCMs for low end systems will have powers ranging up to 100 watts.

18.8 TIMELINESS OF INTRODUCTION

Assuming that the complement of assembly processes, interface designs and system cooling configuration capability exists, MCMs can be designed and built with a turnaround time equivalent to the semiconductor. The prerequisites to the achievement of this goal are:

- Freezing device pad assignments allowing module design to start early (concurrent engineering)
- Good information and design computer aided design systems
- Rapid prototype facilities
- Good simulation systems for electrical and thermal modeling

These techniques all exist today, able to meet requirements for rapid design iterations and performance increases expected in small systems. Following the semiconductor evolution, it is reasonable to expect small systems to double in performance every three years.

Other alternatives should be considered. One approach is for a consortia of companies to develop functional building blocks utilized by its members. Another alternative is for semiconductor houses to develop and sell functional building blocks. Systems houses currently buy the processor device from one supplier and complementary devices and memory from other suppliers. If a semiconductor house supplied MCMs with the whole function, they could supply more of their devices to the system purchaser.

18.9 ALTERNATIVES TO MCMs

Consider a system consisting of nine large chips each with 600 I/Os (400 signal I/Os, 200 power and ground) connected to a large memory array and some other low density parts. First, consider how the system can be partitioned by interconnection needs. Unless the system performance is particularly critical to small improvements in delay to the memory array chips, there is little benefit gained by packaging the memory chips in an MCM. The memory and low density part of the system is slightly larger, slightly slower and, certainly, much more cheaply packaged with plastic single chip packages on a four layer (two signal, power and ground) board.

On the other hand, if the nine chips in single chip packages are spaced 2" apart, the required wiring capacity necessary is about 1,800" per component or 450" per square inch. This could be provided by a PWB with 9 - 10 signal layers, as long as the average line pitch was better than 1 line per 20 mils. This is a credible state of the art for leading PWB suppliers. The critical technology for this board would be providing through hole vias on a 40 -60 mil grid. This will always be a critical technology element for future high performance PWBs, or similar laminate MCMs, the other critical element being coefficient of thermal expansion (CTE) match. Progress in this technology was discussed in Chapters 2 and 5. Despite this, their cost is significantly cheaper than the equivalent MCM.

However, consider the edge connector of this board. If they are all to be on one edge of the PC board (6" in length), then the edge contact density is approximately $\left(\sqrt{9} \times 400 \right)$ = 1200 or 200 per inch, using a Rent's rule relationship. This is equivalent to 10 rows of contacts on a 50 mil grid or 5 rows on each side of the edge. A PWB edge connection system of this type has not yet been qualified. However, a peripheral leaded MCM flex demonstrates the contact technology and was released in the DEC 9000 computer series (see Chapter 17).

For comparison purposes, consider an equivalent nine chip MCM. With modest device spacing, the same nine chips could be mounted on a 2.5" square substrate. Adequate wireability [21] and I/O are achievable with either a multilayer ceramic thick film (MCM-C) or thin film (MCM-D) design with I/O on a staggered 0.100" grid (area array). Conventional spring connectors are available with this spacing. Also, existing designs [19] cool up to 90 watts in air for these size modules.

The single chip PWB approach results in a 6" × 6" (36 square inch) package and a formidable connector challenge. The MCM alternatives result in a 2.5" square package (6.25 square inch) with a readily achievable interconnect solution. One readily can discuss performance differences in the range of 15 - 20% based on the usual rule of thumb that performance is one-third chip, one-third circuit and one-third package delay.

The other issue is cost. The PWB has two advantages: it is low on the manufacturing learning curve, and many pieces can be cut from the large panel processed. The MCM, in comparison, can be six times as expensive per unit area at the substrate level.

Finally, examine the significance of improvements in the cost/performance ratio with the use of MCMs. For example, if mounting the chips on an MCM substrate doubles the performance and also doubles the cost compared with single chip mounting, the cost/performance ratio is unchanged at the mounting level. This may well be sufficient to sell additional applications. However, at a system level, the cost/performance ratio may be improved significantly. Consider a case where the system level cost split is 25% to semiconductor devices, 50% to the package and 25% to assemble and test. Then, assuming that chip mounting is approximately 50% of the total package cost:

Cost/performance = [.25 + (.25+.25) + .25] / 1 = 1/1
for single chips

Cost/performance = [.25 + (.25+.50) + .25] / 2 = 1.25/2
for MCMs

Thus the improvement in the cost/performance ratio at the system level is 1/0.6125 = 1.63 which is substantial.

In conclusion, as performance continues to increase in low end systems, and, especially as multiprocessor applications increase, MCMs will capture a significant fraction of the electronic packaging market. Estimates by Sage and Hartley are in the range of 4% to 12% of an $11.5 million unit total market for just work stations, high end PCs and portable computers by 1995 [14].

18.9.1 Future Growth in Single Chip Packaging

PLCCs, with I/O counts greater than 300, are now available and sample quantities of 500+ I/O are offered. For our conceptual package we chose 2" spacings for the components. Even with 0.3 mm OLB spacings, the achievement of 600 peripheral leads on a flat pack will not be contained in the space allotted. Also, it is recognized in industry that peripheral lead packages with 0.5 mm or smaller spacings are difficult to join at high yield and reasonable cost to the printed circuit carrier. It is barely possible to achieve a 3σ joining yield. A goal of 6σ joining does not appear practical without new processes and paradigms being invented. On the other hand, surface mount array (SMA) packages, with area array I/O, can achieve reasonable sizes at 600 I/O (Figure 18-10) and, at a 0.050" grid, can be joined readily at the 6σ level with current processes. Both pin grid array and land (or pad) grid array ceramic packages are readily available for surface mounting (Figure 18-11).

Although advanced area array packages provide higher packaging density, they also drive higher PWB wiring capability (density of interconnection) [23], because the same number of wires have to be contained in the smaller area allowed by the array. However, the wires are shorter proportionally to the pitch or spacing of the components, if the component spacing on the PWB is decreased proportionally to the component size.

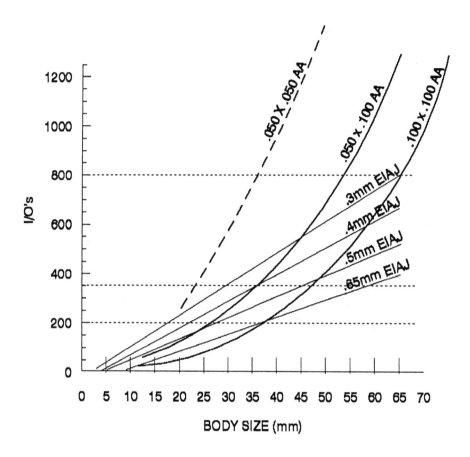

Figure 18-10 Comparison of I/O capability versus size for area array and peripheral lead packages.

PLCC packages increase in size for a given lead pitch as they provide higher and higher I/O. The same number of wires (as compared to area arrays) are contained in a larger area. The wires in this case are longer proportionally to the pitch of the components. If the larger areas are used effectively, some relief in wiring density and area for air cooling is available with the larger components. However, the PWB size is increased.

The system designer has several decisions to make. If the electronic package size is not particularly important (such as a desktop PC), low cost PLCCs can be spread out on a surface mount PWB. The wiring demand on the

Figure 18-11 A surface mount array MLC MCM with 600+ I/O on 0.050" grid.

PWB is not very high; spreading the components out assists in thermal management. If size of the electronics is important, one alternative to an MCM is the SMA package. This very dense package reduces the PWB area required at the next level, at the expense of wiring demand. The wiring demand probably doubles from a 0.5 mm peripheral lead package to a 50 mil grid SMA package, but the required area is reduced substantially. A dense SMA package on a dense PWB should permit up to 100 MHz clock times without resorting to MCMs.

18.10 INEXPENSIVE, LOW COMPLEXITY MCMs

So far our discussion has dealt mainly with the system processor or central electronic complex. However, there is another class of MCMs, generally with only a few chips [22], that are successful. In this breed of MCM, there only may be one to three semiconductor devices and some passive components (resistors, capacitors, etc.). These modules are often associated with unique functions such as analog circuits, sense amplifiers, optical driver/receivers or semipower devices such as hammer drivers. The many passive devices accommodated on a PWB can be placed on an MCM as cheaply. In many cases,

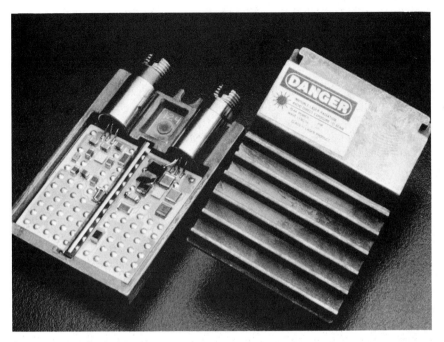

Figure 18-12 A 200 MHz optical driver/receiver MCM.

the circuit performance improvements of the MCM are considerable. In the case of very small signal amplifier circuits, the ability to actively trim resistors and tune the circuits for performance is vital to success of the function.

An optical driver/receiver MCM substrate produced by IBM is shown in Figure 18-12. This MCM design provides both driver and receiver circuits for a 200 Mb/sec optical data communications module and provides interconnection of the fiber optic cable in a simple 30 mm package. A more sophisticated module has been investigated by GTE [23]. This MCM (28 die plus assorted passive components) packaged many functions of a digital signal processor (DSP) board on one MCM. The MCM resulted in a 60% space saving and substantial performance gain.

It is conceivable that the MCM industry will grow, not from a "trickle down" of technology from the very high end, but be driven up by the growing demands for density and performance increases of single chip packaging technology [9].

The other thing that the MCM developer has to keep in mind is that silicon density keeps increasing with little sign of slowing growth. A function that requires an MCM today will, in all likelihood, be designed into a single chip tomorrow. It is feasible that an MCM with three to ten chips will be integrated into a single die three years after the MCM goes into production. For the careful planner, this provides an opportunity. As a single die with the same function requires the same number of I/O, a functional MCM design can be migrated to a single chip design at a lower cost without disturbing the next level of package.

Trends indicate that high end PCs, work stations and large systems will embrace multiway designs to enhance performance in the near future. Assuming that a one chip processor can be achieved, a four way design will require four of these chips plus associated cache. An MCM design can handle the dense interconnections required now; a single die with four processors can replace the MCM when the technology becomes available.

18.11 SUMMARY

The decision to use MCMs in a system, whether small or large, is driven by the need for performance and competitive cost. MCMs have two main performance advantages: their ability to space chips closely, minimizing interconnect delays, and their ability to provide very large amounts of wiring, within the MCM, from the chip to the MCM and through the connector to the next level of packaging. Cost benefits will be achieved, and should be optimized, at the system level. However, consideration needs to be given to a number of potential factors that might inhibit MCM use. In particular, the need for burn-in and testing of complex CMOS systems needs to be considered, along with the higher heat density and the extra design challenge. One solution is to use small, less than 10 chip MCMs. This fortunately is becoming more and more feasible with the growth in semiconductor density.

Looking toward the future, a number of challenges are evident that require new solutions. First and foremost is the need for high I/O count, low insertion force, separable connectors for both PWBs and MCMs. Next, better solutions to the test and burn-in issues are needed if high chip count CMOS MCMs are to be a success. This is partly an infrastructure issue, as the chip manufacturers have to provide qualified bare die for this to happen. Finally, there is the challenge of designing a high speed, high power MCM with first pass success in the manufacturing environment.

We believe these challenges will be met and that the MCM technology will find widespread use in the electronics industry.

References

1   D. P. Seraphim, I. Feinberg, "Electronic Packaging Evolution in IBM," *IBM J. of Res. & Devel.*, Anniversary Issue, vol. 22, no. 5, p. 617, Sept. 1981.

2   R. Johnson, R. K. F. Teng, J. W. Balde, *Multichip Modules: System Advantages, Major Constructions, and Materials Technologies*, New York: IEEE Press, 1991:
    J. W. Balde, "New Packaging Strategy to Reduce System costs," pp. 7-10
    W. H. Knausenberger, L. W. Schaper "Interconnection Costs of Various Substrates—The Myth of Cheap Wire," pp. 11-13.
    M. G. Sage, "Future of Multichip Modules in Electronics," pp. 14-20.
    G. Messner, "Cost Density Analysis of Interconnections," pp. 21-29.

3   Albert J. Blodgett, Jr., "A Multilayer Ceramic Multichip Module," *IEEE CHMT Trans.*, vol CHMT-3, no. 4, pp. 634-637, Dec, 1980. Also *Scientific American*, vol. 249, pp. 86-96, July 1983.

4   L. E. Thomas, F. W. Haining, J. S. Macek, "Second Level Packaging for the IBM ES/9000 Processors", *Internat. Electr. Packaging Soc. Conf.*, (San Diego CA), pp. 1049-1055, Sept. 1991.

5   D. Akihiro, W. Toshihiko, N. Hideki, "Packaging Technology for the NEC SX-3/SX-X Supercomputer," *Proc. of the IEEE 40th Electr. Components and Tech. Conf.*, (Las Vegas, NV), pp. 522-533, May 1990.

6   A. Kaneko, K. Kuwabara, S. Kikuchi, T. Kano, "Hardware Technology for Fujitsu VP2000 Series," *Fujitsu Scientific and Tech. J.*, vol. 2, no. 27, pp. 158-168, June 1991.

7   T. Chikazawa, *et al.*, "Sensor Technology for ZIF Connectors," *ASME J. of Electr. Packaging*, vol. 112, pp. 187-191, Sept. 1990.

8   *Proc. NEPCON West*, (Anaheim CA), 1990.
    L. S. Buchoff, "Solving High Density Electronic Problems with Elastomeric Connectors," pp. 3-7.
    K. Gilleo, "A New Multilayer Circuit Process Based on Anisotropicity," pp. 8-31.
    J. A. Fulton, *et al.*, "The Use of Anisotropically Conductive Polymer Composites for High Density Interconnections," pp. 32-46.
    R. C. Moore, *et al.*, "Testing Bare Boards with an Anisotropically Conductive Elastomer," pp. 47-56.
    H. Yonekura, "Oriented Wire Through Connectors for High Density Contacts," pp. 57-72.

9   "Computer Development and Parallel Processing," *BPA COMPASS Update*, no. 1, 1992.

10   E. Davidson, "The Performance Choice: A Study of Chip and Package Densities," *Proc. of the 40th Electr. Components and Tech. Conf.*, (Las Vegas NV), pp. 147-150, 1990.

11   J. Shaio, D. Nguyen, "Performance Modeling of a Cache System with Three Interconnect Technologies: Cyanate Ester PCB, Chip-on-Board and CuPI MCM," *Proc. of the 1992 IEEE Multichip Module Conf.*, (Santa Cruz CA), pp. 134-137, March 1992.

12   C. Neugebauer, R. D. Carlson, R. A. Fillion, T. R. Holler, "Multi-chip Module Designs for High Performance Applications," *Proc. NEPCON West*, (Anaheim CA), pp. 401-413, vol. 1., 1989.

13 D. E. O'Brien, B. M. Hahne, J. P. Krusius, "Multichip Modules vs. High-density Printed Wiring Board: A Trade-off Study," *Proc. of the IEEE Electr. Components and Tech. Conf.*, (San Diego CA), pp. 31-41, May 1992.

14 M. G. Sage, N. Hartley, "System Issues for Multichip Packaging," *Proc. of the Internat. Soc. Optical Engr.*, (Boston, MA), pp. 302-310, vol. 1390, 1991.

15 A. H. Landzberg, *Microelectronics Manufacturing Diagnostic Handbook*, New York: Van Nostrand Reinhold, 1992, Chapter 8.

16 D. S. Karpenske, "Interconnect Verification of Multichip Modules Using Boundary Scan," *Proc. IEEE 1991 VLSI Test Symp.*, (Atlantic City NJ), pp. 85-91, April 1991.

17 H. Wessely, "Packaging System for High Performance Computer," *Proc. IEEE IEMT Symp.*, (Nara Japan), pp. 85-91, April 1989.

18 J. H. Lau, *Handbook of Tape Automated Bonding*, New York: Van Nostrand Reinhold, 1992.

19 S. Oktay, B. Dessauer, J. I. Horvath, "New Internal and External Cooling Enhancements for the Air Cooled IBM 4381 Module," *IEEE Internat. Conf. on Comp. Design*, (Portchester NY), pp. 2-5, Nov 1983.

20 U. Deshpande, G. Howell, S. Shamouilian, "High Density Interconnect Technology for VAX-9000 System," *Proc. of the SPIE*, (Boston MA), pp. 489-501, vol. 1390, pt. 2, 1990.

21 L. R. Buda, R. W. Gedney, T. F. Kelley, "High Density Packaging: Future Outlook," *Proc. of the IEEE Electr. Components and Tech. Conf.*, (San Diego CA), pp. 36-41, May 1992.

22 K. M. Nair, "Hybrids and Higher Level Integration," *Electronic Materials Handbook: Volume 1, Packaging*, City: ASM International, 1989, Section 3.

23 R. Chew, "A Study on the Feasibility of Using MCMs: A Designer's Perspective," *Proc. of the 1992 IEEE Multichip Module Conf.*, (Santa Cruz CA), pp. 127-129, March 1992.

EPILOGUE

THE CONCLUSION OF THE MATTER

(Ecclesiastes 12:9-14)

Not only was the Teacher wise,
but he also imparted knowledge to the people.
He pondered and searched out and set in order
many proverbs.
The Teacher searched to find just the right words,
and what he wrote was upright and true.

...BUT, BE WARNED, MY SON....

Of making many books there is no end,
and much study is a weariness of the flesh!

NOW ALL HAS BEEN HEARD. HERE IS THE CONCLUSION OF THE MATTER.

Fear God and keep his commandments,
for this is the whole duty of man.
For God will bring every deed into judgment,
including every hidden thing,
whether it is good or evil!

ABOUT THE AUTHORS

Salma Y. Abbasi, Principal Engineer, Digital Equipment Corp., Cupertino CA. Ms. Abbasi has 10 years of experience in the field of advanced computer and semiconductor packaging and interconnect technologies, including high pin count, high performance packaging, thermal management of single chip and multichip packaging, tape automated bonding, die attach and process development and characterization in both multichip module and surface mount technologies. Additionally, she has extensive experience in TQM, DOE and DFM. She has worked on semiconductor packaging and interconnect related technology development for Advanced Micro Devices and Monolithic Memories for four and two years respectively, before joining Digital Equipment in 1988 as a senior process development engineer in the Multichip Unit Development Engineering Group. She has a BS from Westminster University in England, and is currently enrolled in an MS degree in Engineering Management at San Jose State University CA.

Kaveh Azar received his PhD in hydrodynamic stability from University of Connecticut in 1983. He joined AT&T Bell Laboratories in 1985 and has been involved in electronics cooling for the past six years. His activities include development of generic cooling techniques and thermal consulting. Dr. Azar also has been holding an adjunct position at Tufts and Northeastern Universities and had developed graduate courses in electronics cooling and thermal modeling. He was the past general chairman of IEEE SEMITHERM conference, holds five patents in electronics cooling and has many publications.

Robert L. Bone received a BS degree in Industrial Technology, Electronics Option from California State University at Long Beach in 1968. He currently is Manager of Application and Product Development Engineering at Hughes Aircraft Company, Microelectronic Circuits Division, Newport Beach CA, responsible for engineering, program management, marketing and operations.

Steve J. Bezuk is the Manger of Chip Interconnect Technology, Unisys Corp., San Diego CA. He received the BS (Chemistry) from the University of Pittsburgh, the PhD (Chemistry) from the University of Minnesota and the MBA from the University of Redlands. His prior experience includes management of the Advanced Interconnect Technology Group, Sperry Corp. (now UNISYS), and a member of the Energy Systems Group, RCA David Sarnoff Research Center. His experience includes amorphous silicon research, CMOS processing, GaAs microwave processing, as well as microwave, thin film, wire bond, TAB and FCSB packaging. He has published papers on plasma diagnostics, $TaSiO_2$ - polysilicon gate materials for CMOS devices and laser photochemical and pyrolytic deposition of metals.

Allison Casey Dixon is Product Engineering Manager for Multichip Modules in Motorola's ASIC Division located in Chandler AZ. Ms. Dixon has championed the MCM development effort in the ASIC Division since 1988 and currently is responsible for bringing Motorola MCM products to market. Ms. Dixon has 10 years of experience at Motorola in a variety of engineering positions including wafer fabrication, product engineering and RF gallium arsenide development. She has a special interest in creating products from emerging technologies. Ms. Dixon graduated in 1982 from the M.I.T. with a BS in Chemistry. She is currently pursuing an MBA Administration from Arizona State University.

Dennis F. Elwell received a PhD in Physics from Sheffield University, Sheffield, England, and a Bachelor of Physics from Imperial College, London, England. He has served on the physics faculty of Portsmouth Polytechnic, England, researching electrical conduction in magnetic oxides and crystal growth from high-temperature solutions, has published over 100 technical papers and four books, and has received one patent. He currently is Manager of Technology and Senior Scientist/Engineer for the Process Technology Development Department, Microelectronic Circuits Division, Hughes (IEG), responsible for coordinating division internal technical research efforts in materials development processes and analysis, and heads a team developing LTCC technology in support of pilot manufacturing. Earlier, he managed projects in atomic layer epitaxy, high-temperature superconductors, and nonlinear optical materials.

Claudius Feger, Research Staff Member, IBM T. J. Watson Research Center, Yorktown Heights NY, since 1984. His research concerns all aspects of polymeric electronics packaging materials, particularly polyimides. He holds three patents and has published over 50 technical papers and edited two books, He has organized several conferences including two conferences on polyimides. Before joining IBM he was working at the University of Massachusetts, Amherst, and at UFRGS, Porto Alegre, Brazil. He is currently adjunct professor at the Electrical Engineering Department of the University of Maine, Orono. He received the chemistry diploma from the Albert-Ludwigs University at Freiburg Germany in 1976 and from the same university, the PhD in polymer science in 1981.

Christine Feger, Staff Engineer, IBM Corp., East Fishkill NY. Ms. Feger has 15 years experience in all aspects of thin film process technology. She is presently responsible for polyimide process development and technology transfer in IBMs East Fishkill Packaging Laboratory. Prior to joining IBM in 1989, she worked for Rockwell International developing processes for thin film hybrid circuits. She was responsible for setting up the prototype multichip module fabrication facility for Rockwell in Anaheim CA. She received a BS in Chemistry from the University of California, Santa Barbara in 1977 and an MS in Electrical Engineering and Computer Science from UCSB in 1982.

Ronald W. Gedney, Program Manager, IBM Packaging Development Laboratory in Endicott NY. As engineer and manager, he has been involved in electronic packaging and development for some 30 years. He was responsible for the development of IBM's "metallized ceramic" chip carriers from 1974 to 1982, the most widely used single and multi-chip packaging technology in IBM. From 1982 to 1984, he was a consultant at IBM's plant in Havant, England, during the transfer of a substrate manufacturing line. In 1986-87, he established IBM's TAB/flex capabilities. Over the last three years, he developed IBM's low end chip carrier packaging strategy. A senior member of IEEE, he was President of the CHMT Society during 1990 and 1991. He holds a BS degree in Electrical Engineering from Tufts University.

Dean R. Haagenson is a Senior Engineer, Semiconductor Memory and Packaging Organization, Unisys Corp., San Diego CA. Mr. Haagenson received his BSME from the University of Minnesota, Twin Cities. He joined Sperry Corp., Semiconductor Operations (later UNISYS Corp.) in 1985, primarily focusing on TAB assembly development as applied to multichip module programs. He transferred to his present location at the Unisys semiconductor facility in San Diego in 1987, where he joined the surface mount development group charged primarily with developing advanced SMT processes for manufacturing, including TAB bonding. Currently, Mr. Haagenson is working on advanced TAB and flip chip assembly development programs for Unisys. He has one TAB-related patent.

Eugene F. Heimbecher, II is a Senior Engineer, at Motorola Inc. in Chandler AZ. Mr. Heimbecher recently joined the Motorola ASIC Division Multichip Module Group as a system engineer for substrate analysis, signal integrity and design guidelines. Previously, he worked in Motorola Government Electronics for ten years. His most recent experience is three years as a system engineer with responsibility for design guidelines and noise management in hugh speed CMOS digital processing systems. He has written a CMOS system noise management handbook for use at Motorola. He also has seven years experience in high speed analog and frequency circuit design and analysis, primarily in the area of high performance low noise frequency synthesizers and radio receivers. He received his BSEE from the University of Wisconsin in 1981. He has served three years in the U.S. Army as an electronics technician.

Leo M. Higgins, III is Manager of Multichip Module and C4 Packaging and Technology Development, Motorola Inc., Austin TX. Dr. Higgins has 14 years of experience in microelectronic packaging in the areas of super minicomputer system packaging, ECL multichip module development, thermal management of very high power multichip modules and single chip PGAs, high performance/high power PGA and chip carrier development, inner and outer lead tape automated bonding, flip chip assembly, development and manufacture of ceramic packaging systems, etc. At Motorola, he manages a group responsible for SMT application engineering support and assembly technology development, TAB ILB/OLB engineering, and introduction of C4 assembly into Motorola. He is establishing a prototype/pilot assembly facility for C4/wire bond/TAB multichip modules constructed with MCM-L, MCM-D, and MCM-C substrates. Previously he worked on single and multichip ECL and CMOS packaging at Prime Computer (Framingham MA) for three years. Prior to this, he worked on cofired ceramic packaging technology at Cabot/Augat Technical Ceramics (North Attleboro MA) for five years as Manager of Engineering, and Vice President Research and Technology. Before this, he worked on a broad range of ceramic packaging for three years at Kyocera International (San Diego CA) as Senior Research Scientist, then Manager of Applied Technology. He has a BS, MS, and PhD in Ceramic Science and Engineering from Rutgers. He holds five patents, and has published more than 10 papers.

Ronald J. Jensen, Engineering Fellow, Honeywell Solid State Electronics Center, Plymouth MN. Dr. Jensen has more than ten years of experience in multichip module technology. In 1981, he joined Honeywell's Physical Sciences Center, where he developed the materials and processes for fabricating thin film multilayer (TFML) interconnecting substrates for MCMs. He also developed design and testing methods for MCM substrates, and demonstrated the TFML technology in a variety of test vehicles and functional MCMs for Honeywell divisions and other companies. In 1990, he transferred the TFML technology to a production facility at SSEC and implemented SPC and DOE methods for optimizing processes. He is now involved in other aspects of MCM technology, including bare die testing and MCM design and assembly. He received a BS in Chemical Engineering from Iowa State University and a PhD in Chemical Engineering from the University of California at Berkeley. He has published more than 20 papers and book chapters in areas related to MCM technology.

R. Wayne Johnson received the BE and MS degrees in 1979 and 1982 from Vanderbilt University, Nashville, TN, and the PhD degree in 1987 from Auburn University, Auburn, AL. All are in electrical engineering. He is an Associate Professor of Electrical Engineering at Auburn University. At Auburn, he has established teaching and research laboratories for thick and thin film hybrid technology. His thin film research efforts are focused on materials, processing and modeling for multichip modules. Research in power hybrids and high temperature electronics is being directed in the thick film laboratory. Dr. Johnson has published in technical journals and presented numerous papers at workshops and conferences. He also worked in hybrid microelectronics for DuPont, Eaton and Amperex. Dr. Johnson is a member of the IEEE CHMT Society, IEPS and was 1991 President of ISHM.

Alan D. Knight joined IBM after receiving his BS degree in Mechanical Engineering from Clarkson College of Technology in 1965. He has held various positions in Manufacturing Engineering, Industrial Engineering, Process Development Engineering, Product Engineering and currently is a manager of a connector development group. He has co-authored and presented numerous technical papers and holds several patents pertaining to design and fabrication of cables and connectors.

Erik N. Larson received a BS degree in Mechanical Engineering from San Diego State University. In 1987, he joined Unisys Corp. as a Mechanical and Process Development Engineer. His primary responsibilities at Unisys have been in the areas of thermal design and assembly process development for the corporation's Semiconductor, Memory and Packaging Operations division in San

Diego CA. He has been responsible for developing both wire bond and tape automated bonding processes for single and multichip VLSI ASIC packaging. He has also participated in the development of advanced liquid cooling systems for multichip packaging. Erik has received two corporate level achievement awards for enhancing wire bond production yields.

James J. Licari has a degree of Bachelor of Science in Chemistry from Fordham College and a PhD from Princeton University, where he was a DuPont Senior Fellow. At American Cyanimid, Remington-Rand Laboratories and Rockwell International, he has made extensive contributions in the development of new materials and in the study of fundamental materials properties for electronic applications. He is a Senior Scientist at Hughes Aircraft Company, Microelectronic Circuits Division, Newport Beach CA, where he serves as Program Manager for the Navy's VLSIC Packaging Technology (VPT) contract.

Arun Malhotra, Director of Thin Film Head Operations, IOMEGA Corp. He was manufacturing engineering manager for the VAX-9000 MCU Interconnect (High Density Signal Carrier - HDSC) at Digital from 1986 to 1991. Dr. Malhotra was a manufacturing engineering manager at Hewlett-Packard from 1984 to 1986. From 1976 to 1983 he worked on the NMOS III process development and copper-clad Teflon board development for Hewlett-Packard in Ft. Collins CO. He received the BS degree in Electrical Engineering from IIT in New Delhi in 1970, and the MS and PhD degrees in Electrical Engineering from Purdue in 1972 and 1974 respectively.

John D. Marshall, Product Engineer, Digital Equipment Corp., Cupertino CA. He was born in New Haven CT in 1954. He the BSEE degree from the University of Rochester in 1976, the MSEE degree from the California Institute of Technology in 1977 and the PhD degree in Electrical Engineering from Stanford University in 1988. His doctoral research was concerned with modeling field effect transistors for analog and digital circuits and focused on performance limits of silicon E/D MESFET integrated circuits. Since then he has been employed by Digital Equipment Corp. working on the design, modeling and characterization of high density, thin film circuit interconnect structures. Dr. Marshall is a member of Tau Beta Pi and Sigma Xi.

Robert S. Miles received both a BS degree in Electrical Engineering and a Masters degree in Business Administration from California State University, Los Angeles CA. In his 25 years at Hughes Aircraft Company, he has worked on the Maverick, Phoenix, AMRAAM and LEAP programs, and has held program management, CAD, process development, automated manufacture, and electrical test positions at MSG, S&CG, and IEG. Mr. Miles currently works in the Hughes Microcircuits Division Process Technology Development Department, and manages the Hughes Corporate-wide Electronic Packaging Program (HEPP).

Douglas Modlin, Project Manager, Affymax Research Institute. Throughout his professional career, Dr. Modlin has specialized in the development of instrumentation containing integrated electronic, optical, mechanical and computational systems. From 1986 through November of 1991, Dr. Modlin was employed by Digital Equipment Corp. and led the team which developed the testers used to test the VAX-9000 Multichip Unit interconnect substrates. This included both in-process testers based on existing wafer probers and flying probe testers designed specially for testing the VAX-9000's fine pitch copper-polyimide interconnect substrates. Dr. Modlin is currently employed by Affymax Research Institute where he is developing optically based test equipment for medical diagnostics applications. He received his BS degree from California Polytechnical, San Luis Obispo in 1975 and his Masters and PhD degrees from Stanford University in 1978 and 1983, respectively.

Edward G. Myszka, Staff Engineer, Motorola's Corporate Manufacturing Research Center in Schaumburg, Illinois. He currently is a Project Manager responsible for the development of high density interconnect substrates for high performance MCM and PWB applications. He holds several patents and has published several papers on this subject. During his tenure at Motorola, he also was involved in the design of fiber optic transceiver modules and high speed, high power PAC packages. Prior to Motorola, Mr. Myszka worked for Lytel, Inc., where he was responsible for the design of component and subsystem packages for semiconductor lasers, light emitting diodes and photo diodes for telecommunications applications. Mr. Myszka received his Bachelor degree in Mechanical Engineering from the Polytechnic Institute of New York in 1984 and his Master's in Materials and Metallurgical Engineering from the Illinois Institute of Technology in 1990. He is currently pursuing an MBA degree from the Illinois Institute of Technology.

John A. Nelson Staff Engineer in the Product Technology group of the Unisys Semiconductor, Memory and Packaging Operation in Rancho Bernardo CA. His primary responsibility is in the area of advanced packaging. He holds several patents in packaging technology. At Burroughs/Unisys, he has held several management positions in packaging design, package test, ceramic package fabrication and package assembly. Prior to joining Burroughs in 1974, he had worked in avionics and computer systems packaging at Honeywell. He is active in the packaging activities of IEEE, ISHM and IEPS. Society activities have included speaking, writing, chairing conference and session programs. In 1957, he received his BS in Aeronautical Engineering from the University of Minnesota. He also served as an officer in the U.S. Air Force.

Lee Hong Ng, Project Manager at IBIS Associates, Inc. in Wellesley MA. IBIS specializes in the competitive assessment of alternative technologies. Since joining IBIS, Dr. Ng has been involved with the analysis of alternative packaging schemes ranging from single chip packaging to multichip modules, and has published numerous papers on these topics. Dr. Ng also is involved in business planning and strategic analysis for companies in the electronics industry. Dr. Ng earned a PhD from the Materials Science and Engineering Department at M.I.T. She holds an MS degree from the same department at M.I.T. and a Bachelor's degree in Mechanical Engineering from the University of Texas at Austin.

Rajendra D. Pendse received the BS degree in Metallurgical Engineering from Indian Institute of Technology, Bombay, in 1980 and the PhD degree in Materials Science from the University of California, Berkeley, in 1985. In 1985, he joined National Semiconductor Corporation's Package Engineering Department, where he worked in the areas of thermal, stress and electrical modeling of packages, design of test chips for package reliability studies and several aspects of tape automated bonding. He joined Hewlett-Packard Company in 1989 in the Electronics Packaging Laboratory, where he was co-inventor of a proprietary new package called demountable TAB. He was involved in the development of designs, materials and processes for DTAB, including its subsequent transfer to manufacturing. He continues to work on advanced, high lead count and emerging packaging technologies.

Ralph Platz, Program Director, Maxtor Corp. From 1971 to 1992, he worked for Digital Equipment Corp. and managed the support of PDP-11 systems, the development of HSC50 Storage Subsystem, the development of a VAX-based parallel processor research system, and the development of the VAX-9000 MCU assembly process. Mr. Platz holds a BA degree in philosophy and an MA degree in Classics.

Karl J. Puttlitz, Sr., Senior Engineer, IBM Corp., Packaging Laboratory East Fishkill NY. He has been involved with the design, development and evaluation of interconnections for electronic packages since the early 1960s when IBM instituted its solder ball flip chip (SBFC) program. His early work included the development and implementation of SBFC replacement technologies practiced by IBM. More recently, he has been involved with hybrid interconnection applications and also substrate surface mount interconnections, such as the area-array, solder-ball connection (SBC). He holds a BS, MS and PhD degrees in Metallurgy from Michigan State University, published numerous papers and credited with over 30 published inventions. His memberships include Sigma Xi, ASM International and New York Academy of Sciences.

Robert E. Rackerby, Principal Engineer, Unisys Corp., San Diego CA. Mr. Rackerby has been working in packaging and process engineering at Unisys for more than eight years. Early in his career at Unisys, he became involved with multichip packaging, working heavily in reliability, mechanical modeling and testing. Since 1986, he has been involved in the design and assembly of ceramic multichip modules. His primary responsibilities in MCM assembly have been to evaluate adhesives and strategies for die attach, select die attach materials and equipment, engineer attach processes and certify processes to production. He is currently involved with developing TAB and C4 interconnect processes. Mr. Rackerby has been awarded one patent and received his BSE from the departments of Chemical and Mechanical Engineering from California State University at Northridge in 1983.

Randy D. Rhodes, Department Manager, Applications Function of the Product Technology group of the Unisys Semiconductor, Memory and Packaging Operation in Rancho Bernardo CA. His responsibilities include packaging applications, electrical design, IC applications, memory engineering, product engineering and test. Mr. Rhodes has been involved in the packaging development at Burroughs/Unisys since joining Burroughs in 1983. Prior to that time, he worked on the first CMOS microprocessor and associated development systems at RCA. Mr. Rhodes received his BSEE from Georgia Tech in 1973.

Thomas C. Russell, Manager of the Design and Test Development Department, Alcoa Electronic Packaging, Inc., San Diego CA. Dr. Russell has been employed by Alcoa Electronic Packaging since 1989. His department is responsible for the design and test of advanced ceramic and silicon packages. During 1985 to 1989 he was a Member of the Technical Staff at AT&T Bell Laboratories in the Advanced VLSI Packaging Technology Development Department. While at Bell Laboratories, he was responsible for the development of a two probe, high speed robotic test system used in the electrical test of thin film substrates, ceramic hybrids and printed circuit boards. Dr. Russell holds a patent in a methodology for testing interconnect substrates. He received his ScB from Brown University in 1978, his MS in Mechanical Engineering from Stanford University in 1981 and his PhD in Electrical Engineering from Brown University in 1985.

Donald P. Seraphim, is a retired IBM Fellow and consultant. Dr. Seraphim has been developing advanced package technologies since 1965. He directed IBM's early programs on multi-chip bipolar packaging and during the 1970s directed the advanced printed circuit board program. Dr. Seraphim was a member of IBM's Corporate Technology Committee between 1977 and 1979, and was elected an IBM Fellow in 1981. He served as manager of materials science and engineering in IBM Endicott's Packaging Technology Laboratory from 1981 to 1986. He has written more than 50 articles, co-edited the book *Principles of Electronic Packaging*, and has over 30 patents and patent publications. Dr. Seraphim holds bachelor and masters degrees in applied science from the University of British Columbia, and a Dr. of Engineering in metallurgy from Yale University.

Kevin P. Shambrook has an AEA degree from Stanford University, a BSEE from Manchester University in England, an MSEE from University of Pittsburgh, and a PhD in Engineering and Economic Development from University of California, Los Angeles CA. He joined Hughes in 1983 after more than 20 years of experience in the management of engineering projects. At Hughes, he managed an Air Force-sponsored Technology Modernization program involving a paperless hybrid microcircuit manufacturing facility using a fault-tolerant computer for MRP II, and a Corporate-wide High Density Multichip Interconnect (HDMI) program. Currently, Dr. Shambrook is Director of the Center for Applied Competitive Technologies at Orange Coast College.

Stanley M. Stuhlbarg received a BS degree in Electrical Engineering from University of Cincinnati OH and a Management Certificate from Indiana University. At Hughes since 1974, he was responsible for engineering, marketing and strategic planning projects involving hybrid microcircuits, tape chip carrier development, and multichip module packaging. He initiated an Air Force technology modernization program in microcircuit manufacturing and a Mantech program on VLSIC packaging technology. Currently, Mr. Stuhlbarg is President of Engineering Technologies Associates, a consulting firm specializing in assistance to companies desiring expansion in Manufacturing Technology, Industrial Modernization Incentive Programs (IMIP) and Advanced Technology Projects (ATP).

Hongbee Teoh, Reliability Engineer, Digital Equipment Corp., MicroModule Systems, Cupertino CA. Dr. Teoh is currently working on the reliability testing and qualification of multichip modules technology for high performance applications. Since joining Digital in 1987, he has been working in the Quality and Reliability organization in Digital's Semiconductor Operations in

Hudson/Andover MA. He has been active in the area of single chip and multichip packaging, process technology and product qualifications, interconnect reliability and nondestructive testing of IC interconnects. Prior to joining Digital, Dr. Teoh was the Quality Assurance Manager for Kyocera America, Inc., a manufacturer of ceramic IC packages located in San Diego. Dr. Teoh received his BSME from the University of Rochester NY, an MS and PhD in Materials Science and Engineering from the University of California, Los Angeles CA (1984).

Philip A. Trask received a B.S. degree in Engineering from UCLA, an M.S. degree in Solid State Engineering from Arizona State University, and a Masters degree in Business Administration from Pepperdine University. He was responsible for R&D projects involving integration of passive component parts within silicon substrates, and development of a one-metabit MNOS memory using wafer scale integration. He has established a Class-100 fabrication line from High Density Multichip Interconnect (HDMI) Substrates, and has set up deep UV resist lithography in a prototype CMOS line. Currently, Mr. Trask is Manager of HDMI Substrate Engineering within the Hughes Semiconductor Products Center in Newport Beach, California.

William A. Vitriol, Senior Scientist, CTS Corp., Elkhart IN. Dr. Vitriol has more than 21 years experience in thick film ink development, hybrid microcircuit processing, and ceramic packaging technology. Since joining CTS, he has served as a technical resource for the Corp. in the area of electronic packaging, and, in particular, low temperature cofired ceramic technology. Prior to joining CTS, Dr. Vitriol worked for Hughes Aircraft Company, Newport Beach, CA for 10 years, where he was an Assistant Department Manager in the Technology Development Laboratory, for McDonnell Douglas Corp., and for Beckman Instruments. He has a BS and a PhD degree in ceramics from Rutgers University. He also holds six patents, plus two pending, in the area of thick film and ceramic packaging technology and has published over 15 technical papers. Dr. Vitriol is currently the ISHM Materials Division Chairman.

Yenting Wen, Staff Engineer, ALCOA Electronic Packaging Inc., San Diego, CA. Mr. Wen has been employed by Alcoa since January 1990. He has been involved in MCM development including design, test and high speed digital applications. He is also working on interconnect modeling and packaging characterization. Yenting Wen was a test engineer at AT&T Bell Laboratories from 1986 through 1990. From 1989 to 1990, he worked at Electrical Material Research Department and was responsible for MCM functional testing and electrical design and modeling. From 1986 to 1989, he worked in the Advanced VLSI Packaging Technology Department and was responsible for the functional testing of thin-film multichip modules. He received his BS in Physics from National Central University, Taiwan in 1980 and MSEE from Rutgers, the State University of NJ in 1985.

Scott Westbrook, Member of Technical Staff, Design and Product Engineer Supervisor, Digital Equipment Corp., Cupertino CA. Since coming to California in 1980, he has worked in sequence for Hewlett-Packard, Trilogy Systems, and Digital Equipment Corp. During his career he has researched and developed MOS power devices, MOS memory, bipolar wafer scale technology and most recently thin film interconnect for multichip packaging. He was one of the developers of the copper/polyimide interconnect technology for multichip packaging that is used in the VAX-9000 mainframe/super computer. He is currently responsible for New MicroModule Products Engineering. Mr. Westbrook holds a BS (1978) and an MS (1980) in Electrical Engineering from M.I.T.

Chee C. Wong received his BS and PhD in Materials Science from M.I.T. in 1981 and 1986, respectively. Since then he has been with AT&T Bell Laboratories, where he is currently a Member of Technical Staff in the Electronics Packaging Research Department. His research activities include amorphous silicon devices, thin film processing and artificial neural network hardware implementations using amorphous silicon photoconductors. He was responsible for the development and transfer of solder bumping technology into manufacturing. More recently, he has been engaged in advanced interconnection technology for electronic/photonic multichip module applications.

Jimmy Jun-Min Yang, Principal Reliability Engineer, Digital Equipment Corp., MicroModule Systems, Cupertino, CA. Mr. Yang started at the Digital Taiwan Terminal Design and Manufacturing plant as Mechanical Engineer in 1980. He has previously been involved in VAX-9000 Manufacturing program office for Quality and Reliability planning, MCU Process Qualification as principal Quality Engineer in Marlborough MA, DEC disk and tape drives manufacturing plant as a Senior Component/Quality Engineer. He has a BS degree in Mechanical Engineering from Feng-Chia University, Taiwan, and a MS degree in Manufacturing Engineering from Pittsburgh State University. He is also a candidate for an MS degree in Manufacturing System Engineering from NTU, and recognized by the ASQC as both a Certified Quality Engineer, and a Certified Reliability Engineer. Mr. Yang has contributed to DEC internal Quality Planning Handbook and has written several Quality and Reliability technical papers. His current interests include System Quality and Reliability, Process Quality Modeling, System Reliability prediction and modeling techniques.

David F. Zarnow received a BS degree in Computer Engineering from the University of Illinois. He was associated with the Naval Avionics Center in Indianapolis IN, as a member of the Advanced Microelectronics Engineering staff, co-founding their hybrid microcircuit pilot production facility. Joining Hughes Microelectronic Circuits Division (MCD), Newport Beach CA in 1982, he became Manager of the Corporate MultiChip Module Computer Aided-Design (CAD) program, establishing standard design tools and company-wide foundry interfaces for multichip product designs. Currently, Mr. Zarnow heads product design and design information activities for the Hughes Microelectronic Circuits Division.

INDEX

625-634, 636-658, 706, 725-732,
746, 759, 774-778, 786-790,
792-795, 800, 807-809, 829, 836,
840
assembled module, 644
at-speed, 60
bare die, 24, 32, 632, 678, 679, 728,
812, 826, 830
bed-of-nails tester, 619, 624-626
board testers, 649, 654, 657
boundary scan, 362, 633, 635, 636,
648, 649, 655, 657
built-in self test (BIST), 632, 637, 651
cost, 115
device under test (DUT), 638
design for test, 653
dual sided two probe testers, 626
electron beam probing, 628-630
equipment, 645, 654
escape, 116
final performance, 648
fixed probe array tester, 619
full functional, 646
Greedy Algorithm, 628
IC testers, 631, 654
infant mortality, 642, 826
integral method, 595, 598
internal scan, 633, 635, 650
JTAG boundary scan, 76
limited functional, 646, 657
parametric distribution, 826
parametric, 632, 637
patterns, 633
probe card, 638, 728
strategies, 645
single probe testers, 621, 631
substrate test, 615, 616, 631
test access port (TAP), 634, 650
test and burn-in, 83, 351
test, rework and scrap, 828
test-on tape, 400
8 testability, 161, 349, 428, 433, 774,
818, 830
two probe testers, 625-628, 631
vectors, 654

wire pull tests, 386
Texas Instruments, 418, 755-758
TFM (Thin film module), 687
TFML (Thin film multilayer), 255, 256,
260, 262, 263, 266, 272, 276, 300, 301,
305, 706, 708, 712, 738, 740
Thermal analysis, 323, 418, 593-595, 597,
611, 725, 726
Thermal conduction, 18, 19, 41, 98, 109,
226, 228, 236, 269, 289, 300, 301, 402,
419, 587-590, 670, 708, 710, 712, 722,
778
thermal coupling, 580, 581
Thermal conduction module (TCM), 98,
289, 300, 301, 587, 589, 590, 670, 708,
778
Thermal dissipation, 78, 244, 248, 266, 474,
349, 721, 726, 770, 811
Thermal fatigue failures, 19, 462, 468
Thermal performance, 16, 38, 70, 75, 272,
299, 355, 417, 419, 430, 437, 478, 488,
502, 581, 592, 593, 610, 640, 669, 674,
681, 682, 722, 732, 738, 742, 758
thermal mismatch, 318, 712
thermal noise, 561
thermal paths, 15, 571
Thermal stresses, 197, 242, 492, 570
thermal resistance, 577, 578, 580
thermal stability, 18, 316, 319, 321, 340,
365, 716
Thévenin equivalent, 533, 541, 546
Thick film hybrids, 44, 55, 56
thick film layout, 240
thick film multilayer, 215, 244, 245
thick film substrates, 146
thick film technology, 146, 215, 454, 718
Thin film MCM, 98, 113, 120, 126, 127,
142, 150, 153, 156, 250, 448, 545, 560,
562, 537, 544, 686, 738, 750
Thin film module (TFM), 687
Thin film multilayer (TFML), 255, 256,
260, 262, 263, 266, 272, 276, 300, 301,
305, 706, 708, 712, 738, 740
TFML design rules, 262, 263
Thin small outline package (TSOP), 13, 14